Introduction to Communications Engineering

Second Edition

WILEY SERIES IN TELECOMMUNICATIONS

Donald L. Schilling, Editor
City College of New York

Introduction to Communications Engineering

Second Edition

Robert M. Gagliardi
Department of Electrical Engineering
University of Southern California

WILEY

A Wiley-Interscience Publication
JOHN WILEY & SONS
New York • Chichester • Brisbane • Toronto • Singapore

Library of Congress Cataloging in Publication Data:

Gagliardi, Robert M., 1934–
   Introduction to communications engineering.

   (Wiley series in telecommunications)
   "A Wiley-Interscience publication."
   1. Telecommunication.   I. Title.   II. Series.
TK5101.G24   1988   621.38   88-5629
ISBN 0-471-85644-4

Printed in the United States of America

10 9 8 7

*To the memory of my beloved mother and father,
Louise and Michael Gagliardi*

PREFACE

In the first edition of this book I attempted to provide a college text devoted to the engineering aspects of modern communication systems. My objective was to concentrate on the practical system considerations in the end-to-end construction of a typical communication link, allowing a reader to enter the present-day engineering field with some knowledge of the terminology, methodology, and procedures currently being used. In this second edition I have continued with the same objective, but with more emphasis on modern technology and hardware considerations. The reader will benefit from an overall enhancement and updating of the key areas, at the expense of a de-emphasis of some less important topics. While the first edition tended to be segmented into separate discussions of analog and digital systems, this edition integrates these two disciplines, making it more compatible with today's overall design philosophies.

The principal modifications from the first edition can be summarized as follows: (1) introduction of digital systems and waveforms early in the text; (2) insertion of new sections devoted to the important technology areas such as oscillators, frequency generators, mixers, amplifiers, and both digital and switching circuitry, all of which strongly influence system design; (3) the inclusion of fiber links as an important alternative to wireline communications, even though it requires lightwave technology (the reader and instructor need not be familiar with optics to cover these areas); (4) more emphasis on overall link analysis and a corresponding de-emphasis of some highly mathematical topics such as quantization theory and information theory; (5) an updated digital communication discussion, integrating both theory and hardware, with a new appendix to aid the student in analysis; (6) a detailed study (separate chapters) of two of today's most important communication systems—the satellite link and the fiber link; (7) new and upgraded homework problems to augment these changes.

The format of the book is again that of a classroom text. Emphasis is placed on deriving design equations of various subsystems, interpreting their implications, and carrying out tradeoff studies for system comparisons. In the process the student is introduced to modern communication systems through examples and diagrams. The material is arranged to aid an instructor in moving through the topics in logical order, beginning with basic system models, and developing and interfacing the various communication subsystems. Homework problems have been carefully selected to review the text material and give the student practice in design procedures.

In terms of electrical engineering curricula, the book falls between basic undergraduate signal electronics and the more theoretical graduate topics already stated. The text begins at a basic introductory level suitable for college seniors, yet is wide enough in topic coverage to be usable on a graduate level. Prerequisites for the book are courses in electronics, electromagnetic theory, and linear system theory, all core courses in an electrical engineering curriculum. Chapter 1 is intended to establish the necessary background material for the remainder of the book. A previous course in probability would be advantageous to the student, but is not necessary for one guided by an instructor. The important aspects of noise theory required here are covered as simply as possible in the first chapter. Appendixes on random theory and signal analysis are provided for the instructor who wishes to digress further into these areas.

For an undergraduate text Chapters 1–5, and parts of 6 and 7, provide a useful, self-contained introduction to the communication engineering field for both the graduating senior and the future graduate student. The classroom instructor must provide the proper match between the rate of topic coverage and the student's background. On a graduate level Chapters 1–7 can be easily covered in a semester and would serve well as an introductory course for a modern graduate communication curriculum. Such a course would give students an overall view of all aspects of communication system design (signal processing, modulation, link analysis, analog and digital systems, and synchronization). The remainder of the text can then be used as a second semester or follow-on type of course. In addition, Appendix C and Chapters 7–10 alone form a fairly comprehensive one-semester course devoted to modern digital communication system design. The book is also tailored for industrial in-house courses and for one- or two-week short courses covering these aspects of communications engineering.

I wish to personally thank Ms. Georgia Lum of the Electrical Engineering Department at USC for typing the manuscript. I also acknowledge the help of the staff of the Communications Science Institute at USC, Ms. Milly

Montenegro, Ms. Cathy Cassells, and Ms. Neela Sastry, in putting together the new edition. Lastly, I would like to thank the classroom students and instructors who used the first edition over the years for their criticisms, comments, and suggestions that aided me in upgrading this edition.

ROBERT M. GAGLIARDI
University of Southern California

CONTENTS

COMMUNICATION
SYSTEM MODELS

The principal objective of a communication system is to transmit *information signals* (waveforms) from one point to another. The information signals may be the result of a voice message, a television picture, or a meter reading or may take on a variety of other formats depending on the specific application. The points between which the information is to be sent may be located in close proximity or may be continents, or even planets, apart. *Communications engineering* involves the analysis, design, and fabrication of an operating system that satisfactorily performs the communication objective. In this book we are concerned with the development and application of the basic analytical tools and system alternatives necessary to carry out the communications engineering tasks.

The required transfer of information is most often accomplished by *modulating* (superimposing) the desired information signal onto a *carrier* (sine wave) and converting the carrier to an electromagnetic field. This field is then *transmitted* (propagated) to the desired destination, where it is *received* (intercepted) and the information signal *demodulated* (recovered). The design objective is to construct a suitable system that accomplishes these operations with minimal *distortion* (perturbations) of the information signal itself. To produce such a design, it is necessary to understand the basic functional aspects of each component of the system model, along with their various characteristics and properties that influence the eventual performance.

1.1 ELEMENTS OF COMMUNICATION SYSTEMS

A communication system can be divided into three primary parts, as shown in Figure 1.1. The *transmitter* represents the part of the system where the information signal is generated, modulated, and transmitted. Each of these

Figure 1.1. The basic communication system.

operations is not necessarily physically located at the same place, so that the transmitter block does not always represent a single, self-contained unit. The *receiver* portion performs the necessary operation for intercepting the transmitted carrier and recovering the information signal at the desired destination. It also need not represent a single, self-contained unit. The *channel* represents the propagating media or the electromagnetic path interconnecting the transmitter and the receiver. It is from this transmission channel that various anomalies and interference effects enter the system operation. Thus, although the design and fabrication of the transmitter and receiver portions are, for the most part, entirely in the hands of the communication engineer, it is often the properties of the channel that ultimately influence and dictate such design procedures. As we shall see, knowledge of these channel properties is mandatory for successful engineering. The combination of transmitter, channel, and receiver is referred to as a communication *link*. Any part of a communication system is called a *subsystem*, and the overall design is accomplished by properly interfacing the individual subsystems.

The transmitter block can be further subdivided into the subsystems shown in Figure 1.2*a*. The information *source*, whatever its form, generates the electronic signal to be sent to the receiver. Information sources are classified as either *analog* or *digital* sources. An analog source produces a time continuous electronic waveform as its output. A digital source produces sequences of data symbols for its output. Although Figure 1.2*a* shows only a single source output, we may in fact have several different outputs from a single source, and perhaps even more than one distinct source. In such cases the communication system must be capable of handling simultaneously the complete set of source signals.

Source signals are generally converted into a baseband waveform to prepare it for the carrier transmission. The principal objective of this baseband conversion is to insert suitable control and proper formulation of the source output prior to carrier modulation. In some simple systems baseband conversion is not used, and the analog source output is used directly for carrier modulation. In other systems the conversion operation may take on a variety of forms. For example, it may involve only filtering of the source waveform, or it may involve a separate modulation of the source signal onto a secondary carrier (called a *subcarrier*) before modulating onto the main carrier. For digital sources, baseband conversion is used to convert the source symbols into baseband waveforms for transmission, and the conversion operation is called *encoding*. In a multiple source environment

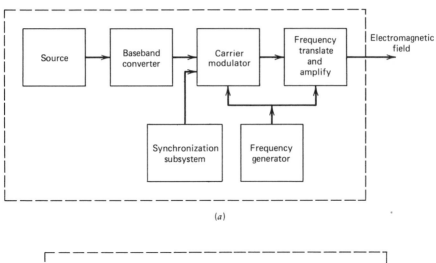

(a)

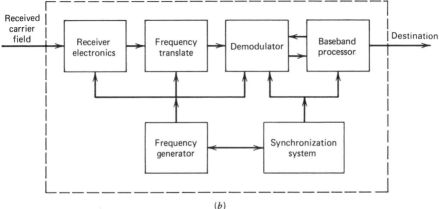

(b)

Figure 1.2. Transmitter and receiver block diagrams: (a) transmitter, (b) receiver.

the converter involves additionally a combining of all the various source outputs into a single baseband waveform, the latter conversion called *multiplexing*. The selection of the proper baseband conversion operation so as to achieve satisfactory communication becomes an important aspect of transmitter design. In later discussions the various types and alternatives for the conversion operation are examined in detail.

The baseband waveform is modulated onto the carrier to form the carrier waveform. The basic objective of performing a modulation is to generate a transmission waveform that is best suited for the electromagnetic propagation and communication reception. These advantages are made more obvious in subsequent chapters. Here again the modulation may take on various forms, each with its own advantages and disadvantages relative to the channel and receiver. Modulation formats are important in determining the extent of the channel interference and combating the associated distortion. Also, the modulation method must interface properly with the baseband conversion operation and the basic parameters of the receiving system.

In addition to the subsystems devoted to the processing of the source signal, it may be necessary to provide auxiliary transmitter waveforms that aid in the transmission and recovery of the desired information. These auxiliary waveforms can be classified as transmitter *synchronization* signals whose basic objective is to keep the link operationally aligned and properly interfaced with the transmitter. These synchronization signals can take on a variety of forms and may be superimposed on the source, baseband, or carrier waveforms. In the sophisticated system characterizing modern design, the ability to maintain adequate synchronization and alignment must be a primary consideration and often becomes the ultimate limit to attainable performance.

At the transmitter the modulated carrier is often frequency translated and amplified to a desired power level. Frequency translation involves a direct shift of the modulated carrier from one frequency band to another. (These bands are discussed in Section 1.2). This frequency shifting is accomplished with the aid of signals produced from devices called *frequency generators*, which may also provide the carrier waveforms needed for the modulation. The majority of modern systems use some form of transmitter frequency translation.

In electronic communication systems, the amplified carrier is used to generate the electromagnetic field for channel propagation. In optical communication systems, the baseband signal is used to directly modulate an optical source to produce an electromagnetic field corresponding to a light beam. The predominate use of light beams is with optical fibers, in which transmitter and receiver are interconnected by a glass "pipe." The use of lightwave fiber systems as a communication link is rapidly becoming one of our most important communication channels.

All propagating electromagnetic fields and optical beams are characterized by their *polarization* (the orientation of their electric field in space) and their transmitted power distribution (beamshape) in the various directions of propagation. Electromagnetic transmission channels can be roughly divided into *guided* and *unguided* channels. In an unguided channel the transmitter propagates the field formed from the carrier freely in a medium with no attempt to control its propagation other than by its field propagation pattern. The prime example of an unguided channel is the so-called space channel, in which the medium involved may be free space, the atmosphere, or the ocean (under water). In a guided channel a waveguide is used to confine the field propagation from transmitter to receiver. The prime example of guided system is the cable, hard-wire, or optical fiber channel. Specific properties of these various channels are discussed (in Chapter 3).

The system receiver can be divided into the subsystems shown in Figure 1.2b. The front end electronics receives and processes the electromagnetic field and attempts to eliminate as much channel interference as possible. The received electromagnetic field is then demodulated to recover the baseband waveform. In an optical receiver a photodetector is used in the front end to regenerate the baseband. Channel effects are generally exhibited at this point and cause the recovered baseband to no longer represent the transmitter baseband identically. The recovered baseband is then deconverted to produce the recovered information signal (or signals, in the multiple source case) for the final destination. In general, this deconversion simply attempts to invert the operation performed at the transmitter. It may, for example, involve baseband filtering, subcarrier demodulation, *decoding* (reconverting the baseband back into digital symbols), or possibly *demultiplexing* (splitting the baseband into its various source waveforms). Over the past several decades there have been increased efforts to improve receiver design to achieve better system performance in terms of both theoretical analysis and hardware fabrication. Much of this effort is concentrated on more sophisticated receiver baseband processing, and the selection of the appropriate processing alternative becomes a vital part of successful system engineering.

If synchronization signals were provided at the transmitter, they must also be recovered and properly used in the receiver processing. In this case a synchronization subsystem must be included within the receiver to separate out the synchronization waveform and reinsert it into the correct receiver subsystem (Figure 1.2b). This synchronization information may be used in the carrier, in the baseband, or in the eventual application of the recovered source signal at the destination. Chapter 10 is devoted to some common synchronization subsystems and their associated design.

The system model shown in Figure 1.2 implies that a single carrier is involved, which indeed represents the most common situation. In discussing more advanced modern communication systems, however, we extend this basic model to include the case where more than one simultaneous carrier is involved in the transmitting operation. Such a system is called a *multiple carrier* system, and its analysis requires modification and extension of many of the results to be derived for the single carrier case.

1.2 COMMUNICATION SYSTEMS

The communication system models depicted in Figures 1.1 and 1.2 are simply generalized forms of some of the basic communication systems with which we are personally familiar. Three of the most common examples in our everyday life are the telephone, television, and teletype systems (sketched in Figure 1.3). In the telephone system the speaker's voice is converted to an electrical signal directly in the hand-held phone set, which therefore acts as an analog source for the link. The voice signal is then sent by wire to a central telephone exchange. Here it is carrier modulated and relayed via a

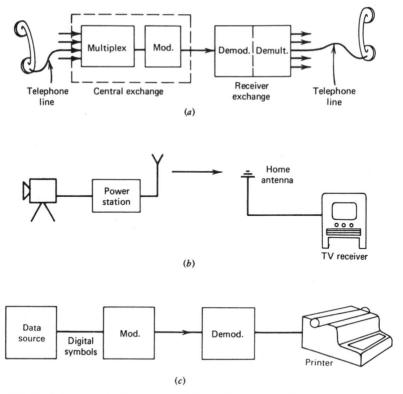

Figure 1.3. Typical communication systems: (*a*) telephone system, (*b*) television system, (*c*) data transmission system.

complicated switching procedure to the central exchange of the intended listener. The latter exchange is determined by the dialing code sent by the speaker, which thus acts as synchronization information interconnecting the speaker and the listener. If the listener's exchange is in a nearby locality, the modulated carrier may simply be cable connected to it. If it is at a distant location, electromagnetic space propagation may be used. At the listener's exchange, the voice signal is demodulated and passed through telephone lines to the listener's phone. Note that in such a system the various subsystems of the transmitter and receiver blocks are separated in their physical location.

In commercial television (Figure 1.3*b*) the camera acts as the system analog source, generating the video (television) signal for transmission. Typically, this signal is relayed via cable to a transmitting power station invariably positioned in some electromagnetically strategic point in a high location. Here the carrier is formed and the resulting field is broadcast to the immediate locality. For reception, the electromagnetic field is intercepted by our rooftop antennas and fed into our receiving subsystem—the home television set. Within the receiver the carrier is demodulated and processed, including voice and picture separation and synchronization alignment. Note that although the source and the transmitter are physically separated, in the television system the receiving subsystem is self-contained at a single location.

In data systems (Figure 1.3c) customer messages, facsimile graphics, computer words, and so on are encoded into digital symbols, producing a synchronized digital source for the system. The symbols are then encoded into modulated carrier signals for long range transmission, either over

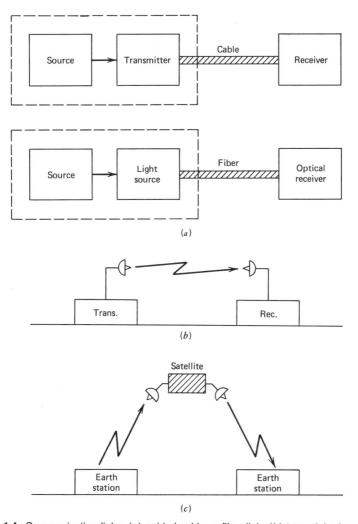

Figure 1.4. Communication links: (a) guided cable or fiber link, (b) terrestial microwave link, (c) satellite link.

telephone lines or by space propagation. At the receiving station the carrier is demodulated, decoded back into symbols, and used to manipulate a synchronized teletype machine for final message printout.

The systems shown in Figure 1.3 can interface with many different types of communication channel. Figure 1.4a shows a common cable link, in which the information carrier, containing voice, video, or data is propagated through a metallic waveguide or transmission line to the receiver. Equivalently, an optical fiber version of the same link can be used, with the information signal used to generate a light beam that shines through the fiber to the receiver. Although both channels perform as basic guided channels, their auxiliary hardware and transmission characteristics differ considerably. For this reason fiber systems will be treated as a special guided link and their discussion postponed to Chapter 12.

Cable or fiber links can be interconnected to form *communication networks*, in which a multiple set of stations can communicate mutually. Each station acts as a network node where information signals are transmitted, received, or relayed. Communication networks can be local area networks (within a building), metropolitan (within a city), or continental (countrywide). Networks allow video, voice, or data to be routed between sources, computers, terminals, and displays.

Figure 1.4b shows a terrestial (ground–ground) link, in which the transmitter signal is sent as unguided propagation to the receiver. Such links are generally short range (tens of miles) and usually confined to line-of-sight transmissions. Figure 1.4c shows a satellite channel in which modulated carriers from transmitting Earth stations are transmitted to orbiting satellites and relayed to receiving Earth stations. This allows communication to be completed between stations that are not in line of sight and may be separated by mountainous terrain, or may even be continents apart. Today satellites are an inherent part of our everyday communications and will continue to play an important role as each new generation of satellites become bigger and more electrically sophisticated. Because of the importance and uniqueness of the satellite channel, Chapter 11 has been devoted to a detailed study of its specific characteristics.

1.3   THE ELECTROMAGNETIC FREQUENCY SPECTRUM

Carrier modulation involves the embedding of baseband or information waveforms onto a carrier sine wave of a fixed frequency. A carrier sine wave has the mathematical form

$$A \cos (2\pi f t + \psi) \tag{1.3.1}$$

where A is the carrier *amplitude*, f is the carrier *frequency* in hertz (Hz), and ψ is its *phase angle* in radians. The modulated carrier is formed from this sine wave and becomes the time variation of the electric field component of the propagated eletromagnetic field. The electromagnetic carrier is often labeled by the location of its frequency f within the electromagnetic frequency spectrum (Figure 1.5). The electromagnetic frequency spectrum is divided into groupings in which the frequency bands are denoted as shown. A carrier whose frequency lies in a given band is labeled according to that designation. Hence a carrier selected from the RF (radio frequency) portions of the spectrum is called an *RF carrier*, one from the optical region is an *optical carrier*, and so on. For high frequencies we find it more convenient to deal with carrier wavelength than with its frequency. The carrier wavelength λ is obtained from the frequency f in hertz by

$$\lambda = \frac{3 \times 10^8}{f} \text{ meters} \tag{1.3.2}$$

Thus millimeter wavelength carriers are carriers whose frequencies are in the range 3×10^{10} to 3×10^{12} Hz. Optical frequencies have wavelengths on the order of microns (10^{-6} meters) and correspond to frequencies in the 10^{14} Hz range.

Conventional electronic communication systems use RF carriers, for reasons discussed in Chapter 3. However, increasing use of millimeter and optical carriers is expected as future communication demands increase and the associated technology advances. This extensive use of RF carriers has led to a further subdivision of the RF portion of the spectrum into the designations shown in Table 1.1. In addition, the SHF and EHF bands are further divided as shown in Table 1.2. To control the excessive demands for terrestial and satellite systems in the S-band, X-band, and K-band carrier frequencies, specific assignments for frequency usage within the United States have been made. Table 1.3 lists some of the important frequency band restrictions and their designated uses.

Selection of the proper carrier frequency is one of the initial decisions that must be made in system design. This selection is generally based on three main considerations: (1) the ease with which the baseband waveform

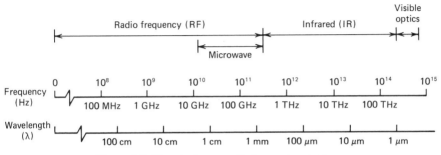

Figure 1.5. The electromagnetic frequency spectrum.

TABLE 1.1 Frequency Designations

Frequency	Wavelength	Designation
3–30 MHz	100–10 m	HF
30–300 MHz	10–1 m	VHF
300–3000 MHz	1–0.1 m	UHF
(3 GHz)		
3–30 GHz	100–10 mm	SHF
30–300 GHz	10–1 mm	EHF

TABLE 1.2 Frequency Band Designations

Bands	Frequency
L	0.39– 1.55 GHz
S	1.55– 5.2
C	3.9 – 6.2
X	5.2 –10.9
K	10.9 –36
Ku	11.7 –14.5
Ka	17 –31
Q	36 –46
V	46 –56

TABLE 1.3 Assigned Frequencies for United States

Application	Frequency Band
Commerical AM radio	0.53– 1.6 MHz
Amateur radio and citizen's band	1.6 – 10
Mobile radio	10 – 20
Citizen band radio (40 channels)	26.9 – 27.5
VHF commercial television	54 – 88
Commercial FM radio	88 –108
VHF commercial television	178 –216
UHF commercial television	470 –890
Military	1.7 – 1.85 GHz
Studio transmitter link	1.9 – 2.1
Commercial satellite (downlink)	3.7 – 4.2
Military	4.4 – 5.0
Commercial satellite (uplink)	5.9 – 6.4
Military	7.1 – 8.4
Community antenna television (CATV)	12.7 – 12.9
Military	14.4 – 15.2

can be modulated onto a carrier of a given frequency, (2) the dependence of system hardware (component size and weight) on carrier frequency, and (3) the dependence of channel characteristics on carrier frequency. Each of these factors must be properly taken into account (as is discussed in Chapter 3) before judicious carrier frequency selections can be made. It is precisely these factors that led to the frequency designations in Table 1.3, although further frequency selection must usually be made within a given category.

A basic property of a modulated carrier waveform is the limitation of the amount of information that can be physically modulated onto a carrier of a given frequency. The conditions involved are developed later; it suffices to point out that the maximum modulating waveform frequency that can be superimposed on a carrier of a fixed frequency is often limited to a certain percent of that carrier frequency. Typical values are 1–10%, depending on the modulation technique. We immediately see a basic advantage of increasing the carrier frequency being used—higher frequency modulating waveforms can be transmitted. For example, a carrier from the millimeter frequency range can incorporate modulating frequencies 1000 times larger than a carrier in the RF range, and optical carrier can handle frequencies 10^5 times as large. The ability to transmit higher modulating frequencies (i.e., more baseband information) is a prevailing property demanded of modern systems and is the force that continually drives engineers to extend technology into the millimeter and optical carrier frequencies. Design of electronic components and devices for such systems, however, with their accompanying small wavelengths and component limitations, often becomes a major hurdle in this advancement.

1.4 WAVEFORMS AND SIGNALS

It is obvious from our basic system model that communication engineering requires us to deal extensively with time waveforms. In general, time waveforms may be classified as either *deterministic* or *random*. A deterministic waveform is one whose value at all time instants is known precisely. Examples of deterministic waveforms are sine waves or square waves of known amplitude, phase, and frequency, exponential functions of known amplitude and decay rate, and pulses of fixed height and width. A random waveform is one whose unobserved values at any instant in time can only be described statistically (i.e., treated as a random variable). Random waveforms are usually referred to as random, or *stochastic*, processes and model the noise and interference waveforms encountered in system operation. In addition, analog source waveforms are in fact random and noiselike in nature, although their basic properties are often difficult to measure and characterize. For this reason it is sometimes analytically convenient to model source outputs as deterministic waveforms in order to gain some initial insight into design and performance. Care must be used in extending results derived under these conditions to actual operating systems. It is also common to deal with a waveform of known waveshape but containing a random parameter. Such a waveform is actually a random process, requiring statistical description, even though it may not be completely noiselike in appearance.

In dealing with deterministic waveforms it is convenient to introduce the *Fourier transform*. If $x(t)$ is the time waveform involved, its Fourier transform is defined as

$$X(\omega) = \int_{-\infty}^{\infty} x(t) e^{-j\omega t} \, dt \qquad (1.4.1)$$

where ω is the radian frequency variable in radians per second (rps). The radian frequency is related to the frequency f in hertz by

$$\omega = 2\pi f \qquad (1.4.2)$$

Note that the Fourier transform $X(\omega)$ is a complex function (i.e., has a real and imaginary part) in the variable ω. It is well known that $x(t)$ and $X(\omega)$ form a transform pair and each can be uniquely derived from the other. The *inverse* Fourier transform is given by

$$x(t) = \frac{1}{2\pi} \int_{-\infty}^{\infty} X(\omega) e^{j\omega t} \, d\omega \qquad (1.4.3)$$

A tabulation of some useful transform pairs and some basic identities and properties of the Fourier transform are summarized in Appendix A. More extensive tabulations are available [1–3].

The important frequency characteristics of a deterministic waveform are contained in its Fourier transform function. The complex transform function $X(\omega)$ can be written in polar coordinates as

$$X(\omega) = |X(\omega)| \exp\left[j\angle X(\omega)\right] \qquad (1.4.4)$$

The function $|X(\omega)|$ is called the transform *amplitude* function, and $\angle X(\omega)$ is the *phase* function. Deterministic waveforms are therefore uniquely described by their amplitude and phase functions. It is easy to show (Problem 1.1) that since $x(t)$ is a real waveform its amplitude function is always even in ω [i.e., $|X(-\omega)| = |X(\omega)|$] and its phase is always odd [$\angle X(-\omega) = -\angle X(\omega)$]. Thus these functions need be described by only their positive frequency behavior. Frequency functions defined over the positive frequencies only are called *one-sided* functions.

Transform amplitude functions associated with typical communication waveforms are generally of a form in which the functions are concentrated over a particular frequency range and tend to roll off (decrease in mag-

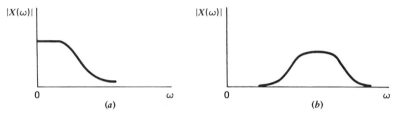

Figure 1.6. Transform amplitude functions: (a) low pass waveform, (b) bandpass waveform.

nitude) at other frequencies (Figure 1.6). If this concentration along the positive frequency axis begins at zero frequency and extends to some higher frequency before roll-off begins, the waveform is said to be a *low pass waveform*. If the concentration is centered along the positive frequency axis and rolls off toward zero and higher frequencies, the waveform is said to be a *bandpass waveform*. As an example, consider the waveform

$$x(t) = \begin{cases} A, & 0 \le t \le \tau \\ 0, & \text{elsewhere} \end{cases} \qquad (1.4.5)$$

corresponding to a voltage pulse (Figure 1.7a). The Fourier transform is given by (Section A.1)

$$X(\omega) = A \left[\frac{1 - e^{-j\omega\tau}}{j\omega} \right] \qquad (1.4.6)$$

The amplitude function $|X(\omega)|$ is then

$$|X(\omega)| = A \frac{|1 - \cos \omega\tau - j \sin \omega\tau|}{|j\omega|}$$

$$= A \frac{[(1 - \cos \omega\tau)^2 + \sin^2 \omega\tau]^{1/2}}{\omega}$$

$$= A\tau \left| \frac{\sin (\omega\tau/2)}{(\omega\tau/2)} \right| \qquad (1.4.7)$$

and is sketched in Figure 1.7b. Thus the pulse waveform has a low pass frequency function extended over the entire frequency axis but concentrated primarily in the range $|f| \le 1/\tau$. This means that as the pulse is made narrower in time, its frequency function is spread in frequency, and an inverse relation exists between pulse time width and frequency width. This inverse time-frequency behavior is characteristic of transform pairs in general (Problem 1.4).

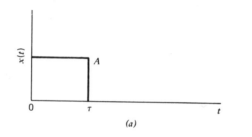

(a)

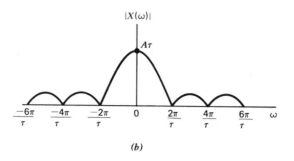

(b)

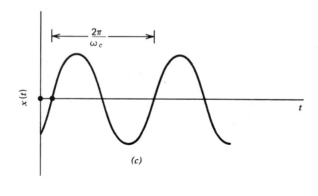

(c)

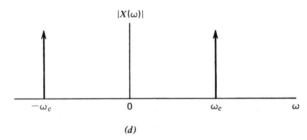

(d)

Figure 1.7. Examples of waveforms and transforms: (a) pulse waveform, (b) pulse transform, (c) sine wave, (d) sine wave transform.

As a second example, consider the sine wave of (1.3.1) (Figure 1.7c):

$$x(t) = A \cos (\omega_c t + \psi) \qquad (1.4.8)$$

This has the frequency function

$$X(\omega) = \pi A e^{j\psi} \delta(\omega - \omega_c) + \pi A e^{-j\psi} \delta(\omega + \omega_c) \qquad (1.4.9)$$

where $\delta(\omega - \omega_c)$ is the *delta*, or *impulse*, *function* located at ω_c. The amplitude function for this transform is shown in Figure 1.7d. Thus a sine wave at frequency ω_c has its frequency function concentrated at the points $\pm \omega_c$ along the frequency axis. Conversely, we can say that delta functions in the frequency domain imply purely sinusoidal components in the corresponding time waveform. Delta functions along the frequency axis are often called frequency *lines*.

In order to discuss noise and random signals in communication systems, we are forced to deal with random processes. Although a rigorous development of the theory of random analysis is well beyond our scope here,[†] we should point out several basic definitions that continue to arise in a later analysis. Random waveforms can be described statistically only in time. That is, at each point in time t, an unobserved value of the waveform is a random variable described by its probability density at that t. From this density, probabilities, moments, and averages of the process at that t can be determined. For example, if $x(t)$ is a random process, then at $t = t_1$, $x(t_1)$ is a random variable having probability density $p_{x_1}(\xi)$, which statistically describes the process at t_1 (Figure 1.8). The probability that the process $x(t)$ will have a value in the range (a, b) at time t_1 is then given by

$$\text{Prob} [a \le x(t_1) \le b] = \int_a^b p_{x_1}(\xi) \, d\xi \qquad (1.4.10)$$

Similarly, the mean and mean squared value (first two moments) of the process $x(t)$ at t_1 are given by

$$\text{mean of } x(t_1) = \mathscr{E}[x(t_1)] = \int_{-\infty}^{\infty} \xi p_{x_1}(\xi) \, d\xi \qquad (1.4.11a)$$

$$\text{mean squared of } x(t_1) = \mathscr{E}[x^2(t_1)] = \int_{-\infty}^{\infty} \xi^2 p_{x_1}(\xi) \, d\xi \qquad (1.4.11b)$$

Here we have introduced the expectation operator $\mathscr{E}[x(t_1)]$, which represents the statistical average of the random variable $x(t_1)$ over its density.

[†]A short review of probability, random variables, and random processes is presented in Appendix B. The reader may wish to review this appendix before continuing.

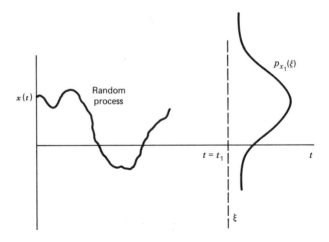

Figure 1.8. Random process and its probability density.

Any other average or moment at t_1 can be computed in a similar manner. In general, the probability density of a random process depends on t, and $p_{x_1}(\xi)$ will be different for different points t_1. This means that the probabilities and moments in (1.4.10) and (1.4.11) will be a function of t_1. However, if the process is *stationary*,[†] then $p_{x_1}(\xi)$ is the same for any point t_1, and the statistical description of the process is the same at all points in time.

Two points in time, t_1 and t_2, define a pair of random variables, $x(t_1)$ and $x(t_2)$, statistically described by their joint second order density $p_{x_1 x_2}(\xi_1, \xi_2)$. The probability that the process $x(t)$ lies in the interval (a, b) at time t_1 and lies in the interval (c, d) at time t_2 is then given by

$$\text{Prob}\left[a \leq x(t_1) \leq b \text{ and } c \leq x(t_2) \leq d\right] = \int_a^b \int_c^d p_{x_1 x_2}(\xi_1, \xi_2)\, d\xi_2\, d\xi_1$$

(1.4.12)

Likewise, a sequence of time points $(t_1, t_2, \ldots, t_n)$ defines a sequence of random variables $x(t_1), x(t_2), \ldots, x(t_n)$, described by their joint nth order probability density. A complete description of a random process requires knowledge of all such joint densities for all possible point locations and orders. The most basic type of noise process encountered in communications is the *Gaussian process*, in which $x(t)$ at any t is a Gaussian random variable.

[†]There are various degrees of stationarity that can be formally defined for a random process (see Section B.3). For simplicity in discussion we do not distinguish precisely between the specific forms of stationarity in subsequent analysis. Noise processes typically encountered in communication systems are generally considered stationary processes.

Gaussian processes have the basic property that all joint densities of any order are described by Gaussian probability densities (Section B.5). The modeling of noise and signal processes as Gaussian processes in communication analysis is quite prevalent, and fortunately, in most cases, this assumption can be physically justified as well as being mathematically convenient.

Rather than deal with probability densities in analysis, we often choose to work with simpler descriptions of random processes. One simple measure of the degree of randomness of a process $x(t)$ is indicated by the process *autocorrelation* function

$$R_x(t, \tau) = \mathcal{E}[x(t)x(t + \tau)] \tag{1.4.13}$$

where $\mathcal{E}$ now denotes a statistical average over the joint density of the process $x(t)$ at time t and $t + \tau$. Note that $R_x(t, \tau)$ is simply the correlation moment of the pair of random variables $x(t)$ and $x(t + \tau)$. For stationary processes the autocorrelation in (1.4.13) does not depend on t, but only on the time separation τ between the points. That is, all pairs of values of $x(t)$ at points in time separated by τ sec have the same correlation value. We can therefore denote this stationary correlation function as simply $R_x(\tau)$. It can be easily shown (see Appendix B, Section B.4) that stationary autocorrelation functions $R_x(\tau)$ are always real, even in τ [i.e., $R_x(-\tau) = R_x(\tau)$], and always take on their maximum value at $\tau = 0$. For zero mean processes $R_x(\tau)$ indicates the extent to which the random values of the process separated by τ sec in time are statistically correlated. If $R_x(\tau)$ is highly positive at a particular τ, it implies that values of the process separated by this τ are positively correlated. This means that if the process takes on a positive (negative) value at some point in time, then the value of the process τ sec away will, with a high probability, also be positive (negative). Similarly, if $R_x(\tau)$ is negative at a particular τ, it implies points separated by this τ are negatively correlated, and therefore are most likely of opposite sign. If $R_x(\tau) = 0$, the points involved are uncorrelated, and no statistical inference about one can be inferred from the other. Thus autocorrelation functions present a descriptive indication of the statistical relation between points of the process. A stationary Gaussian process has the further advantage that all its joint densities (and therefore a complete statistical description) are derivable solely from its autocorrelation function.

The frequency characteristics of a stationary random process $x(t)$ are exhibited by its *spectral density*, $S_x(\omega)$, defined as the Fourier transform of the process autocorrelation function

$$S_x(\omega) = \int_{-\infty}^{\infty} R_x(\tau)e^{-j\omega\tau} \, d\tau \tag{1.4.14}$$

where ω is again the frequency variable in rps. Similarly, by Fourier inversion

$$R_x(\tau) = \frac{1}{2\pi} \int_{-\infty}^{\infty} S_x(\omega)e^{j\omega\tau} \, d\omega \qquad (1.4.15)$$

Thus the correlation function and spectral density of a stationary random process form a transform pair, and each is uniquely derivable from each other. These two functions are two of the more important descriptive functions of noise processes and are used extensively in communication analysis. We point out that since $R_x(\tau)$ is both real and even in τ, the spectral density $S_x(\omega)$ in (1.4.14) must necessarily be real and even in ω (Problem 1.7).

Noise processes are often given special names based on the characteristics of its spectral density. If

$$S_x(\omega) = S_0 \qquad (1.4.16)$$

the process is referred to as a *white noise process*, and the parameter S_0 is called the *spectral level* of the process (Figure 1.9a). A white noise process has an autocorrelation function given by the transformation of (1.4.16). From Section A.1 this transform is

$$R_x(\tau) = S_0 \, \delta(\tau) \qquad (1.4.17)$$

where $\delta(\tau)$ is the delta function, as shown in Figure 1.9b. Note that this correlation function indicates that any point on the process is uncorrelated with any other point of the process. For this reason white noise processes are considered to be the ultimate in randomness. If

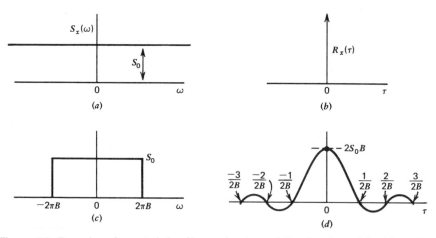

Figure 1.9. Examples of spectral densities and autocorrelation functions: (a) white noise spectrum, (b) white noise autocorrelation, (c) bandlimited white noise spectrum, (d) bandlimited white noise autocorrelation.

$$S_x(\omega) = \begin{cases} S_0, & |\omega| \le 2\pi B \\ 0, & \text{elsewhere} \end{cases} \qquad (1.4.18)$$

the process is called a *bandlimited white noise process* (Figure 1.9c). Such a process has the corresponding autocorrelation function

$$R_x(\tau) = 2BS_0 \frac{\sin(2\pi B\tau)}{2\pi B\tau} \qquad (1.4.19)$$

which is shown in Figure 1.9d. Note that the autocorrelation oscillates between positive and negative correlation values and decays asymptotically to zero as $\tau \to \infty$. As an approximation we can consider $R_x(\tau)$ in (1.4.19) to be approximately zero after several multiples of $1/B$ sec. Hence a band-limited white noise process, having a spectral extent of B Hz, will have a correlation extent of several times $1/B$ sec. That is, values of the process separated by this correlation extent are approximately uncorrelated. We make use of this property in later studies of receiver noise.

Two different random processes, $x(t)$ and $y(t)$, are described by the joint probability densities between their values at specific points in time. The *cross-correlation* of the two processes can be similarly defined as

$$R_{xy}(t, \tau) = \mathscr{E}[x(t)y(t + \tau)] \qquad (1.4.20)$$

The cross-correlation function therefore indicates the statistical correlation between the value of one process at time t and the other at time $t + \tau$. If the processes are jointly stationary, then (1.4.20) does not depend on t, and we can simply write $R_{xy}(\tau)$. Note that the ordering of the indexes is now important (i.e., which is written first), and the function $R_{xy}(\tau)$ is not the same as $R_{yx}(\tau)$ (Problem 1.8). If the processes are zero mean and if $R_{xy}(\tau) = 0$ at $\tau = 0$, values of the two processes are uncorrelated at the same t. If $R_{xy}(\tau) = 0$ for all τ, the two processes are everywhere uncorrelated and are referred to as a pair of *uncorrelated random processes*. Two processes generated from two completely independent sources are always uncorrelated, but the converse is not always true. Again $R_{xy}(\tau)$ serves as an indication of the relative randomness between the two processes. The Fourier transform of $R_{xy}(\tau)$ can be defined as a *cross-spectral density*, but it is difficult to associate physical meaning to it as a frequency function.

1.5 POWER AND BANDWIDTH

Two of the most significant communication parameters associated with a waveform are its *power* and *bandwidth*. The power is an indication of the strength of the signal, whereas the bandwidth indicates its frequency extent. Both of these parameters play a key role in system analysis. For a deterministic waveform $x(t)$ defined over an interval $(0, T)$, the power is given by

$$P_x = \frac{1}{T} \int_0^T x^2(t) \, dt \qquad (1.5.1)$$

If the waveform is defined over all time (i.e., it began an infinite time in the past and will continue an infinite time into the future), P_x must be evaluated as a limiting operation as $T \to \infty$. In this case

$$P_x = \underset{T \to \infty}{\text{Lim}} \frac{1}{2T} \int_{-T}^T x^2(t) \, dt \qquad (1.5.2)$$

For an infinitely long periodic waveform of period t_0, we can let $T = nt_0$ in (1.5.2) and interpret the limit as

$$P_x = \underset{n \to \infty}{\text{Lim}} \frac{1}{2nt_0} \int_{-nt_0}^{nt_0} x^2(t) \, dt$$

$$= \underset{n \to \infty}{\text{Lim}} \left(\frac{1}{2nt_0} \right) 2n \int_0^{t_0} x^2(t) \, dt$$

$$= \frac{1}{t_0} \int_0^{t_0} x^2(t) \, dt \qquad (1.5.3)$$

Thus the power of a periodic waveform is that occurring in a single period. Since $x(t)$ is assumed to be deterministic, P_x in either (1.5.1), (1.5.2), or (1.5.3) can be evaluated by straightforward integration, and application of the limit, if required. For example, (1.5.1) shows that the pulse in (1.4.5) has a power of $P_x = A^2 \tau / \tau = A^2$, whereas (1.5.2) yields the sine wave power in (1.4.8) as

$$P_x = \underset{T \to \infty}{\text{Lim}} \frac{1}{2T} \int_{-T}^T A^2 \cos^2 (\omega_c t + \psi) \, dt$$

$$= \underset{T \to \infty}{\text{Lim}} \frac{1}{2T} \left[\frac{A^2}{2} \left(2T + \frac{2 \sin (2\omega_c T) \cos (2\psi)}{2\omega_c} \right) \right]$$

$$= \underset{T \to \infty}{\text{Lim}} \left[\frac{A^2}{2} + \left(\frac{A^2}{2} \right) \frac{\sin (2\omega_c T) \cos (2\psi)}{2\omega_c T} \right] \qquad (1.5.4)$$

Since $\sin (2\omega_c T)$ is bounded by one for any T, the limit of the second term is zero, and

$$P_x = \frac{A^2}{2} \qquad (1.5.5)$$

Thus the power in a sine wave is always one-half the amplitude squared. This same result could have been derived by treating the sine wave as a periodic function and using (1.5.3). In the laboratory, power in a waveform is measured by a power meter, which essentially performs the operation in (1.5.2).

Note that in the preceding power definitions P_x has units corresponding to the square of the waveform units (volts^2, amp^2, etc.). These units are converted to watts only if $x(t)$ is properly normalized by impedance units. In analyses we find it convenient to deal with power in terms of *decibels*. The power in decibels (dB) is given by

$$(P_x)_{\text{dB}} = 10 \log_{10} P_x \tag{1.5.6}$$

If P_x has the units of watts, $(P_x)_{\text{db}}$ has the units of decibel-watts, denoted dBW. If P_x has the units of milliwatts, $(P_x)_{\text{dB}}$ has the units of decibel-milliwatts, or dBM. Thus 100 W is equivalent to 20 dBW or 50 dBM. Similarly, 0.01 W is equivalent to -20 dBW or 10 dBM.

For random processes, the power of the process at a point t is defined by

$$P_x = \mathscr{E}[x^2(t)] \tag{1.5.7}$$

which is simply the mean squared value of $x(t)$ at time t. As such, P_x depends on the choice of the point t. If the process is stationary, $\mathscr{E}[x^2(t)]$ does not depend on t, and the power of the process is simply taken to be its statistical second moment. Note that for a stationary process, it follows from a comparison of (1.5.7) to (1.4.13) that

$$P_x = R_x(0) \tag{1.5.8}$$

Thus the power is also given by the value of the autocorrelation function at $\tau = 0$. Furthermore, from (1.5.8) and (1.4.15) we can also write

$$P_x = \frac{1}{2\pi} \int_{-\infty}^{\infty} S_x(\omega) e^{j\omega\tau} \, d\omega \Big|_{\tau=0}$$

$$= \frac{1}{2\pi} \int_{-\infty}^{\infty} S_x(\omega) \, d\omega \tag{1.5.9}$$

Hence the power in the process can also be determined by integrating over the spectral density of the process. For this reason $S_x(\omega)$ is also referred to as the *power spectrum* of the process and indicates the distribution of power over the frequency axis. Note the integration in (1.5.9) is over the entire frequency axis, and $S_x(\omega)$ is referred to as a *two-sided spectral density*. However, since $S_x(\omega)$ is even in ω, we can also write

$$P_x = \frac{2}{2\pi} \int_0^\infty S_x(\omega)\, d\omega$$

$$= \frac{1}{2\pi} \int_0^\infty \hat{S}_x(\omega)\, d\omega \qquad (1.5.10)$$

where we have now defined

$$\hat{S}_x(\omega) = \begin{cases} 2S_x(\omega), & \omega \geq 0 \\ 0, & \omega < 0 \end{cases} \qquad (1.5.11)$$

Here $\hat{S}_x(\omega)$ is referred to as a *one-sided spectral density* and is generated from $S_x(\omega)$ by folding the negative frequency part onto the positive frequency part, as indicated in (1.5.11). Note that the one-sided density is integrated over positive frequencies only to obtain total power. For this reason we must carefully distinguish in later analysis between one-sided and two-sided spectral densities in power calculations. It is important to recognize that the power of a random process can be determined directly from its autocorrelation function, using (1.5.8), or from its spectrum, using (1.5.9) or (1.5.10) without the necessity of specifying the actual statistics of the process, that is, using (1.5.7).

We see from Figure 1.9a that the white noise process has its power spread evenly over all frequencies. This means it has infinite power and is therefore not truly a physically meaningful process. Nevertheless, the simpicity of its form makes it a mathematically convenient function to deal with in system analysis. The bandlimited white noise process (Figure 1.9c) has a finite power of

$$P_x = \frac{1}{2\pi} \int_{-2\pi B}^{2\pi B} S_0\, d\omega$$

$$= 2S_0 B \qquad (1.5.12)$$

where S_0 is again the two-sided spectral level. Thus the power is proportional to the spectral level and to twice the positive frequency spectral width.

If two jointly stationary processes $x(t)$ and $y(t)$ are summed to form a new process

$$z(t) = x(t) + y(t) \qquad (1.5.13)$$

the autocorrelation function of the sum process is then

$$R_z(\tau) = \mathscr{E}[z(t)z(t+\tau)]$$

$$= \mathscr{E}\{[x(t) + y(t)][x(t+\tau) + y(t+\tau)]\}$$

$$= R_x(\tau) + R_y(\tau) + R_{xy}(\tau) + R_{yx}(\tau) \qquad (1.5.14)$$

The power of the sum processes $z(t)$ then follows as

$$P_z = R_z(0) = P_x + P_y + 2R_{xy}(0) \tag{1.5.15}$$

If the processes are uncorrelated at the same t [i.e., $R_{xy}(0) = 0$], then the power of the sum is simply the sum of the powers. On the other hand, if the processes are uncorrelated processes, then $R_{xy}(\tau) = R_{yx}(\tau) = 0$, $P_z = P_x + P_y$, and $R_z(\tau) = R_x(\tau) + R_y(\tau)$. That is, the autocorrelation of $z(t)$ is the sum of the individual autocorrelation functions. This also means that the power spectrum of $z(t)$ is the sum of the individual spectra,

$$S_z(\omega) = S_x(\omega) + S_y(\omega) \tag{1.5.16}$$

In later analyses, when dealing with interference composed of the sum of uncorrelated noise processes, (1.5.16) shows that the spectrum of each can be simply summed to obtain the total interference spectrum.

In analyzing deterministic waveforms it is often necessary to deal with the waveform *energy*, defined as the integral portion of the power definition. That is,

$$E_x = \int_{-\infty}^{\infty} x^2(t)\, dt \tag{1.5.17}$$

Note that a waveform with finite energy may have zero power, according to (1.5.2), whereas a periodic waveform must have infinite energy. Thus a waveform in a system may be constrained in either its power or energy values. From Parceval's theorem (Appendix A, Section A.2), (1.5.17) can be equivalently written as

$$E_x = \frac{1}{2\pi} \int_{-\infty}^{\infty} |X(\omega)|^2\, d\omega \tag{1.5.18}$$

where $X(\omega)$ is the Fourier transform of $x(t)$. We see that $|X(\omega)|^2$ plays the role of a waveform *energy spectrum* over which we integrate to get total energy. Thus the energy spectrum of a deterministic waveform is related to waveform energy in the same way that a power spectral density of a random process is related to the process power.

Although the power and energy of a waveform can be defined precisely, the notion of a waveform bandwidth can sometimes be vague and misleading. The basic objective for specifying a waveform bandwidth is to indicate the frequency range over which most of the waveform power is distributed. For deterministic waveforms, this bandwidth information is contained within the waveform energy spectral function $|X(\omega)|^2$. For random processes, it is contained within the power spectral function, $S_x(\omega)$. However, these are many different ways in which the frequency extent may be specified from these frequency functions (see Figure 1.10). The most common definition of bandwidth is the *3 dB bandwidth*. The 3 dB bandwidth is defined as the

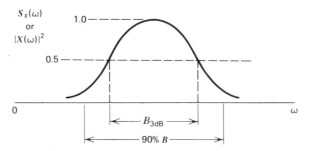

Figure 1.10. Bandwidth definitions.

frequency width along the positive real frequency axis between the frequencies where the frequency function [either $|X(\omega)|^2$ or $S_x(\omega)$] has dropped to one-half of its maximum value. This definition of spectral bandwidth has the advantage that it can be read directly from the frequency function, but has the disadvantage that it may be misleading if the function has slowly decreasing tails. Furthermore, it may become ambiguous and nonunique with multiple peak functions. This definition can be extended to a 6 dB bandwidth (width between 25% points) or a 10 dB bandwidth (width between 10% points), and so on.

Another definition of bandwidth is the *90% bandwidth*, defined as the frequency width along the positive axis within which 90% of the one-sided spectrum lies. Note that this bandwidth is always unique but again may yield misleading values for spectra that have somewhat slowly decaying spectral tails. Although it can sometimes be accurately estimated from the frequency function plot, the 90% bandwidth must be determined by integration for a precise evaluation. Fortunately, most of the waveform spectra encountered in typical communication systems are of a single hump type with relatively fast falloff, much like those in Figure 1.6. For such spectra all the preceding bandwidth definitions yield approximately the same value, and none of these ambiguities occur. Hence it is customary to speak of a general waveform bandwidth without necessarily resorting to a specific definition. We point out that each of the preceding bandwidth definitions is defined along the positive frequency axis only and always defines positive frequency, or *one-sided*, *bandwidths* only.

1.6 CARRIER WAVEFORMS

An important type of waveform encountered in communication systems is the carrier waveform. The *pure*, or unmodulated, carrier was given earlier in (1.4.8) as

$$c(t) = A \cos(\omega_c t + \psi) \qquad (1.6.1)$$

where A, ω_c, and ψ are constants. The pure carrier is therefore a sinusoidal waveform with fixed amplitude A, frequency ω_c in rps selected from the electromagnetic chart in Figure 1.5, and phase angle ψ. Pure carriers are generated by some form of harmonic oscillator, although in practice generation of the ideal carrier of (1.6.1) is hindered by internal oscillator noise and instabilities that cause parameter variations. The assumptions concerning the phase angle ψ become an extremely important part of the carrier waveform characterization. If ψ is considered to be a fixed, known parameter, then the phase of the carrier sinusoid is assumed to be known exactly at any t, and the pure carrier is completely deterministic. Such a waveform has a power of $A^2/2$ and is described in the frequency domain by its Fourier lines at $\pm\omega_c$ (Figure 1.7d). In practice it is necessary to ensure that a known phase is indeed a practical assumption. Knowledge of a carrier phase implies that we know exactly where on a sine wave cycle the carrier will be at the instant it is turned on. For this reason it is often more meaningful to consider ψ in (1.6.1) as a random variable, uniformly distributed over $(0, 2\pi)$. We are then admitting that carrier phase is completely unknown, and subsequent processing must take this into account. Note that a carrier with a random phase ψ becomes a random waveform in time, since at any t its value can only be statistically described. Thus all the averaging operations denoted earlier now involve averages over the probability density of this phase angle. In particular, if ψ is assumed a uniform variable, then $c(t)$ in (1.6.1) has a mean

$$\mathcal{E}[c(t)] = \int_0^{2\pi} A \cos(\omega_c t + \psi)\left(\frac{1}{2\pi}\right) d\psi = 0 \qquad (1.6.2)$$

and an autocorrelation given by

$$R_c(\tau) = \mathcal{E}[c(t)c(t + \tau)]$$

$$= \int_0^{2\pi} A \cos(\omega_c t + \psi)A \cos[\omega_c(t + \tau) + \psi]\left(\frac{1}{2\pi}\right) d\psi \quad (1.6.3)$$

Expanding the cosine product allows us to write this as

$$R_c(\tau) = \frac{A^2}{2}\int_0^{2\pi} \cos(\omega_c \tau)\left(\frac{1}{2\pi}\right) d\psi + \frac{A^2}{2}\int_0^{2\pi} \cos(2\omega_c t + \omega_c \tau + 2\psi)\left(\frac{1}{2\pi}\right) d\psi$$

$$(1.6.4)$$

The second integral is zero for any t and τ, and (1.6.4) reduces to

$$R_c(\tau) = \frac{A^2}{2}\cos(\omega_c \tau) \qquad (1.6.5)$$

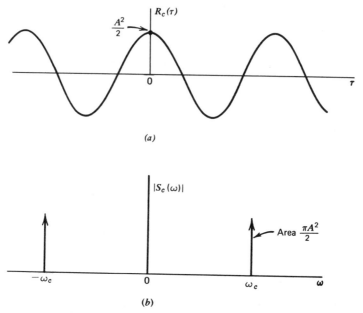

(a)

(b)

Figure 1.11. Carrier waveforms with random phase: (a) autocorrelation function, (b) power spectrum.

Thus random phased pure carriers have autocorrelation functions that are cosinusoidal in τ (Figure 1.11a) exhibiting periodic positive and negative correlation regions. The power spectral density follows as the transform of $R_c(\tau)$,

$$S_c(\omega) = \frac{\pi A^2}{2}\,\delta(\omega + \omega_c) + \frac{\pi A^2}{2}\,\delta(\omega - \omega_c) \qquad (1.6.6)$$

corresponding to spectral lines in the frequency domain[†] (Figure 1.11b). These lines exhibit the concentration of power at the specific frequency ω_c, a result somewhat obvious from (1.6.1) but shown now to be rigorously true. Conversely, delta functions in the spectral density imply random phased sinusoidal components in the corresponding time process. Note that a pure carrier has a theoretical bandwidth of zero width and is often referred to as a *monochromatic* (single frequency) waveform. The power of the random phased carrier follows from (1.5.8) as $P_c = R_c(0) = A^2/2$, which is identical to (1.5.5) for the deterministic sine wave. Thus the pure carrier has the same power whether the phase angle is assumed random or not.

[†]The reader should be careful here to distinguish between delta functions in the waveform transform of a deterministic carrier (Figure 1.6d) and delta functions in the power spectrum of a random phased carrier, as in Figure 1.11b.

The modulated carrier is formed by superimposing a baseband waveform onto a pure carrier waveform. The most prevalent type of modulated carriers are the *amplitude modulated* (AM) carrier, the *phase modulated* (PM) carrier, and the *frequency modulated* (FM) carrier. Understanding the properties of these particular waveforms is therefore a prerequisite for later analysis. The general form of a modulated carrier is given by

$$c(t) = a(t) \cos \left[\omega_c t + \theta(t) + \psi \right] \tag{1.6.7}$$

where $a(t)$ is the nonnegative amplitude modulation of the carrier and $\theta(t)$ is the phase modulation. The AM, PM, and FM carriers are simply special cases of this general carrier waveform. The electronic circuitry and hardware for generating the modulated carriers has been adequately discussed elsewhere [4–8] and need not be rigorously pursued here. Note that a modulated carrier is no longer a pure carrier, and its frequency function can no longer be represented by a pair of delta functions. This means that the modulated carrier tends to spread its power over a frequency band rather than concentrating it at a single frequency. Frequency functions of the specific modulated carriers are considered in detail in Chapter 2.

It is sometimes convenient to rewrite the modulated carrier of (1.6.7) in the alternate form

$$c(t) = \text{Real} \left\{ a(t) e^{j\theta(t)} e^{j\omega_c t + j\psi} \right\} \tag{1.6.8}$$

where Real $\{\cdot\}$ means the real part of $\{\cdot\}$. The time function in braces is now a complex time function, and (1.6.8) is called the *complex representation* of $c(t)$. The term $\exp j(\omega_c t + \psi)$ is called the *complex carrier*, and $a(t) \exp j\theta(t)$ is the *complex envelope function*. By writing the envelope function in terms of its real and imaginary part, $a(t) \cos [\theta(t)] + ja(t) \sin [\theta(t)]$, (1.6.8) can also be expanded as

$$c(t) = a(t) \cos [\theta(t)] \cos (\omega_c t + \psi) - a(t) \sin [\theta(t)] \sin (\omega_c t + \psi) \tag{1.6.9}$$

The preceding equation is an expansion of the general modulated carrier into cosine and sine components and is called a *quadrature expansion* of $c(t)$. At certain points in later analysis it is convenient to rewrite the modulated carrier in (1.6.7) in one of the alternate forms in (1.6.8) or (1.6.9).

When dealing with carrier waveforms that are in the form of a quadrature expansion, it is often necessary to determine the envelope and phase of the resulting waveform. This can be obtained by reconverting (1.6.9) back to (1.6.7). If a quadrature carrier has the form

$$c(t) = A m_c(t) \cos (\omega_c t + \psi) + A m_s(t) \sin (\omega_c t + \psi) \tag{1.6.10}$$

it is equivalent to the carrier waveform in (1.6.7) with envelope

$$a(t) = A[m_c^2(t) + m_s^2(t)]^{1/2} \tag{1.6.11}$$

and phase

$$\theta(t) = \tan^{-1}\left[\frac{-m_s(t)}{m_c(t)}\right] \tag{1.6.12}$$

The *instantaneous frequency* of the carrier waveform $c(t)$ in (1.6.7) is defined as the time derivative of the sinusoidal argument. Hence

$$\omega(t) = \frac{d}{dt}[\omega_c t + \theta(t) + \psi]$$

$$= \omega_c + \frac{d\theta}{dt} \tag{1.6.13}$$

Thus the instantaneous frequency of the modulated carrier does not depend on ψ and varies about the carrier frequency ω_c according to the derivative of the function $\theta(t)$. This phase function $\theta(t)$ may have been intentionally imposed, because of phase or frequency modulation, or may occur unintentionally, because of undesired phase variations of the carrier waveform. If $\theta(t) = 0$, the modulated carrier will have the same instantaneous frequency as the pure carrier. The specific forms of modulated carriers are discussed below.

Amplitude Modulation

The amplitude modulated carrier waveform is produced by linearly multiplying the pure carrier sinusoid by an amplitude modulating waveform $s(t)$. If $m(t)$ represents the baseband waveform to be modulated onto the carrier, the amplitude modulating signal can be written in general form as

$$s(t) = a_0 + b_0 m(t) \tag{1.6.14}$$

where a_0 and b_0 are fixed constants. The AM carrier then has the form

$$c(t) = s(t) \cos(\omega_c t + \psi)$$

$$= [a_0 + b_0 m(t)] \cos(\omega_c t + \psi) \tag{1.6.15}$$

This can be put into the form of (1.6.7) by rewriting it as

$$c(t) = |a_0 + b_0 m(t)| \cos[\omega_c t + \theta(t) + \psi] \tag{1.6.16}$$

where now $\theta(t)$ has the specific form

$$\theta(t) = \begin{cases} 0 & \text{if } s(t) \geq 0 \\ \pi & \text{if } s(t) < 0 \end{cases} \tag{1.6.17}$$

If the modulating waveform in (1.6.14) is always nonnegative, $\theta(t) = 0$, and the AM carrier reduces to simply

$$c(t) = [a_0 + b_0 m(t)] \cos(\omega_c t + \psi)$$

$$= a_0 \left[1 + \frac{b_0}{a_0} m(t) \right] \cos(\omega_c t + \psi) \tag{1.6.18}$$

This corresponds to a sinusoid whose amplitude varies about some average value a_0 in accordance with the modulating waveform $m(t)$ (Figure 1.12a). Such modulation is referred to as *standard* AM. If $s(t)$ becomes negative for any t, the negative sign can be incorporated into the carrier phase as a shift of π radians. Thus the carrier phase, relative to ψ, shifts π radians every time the amplitude modulating waveform changes polarity. If the latter waveform becomes negative for some t, the carrier is said to be *overmodulated*, and the resulting AM waveform is both amplitude and phase modulated. This unintentional phase modulation of an AM carrier caused by overmodulation may or may not be important, depending on the receiver processing. Proper selection of the constants a_0 and b_0 prevents this overmodulation from occurring. If the most negative value of $m(t)$ is normalized to unity, then the required condition is that the parameter b_0/a_0 in (1.6.18) (referred to as the *amplitude modulation index*) be less than one. If $a_0 = 0$, the carrier is overmodulated whenever the modulating waveform $m(t)$

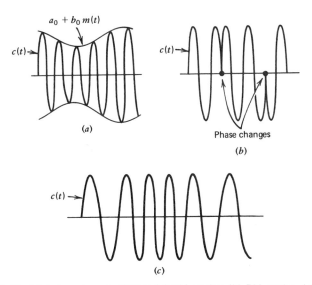

Figure 1.12. Modulated carrier waveforms: (*a*) AM carrier, (*b*) PM carrier, (*c*) FM carrier.

becomes negative. The resulting AM carrier is called a *carrier suppressed AM waveform*. This nomenclature arises since the a_0 term in (1.6.18) corresponds to a pure sine wave carrier within the modulated waveform.

Phase Modulation

A phase modulated carrier has the form

$$c(t) = A \cos \left[\omega_c t + \Delta m(t) + \psi \right] \qquad (1.6.19)$$

where A is the carrier amplitude, $m(t)$ is again the modulating waveform, and Δ is a *phase deviation coefficient*. This coefficient indicates the amount of phase deviation imparted to the carrier and has units of radians per volt. If the maximum value of $m(t)$ is normalized to one, then Δ represents the maximum phase deviation of the carrier. Note that the modulation directly varies the phase of the carrier, and the carrier phase ψ adds directly to the desired modulation. Phase modulated carriers are generated by voltage control of phase shifting networks into which the carrier is inserted. PM carriers have fixed amplitudes and exhibit the modulation only through their phase behavior (Figure 1.12*b*). PM is often called *angle modulation*, as distinguished from the linear modulation associated with AM.

Frequency Modulation

A frequency modulated carrier is a carrier whose frequency is varied according to the modulating waveform and has the form

$$c(t) = A \cos \left[\omega_c t + \Delta_\omega \int m(t)\, dt + \psi \right] \qquad (1.6.20)$$

Hence an FM carrier is simply a phase modulated carrier with the phase modulation proportional to the integral of the modulating waveform. The preceding carrier has an instantaneous frequency defined in (1.6.13) of

$$\omega(t) = \omega_c + \frac{d}{dt} \left[\Delta_\omega \int m(t)\, dt + \psi \right]$$

$$= \omega_c + \Delta_\omega m(t) \qquad (1.6.21)$$

which therefore varies in proportion to the modulating signal $m(t)$ as desired. The proportionality constant Δ_ω is the *frequency deviation coefficient* with units of rps/volt and determines the amount of frequency variation given to the carrier. Again, if $m(t)$ is normalized to a unit peak value, then Δ_ω is the maximum frequency deviation in rps that the carrier will undergo.

FM carriers are produced by varying the tuning frequencies of a resonant oscillator. Such a device is characterized by the fact that its output frequency can be made to vary in accordance with an applied control (modulating) waveform. Unfortunately, however, there is a physical limit to the maximum amount by which an oscillator sine wave of a given frequency can have its frequency altered. This means that in (1.6.20) a condition is imposed of the form

$$\Delta_\omega \le \eta \omega_c \tag{1.6.22}$$

where η is a constant, typically in the range 0.05–0.1. This limit to the amount of frequency modulation that can be inserted on a carrier could become a fundamental constraint in FM carrier systems. Note that the FM carrier also has a fixed amplitude, with the modulation information appearing only in the varying frequency (Figure 1.12c) and therefore represents a form of angle modulation.

In summary, we have reviewed specific forms of the general modulated carrier of (1.6.7). The AM, PM, and FM carriers are generated by superimposing the modulating signal onto the parameters (amplitude, phase, or frequency) of the basic sine wave carrier. These particular carrier formats are the most prevalent in communication systems and are examined in more detail in Chapters 2–4. Some modifications of these basic canonic carrier forms can be made so as to achieve some advantage in specific applications. We later comment on some of these alterations.

1.7 FILTERING OF WAVEFORMS

A common operation in communication signal processing is to filter a waveform, that is, pass the waveform through a linear filter (Figure 1.13). Linear filters have their output and input related through the filter *impulse response* $h(t)$ or the filter *transfer function* $H(\omega)$, the latter the Fourier transform of the former. [If the filter transfer function is stated in terms of the Laplace variable, then $H(\omega)$ is obtained by replacing this variable by $j\omega$.] The filter output time waveform $y(t)$ is related to the input waveform $x(t)$ by either of the forms

$$y(t) = \int_{-\infty}^{\infty} h(t - \rho)x(\rho)\, d\rho = \int_{-\infty}^{\infty} h(\rho)x(t - p)\, d\rho \tag{1.7.1}$$

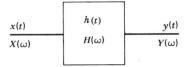

Figure 1.13. Linear filtering model.

These integrals are called *convolutions* of $x(t)$ and $h(t)$. The function $h(t)$ is called the *impulse response* since $y(t) = h(t)$ if an impulse function is applied at $t = 0$ at the input. For a filter to be physically realizable, it is necessary that $h(t) = 0$ for $t < 0$. Similarly, the output waveform transform $Y(\omega)$ is related to the transform $X(\omega)$ of the input by

$$Y(\omega) = H(\omega)X(\omega) \tag{1.7.2}$$

Hence filter output waveforms can be determined by convolving with the impulse response, or by multiplying with the filter transfer function and inverse transforming. By using polar notation we can write

$$Y(\omega) = |H(\omega)| \exp[j\angle H(\omega)]|X(\omega)| \exp[j\angle X(\omega)]$$
$$= |H(\omega)||X(\omega)| \exp\{j[\angle H(\omega) + \angle X(\omega)]\} \tag{1.7.3}$$

The output transform amplitude function follows as

$$|Y(\omega)| = |H(\omega)||H(\omega)| \tag{1.7.4}$$

and the output phase function, as defined in (1.4.4), is

$$\angle Y(\omega) = \angle X(\omega) + \angle H(\omega) \tag{1.7.5}$$

The filter therefore modifies the input transform amplitude by multiplying by $|H(\omega)|$ and adds its phase function to that of the input. The function $|H(\omega)|$ is called the *gain function* of the filter. Thus the filter amplifies the frequencies of the input where its gain function is high and attenuates those frequencies where the gain is low. Undesirable modification of the input amplitude function by the filter is called filter *amplitude distortion*. Addition of a nonlinear phase function by the filter produces *phase distortion*. To prevent significant amplitude distortion by the filter, it is necessary that $|H(\omega)|$ be constant over all ω within the bandwidth of the input waveform. To prevent significant phase distortion by the filter, it is necessary that $\angle H(\omega)$ be linear in ω over the bandwidth of the input.

Typical filters have gain functions that are constant over a specific band of frequencies (called the filter *passband*) and fall off fairly rapidly outside the band, as shown in Figure 1.14. This means filters have bandwidths that can be defined just as with waveform frequency functions. To reduce amplitude distortion it is, therefore, necessary that the filter passband equal or exceed the bandwidth of the input waveform. The usual filter bandwidth definition is the 3 dB bandwidth, defined from the squared gain function $|H(\omega)|^2$. Filter phase functions are generally linear over the passband of the filter gain function and depart from linearity during filter roll-off. Hence maximum phase distortion occurs for values of ω at band edges. If the passband

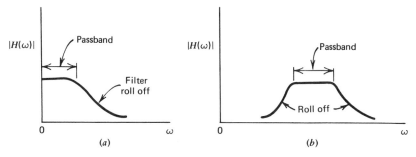

Figure 1.14. Filter gain functions: (a) low pass filter, (b) bandpass filter.

of the filter begins at zero frequency and extends to a higher frequency, the filter is said to be a *low pass filter* (Figure 1.14a). If the passband extends over a range of frequencies away from zero, the filter is said to be a *bandpass filter* (Figure 1.14b).

An *ideal*, or *rectangular*, filter is one having a constant passband gain function with perfect rejection, as shown in Figure 1.15a. Although truly ideal filter functions are unrealizable (Problem 1.25), they can be approximated arbitrarily closely by realizable frequency functions. The two most common are the Butterworth and Chebychev filter functions. The Butterworth function (Figure 1.15b) is given by

$$|H(\omega)|^2 = \frac{1}{1 + [(\omega - \omega_0)/\omega_3]^{2\mu}} \qquad (1.7.6)$$

where ω_3 is the 3 dB bandwidth, ω_0 is the filter center frequency, and μ is the order of the filter, the latter determining its falloff rate. For low pass filters, $\omega_0 = 0$. As the order of the filter increases, the Butterworth function approaches the ideal bandpass (low pass, if $\omega_c = 0$) filter function. Chebychev filter functions (Figure 1.15c) have the form

$$|H(\omega)|^2 = \frac{1}{1 + \epsilon^2 C_\mu^2 [(\omega - \omega_0)/\omega_3]} \qquad (1.7.7)$$

where $C_\mu(\omega)$ is the Chebychev polynomial [9] of order μ. Chebychev filter functions exhibit a passband ripple of height $\epsilon^2/2$ and a falloff rate that increases with order for improved out-of-band rejection. For a specific value of μ and the same 3 dB bandwidth, a Butterworth filter has flatter inband gain, whereas a Chebychev filter has slightly better out-of-band rejection. Both Butterworth and Chebychev filter functions can be constructed by passive networks with properly located filter poles [9, 10].

If the input to the filter is a random process, then the output in (1.7.1) will also evolve as a random process. The statistics of the output process,

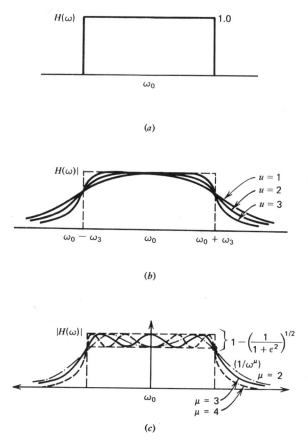

Figure 1.15. The ideal filter function: (*a*) the exact form, (*b*) approximation by Butterworth filters, (*c*) approximation by Chebychev filters.

however, are related in a rather complicated way to those of the input process (see, e.g., Reference 11, Chapter 7). Two important exceptions are when the input is a Guassian process, in which case the output is also a Gaussian process, and when the bandwidth of the input process is much larger than the filter bandwidth. In the latter case the output has been shown to be approximately Gaussian no matter what the input statistics [12]. There are, however, relatively straightforward expressions relating input and output correlation and spectral densities for any process.

The autocorrelation of the output of a linear filter with impulse response $h(t)$ is shown in Appendix B.4 to be

$$R_y(\tau) = \int\limits_{-\infty}^{\infty}\!\!\int R_x(\tau + \rho - \alpha)h(\rho)h(\alpha)\, d\rho\, d\alpha \qquad (1.7.8)$$

where $R_x(\tau)$ is the input autocorrelation function. The output power spectrum is obtained by Fourier transforming (1.7.8) (see B.4.22) to produce

$$S_y(\omega) = S_x(\omega)|H(\omega)|^2 \qquad (1.7.9)$$

where $S_x(\omega)$ is the power spectrum of the input and $H(\omega)$ is the filter transfer function. Hence the square of the filter gain function filters the power spectral density of the input to produce the spectral density of the output. Note that the resulting power in the output process follows as

$$P_y = \frac{1}{2\pi} \int_{-\infty}^{\infty} S_x(\omega)|H(\omega)|^2 \, d\omega \qquad (1.7.10)$$

If the input process is a white noise process of spectral level S_0 W/Hz, then (1.7.10) becomes

$$P_y = S_0 \left(\frac{1}{2\pi} \int_{-\infty}^{\infty} |H(\omega)|^2 \, d\omega \right) \qquad (1.7.11)$$

It is conventional to define

$$B_n = \frac{1}{2\pi} \int_{0}^{\infty} |H(\omega)^2| \, d\omega \qquad (1.7.12)$$

as the *noise bandwidth* of the filter. The output power in (1.7.10) can then be written as simply

$$P_y = 2S_0 B_n \qquad (1.7.13)$$

Thus the filter noise bandwidth is the effective bandwidth determining the output noise power when a white noise process is inserted at the input. Note that if a filter has a constant transfer function, $H(\omega) = \sqrt{G}$, the output power is given by

$$P_y = GP_x \qquad (1.7.14)$$

The parameter G is referred to as the *power gain* of the filter. Power gain is often stated in decibels, with

$$(G)_{\text{dB}} = 10 \log_{10} G \qquad (1.7.15)$$

If $G < 1$, representing a power loss or an attenuation, $(G)_{\text{dB}}$ will be negative.

If two filters are placed in cascade (Figure 1.16a), their overall filter function $H(\omega)$ is given by the product of their individual functions. Thus

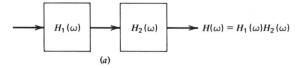

(a)

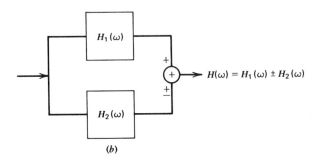

(b)

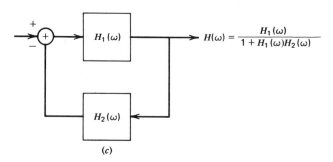

(c)

Figure 1.16. Filter combinations: (a) cascade, (b) parallel, (c) feedback.

$$H(\omega) = H_1(\omega)H_2(\omega) \tag{1.7.16}$$

where the $H_i(\omega)$ are the individual filter functions. If a filter function has a peak value of $\sqrt{G}$, it is convenient to factor its filter function as $\sqrt{G}H(\omega)$, where $H(\omega)$ is now a normalized filter function of unit peak value. Using (1.7.16), we can then interpret the filter as the cascade of a device with constant power gain G and a unit gain filter function $H(\omega)$. This separation allows us always to deal with unit level frequency functions in our subsequent analysis. If the filters are placed in parallel (Figure 1.16b),

$$H(\omega) = H_1(\omega) \pm H_2(\omega) \tag{1.7.17}$$

whereas if the filters are placed in a negative feedback arrangement (Figure 1.16c),

$$H(\omega) = \frac{H_1(\omega)}{1 + H_1(\omega)H_2(\omega)} \qquad (1.7.18)$$

These results can be easily derived by basic linear system theory (Problem 1.30). Their basic advantage is they allow us to build up complicated filter functions from relatively simple filters.

1.8 SIGNAL TO NOISE RATIO

In the actual operation of a communication system the recovered waveforms at the receiver do not generally correspond exactly to the desired waveforms produced in the transmitter. This is due to the anomalies occurring during transmission and reception that cause signal distortion and the insertion of interference and noise waveforms. These effects cause a basic deterioration of the desired waveform and a degradation in the overall communication operation. To assess the capability of the link, it is therefore necessary to account for these effects in system analysis and design. Typically, a specific performance criterion relating desired and actual operation is first decided on. Subsequent design or system comparison is then based on satisfying the criterion. When more than one system design is being considered, a comparison can be made with respect to the decided criterion and the most favorable systems can be determined.

One of the most convenient and widely used measures of performance in communication analysis is the concept of *signal to noise ratio* (SNR). When dealing with a situation in which a desired waveform, whether it be an information signal, baseband waveform, or modulated carrier, is contaminated with an additive, interfering waveform, the SNR of the combined waveform at that point in the system is defined as

$$\text{SNR} = \frac{\text{power in the desired waveform}}{\text{power of the interfering waveform}} \qquad (1.8.1)$$

Thus SNR indicates how much stronger the desired waveform is relative to the interference at the point in question. If SNR is greater than one, there is more power in the desired signal than in the interference, and the opposite is true if SNR is less than one. Adequate system design generally requires the desired signal power to be many times larger than that of the interference. Thus $\text{SNR} \gg 1$ is generally required in design. Note that the SNR parameter depends only on power levels of the signals involved, which is the basic reason for its wide usage. It is, therefore, easily measured, and applicable whether dealing with deterministic waveforms or random processes. Furthermore, its computation does not require knowledge of the properties and statistics of the random waveforms involved.

It is common to express SNR in decibels. Hence if P_d is the power of the desired signal and P_n is the power of the interference, then

$$(\text{SNR})_{\text{dB}} = 10 \log_{10} \frac{P_d}{P_n}$$
$$= 10 \log_{10} P_d - 10 \log_{10} P_n$$
$$= (P_d)_{\text{dB}} - (P_n)_{\text{dB}} \qquad (1.8.2)$$

Thus $(\text{SNR})_{\text{dB}}$ is the difference of the two powers expressed in decibels. When the interference is composed of the sum of two separate uncorrelated, interfering waveforms, say, $n_1(t)$ and $n_2(t)$, the SNR in (1.8.1) is then

$$\text{SNR} = \frac{\text{power of the desired signal}}{\text{power of } n_1(t) + \text{power of } n_2(t)} \qquad (1.8.3)$$

The result can be easily extended to more than two interfering waveforms as well. We emphasize that the SNR is associated with a particular point or location within the system. Different points often generate different values for SNR since the interference may vary throughout the link. When the desired signal involved is a carrier waveform, we often speak instead of the *carrier to noise ratio* (CNR) instead of SNR.

Care must be used in dealing with SNR when signal distortion has occurred. The recovered signal waveform may no longer correspond to the desired form, even though its power remains the same. To include distortion effects, we define a desired signal error waveform

$$e(t) = s_d(t) - s(t) \qquad (1.8.4)$$

between the recovered signal $s(t)$ and the desired signal $s_d(t)$. By rewriting this as

$$s(t) = s_d(t) - e(t) \qquad (1.8.5)$$

it appears as if the recovered signal has been formed by adding an error "noise" to the desired waveform. Neglecting any additive interference, we can treat the error waveform as an equivalent noise and define a *signal to distortion ratio* (SDR) as

$$\text{SDR} = \frac{\text{power in desired signal}}{\text{power in the error waveform}} \qquad (1.8.6)$$

Hence SDR is comparable in interpretation to SNR and is generally treated as such. However, it must be remembered that the "noise" is actually a distortion error, and in fact may be itself dependent on the desired signal.

When the distorting system corresponds to a linear filter and $s(t)$ and $s_d(t)$ are the output and input signals, respectively, the error waveform in (1.8.4) has the transform

$$E(\omega) = F_{s_d}(\omega) - F_{s_d}(\omega)H(\omega)$$

$$= [1 - H(\omega)]F_{s_d}(\omega) \qquad (1.8.7)$$

where $H(\omega)$ is the frequency transfer function of the linear system and $F_{s_d}(\omega)$ is the transform of the desired waveform $s_d(t)$. When written as in (1.8.7), it appears as if the error waveform is generated by an effective filtering of $S_d(t)$ with the function $[1 - H(\omega)]$. If $s_d(t)$ is a random process with power spectral density $S_d(\omega)$, then the error process has the equivalent spectrum $|1 - H(\omega)|^2 S_d(\omega)$. The corresponding SDR in (1.8.6) then becomes

$$\text{SDR} = \frac{P_d}{(1/2\pi)\int_{-\infty}^{\infty}|1 - H(\omega)|^2 S_d(\omega)\,d\omega} \qquad (1.8.8)$$

If there is additional interference to the distorted waveform, the overall SNR can be evaluated as in (1.8.3), using SDR to replace one of the individual SNR_i terms. In effect, this combines the waveform error and the waveform interference into a total interference superimposed on the desired waveform.

1.9 DIGITAL COMMUNICATION SYSTEMS

The continual development of high speed, low cost, low weight switching circuitry has fostered a predominate trend toward digital communications. A digital version of the basic communication link is shown in Figure 1.17. Recall that in a digital system, source information corresponds to binary symbols (bits) that are to be transmitted by the link to the receiver. The most common types of digital source are (1) command generators, in which the bit sequence represents instruction words to be executed at the receiver; (2) numerical symbol generators, in which the source sequence represents

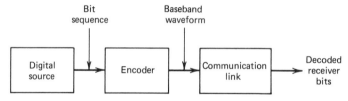

Figure 1.17. The digital communication model.

outputs of computers, graphics, terminals, digital readout, and so on; and (3) analog-to-digital (A-D) convectors, in which original source waveforms are converted into a bit sequence to represent the waveform. The bit sequence is then transmitted to the receiver instead of the actual source waveform, where it is converted back (digital-to-analog (D-A) conversion) to the source waveform. Hence bit sequences from digital sources can represent many different types of information.

In a digital system, the task of the communication link is to transmit the source bits as accurately as possible to the receiver destination. Instead of accurately reproducing a baseband waveform, the design objective in a digital link is to reproduce the bit sequence of the source. Therefore, rather than a waveform quality parameter such as SNR or SDR, a digital communication link is instead constrained by bit reliability. The accepted measure of the bit integrity is the probability that a transmitted bit will be received in error, often simply called the bit *error probability* (PE) or the *bit error rate* (BER). [The latter follows since if N bits are transmitted, then, on the average, $N(PE)$ receiver bits will be in error.] Hence a digital communication link is generally specified to operate with a required PE. Typical values for PE are in the range from 10^{-3} to 10^{-7}, depending on the type of source. For example, a command link may require better PE performance (withstand fewer bits in error) than an A-D conversion system.

Since PE is, by definition, an event probability it can be determined only from a true statistical analysis at the receiver. In general, this involves more than simply a power analysis, as is used to determine SNR. Hence PE values in digital systems are more model oriented, and depend strongly on both the statistical characteristics of the interference and the receiver processing inserted.

In the digital systems in Figure 1.17, source bits are transmitted by first converting (encoding) the bit sequence into a baseband waveform. The generation of this digital waveform (the encoded waveform carrying the source bits) permits the bits to be transmitted by basic carrier modulation formats. At the receiver the recovered digital waveform is converted back (decoded) into the bit sequence. In the simplest bit encoding format each source bit is encoded separately by using a specific waveform to represent each bit. For example, let $s_1(t)$ and $s_0(t)$ be arbitrary waveforms, each T_b sec long, as shown in Figure 1.18a. We now encode each source bit as either $s_1(t)$ or $s_0(t)$. That is, each one-bit from the source is sent as $s_1(t)$, and each zero-bit is sent as $s_0(t)$. A sequence of source bits is therefore converted into a baseband waveform corresponding to a sequence of $s_1(t)$ and $s_0(t)$ as shown in Figure 1.18b. This encoded waveform can be mathematically represented by the general binary digital waveform[†]

[†] Here, and throughout the book, a summation indexed by k alone (Σ_k) represents a sum over all integer values of k, from $-\infty$ to $+\infty$. Restricted sums on k will be so indicated by specifically indexing the summation sign.

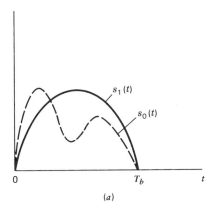

(a)

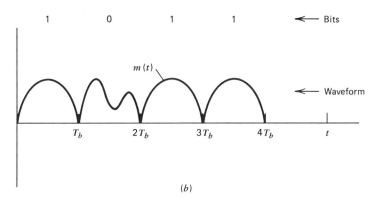

(b)

Figure 1.18. Binary encoding: (a) bit waveforms, (b) encoded bit sequence.

$$m(t) = \sum_k [b_k s_1(t - kT_b) + (1 - b_k)s_0(t - kT_b)] \qquad (1.9.1)$$

where b_k is the kth source bit ($b_k = 0, 1$). Thus $\{b_k\}$ represents the sequence of bits as they are produced by the source. The waveforms $s_i(t)$ in Figure 1.18 are referred to as *bit waveforms*. Note that $m(t)$ is actually a random waveform (because of the random data bits) and carries one source bit every T_b sec, or equivalently has a bit rate of

$$R_b = \frac{1}{T_b} \quad \text{bits/sec} \qquad (1.9.2)$$

for any set of bit waveforms. Baseband bit rate is therefore determined by the time length of the bit waveforms. This, in turn, is inherently related to the switching speed of the encoder.

An important set of bit waveforms are the *antipodal* signal pair, where $s_0(t)$ is selected as $-s_1(t)$. The bit waveforms are negatives of each other, and source bits are encoded into sequences of positive and negative versions of the same waveform (Figure 1.19). The binary digital waveform in (1.9.1) reduces to simply

$$m(t) = \sum_k d_k s(t - kT_b) \tag{1.9.3}$$

where $s(t)$ is the bit waveform, and $d_k = +1$ if $b_k = 1$, and $d_k = -1$ if $b_k = 0$. In antipodal signaling, the encoder uses the same bit waveform but only encodes the polarity of the signal.

For ease of implementation, particular antipodal bit waveforms $s(t)$ may be more advantageous than others. For example, the pulse waveform in (1.4.5)

$$s(t) = \begin{cases} A, & 0 \le t \le T_b \\ 0, & \text{elsewhere} \end{cases} \tag{1.9.4}$$

when used in (1.9.3) generates a digital waveform corresponding to a pulse

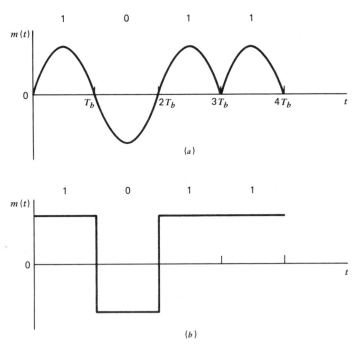

Figure 1.19. Encoding with antipodal waveforms: (a) arbitrary bit waveforms, (b) pulse waveform.

sequence (Figure 1.19b) that flips in polarity according to the data sequence. These digital pulse waveforms are easy to generate from simple two-state, flip-flop (multivibrator) circuits, with the data bit selecting the device state. Since high speed, fast-switching flip-flops are readily available, this particular waveform is conducive to high rate binary signaling. Note that the digital pulse waveform $m(t)$ in Figure 1.19b has the important property that it is a true binary waveform in that it has only the values $\pm A$ at any time t.

Another important waveform pair is that for which $s_0(t) = 0$; that is, a zero bit is transmitted as the absence of a signal during its $(0, T_b)$ interval. Encoding with this bit waveform is referred to as on–off keying, and source bits are encoded by switching the waveform $s_1(t)$ on or off. This particular signaling format is commonly used in fiberoptic links, in which light pulses are keyed on or off to represent source bits.

The encoding scheme in (1.9.1) is referred to as *binary*, or *bit-by-bit*, encoding. The concept can be easily extended to block encoding, in which an entire block of source bits is encoded into one of a set of waveforms $\{s_i(t)\}$. Further discussions of block and binary encoding, as well as best choices of the encoding waveforms $s_i(t)$, is postponed until Chapters 7–9. In fact, a comprehensive theory of digital encoding and decoding has been well developed and will be presented somewhat in detail in those chapters. Prior to applying that theory, however, it is first necessary to understand how baseband waveforms are modulated, transmitted, and received and how the various sources of noise, interference, and distortion enter the link (see Chapters 2–6).

In summary, this first chapter has presented an overview of a typical communication system, from information source to final destination. Several examples of familiar systems were discussed with respect to this model. Our prime objective was to delineate the overall system structure and indicate the various subsystems and functional operations comprising the typical link. A review of waveform analysis, noise theory, and linear system theory was presented, along with some key system definitions that will be used extensively. We now begin to examine somewhat more rigorously the component subsystems and to indicate their physical and mathematical characterization. It is often convenient to examine these subsystems in groups so as to portray their interconnections more clearly. We begin in the next chapter by considering baseband waveforms and the modulated carriers they create.

REFERENCES

1. Bracewell, R. *The Fourier Transform*, McGraw-Hill, New York, 1965.
2. Abramowitz, M. and Stegun, I. *Handbook of Mathematical Functions*, National Bureau of Standards, Washington, D.C., 1965, Chap. 29.

3. Sneddon, I. *Fourier Transforms*, McGraw-Hill, New York, 1958.

4. Terman, F. *Electronic and Radio Engineering*, McGraw-Hill, New York, 1955.

5. Panter, P. *Modulation, Noise, and Spectral Analysis*, McGraw-Hill, New York, 1965.

6. Gregg, W. *Analog and Digital Communications*, Wiley, New York, 1977.

7. Smith, J. *Modern Communication Circuits*, McGraw-Hill, New York, 1986.

8. Carlson, B. *Communication System*, 3rd ed., McGraw-Hill, New York, 1986.

9. Van Kalkenberg, M.E. *Introduction to Modern Network Synthesis*, Wiley, New York, 1960, Chap. 13.

10. Balabanian, N. *Network Synthesis*, Prentice-Hall, Englewood Cliffs, N.J., 1958.

11. Papoulis, A. *Probability, Random Variables, and Stochastic Processes*, McGraw-Hill, New York, 1965.

12. Middleton, D. *Statistical Communication Theory*, McGraw-Hill, New York, 1960.

PROBLEMS[†]

1.1   (1.4)   Show that if $x(t)$ is real, then $|X(\omega)|$ is even in $\omega [|X(-\omega)| = |X(\omega)|]$ and $\angle X(\omega)$ is odd $[\angle X(-\omega) = -\angle X(\omega)]$.

1.2   (1.4)   Determine the amplitude function $|X(\omega)|$ of the waveform

$$x(t) = \begin{cases} A \cos(\omega_0 t + \psi), & 0 \le t \le T \\ 0, & \text{elsewhere} \end{cases}$$

[*Hint:* Treat $x(t)$ as the product of a sinusoid and a pulse and use frequency convolution (Section A.1).]

1.3   (1.4)   Using transforms, verify the frequency spreading that occurs as pulses are made narrower, using pulses that are time waveforms composed of Gaussian-shaped time pulses.

1.4   (1.4)   Starting with the fact that the waveform energy is finite, show that for any transform pair with bounded time and frequency functions

$$\int_{-\infty}^{\infty} |x(\omega)| \, d\omega \int_{-\infty}^{\infty} |x(t)| \, dt = \text{constant}$$

Comment on what this means in terms of the relation between the time and frequency extent of a waveform.

[†]Numbers in parentheses after problem numbers refer to the section required for their solution.

1.5   (1.4)   A random voltage $x(t)$ has a first order density

$$p_{x_t}(x) = \tfrac{1}{20}, \qquad -10 \le x \le 10$$

at any t. What is the probability that $x(t)$ will have a voltage value between 5 and 7 V at time $t = 5$ sec?

1.6   (1.4)   Given a stationary Gaussian random process with zero mean and correlation $R_x(\tau) = 2e^{-5|\tau|}$, set up the integral for determining the probability that the process is in the interval $(4, 6)$ at $t = 5$ sec and in the interval $(5, 8)$ at $t = 10$ sec.

1.7   (1.4)   Show that since $R_x(\tau)$ is real and even in τ, the spectral density $S_x(\omega)$ must necessarily be real and even in ω [i.e., $S_x(-\omega) = S_x(\omega)$].

1.8   (1.4)   Let $x(t)$ and $y(t)$ be jointly stationary random processes. Show that (a) $R_{xy}(-\tau) = R_{yx}(\tau)$, (b) $R_{xy}(\tau) + R_{yx}(\tau)$ is even in τ.

1.9   (1.4)   Using (B.2.16) of Appendix B, show that the second order characteristic function of a zero mean stationary Gaussian process at times t and $t + \tau$ is given by

$$\Psi(\omega_1, \omega_2, \tau) = \exp \tfrac{1}{2}[R_x(0)\omega_1^2 + R_x(0)\omega_2^2 + 2R_x(\tau)\omega_1\omega_2]$$

1.10   (1.5)   Convert the following power values to decibels: (a) 3 W, (b) 0.25 W, (c) 25 mW, (d) 230 W. Convert the following to watts: (e) 3 dBW, (f) 70 dBM, (g) -15 dBW.

1.11   (1.5)   Convert the following numbers to decibels:
(a) 200   (b) 0.01   (c) 1/8   (d) 27   (e) 10^4
Convert the following to numbers:
(a) 37 dB   (b) 6 dB   (c) -6 dB   (d) 20 dB   (e) -190 dB

1.12   (1.5)   Determine the 3 dB, 6 dB, and 90% bandwidths of the frequency function

$$|X(\omega)|^2 = \frac{2}{1 + 0.25\omega^2}$$

1.13   (1.5)   The *rms bandwidth* of a low pass power spectral density $S_x(\omega)$ is defined by

$$B_{\text{rms}} = \left[\frac{1}{2\pi P_x} \int_{-\infty}^{\infty} \omega^2 S_x(\omega)\, d\omega \right]^{1/2}$$

where P_x is the power in $S_x(\omega)$. Show that the power in a bandwidth of kB_{rms} is bounded by $[1 - (1/k^2)]P_x$.

1.14 (1.5) Show that the variable transformation

$$jw \rightarrow \frac{\omega_c}{B} \left[\frac{jw}{\omega_c} + \frac{\omega_c}{jw} \right]$$

converts a low pass function to a bandpass function.

1.15 (1.6) Let $x(t) = A \cos(\omega_c t + \psi)$, where ψ is a random phase variable with probability density $p_\psi(\xi)$. (a) Find the mean of $x(t)$ if ψ has a Gaussian distribution with mean zero and variance σ_ψ^2. (b) Find the condition on $p_\psi(\xi)$ needed for the mean of $x(t)$ to be zero.

1.16 (1.6) Let $q(t)$ be a $(0, 1)$ square wave. Show that phase modulating with $\pi q(t)$ is identical to amplitude modulating with $[1 - 2q(t)]$.

1.17 (1.6) Three carriers at the same frequency and amplitude but with different phases are summed, forming

$$c(t) = \sum_{i=1}^{3} A \cos(\omega_c t + \psi_i)$$

Determine the envelope and phase of the resulting $c(t)$.

1.18 (1.6) Using the complex envelope representation of a modulated carrier, show how its imaginary part is related to the instantaneous frequency of the carrier.

1.19 (1.6) Show that a filter with a unit gain and linear phase function $H(\omega) = e^{-j\alpha\omega}$, does not distort, but only delays, any input waveform.

1.20 (1.7) Given the RC network in Figure P1.20: (a) find $|H(\omega)|$ and $\angle H(\omega)$; (b) repeat for N cascaded stages of this type (neglect network loading).

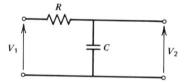

Figure P1.20.

1.21 (1.7) Consider the network in Figure P1.20, with $R = 10,000$ ohms and $C = 10^{-6}$ F. The sine wave $10 \sin(2\pi10t + \pi/9)$ is applied. (a) State the output time function. (b) If the output has an amplitude of 5 when the input sine wave amplitude is 10, what is the frequency of the input sine wave?

1.22 (1.7) Given the linear system described by the differential equation

$$\frac{dy}{dt} + 2y(t) = 3x(t)$$

where $x(t)$ is the input and $y(t)$ the output. Determine $h(t)$ and $H(\omega)$ for this system.

1.23 (1.7) A square pulse of width T sec and height A is applied to a first order low pass filter with impulse response $h(t) = 2\pi B e^{-2\pi B t}$, $t \geq 0$, where B is the filter 3 dB bandwidth. (a) Compute and sketch the output response. (b) Assume $T \gg 1/B$ and find the time in terms of B for the output to rise to 90% of its peak amplitude.

1.24 (1.7) The impulse response of a network is $h(t)$. Show that its response to a unit pulse of width τ is given by

$$\int_{-\infty}^{t} h(\rho)\, d\rho - \int_{\tau}^{t} h(\rho - \tau)\, d\rho$$

(*Hint:* Treat the pulse as two step functions τ sec apart.)

1.25 (1.7) A realizable filter must have an impulse response that is zero for negative t. Show that the ideal, rectangular low pass filter is unrealizable.

1.26 (1.7) What must the order of a low pass Butterworth filter be in order that $H(\omega)$ is down by at least 30 dB at $\omega = 10\omega_3$?

1.27 (1.7) Determine the noise bandwidth of a low pass Butterworth filter function with a 3 dB bandwidth of ω_3 and order μ.

1.28 (1.7) Given a linear system with impulse response $h(t)$, show that the input–output cross-correlation is related to the input autocorrelation function $R_x(\tau)$ of a stationary random process by

$$R_{yx}(\tau) = \int_{-\infty}^{\infty} h(\rho) R_x(\tau + \rho)\, d\rho$$

1.29 (1.7) A differentiator has the filter transform function $H(\omega) = j\omega$. If the input is a random process $x(t)$, show that the output autocorrelation is the negative second derivative of the input autocorrelation, that is,

$$R_y(\tau) = R_{dx/dt}(\tau) = -\frac{d^2}{d\tau^2} R_x(\tau)$$

(*Hint:* First determine the output spectral density.)

1.30 (1.7) Prove the overall transfer function result for the cascade, parallel, and feedback networks in (1.7.16)–(1.7.18). (*Hint:* Apply input inpulse functions, trace the response through the system, and write the transform of the output.)

1.31 (1.8) (a) A dc signal of 5 V is desired at a point in a system. Instead the waveform observed is $5 + 0.2 \sin (2\pi 10t + \pi/3)$. Determine the SDR of the waveform at that point. (b) Repeat with the observed waveform at $5 + n(t)$ V, where $n(t)$ is a stationary, Gaussian random process with correlation function $R_n(\tau) = 0.02e^{-\tau/10}V^2$.

1.32 (1.8) The input to a linear system $H(\omega)$ is a random signal with spectral density $S_d(\omega)$ and additive independent noise with spectral density $S_n(\omega)$. Show that the output SDR is

SDR

$$= \frac{P_d}{(1/2\pi)\int_{-\infty}^{\infty}|1 - H(\omega)|^2 S_d(\omega)\, d\omega + (1/2\pi)\int_{-\infty}^{\infty}|H(\omega)|^2 S_n(\omega)\, d\omega}$$

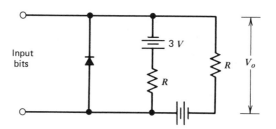

Figure P1.33.

1.33 (1.9) A flip-flop circuit is shown in Figure P1.33. The diode is either forward biased (short circuit) or reversed bias (open circuit). Assuming that one-bit forward biases the diode while a zero-bit reverse biases it, sketch the output voltage as a function of the input bit sequence 1 0 1 1 0.

2

CARRIER MODULATION

In most communication systems, information is transmitted by first modulating the source baseband waveforms onto a carrier to form the modulated carrier. In order to analyze and design the communication system properly, it is first necessary to understand the properties of the modulated carrier waveforms involved. In this chapter we present a more detailed modulated carrier description, in terms of power levels and spectral properties.

The specific modulated carriers introduced in Section 1.6 are reconsidered. We first examine some basic source waveforms and then extend to the modulated carriers they create. Later, in Chapter 6, more complicated forms for the modulating waveforms are considered.

2.1 BASEBAND WAVEFORMS

Carrier modulation is achieved by a modulation subsystem, as depicted in Figure 2.1. The baseband waveform generated from the information sources is modulated onto a pure carrier waveform to form the modulated carrier. Modulation can be achieved by any of the standard formats discussed in Chapter 1, with modulation occurring in either amplitude, frequency, or phase. In order to determine the properties of the modulated carrier, it is first necessary to characterize the baseband signal itself. The baseband signal can be the direct output of an audio, video, message, or other source or can be formed as a combination of these via baseband multiplexing.

One of the most important baseband waveforms is the common voice signal—the electronic waveform generated from a telephone headset, microphone, or audio amplifier. A typical voice signal has the power spectral density shown in Figure 2.2a. Although voice frequencies exist out to about 10 kHz, most of the speaker's energy is confined to a bandwidth out to approximately 4 kHz, with the power scaling dependent on the actual speaker (male, female, high pitch, low pitch, etc.). A low pass 4 kHz basebandwidth is generally sufficient to adequately reproduce the human

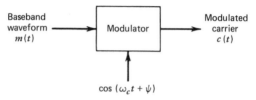

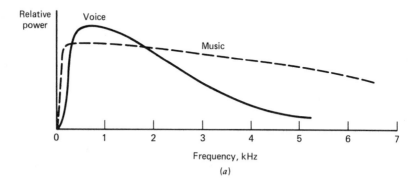

Figure 2.1. Carrier modulation model.

Figure 2.2. Source power spectra: (a) audio, (b) television.

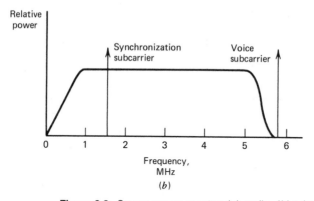

voice. Music bandwidths on the other hand are significantly wider as shown, with audio frequency components out to about 15 kHz. High quality (stereo) music reproduction therefore requires bandwidths at least this large to adequately represent the source waveform.

Video (television) waveforms are generated as electronic signals from video cameras or imaging devices. The camera forms its electronic signal by scanning the imaged scene and producing a voltage proportional to the light intensity at each point during its scan. In standard TV cameras, the scene is scanned once every $1/60$ sec to form a field of 263 lines. The interlacing of two such fields forms a scene frame of 525 lines every $2/60$ sec. A line scan is therefore $(2/60)/525 = 63.5\ \mu$sec of which $10\ \mu$sec is needed for horizontal retrace. The line sweeps produce a vertical resolution of about 365 vertical scene samples per frame. With a scene aspect ratio of $4/3$, a television picture is designed to produce $(4/3)(365) = 485$ horizontal resolution samples (pixels) per line. If a scene intensity change occurs from one horizontal pixel to the next, the system must have the capability of discerning light changes every $(63.5 - 10)/(485/2) = 0.22\ \mu$sec. Thus a TV camera produces an electronic video signal having frequency components as high as $1/0.22\ \mu$sec $= 4.5$ MHz. Figure 2.2b shows the power spectrum of a typical video signal. In the United States a commercial television channel is allowed an overall bandwidth of 6 MHz for its transmission. The delta functions within the spectrum signify the presence of sinusoidal components used for TV synchronization and audio transmission. We see that the basebandwidth of a video signal is about 10^3 times wider than a typical audio signal.

A facsimile system also transmits an image by scanning a written document or graphic. Since there is no scene motion involved, there is no requirement for rapid scanning, and the line scan rate can be slowed to produce a smaller bandwidth. This permits the facsimile information to be sent over low rate (voice) channels. If the scanning produces n_v horizontal lines, each scanned at the rate of T_{line} sec per line, the facsimile will be printed out in $T_{\text{frame}} = n_v T_{\text{line}}$ sec. If the image is reconstructed with the same horizontal resolution as a television scene, a line will have $(4/3)n_v$ horizontal resolution samples. This can produce a light intensity change every $T_{\text{line}}/(4/3)n_v$ sec and a bandwidth of $B = (1.33)n_v^2/T_{\text{frame}}$. Thus an inverse relation exists between the required facsimile bandwidth and its printout rate. For example, if there are 30 line scans per inch, then an 8×10 inch page, transmitted by facsimile over a 4 kHz link, will require about 30 sec to print out.

Another important baseband signal is the binary digital waveform in (1.9.1) generated from digital sources. Since digital communication systems are of prime importance, it is mandatory that we understand the properties of these waveforms. Recall the antipodal digital waveform, which has the form

$$m(t) = \sum_k d_k s(t - kT_b) \qquad (2.1.1)$$

where T_b is the bit time, $s(t)$ is the bit waveform over $(0, T_b)$, and $d_k = \pm1$ depending on the kth data bit from the source. From (1.9.2) the waveform carries bits at a rate of $1/T_b$ bits per sec for any bit waveform $s(t)$.

The average power of the digital waveform is computed directly from (1.5.3) as

$$P_m = \lim_{n \to \infty} \sum_{k=-n}^{n} \frac{1}{2nT_b} \int_{kT_b}^{(k+1)T_b} d_k^2 s^2(t - kT_b) \, dt$$

$$= \lim_{n \to \infty} \frac{2nd_k^2}{2nT_b} \int_0^{T_b} s^2(t) \, dt$$

$$= \frac{1}{T_b} \int_0^{T_b} s^2(t) \, dt$$

$$= P_s \tag{2.1.2}$$

The data modulated digital waveform therefore has the same power as the bit waveform $s(t)$.

To determine the spectral characteristics of the binary waveform, it is first important to recall that $m(t)$ is a random process, as a result of the random data bits occurring during each bit interval. The sequence $\{d_k\}$ is a sequence of binary random variables, and we assume that the sequence correlation function is known. To determine the power spectrum of $m(t)$, it is necessary to compute its autocorrelation function as in (1.4.13),

$$R_m(t, \tau) = \mathcal{E}[m(t)m(t + \tau)]$$

$$= \mathcal{E}\left[\sum_k \sum_j d_k d_j s(t - kT_b) s(t + \tau - jT_b) \right]$$

$$= \sum_k \sum_j \mathcal{E}(d_k d_j) s(t - kT_b) s(t + \tau - jT_b)$$

$$= \sum_k \sum_j \mu_d(k - j) s(t - kT_b) s(t + \tau - jT_b) \tag{2.1.3}$$

Here the expectation operator $\mathcal{E}$ averages over the statistics of the data sequence. We have assumed the random d_k sequence is stationary and denoted its correlation as

$$\mathcal{E}(d_k d_j) = \mu_d(k - j) \tag{2.1.4}$$

It is apparent that even though the bit sequence is stationary, the correlation function in (2.1.3) of the waveform $m(t)$ is not (i.e., it depends on t). However, the latter correlation is periodic in t, since its value at $(t + T_b)$ is the same as at t. To remove the dependence on t, we compute the time average correlation by averaging over a period of t. Hence

$$R_m(\tau) = \frac{1}{T_b} \int_0^{T_b} R_m(t, \tau) \, dt$$

$$= \frac{1}{T_b} \sum_k \sum_j \mu_d(k - j) \int_0^{T_b} s(t - kT_b) s(t + \tau - jT_b) \, dt$$

$$= \frac{1}{T_b} \sum_k \sum_j \mu_d(k - j) \int_{-kT_b}^{(1-k)T_b} s(z) s(z + \tau + (k - j)T_b) \, dz \tag{2.1.5}$$

where the last equation is obtained by changing the integration variable to $z = t - kT_b$. Since the terms involve only the integer difference $(k - j)$, it is convenient to rewrite the double sum by substituting $q = k - j$ and writing

$$R_m(\tau) = \frac{1}{T_b} \sum_k \sum_q \mu_d(q) \int_{-kT_b}^{(1-k)T_b} s(z)s(z + \tau + qT_b)\, dz$$

$$= \frac{1}{T_b} \sum_q \mu_d(q) \sum_k \int_{-kT_b}^{(1-k)T_b} s(z)s(z + \tau + qT_b)\, dz \qquad (2.1.6)$$

Applying the summation in k to the integral results in

$$R_m(\tau) = \sum_q \mu_d(q)f_s(\tau + qT_b) \qquad (2.1.7)$$

where we have introduced the function

$$f_s(\tau) = \frac{1}{T_b} \int_{-\infty}^{\infty} s(z)s(z + \tau)\, dz \qquad (2.1.8)$$

Note that $f_s(\tau)$ is a completely deterministic function that depends only on the waveshape $s(t)$. The result in (2.1.7) therefore gives the general correlation function of the antipodal digital waveform in (2.1.1). We see that this correlation depends on both the bit waveform $s(t)$ [through the $f_s(\tau)$ function] and the correlation of the data bit sequence produced by the digital source.

Let us first consider the important case where the source bits are produced independently, with a probability of p that $d_k = +1$ and probability $1 - p$ that $d_k = -1$. The sequence $\{d_k\}$ therefore represents independent, identically distributed binary variables, each with mean

$$\mathcal{E}(d_k) = (+1)p + (-1)(1 - p)$$
$$= 2p - 1 \qquad (2.1.9)$$

and mean squared value

$$\mathcal{E}(d_k^2) = (+1)^2 p + (-1)^2 (1 - p)$$
$$= 1 \qquad (2.1.10)$$

The sequence correlation in (2.1.4) is then

$$\mu_d(q) = \begin{cases} 1, & q = 0 \\ (2p - 1)^2, & q \neq 0 \end{cases} \qquad (2.1.11)$$

and the resulting waveform correlation in (2.1.7) reduces to

$$R_m(\tau) = f_s(\tau) + (2p - 1)^2 \sum_{q \neq 0} f_s(\tau + qT_b) \qquad (2.1.12)$$

Since the summation deletes the $q = 0$ term, it is convenient to insert (and subtract) this term and write $R_m(\tau)$ more compactly as

$$R_m(\tau) = [1 - (2p - 1)^2]f_s(\tau) + (2p - 1)^2 \sum_q f_s(\tau + qT_b)$$

$$= 4p(1 - p)f_s(\tau) + (2p - 1)^2 \sum_q f_s(\tau + qT_b) \qquad (2.1.13)$$

This represents the time averaged correlation function of an antipodal digital waveform with independent data bits, each having probability p of being a one-bit.

The spectral density is obtained by transforming $R_m(\tau)$. If $F_s(\omega)$ is the Fourier transform of $s(t)$, it follows that (Problem 2.3)

$$\text{Fourier transfer of } f_s(\tau) = \frac{1}{T_b} |F_s(\omega)|^2 \qquad (2.1.14a)$$

$$\text{Fourier transfer of } \sum_q f_s(\tau + qT_b) = \frac{2\pi}{T_b^2} \sum_n \left| F_s\left(\frac{n2\pi}{T_b}\right) \right|^2 \delta\left(\omega - \frac{n2\pi}{T_b}\right)$$

$$(2.1.14b)$$

The power spectrum of the digital waveform is then

$$S_m(\omega) = \frac{4p(1 - p)}{T_b} |F_s(\omega)|^2 + \frac{(2p - 1)^2}{T_b^2} \sum_n \left| F_s\left(\frac{n2\pi}{T_b}\right) \right|^2 2\pi\delta\left(\omega - \frac{n2\pi}{T_b}\right)$$

$$(2.1.15)$$

The spectrum is composed of two basic parts. The first term is the continuous portion, and its spectral shape depends solely on the frequency characteristics (Fourier transform) of the bit waveform $s(t)$ itself. The second term corresponds to discrete spectral lines occurring at the harmonics of the data rate $1/T_b$, each with a strength dependent on the value of $F_s(\omega)$ at each frequency. Hence a binary digital waveform, in general, has both a continuous and a discrete part. The discrete part signifies that the binary wave contains harmonic energy at the data rate, in spite of the random modulation.

Note that the power in each discrete line also depends on the bit probability p. Thus the distribution of ones and zeros in the source bit stream determines the harmonic power. In particular, we note that the discrete lines all vanish if $p = \frac{1}{2}$, that is, if the bit symbols are equal likely. The more evenly distributed are the individual bits within the source, the less the power of the harmonic lines in the spectrum. Since the harmonic lines are useful for aiding digital synchronization (we will explore this point in more detail in Chapter 10), a system designer may wish to artificially control the source bit distribution by purposely inserting auxiliary bits so as to maintain the discrete lines in the baseband spectrum.

The continuous portion of $S_m(\omega)$ depends on the frequency charcteristics of the bit signal $s(t)$ and therefore can be controlled by waveform selection. For example, an antipodal pulsed waveform uses rectangular pulses for $s(t)$,

$$s(t) = \begin{cases} A, & 0 \leq t \leq T_b \\ 0, & \text{elsewhere} \end{cases} \qquad (2.1.16)$$

Its transform is given in (1.4.7), so

$$|F_s(\omega)|^2 = (AT_b)^2 \left(\frac{\sin (\omega T_b/2)}{\omega T_b/2} \right)^2 \qquad (2.1.17)$$

This transform contains nulls at the harmonics of $1/T_b$ and spectral humps distributed out along the ω axis with decreasing amplitudes. Because of the nulls in $F_s(\omega)$ all discrete harmonics of $S_m(\omega)$ in (2.1.15) are zero, except possibly at $\omega = 0$. The spectrum of the binary pulsed waveform therefore simplifies to

$$S_m(\omega) = 4p(1-p)A^2 T_b \left(\frac{\sin (\omega T_b/2)}{\omega T_b/2} \right)^2 + A^2(2p-1)^2 2\pi\delta(\omega) \quad (2.1.18)$$

This continuous spectrum is illustrated in Figure 2.3. The pulsed binary waveform is extremely important in digital communications, since it represents one of the more popular types of digital waveform. It can be generated from a simple binary, two-state flip-flop device that merely switches states according to the source bits and produces an antipodal voltage output at each state.

Also shown in Figure 2.3 is the waveform spectrum for a rounded pulse, the so-called raised cosine pulse. The latter has the transform stated and produces the continous spectrum shown. Since its transform has nulls at $\omega = \pm n2\pi/T_b$ for all $n \geq 2$, the spectrum can have discrete components at only $\omega = 0$ and $\omega = \pm 2\pi/T_b$. The raised cosine pulse produces a slightly wider main hump, but the spectral tails decrease at a faster rate than for the rectangular pulse case. This spectral shaping by proper selection of the bit waveform will play an important role in digital communication system design.

When the source bits are not independent, the power spectral density can still be computed by transforming (2.1.7), except the sequence correlation $\mu_d(q)$ must be considered a general function of q. This leads to

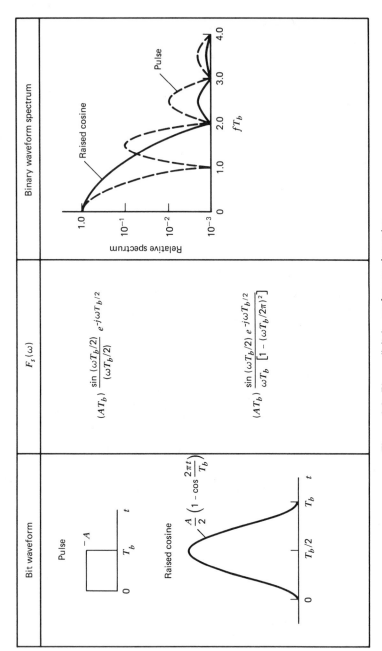

Figure 2.3. Binary digital waveform and spectra.

56

$$S_m(\omega) = \int_{-\infty}^{\infty} R_m(\tau)e^{-j\omega\tau}\,d\tau$$

$$= \sum_q \mu_d(q) \int_{-\infty}^{\infty} f_s(\tau + qT_b)e^{-j\omega\tau}\,d\tau \qquad (2.1.19)$$

Rewriting the integral, and using (2.1.14), produces

$$S_m(\omega) = \frac{1}{T_b}|F_s(\omega)|^2 \sum_q \mu_d(q)e^{j\omega qT_b} \qquad (2.1.20)$$

The summation represents the discrete Fourier transform of the sequence $\mu_d(q)$, and the waveform spectrum is obtained by multiplying $|F_s(\omega)|^2$ by this transform. This again illustrates that the properties of the binary waveform depend on the characteristics of the data bit sequence as well as the frequency characteristics of the bit waveform $s(t)$. This further shows that by control of the sequence statistics, one can effectively shape the waveform spectrum. For example, suppose equally likely, independent binary bits d_k are produced from a source, and a new digital variable is generated as

$$a_k = d_k + d_{k-1} \qquad (2.1.21)$$

That is, by numerically summing the present bit with the previous bit. We now use the a_k variables as the waveform coefficients in the digital waveform in (2.1.1). The sequence of the new variables a_k, which are no longer binary value, has a correlation

$$\mu_a(q) = \begin{cases} 0, & q \geq \pm 2 \\ 1, & q = \pm 1 \\ 2, & q = 0 \end{cases} \qquad (2.1.22)$$

Its discrete transform is then

$$\sum_q \mu_a(q)e^{j\omega T_b} = 2 + e^{-j\omega T_b} + e^{j\omega T_b}$$

$$= 2(1 + \cos \omega T_b)$$

$$= 4\cos^2 \frac{\omega T_b}{2} \qquad (2.1.23)$$

and the resulting spectrum reduces to

$$S_m(\omega) = \frac{4}{T_b}|F_s(\omega)|^2 \cos^2 \frac{\omega T_b}{2} \qquad (2.1.24)$$

Note that there are no discrete lines, and the sequence in (2.1.21) has effectively low pass filtered the continuous spectrum as a result of the bit waveform.

Lastly, we should point out that although we confined our attention to digital waveforms generated from positive and negative versions of the same waveform $s(t)$, the results can be easily extended to the case where the data bits are encoded onto arbitrary waveform pairs $s_1(t)$ and $s_0(t)$, $0 \le t \le T_b$, producing the data modulated waveform

$$m(t) = \sum_k [b_k s_1(t - kT_b) + (1 - b_k) s_0(t - kT_b)] \qquad (2.1.25)$$

where $b_k = 1$ if a one-bit was sent in the kth interval, and $b_k = 0$ if a zero-bit was sent. The correlation function of $m(t)$ can be computed similar to (2.1.7), and for independent data bits the spectrum can be shown to be (see Problem 2.5)

$$S_m(\omega) = \frac{p(1 - p)}{T_b} [|F_{s_1}(\omega) - F_{s_0}(\omega)|^2]$$
$$+ \frac{2\pi}{T_b^2} \sum_n \left| pF_{s_1}\left(\frac{n2\pi}{T_b}\right) + (1 - p)F_{s_0}\left(\frac{n2\pi}{T_b}\right) \right|^2 \delta\left(\omega - \frac{2\pi n}{T_b}\right)$$

$$(2.1.26)$$

where p is the probability of a one bit and $F_{s_1}(\omega)$ and $F_{s_0}(\omega)$ are the Fourier transforms of $s_1(t)$ and $s_0(t)$, respectively. The first term is the continuous part, now dependent on the sum of the individual waveform spectra. The discrete lines again occur at the bit rate harmonics, but no discrete lines will occur if

$$pF_{s_1}\left(\frac{2\pi n}{T_b}\right) + (1 - p)F_{s_0}\left(\frac{2\pi n}{T_b}\right) = 0, \qquad n = 1, 2, \ldots \qquad (2.1.27)$$

Again waveform selection and bit probabilities will determine the actual waveform spectra. For antipodal signals, $s_0(t) = -s_1(t)$, and (2.1.26) reduces to our previous result in (2.1.15).

2.2 THE AMPLITUDE MODULATED CARRIER

Let us next consider the amplitude modulated carrier. The standard AM carrier was given in (1.6.15) as

$$c(t) = a(t) \cos(\omega_c t + \psi) \qquad (2.2.1)$$

where $a(t)$ is the amplitude function, ω_c is the carrier frequency in rps, and ψ

is an arbitrary carrier phase. The amplitude function $a(t)$ is related to the baseband modulating function $m(t)$ by

$$a(t) = a_0 + b_0 m(t) \qquad (2.2.2)$$

where we assume that the waveform $m(t)$ has a power of P_m and the parameters a_0 and b_0 are selected to prevent overmodulation. We see from (2.2.1) that AM carriers are formed by multiplying circuits that electronically multiply the baseband $a(t)$ by the carrier waveform (Figure 2.4). Hence amplitude modulated carriers have the advantage of being relatively easy to generate.

To determine the power in a standard AM carrier we again apply the results of Section 1.5. For deterministic carrier waveforms, we must evaluate (1.5.2). Hence

$$P_c = \operatorname*{Lim}_{T \to \infty} \frac{1}{2T} \int_{-T}^{T} a^2(t) \cos^2 (\omega_c t + \psi)\, dt \qquad (2.2.3)$$

By expanding the squared cosine term, we can rewrite this as

$$P_c = \operatorname*{Lim}_{T \to \infty} \left[\frac{1}{2T} \int_{-T}^{T} \frac{a^2(t)}{2}\, dt + \frac{1}{2T} \int_{-T}^{T} \frac{a^2(t)}{2} \cos (2\omega_c t + 2\psi)\, dt \right] \qquad (2.2.4)$$

The second integral requires specification of $a(t)$ in order to be evaluated exactly. However, we can upper bound the magnitude of this integral by

$$\left| \frac{1}{2T} \int_{-T}^{T} \frac{a^2(t)}{2} \cos (2\omega_c t + 2\psi)\, dt \right| \le \frac{\max a^2(t)}{2} \left| \frac{1}{2T} \int_{-T}^{T} \cos (2\omega_c t + 2\psi)\, dt \right|$$

$$= \frac{\max a^2(t)}{2} \left| \frac{\sin (2\omega_c T) \cos (2\psi)}{2\omega_c T} \right|$$

$$(2.2.5)$$

where $\max a^2(t)$ is the maximum value of $a^2(t)$. Since the sine function is bounded in magnitude by one, the right-hand side is bounded by

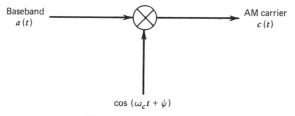

Figure 2.4. AM modulator.

[max $a^2(t)$]$/4\omega_c T$. In the limit as $T \to \infty$ this magnitude bound approaches zero for all $\omega_c > 0$, so long as $a(t)$ is finite for all t. Hence the second integral (2.2.4) must necessarily approach zero in the limit. The power is therefore given by the first term,

$$P_c = \lim_{T \to \infty} \frac{1}{2T} \int_{-T}^{T} \frac{a^2(t)}{2} \, dt$$

$$= \tfrac{1}{2} P_a \qquad\qquad (2.2.6)$$

where P_a is the power in the amplitude function $a(t)$. From (2.2.2)

$$P_a = \lim_{T \to \infty} \frac{1}{2T} \int_{-T}^{T} [a_0 + b_0 m(t)]^2 \, dt$$

$$= a_0^2 + b_0^2 P_m \qquad\qquad (2.2.7)$$

where P_m is the power in the modulation $m(t)$, and we have assumed the limiting time average of $m(t)$ is zero. Hence for the deterministic AM carrier, the power is given by

$$P_c = \tfrac{1}{2}(a_0^2 + b_0^2 P_m) \qquad\qquad (2.2.8)$$

Note that the total carrier power is composed of a contribution from the modulation power P_m and from the carrier component a_0. Conversely, we can state that of the total available carrier power P_c, only a portion is available as modulation power. Of course, it must be remembered that the overmodulation condition requires $b_0^2 P_m \le a_0^2$.

To determine the frequency function of a deterministic AM carrier, we compute the frequency transform of (2.2.1). For a deterministic waveform, $a(t)$ will have the Fourier transform

$$A(\omega) = 2\pi a_0 \, \delta(\omega) + b_0 M(\omega) \qquad\qquad (2.2.9)$$

where $M(\omega)$ is the transform of $m(t)$. The carrier $c(t)$ in (2.2.1) therefore has the transform

$$C(\omega) = A(\omega) \oplus F_c(\omega) \qquad\qquad (2.2.10)$$

where $\oplus$ denotes frequency convolution and $F_c(\omega)$ is the Fourier transform of $\cos(\omega_c t + \psi)$, given in (1.4.9). Integrating out the delta functions in the frequency convolution integral (Appendix A, A.1) yields

$$C(\omega) = \frac{e^{j\psi}}{2} A(\omega - \omega_c) + \frac{e^{-j\psi}}{2} A(\omega + \omega_c) \qquad\qquad (2.2.11)$$

The Fourier transform of the AM carrier is therefore simply a shift of the amplitude transform in (2.2.9) to the carrier frequency $\pm\omega_c$, accompanied by a phase shift and a division by 2. We see that AM carrier modulation causes a shift of the amplitude frequency function up to the carrier frequency. Since $M(\omega)$ is generally low pass in nature (or at least confined to low frequencies), $C(\omega)$ will have its spectrum concentrated in the vicinity of ω_c, as shown in Figure 2.5. Thus AM waveforms are typically bandpass waveforms with a center frequency at the carrier frequency. Note that since $|A(\omega)|$ is necessarily symmetric about $\omega = 0$, $|C(\omega)|$ will be symmetric about ω_c. The frequency components of $C(\omega)$ above ω_c are called *upper sideband* frequencies, and those below are *lower sideband* frequencies. We also see that if $m(t)$ has a one-sided bandwidth of B_m Hz, then $C(\omega)$ has a bandwidth of $2B_m$ Hz around its center frequency. The spectral line (delta function) at $\omega = \pm\omega_c$ indicates a sinusoidal component at the carrier frequency within the AM spectrum. This is called the *carrier component* of the modulated signal and is often important in receiver processing.

In suppressed carrier AM signals, $a_0 = 0$, and the carrier is simply formed as

$$c(t) = Am(t)\cos(\omega_c t + \psi)$$
$$= A|m(t)|\cos(\omega_c t + \psi + \theta(t)) \qquad (2.2.12)$$

where $\theta(t)$ is given in (1.6.17). This AM carrier has the same spectral extent as in Figure 2.5 but contains no carrier component, and the carrier shifts phase every time the baseband changes sign. The suppressed carrier AM has

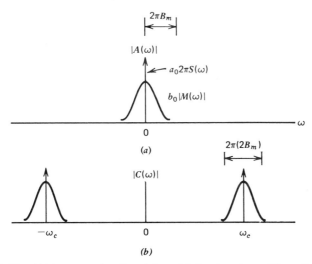

Figure 2.5. The AM frequency function: (*a*) modulating transform, (*b*) carrier transform.

more exaggerated amplitude variations (Figure 2.6), and its power in (2.2.12) is given by

$$P_c = \frac{A^2 P_m}{2} \tag{2.2.13}$$

Thus all the carrier power is devoted to the modulation sidebands, and no power is used for the carrier component.

The ability to shift the baseband function to higher frequencies via amplitude modulation begins to illustrate an advantage of the carrier

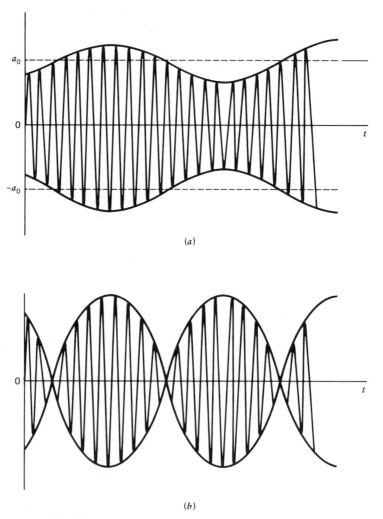

(a)

(b)

Figure 2.6. AM waveforms: (a) standard AM, (b) suppressed carrier AM.

modulation operation, and why we use it at all in communication systems. By modulating we can arbitrarily select (to within certain constraints) the set of frequencies to be transmitted, while still sending the baseband information. For example, rather than having to transmit a frequency ω of $M(\omega)$ we can, by use of AM, instead be concerned with transmitting frequency $\omega_c \pm \omega$ of $C(\omega)$. As we shall see, it is generally easier to transmit higher frequencies over long distances than lower frequencies. Thus modulation allows us to match together properly two basically conflicting conditions—that baseband information is generally low frequency, whereas communication transmission is best accomplished at high frequencies. It is precisely this match-up that governs most of our transmitter and receiver design.

Consider now the case where the carrier amplitude function $a(t)$ is a stationary random waveform, with the carrier phase angle ψ a random phase angle uniformly distributed over $(0, 2\pi)$ and statistically independent of the $a(t)$ process. The standard AM carrier $c(t)$ becomes a random carrier process, dependent on the randomness in $a(t)$ and ψ. Its autocorrelation function is

$$R_c(\tau) = \mathscr{E}[c(t)c(t+\tau)]$$
$$= \mathscr{E}\{a(t)a(t+\tau)\cos(\omega_c t + \psi)\cos[\omega_c(t+\tau) + \psi]\} \quad (2.2.14)$$

where $\mathscr{E}$ represents the statistical average over the statistics of $a(t)$ and ψ. Expanding the product of the cosine terms into sum and difference angle terms, and making use of the independence of $a(t)$ and ψ, yields

$$R_c(\tau) = \tfrac{1}{2}\mathscr{E}[a(t)a(t+\tau)]\cos\omega_c\tau$$
$$+ \tfrac{1}{2}\mathscr{E}[a(t)a(t+\tau)]\mathscr{E}[\cos(2\omega_c t + \omega_c\tau + 2\psi)] \quad (2.2.15)$$

The last average term is zero because of the uniform assumption on ψ, just as in (1.6.4). The first average is precisely the definition of the autocorrelation function $R_a(\tau)$ of the $a(t)$ process and

$$R_c(\tau) = \tfrac{1}{2}R_a(\tau)\cos\omega_c\tau \quad (2.2.16)$$

Thus the autocorrelation of the random AM carrier is one-half the product of the autocorrelation of the amplitude and pure carrier processes. The power in $c(t)$ follows as

$$P_c = R_c(0) = \tfrac{1}{2}P_a \quad (2.2.17)$$

identical to (2.2.6). Hence the power in the standard AM carrier is always given by one-half the power in the amplitude waveform, whether it is considered as a deterministic or random waveform.

The power spectral density of the random AM carrier is obtained by

Fourier transforming $R_c(\tau)$. We again see that this involves a frequency convolution of the spectrum $S_a(\omega)$ of $a(t)$ and the delta function spectrum in (1.6.6). Thus

$$S_c(\omega) = \tfrac{1}{2}\{S_a(\omega) \oplus [\pi\,\delta(\omega - \omega_c) + \pi\,\delta(\omega + \omega_c)]\}$$
$$= \tfrac{1}{4}S_a(\omega - \omega_c) + \tfrac{1}{4}S_a(\omega + \omega_c) \qquad (2.2.18)$$

The power spectrum of $c(t)$ is therefore again obtained by shifting the power spectrum of the process $a(t)$ to $\pm\omega_c$, just as in (2.2.11). For $a(t)$ in (2.2.2),

$$S_a(\omega) = 2\pi a_0^2\,\delta(\omega) + b_0^2 S_m(\omega) \qquad (2.2.19)$$

where $S_m(\omega)$ is now the power spectrum of $m(t)$. When (2.2.19) is used in (2.2.18), we see again that $S_c(\omega)$ occupies a one-sided spectral bandwidth equal to twice the bandwidth of the power spectrum of $m(t)$. Whether dealing with Fourier transforms or power spectra, the AM carrier frequency spectrum always appears as a shifted version of the baseband spectrum to the carrier frequency. This is all, of course, simply a consequence of the fact that amplitude modulation produces a carrier waveform that is linearly proportional to the modulating signal.

An important case is when the carrier is amplitude modulated by a binary digital waveform. This would correspond, for example, to the direct multiplication of the waveform in (2.1.1) by a pure carrier. The AM carrier would be formed as

$$c(t) = A\left[\sum_k d_k s(t - kT_b)\right]\cos(\omega_c t + \psi) \qquad (2.2.20)$$

The carrier has power $A^2 P_s/2$, and the carrier spectrum is obtained by combining (2.1.15) with (2.2.18). The resulting carrier spectrum corresponds to the shift of the digital spectrum in Figure 2.3 up to the carrier frequency ω_c. Figure 2.7 plots the AM spectrum for a pulse binary waveform $s(t)$. Note that the carrier component may be present (if the data waveform has a delta function at $\omega = 0$) but no other spectral lines will appear. The main-hump spectral width extends over a bandwidth of $2/T_b = 2R_b$ Hz. This means higher bit rate sources require wider AM bandwidths. It should be emphasized that frequency components actually extend over an infinite bandwidth, as a result of the high frequency content of the modulating pulse waveforms. Out-of-band energy in a modulated carrier can cause interference to an adjacent carrier channel. These spectral tails can be reduced by rounding off the pulse edges of the modulating waveform, or by bandpass filtering of the carrier. This latter method is considered in Section 2.6.

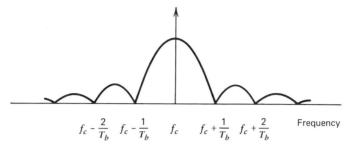

Figure 2.7. AM carrier spectrum with binary pulse digital modulation.

2.3 SINGLE SIDEBAND AM

The fact that the upper and lower sidebands of the AM carrier are related suggests the possibility of transmitting only one such part without loss of baseband information. This notion is the basis of *single sideband AM* (SSB-AM) modulation and is quite commonly used in modern voice communications. The objective of SSB-AM is to generate a carrier waveform in which only an upper (or lower) sideband appears. The resulting carrier bandwidth is only one-half of the standard AM bandwidth.

The obvious way to generate the SSB-AM carrier is to simply follow an AM modulator with a filter that removes the unwanted sideband (Figure 2.8). This is somewhat difficult to do in practice since a sharp filter is needed to ideally eliminate one sideband, especially if the baseband modulating waveform has predominate low frequency content. The presence of an imperfectly removed sideband can lead to distortion during demodulation. SSB-AM is easier to generate by filtering if the modulation doesn't have significant low frequencies, producing a window around the carrier frequency (Figure 2.8*b*). This makes it easier to insert the necessary filtering. Since baseband audio in Figure 2.2*a* tends to have this spectral property, voice transmission by SSB-AM is commonly generated by sideband filtering. In addition, normal voice can withstand some degree of distortion without significantly losing fidelity, which is also favorable for this method.

If the sideband filtering is applied to a standard AM carrier, then invariably some portion of the carrier component is retained, as shown in Figure 2.8*b*. The resulting SSB-AM therefore contains one sideband plus a portion of the carrier component. This is referred to as *vestigial* SSB. As we shall see shortly, this carrier component is useful for aiding in the eventual demodulation of the SSB-AM carrier.

An exact procedure for producing perfect SSB-AM is implemented by

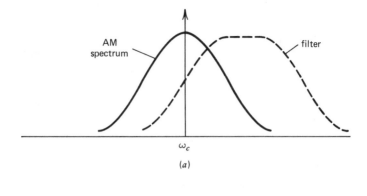

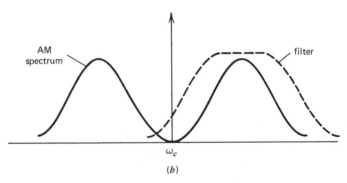

Figure 2.8. SSB-AM generation by filtering: (a) general AM spectrum, (b) displaced AM sidebands with low frequency window.

forming at the modulator the quadrature carrier signal

$$c(t) = a(t) \cos(\omega_c t + \psi) + \hat{a}(t) \sin(\omega_c t + \psi) \qquad (2.3.1)$$

This is simply a modified AM signal in which each quadrature component is formed separately as an AM signal, with $\hat{a}(t)$ obtained by passing $a(t)$ through a linear filter $H_a(\omega)$ with transfer function (Figure 2.9a)

$$H_a(\omega) = \begin{cases} -j, & \omega \geq 0 \\ j, & \omega < 0 \end{cases} \qquad (2.3.2)$$

The transform of $c(t)$ is then

$$C(\omega) = \frac{1}{2}\left[A(\omega - \omega_c) + A(\omega + \omega_c)\right]$$

$$+ \frac{1}{2j}\left[A(\omega - \omega_c)H_a(\omega - \omega_c) - A(\omega + \omega_c)H_a(\omega + \omega_c)\right] \quad (2.3.3)$$

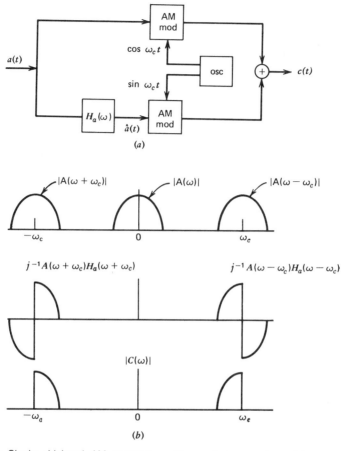

Figure 2.9. Single sideband AM generation with quadrature carriers: (a) modulator, (b) spectrum.

Now since $H_a(\omega - \omega_c) = -j$ for $\omega > \omega_c$ and j for $\omega < \omega_c$, whereas $H_a(\omega + \omega_c) = -j$ for $\omega > -\omega_c$ and j for $\omega < -\omega_c$, the terms in (2.3.3) sum to eliminate one sideband, as shown in Figure 2.9b. Thus the quadrature modulation of (2.3.1) generates a pure SSB-AM carrier. The prime difficulty is to construct the filter $H_a(\omega)$ in (2.3.2) that requires basically a constant phase network over the bandwidth of $a(t)$. Fortunately, such networks can be adequately approximated over relatively small bandwidths, which is again conducive to voice transmission.

As stated earlier, there is an advantage in transmitting a carrier component with SSB-AM, by either a vestigial system or reinserting the carrier after the quadrature system shown in Figure 2.9. Consider adding a carrier to the pure SSB-AM carrier in (2.3.1) to produce the new carrier

$$c(t) = [A + a(t)] \cos (\omega_c t + \psi) + \hat{a}(t) \sin (\omega_c t + \psi) \qquad (2.3.4)$$

The envelope of this carrier, from (1.6.10), is then

$$e(t) = \{[A + a(t)]^2 + [\hat{a}(t)]^2\}^{1/2}$$

$$= [A + a(t)]\left\{1 + \left[\frac{\hat{a}(t)}{A + a(t)}\right]^2\right\}^{1/2} \qquad (2.3.5)$$

If the modulation index is low, $A > a(t)$,

$$e(t) \approx A + a(t) \qquad (2.3.6)$$

Thus the vestigial SSB-AM carrier has an envelope that is approximately the same as that of a standard AM carrier. That is, the SSB-AM carrier resembles a standard AM carrier in terms of its amplitude variation, even though half of its spectrum has been removed. This means demodulation circuitry for SSB-AM can be identical to that of standard AM. This is discussed in Chapter 5.

2.4   THE FREQUENCY MODULATED CARRIER

In frequency modulation the baseband waveform $m(t)$ is used to modulate the frequency of the RF carrier. The general FM carrier has the form

$$c(t) = A \cos \left[\omega_c t + \Delta_\omega \int m(t)\, dt + \psi \right] \qquad (2.4.1)$$

where A is the carrier amplitude, Δ_ω is the frequency deviation coefficient in rps/V, and ψ is again the carrier phase. The power in a deterministic FM carrier can be determined by a procedure similar to (2.2.3)–(2.2.6). By substitution into (2.2.3) we have

$$P_c = \lim_{T \to \infty} \left[\frac{1}{2T} \int_{-T}^{T} \frac{A^2}{2}\, dt + \frac{1}{2T} \int_{-T}^{T} \frac{A^2}{2} \cos [2\omega_c t + 2\theta(t) + 2\psi]\, dt \right] \qquad (2.4.2)$$

where

$$\theta(t) = \Delta_\omega \int m(t)\, dt \qquad (2.4.3)$$

To evaluate the second term, we expand and bound the magnitude of the second integral as

$$\left| \frac{A^2}{4T} \int_{-T}^{T} \cos 2\theta \, \cos 2(\omega_c T + \psi) \, dt + \frac{A^2}{4T} \int_{-T}^{T} \sin 2\theta \, \sin 2(\omega_c t + \psi) \, dt \right|$$

$$\leq \frac{A^2}{4} \left| \max \cos 2\theta(t) \right| \left| \frac{\sin (2\omega_c T) \cos 2\psi}{\omega_c T} \right|$$

$$+ \frac{A^2}{4} \left| \max \sin 2\theta(t) \right| \left| \frac{\cos (2\omega_c T) \sin 2\psi)}{\omega_c T} \right|$$

$$\leq \frac{A^2}{2\omega_c T} \tag{2.4.4}$$

In the limit as $T \to \infty$, the magnitude bound approaches zero, so that in (2.4.2),

$$P_c = \frac{A^2}{2} \tag{2.4.5}$$

for any deterministic frequency modulating waveform $m(t)$. Thus the power in the FM carrier depends only on the amplitude A in (2.4.1) and is not a function of the modulation power P_m, as was the case in AM.

Computation of the FM frequency function is, unfortunately, more difficult, because of the nonlinearity of this type of modulation, and no simple relation exists for relating the frequency spectrum of $c(t)$ to that of $m(t)$ as in the AM case. However, we can investigate several special cases and use these results to approximate more general situations.

Sine Wave Modulation

Consider the case where the baseband modulating signal is a pure sinusoid at frequency ω_m and phase θ_m. That is, let

$$m(t) = \cos (\omega_m t + \theta_m) \tag{2.4.6}$$

We then have

$$c(t) = A \cos [\omega_c t + \beta \sin (\omega_m t + \theta_m) + \psi] \tag{2.4.7}$$

where

$$\beta = \frac{\Delta_\omega}{\omega_m} \tag{2.4.8}$$

is referred to as the *FM modulation index*. Since the peak of $m(t)$ is one, we see that Δ_ω is also the peak frequency deviation from ω_c imported to the carrier by the modulation. We wish to determine the frequency transform of $c(t)$. However, rather than formally transforming, we can instead write

$$c(t) = A \operatorname{Real} \{\exp\left(j\omega_c t + j\psi\right) \exp\left[j\beta \sin\left(\omega_m t + \theta_m\right)\right]\} \tag{2.4.9}$$

We now use the identity

$$\exp\left[j\beta \sin\left(\omega_m t + \theta_m\right)\right] = \sum_{k=-\infty}^{\infty} J_k(\beta) \exp\left[jk(\omega_m t + \theta_m)\right] \tag{2.4.10}$$

where $J_k(\beta) = (-1)^k J_k(\beta)$ and $J_k(\beta)$ is the kth order Bessel function in the argument β [1, 2]. Several of these Bessel functions are shown in Figure 2.10. Substituting into (2.4.9) and taking the real part allows us to write

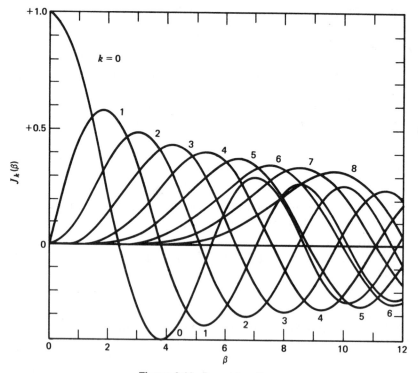

Figure 2.10. Bessel functions.

$$c(t) = A \sum_{k=-\infty}^{\infty} J_k(\beta) \cos\left[(\omega_c + k\omega_m)t + k\theta_m + \psi\right] \qquad (2.4.11)$$

Note that (2.4.11) appears as an infinite sum of sine waves, with amplitudes $AJ_k(\beta)$ and having frequency components above and below the RF carrier frequency ω_c. These sine waves transform to delta functions in frequency, as shown in Figure 2.11. Again, the frequency lines about the carrier frequency form upper and lower sideband frequencies of the modulated carrier. The actual shape of the spectrum (strength of the delta functions) varies with β, as is evident from (2.4.11). As β is changed, the various carrier and sideband components vary in their relative strengths, some increasing while others are decreasing. Contrast this result with the AM case, where a single modulating frequency produces only a single upper and lower sideband. Thus FM, being a nonlinear type of modulation, produces a carrier spectrum significantly different from a simple shift of the baseband spectrum. Note that the carrier phase angle ψ is added to every sideband, whereas multiples of the phase of the modulating sinusoids are added to each corresponding sideband.

Although (2.4.11) theoretically has an infinite number of sine terms, we note from Figure 2.11 that many higher order Bessel terms produce relatively insignificant amplitude values. In particular, we see that $J_k(\beta) \ll 1$ for $\beta \ll k$. Hence only significant terms exist up to values of k slightly greater than the value of β. Roughly speaking, $c(t)$ can be considered to have $\beta + 1$ significant harmonics on each side of the carrier, corresponding to frequency components out to $[\omega_c \pm (\beta + 1)\omega_m]$. This means that $c(t)$ can be considered to occupy a significant frequency bandwidth of approximately

$$\beta_c \cong 2(\beta + 1)f_m \quad \text{Hz} \qquad (2.4.12)$$

where $f_m = \omega_m/2\pi$. The bandwidth B_c is called the *Carson rule* bandwidth for $c(t)$ and is used as a rough rule of thumb for assessing the spectral width of a sinusoidally modulated FM carrier. This bandwidth approximation is convenient because of its simplicity and has been found to be fairly satisfactory in FM analysis. Note that if $\beta \gg 1$, then B_c occupies a bandwidth

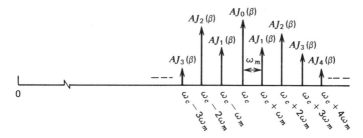

Figure 2.11. Spectrum of FM carrier with sine wave modulation.

$\beta + 1$ times larger than that that would be produced by AM. Thus FM systems with large values of β produce wideband carriers and are often referred to as *wideband FM* systems. On the other hand, *narrowband FM* occurs if $\beta \le 1$, in which case the FM bandwidth is comparable to that of an AM system.

To assess the meaning of the Carson rule bandwidth we might consider the effect of total power of using only M of the sidebands on each side of ω_c. The harmonics excluded therefore constitute lost carrier power. We can therefore define a normalized power factor as

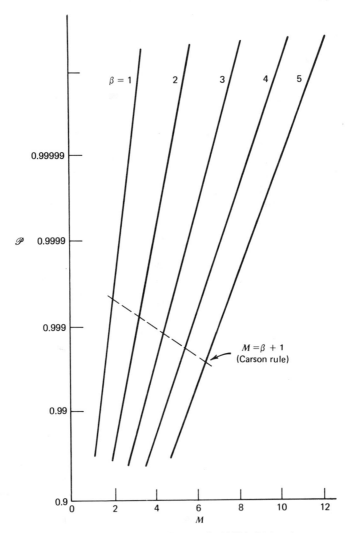

Figure 2.12. Normalized power in M FM sidebands.

$$\mathscr{P} = \frac{\text{power in } M \text{ sidebands of } c(t)}{\text{total power of } c(t)}$$

$$= \frac{(A^2/2)\sum_{k=-M}^{M} J_k^2(\beta)}{A^2/2}$$

$$= J_0^2(\beta) + 2\sum_{k=1}^{M} J_k^2(\beta) \qquad (2.4.13)$$

A plot of $\mathscr{P}$ in (2.4.13) is shown in Figure 2.12 for several β as a function of $2M$, the latter being the total number of sideband components used. We note a continual increase in $\mathscr{P}$ as M is increased, and appropriate bandwidths can be selected for any desired value of $\mathscr{P}$. Superimposed are the values obtained if the filter has the Carson rule bandwidth in (2.4.12), that is, if $M = \beta + 1$. We see that the Carson rule bandwidth produces power values of more than 99% of the total power. If larger values of $\mathscr{P}$ are desired, more sidebands and therefore larger bandwidths must be selected. Plots as in Figure 2.12 help us to specify the bandwidths needed.

Sum of Sine Waves

Consider now the case where $m(t)$ is the sum of K separate sine waves. That is, let

$$m(t) = \sum_{i=1}^{K} C_i \cos(\omega_i t + \theta_i) \qquad (2.4.14)$$

where C_i, ω_i, and θ_i are the corresponding deviations, frequency, and phase angles. We then have

$$c(t) = A \cos\left[\omega_c t + \sum_{i=1}^{K} \beta_i \sin(\omega_i t + \theta_i) + \psi\right] \qquad (2.4.15)$$

where $\beta_i = C_i/\omega_i$ is the modulation index of the ith sine wave. When written as in (2.4.9), and expanded using (2.4.10), we derive

$$c(t) = A \sum_{k_K=-\infty}^{\infty} \cdots \sum_{k_1=-\infty}^{\infty} \left[\prod_{i=1}^{K} J_{k_i}(\beta)\right] \cos\left[\omega_c t + \sum_{i=1}^{K} k_i(\omega_i t + \theta_i) + \psi\right] \qquad (2.4.16)$$

The preceding equation represents the general expression for the FM carrier modulated by K sinusoids. Note that it corresponds to a collection of harmonic frequencies at all the sidebands $\sum_{i=1}^{K} k_i \omega_i$, where all combinations of integers for the $\{k_i\}$ must be considered. Each such combination $\{k_1, k_2, \ldots, k_K\}$ yields a different sinusoid, each with its own phase,

$\sum_{i=1}^{K}(k_i\theta_i + \psi)$ and its own amplitude $\{A\prod_i^K J_{k_i}(\beta_i)\}$. In particular, we note that the carrier component at ω_c corresponds to $k_1 = k_2 = \cdots = k_K = 0$ and has amplitude $\{A\prod_i^K J_0(\beta_i)\}$, whereas the frequency component at frequency $(\omega_c + \omega_j)$ corresponds to $k_1 = 0, k_2 = 0, \ldots, k_j = 1, k_{j+1} = 0, \ldots, k_K = 0$ and has amplitude $\{AJ_1(\beta_j)\prod_{i=1,i\neq j}^{K} J_0(\beta_i)\}$. We also note that the component at $(\omega + \omega_j)$ contains the exact phase angle of the jth sine wave in (2.4.14) added to that of the carrier.

The bandwidth of the FM carrier in (2.4.16) can be estimated in several ways. One way is to simply use the largest bandwidth obtained by applying the Carson rule to each (β_i, ω_i) combination in (2.4.15). Another way is to use ω_K, the highest sine wave frequency of $m(t)$, as the upper frequency bandwidth of the modulation, and define an average index as

$$\beta = \frac{[(1/K)\sum_{i=1}^{K} C_i^2/2]^{1/2}}{\omega_K} \tag{2.4.17}$$

Although these procedures indicate roughly the frequency extent of (2.4.16), they do not give the true spectral shape of the carrier spectrum.

Arbitrary Modulation

When the modulating signal $m(t)$ is more complicated than simply sine waves, an exact expression for the spectrum of $c(t)$ is indeed difficult to generate. Bandwidth extent can be estimated by using Carson rule bandwidths with ω_m as the largest significant frequency of $m(t)$, and by using an effective index of

$$\beta = \frac{\Delta_\omega\sqrt{P_m}}{\omega_m} \tag{2.4.18}$$

where Δ_ω is the peak frequency deviation and P_m is the power of $m(t)$. Although the rms deviation $\Delta_\omega\sqrt{P_m}$ can be accurately measured, the selection of the "highest significant frequency" is often difficult to designate, generally leading to inaccurate band specification.

Another method is to approximate the modulation frequency function with a sum of sine waves at appropriately selected frequencies within the band, each with proper amplitude so as to represent the spectrum in its vicinity, and using (2.4.17). This allows the analysis to account for portions of the modulation frequency spectrum within the band that may have significantly more power, and therefore more effect on carrier spectrum, than the band edge frequencies.

Invariably the most accurate method, in this day of high speed simulation, is to simply compute the carrier spectrum via direct spectral analysis. Figure 2.13 shows measured carrier spectra with different rms indices

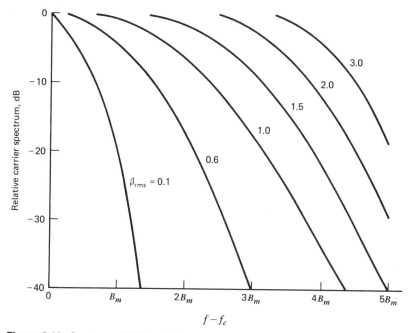

Figure 2.13. Spectrum of FM carrier with video modulation: β_{rms} = rms deviation/B_m.

obtained by frequency modulating a video signal onto a carrier and performing real-time spectral analyses. Frequency modulation by video signals is an important example, since it represents the way in which television signals are commonly transmitted through satellites. Real-time spectral analysis, which produced Figure 2.13, is extremely valuable for accurate estimation of required bandwidth and avoids the difficulties of predicting via Carson rule substitution.

To analytically account for random frequency modulation, we must reconsider (2.4.1) with $m(t)$ modeled as a random process of known statistics. In this case it is more convenient mathematically to treat $c(t)$ as a carrier that is phase modulated with the integral of the baseband process and to derive spectral information in this way. This we do in Section 2.5.

Binary Digital FM

In transmitting digital information, we often frequency modulate with a binary digital waveform. Using the binary waveform in (2.2.1), the carrier phase in (2.4.1), after frequency modulating, becomes

$$\theta(t) = \omega_c t + \Delta_\omega \int_{-\infty}^{t} \sum_k d_k s(\rho - kT_b)\, d\rho + \psi \qquad (2.4.19)$$

For the case of a pulsed waveform $[s(t) = 1, \; 0 \le t \le T_b]$, this phase integrates to

$$\theta(t) = \omega_c t + \Delta_\omega \, d_n q(t) + \sum_{k=-\infty}^{n-1} d_k \Delta_\omega T_b + \psi$$

$$nT_b \le t \le (n+1)T_b \qquad\qquad (2.4.20)$$

with

$$q(t) = \begin{cases} 0, & t \le nT_b \\ t, & t \le (n+1)T_b \\ T_b, & t \ge (n+1)T_b \end{cases} \qquad\qquad (2.4.21)$$

Combination of (2.4.21) with (2.4.20) shows that the phase is actually varying as

$$\theta(t) = (\omega_c + d_n \Delta_\omega)t + \sum_{k=-\infty}^{n-1} d_k \Delta_\omega T_b + \psi$$

$$nT_b \le t \le (n+1)T_b \qquad\qquad (2.4.22)$$

This corresponds to a frequency that shifts between $(\omega_c + \Delta_\omega)$ and $(\omega_c - \Delta_\omega)$ according to the bit stream. This is referred to as *frequency shift keying* (FSK) and is a common modulation method for transmitting binary information. Note that the phase function $\theta(t)$ in (2.4.20) traces out a phase trajectory in the phase domain according to the bit sequence $\{d_k\}$. Hence one can consider the binary digital carriers as either as FSK waveform in (2.4.22) or as a carrier with a phase modulation $\theta(t)$ in (2.4.20). These alternate interpretations lead to different decoding structures, as we shall see in Chapter 9.

To determine the FSK carrier spectrum, it is necessary to treat the modulation as a random digital waveform. Since the carrier involves angle modulation, its spectrum does not correspond to a simple shift of the binary waveform spectrum as was true for AM. Although detailed spectral analyses for FSK has been presented [3], it can be approximately estimated by noting that for FSK, $c(t)$ corresponds to sequences of bursts of a carrier at each of the frequencies $\omega_c \pm \Delta_\omega$. Since each bit lasts for T_b sec, each such burst produces a spectrum centered at the modulation frequency, with a spectral width corresponding to that of a switching interval of T_b sec. Hence the spectrum of an FSK carrier will appear as the combination of two such spectra, each centered at the modulation frequencies. The result depends on the separation between the frequencies [i.e., on the deviation coefficient Δ_ω in (2.4.19)].

Some resulting binary FM spectra are shown in Figure 2.14. For low deviations, the two spectra combine into a single-hump carrier spectrum

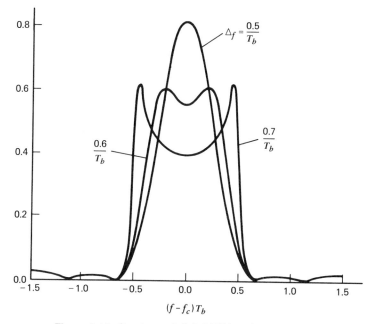

Figure 2.14. Spectrum of digital FSK carrier: $\Delta_f = \Delta_\omega / 2\pi$.

about the carrier frequency. As the deviation is increased, the two modulation frequencies are separated further, and the presence of the two frequencies becomes more apparent. Of course, the carrier bandwidth likewise increases as the deviation increases. The required FSK bandwidth can therefore be estimated as

$$B_c \approx 2\left(\frac{\Delta_\omega}{2\pi} + \frac{1}{T_b}\right) = 2R_b\left(\frac{\Delta_\omega}{2\pi R_b} + 1\right) \quad \text{Hz} \qquad (2.4.23)$$

This is roughly equivalent to a Carson rule bandwidth with an effective FM index of $\Delta_\omega / 2\pi R_b$. For deviations greater than the bit rate frequency, the FSK bandwidth is larger than the main-hump bandwidth for digital AM in Figure 2.7.

2.5   THE PHASE MODULATED CARRIER

If the baseband signal $m(t)$ phase modulates the RF carrier, the carrier waveform is

$$c(t) = A \cos\left[\omega_c t + \theta(t) + \psi\right]$$

$$\theta(t) = \Delta m(t) \qquad (2.5.1)$$

where $\theta(t)$ is the resulting baseband phase modulation. In this case the phase of the carrier varies directly with the desired modulation, and ψ represents an added phase shift that may appear with the carrier. In later work we also are interested in the situations where ψ itself is a function of t, representing extraneous or unintentional phase modulations superimposed on the carrier waveform. The deterministic PM carrier has a power of

$$P_c = \frac{A^2}{2} \tag{2.5.2}$$

obtained by the same development as in (2.4.2)–(2.4.5). To determine the frequency transform we again examine specific types of modulating waveform.

Sine Wave

For the case of a single sine wave modulation at ω_m,

$$\theta(t) = \Delta \sin(\omega_m t + \theta_m) \tag{2.5.3}$$

where Δ is now the phase modulation index in radians per volt. When (2.5.3) is substituted into (2.5.1), the carrier waveform is identical to (2.4.7) with β replaced by Δ. The resulting carrier expansion in (2.4.11), and Carson rule bandwidth in (2.4.12), can therefore be used directly with this substitution. We point out, however, that (2.4.11) is also valid for PM even if θ_m depended on t [i.e., if θ_m represented phase or frequency modulation of the modulating sine wave]. In this case the sidebands in (2.4.11) now become phase modulated sidebands located at the same sideband frequencies. This was not true for the FM case, because the integral of a frequency modulating sine wave containing phase modulation is not equivalent to phase modulating with a sine wave having the same modulation. Often this problem is circumvented in FM, however, by arguing that the time modulation of the sine wave is so much slower than that of the sine wave frequency itself that it can be considered a constant. This assumption allows us to replace the frequency modulating sine wave by a phase modulating sine wave with the same angle modulation, and by using our earlier results in Section 2.4.

Sum of Sine Waves

The results for the case of phase modulating with a sum of sine waves also carry over from the FM case. Let the phase modulation be

$$\theta(t) = \sum_{i=1}^{K} \Delta_i \sin(\omega_i t + \theta_i) \tag{2.5.4}$$

where Δ_i is the phase modulation index of the ith sine wave. The carrier $c(t)$ in (2.5.1) therefore has the same form as (2.4.15) with β_i replaced by Δ_i. The expansion of (2.4.16) is again valid with this substitution. The resulting carrier spectrum therefore appears as a collection of frequency components at all the beat frequencies of the modulation in (2.5.4). Again we point out that the result is valid even if the individual sine waves were themselves modulated [i.e., each θ_i in (2.5.4) depended on t]. The spectral lines in (2.4.16) now broaden into individual PM spectra at the same center frequencies with a bandwidth determined by the modulation on the phase of each component. Bandwidth extent can again be estimated by the Carson rule bandwidth with Δ replacing β and Δ approximated as $[(\Sigma_{i=1}^{K} \Delta_i^2/2K]^{1/2}$.

Binary Digital Waveform

Consider phase modulating with a binary pulse waveform in (2.1.1), so that

$$\theta(t) = \Delta m(t)$$

$$= \Delta \sum_k d_k s(t - kT_b) \qquad (2.5.5)$$

The PM carrier is then formed as

$$c(t) = A \cos[\omega_c t + \Delta m(t) + \psi] \qquad (2.5.6)$$

corresponding to a carrier that is phase shifted between $(\psi + \Delta)$ and $(\psi - \Delta)$ with each bit. The waveform is referred to as a *phase shift keyed* (PSK) carrier. Expanding the argument in (2.5.6) shows the PSK carrier is equivalent to

$$c(t) = A \cos[\Delta m(t)] \cos(\omega_c t + \psi) - A \sin[\Delta m(t)] \sin(\omega_c t + \psi) \qquad (2.5.7)$$

Now using the fact that

$$\sin[\Delta m(t)] = m(t) \sin \Delta$$

$$\cos[\Delta m(t)] = \cos \Delta \qquad (2.5.8)$$

we can rewrite (2.5.7) as

$$c(t) = (A \cos \Delta) \cos(\omega_c t + \psi) - (A \sin \Delta)m(t) \sin(\omega_c t + \psi) \qquad (2.5.9)$$

The first term represents a carrier component with amplitude $A \cos \Delta$, whereas the second represents a carrier amplitude modulated by the binary waveform $m(t)$. The results in Section 2.2 can therefore be applied, and the second term will have a frequency spectrum given by the shift of the spectrum of $m(t)$ to the frequency ω_c. The PSK carrier therefore will have the same spectral distribution as the AM digital carrier with a carrier component inserted. In fact, the result in (2.5.9) shows that a PSK carrier can be created by a direct multiplication, just as in AM. It should be emphasized, however, that the equations in (2.5.8), and therefore the equivalence to AM, holds only for pulsed waveforms $[m(t) = \pm 1$ at every $t]$ in (2.5.6).

Random Modulation

We now want to examine the spectral density of a PM wave when phase modulated by a general stationary random process. Consider again the PM carrier in (2.5.1) where now $\theta(t)$ is a random modulation waveform having autocorrelation $R_\theta(\tau)$ and ψ is a random phase angle, uniformly distributed over $(0, 2\pi)$ and independent of $\theta(t)$. The carrier $c(t)$ has autocorrelation

$$R_c(\tau) = \frac{A^2}{2} \, \mathcal{E} \{\cos [\omega_c \tau + \theta(t + \tau) - \theta(t)]\}$$

$$+ \frac{A^2}{2} \, \mathcal{E} \{\cos [2\omega_c t + \omega_c \tau + \theta(t) + \theta(t + \tau) + 2\psi]\} \quad (2.5.10)$$

The average over ψ in the second term is zero, because of the uniform assumption on ψ, which means

$$R_c(\tau) = \frac{A^2}{2} \, \mathcal{E} \{\cos [\omega_c \tau + \theta(t + \tau) - \theta(t)]\} \quad (2.5.11)$$

Since the average of a real part of a complex random variable is equal to the real part of its complex average, we can instead write

$$R_c(\tau) = \frac{A^2}{2} \, \text{Real} \, \{e^{j\omega_c \tau} \mathcal{E} [e^{j[\theta(t+\tau) - \theta(t)]}]\}$$

$$= \frac{A^2}{2} \, \text{Real} \, \{e^{j\omega_c \tau} \Psi_\theta(1, -1, \tau)\} \quad (2.5.12)$$

where now we have defined

$$\Psi_\theta(\omega_1, \omega_2, \tau) = \mathcal{E} [e^{j\omega_1 \theta(t+\tau) + j\omega_2 \theta(t)}] \quad (2.5.13)$$

and the average is over the random phase $\theta(t)$. We recognize (2.5.13) as the second order *characteristic function* of the process $\theta(t)$ (see Appendix B, B.2.5). Since the modulating process $\theta(t)$ is stationary, its characteristic function depends only on τ. Equation (2.5.12) represents a general expression for the autocorrelation function of any randomly modulated PM carrier with uniform phase variable and requires only the knowledge of the characteristic function of the modulation. Nevertheless, the carrier power follows as in (2.5.2) since $P_c = R_c(0) = A^2/2$ for any random modulation. Thus the PM carrier power is always $A^2/2$ whether the modulation is deterministic or random, just as in the FM case.

The result in (2.5.12) can be expanded further by noting that if we define this new random variable at t and $t + \tau$ by

$$z_1 = e^{j\theta(t+\tau)}$$
$$z_2 = e^{-j\theta(t)} \tag{2.5.14}$$

then

$$\Psi_\theta(1, -1, \tau) = \mathscr{E}[z_1 z_2]$$
$$= \mathscr{E}[(z_1 - \bar{z}_1)(z_2 - \bar{z}_2)] + \bar{z}_1 \bar{z}_2 \tag{2.5.15}$$

where

$$\bar{z}_1 = \mathscr{E}[z_1] = \Psi_\theta(1, 0, 0)$$
$$\bar{z}_2 = \mathscr{E}[z_2] = \Psi_\theta(0, -1, 0) = \Psi_\theta^*(1, 0, 0) \tag{2.5.16}$$

and $*$ denotes the complex conjugate. The last identity follows from the stationarity of the process. Note that the first term in (2.5.15) depends on τ and the second term does not. If we denote the first term as $\tilde{\psi}_\theta(\tau)$, then (2.5.12) can be rewritten as

$$R_c(\tau) = \frac{A^2}{2} \text{ Real } \{e^{j\omega_c \tau}[|\Psi_\theta(1, 0, 0)|^2 + \tilde{\Psi}_\theta(\tau)]\}$$

$$= \frac{A^2}{2} |\Psi_\theta(1, 0, 0)|^2 \cos \omega_c \tau + \frac{A^2}{2} \text{ Real } \{\tilde{\Psi}_\theta(\tau)e^{j\omega_c \tau}\} \tag{2.5.17}$$

In this way we have separated out the constant portion of the function $\Psi_\theta(1, -1, \tau)$. The first term will transform to a delta function (spectral line) at $\omega = \pm\omega_c$ in the power spectrum. This therefore represents the carrier component of the spectrum, and it is interesting that it depends only on the factor $|\Psi_\theta(1, 0, 0)|^2$ in (2.5.16). Since $A^2/2$ is the total carrier power and since $|\Psi(1, 0, 0)|^2 \le 1$, the strength of the carrier component is a computable fraction of the total carrier power. For this reason the factor $|\Psi(1, 0, 0)|^2$ is often called a carrier component *suppression factor* and can be determined directly from the modulation statistics. The second term in (2.5.17) transforms to a continuous spectrum centered at $\pm\omega_c$, the actual shape dependent on the form of the function $\tilde{\Psi}_\theta(\tau)$.

When the phase modulation process $\theta(t)$ is assumed to be a Gaussian random process, the general autocorrelation in (2.5.17) can be expanded further. In this case it is known that the second order characteristic function of a Gaussian process is

$$\Psi_\theta(\omega_1, \omega_2, \tau) = \exp - \left[\frac{\Delta^2\omega_1^2}{2} + \frac{\Delta^2\omega_2^2}{2} + R_\theta(\tau)\omega_1\omega_2\right] \qquad (2.5.18)$$

where $\Delta^2 = R_\theta(0)$. We therefore have

$$|\Psi_\theta(1, 0, 0)|^2 = (e^{-\Delta^2/2})^2 = e^{-\Delta^2} \qquad (2.5.19a)$$

$$\tilde{\Psi}_\theta(\tau) = e^{-\Delta^2 + R_\theta(\tau)} - e^{-\Delta^2} = e^{-\Delta^2}(e^{R_\theta(\tau)} - 1) \qquad (2.5.19b)$$

and

$$R_c(\tau) = \frac{A^2}{2} e^{-\Delta^2} \cos \omega_c\tau + \frac{A^2}{2} e^{-\Delta^2}(e^{R_\theta(\tau)} - 1) \cos \omega_c\tau \qquad (2.5.20)$$

By expanding

$$e^{R_\theta(\tau)} = 1 + \sum_{i=1}^{\infty} \frac{R_\theta^i(\tau)}{i!} \qquad (2.5.21)$$

we can write (2.5.20)

$$R_c(\tau) = \left(\frac{A^2}{2} e^{-\Delta^2}\right)\left[\cos \omega_c\tau + R_\theta(\tau) \cos \omega_c\tau + \frac{R_\theta^2(\tau)}{2!} \cos \omega_c\tau \cdots\right] \qquad (2.5.22)$$

The transform of (2.5.22) yields the power spectrum. Hence

$$S_c(\omega) = \frac{A^2 e^{-\Delta^2}}{2} \left[\pi \delta(\omega \pm \omega_c) + \tfrac{1}{2}S_\theta(\omega \pm \omega_c) + \tfrac{1}{4}[S_\theta(\omega) \oplus S_\theta(\omega)]_{\omega = \omega \pm \omega_c} \cdots\right] \qquad (2.5.23)$$

where $S_\theta(\omega)$ is the power spectrum of $\theta(t)$ and $\oplus$ denotes frequency convolution. The first term is the carrier component, the second is a shift of the modulating spectrum to $\pm\omega_c$, the third is a shift of the convolved modulating spectrum, and so on. The remaining terms involve higher order convolutions of the baseband spectrum, each shifted to $(\pm\omega_c)$. (Recall that convolving two functions spreads the resulting function while reducing its amplitude.) The resulting PM spectrum appears as sketched in Figure 2.15. We see that the spectrum is actually built up from an infinite sequence of overlapping convolved spectra. Hence the phase modulated spectrum is much more complicated than that of the modulation itself and does not simply correspond to a shift of the baseband up to the carrier frequency.

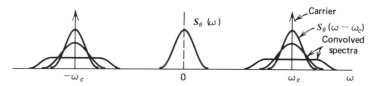

Figure 2.15. PM spectrum due to random Gaussian phase modulation.

The bandwidth extent can usually be estimated by superimposing the first few terms of (2.5.23). More quantitative estimates can often be derived (Problem 2.14).

In summary, these first sections have attempted to outline the intrinsic properties of the modulated carrier waveform generated at the transmitter. We have concentrated particularly on the power and on the modulated carrier frequency functions and the manner in which they are derived from the baseband waveform. We have shown that in all cases the modulated carrier has a bandpass frequency axis by proper choice of carrier center frequency. As stated earlier, this basic property of frequency spectral control represents the primary reason for utilizing a transmitter modulation operation in communication systems. In subsequent discussions, these advantages become more apparent.

2.6 FILTERING OF MODULATED CARRIERS

Modulated carriers will invariably have to pass through bandpass filters at both the transmitter and receiver ends of the system. Filtering is generally inserted to remove extraneous interference and noise in the vicinity of the modulated carrier spectrum. The filter designer must always be aware of the effect of the filtering on the carrier waveform itself.

Simply constructing a filter with a bandwidth that passes the major portion of the carrier spectrum may not be adequate in terms of eventual waveform reconstruction. Of particular concern is the effect of the filter on the modulating baseband waveform embedded within the carrier, which will be subsequently extracted by the demodulation process. In this section we examine the basic effect of a bandpass carrier filter on the various carrier waveforms $c(t)$.

Consider the system in Figure 2.16, where we let $c(t)$ be a general modulated carrier at the input of a bandpass filter, let $c_0(t)$ be the output carrier signal, and represent the filter by its transfer function $H_c(\omega)$ and impulse response $h_c(t)$. We then have

$$c_0(t) = \int_{-\infty}^{\infty} h_c(x)c(t - x)\, dx \qquad (2.6.1)$$

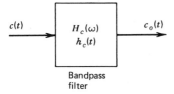

Bandpass
filter

Figure 2.16. Bandpass filtering of a carrier waveform.

For the class of modulated carrier signals of interest, we recall that $c(t)$ can be written as

$$c(t) = a(t) \cos [\omega_c t + \theta(t)]$$
$$= \text{Real} \{a(t)e^{j\theta(t)}e^{j\omega_c t}\} \tag{2.6.2}$$

where the form of $a(t)$ and $\theta(t)$ depends on whether it is amplitude or angle modulated. This means (2.6.1) takes the general form

$$c_0(t) = \int_{-\infty}^{\infty} h_c(x) \, \text{Real} \{a(t-x)e^{j\theta(t-x)}e^{j\omega_c t - j\omega_c x}\} \, dx$$
$$= \text{Real} \left\{ e^{j\omega_c t} \int_{-\infty}^{\infty} h_c(x)a(t-x)e^{j\theta(t-x)}e^{-j\omega_c x} \, dx \right\} \tag{2.6.3}$$

Thus a study of filtering effects on carrier signals reduces to an examination of this particular equation. We examine the preceding for specific types of carrier waveform.

Filtering AM Signals

Let $\theta(t) = \psi$, a constant, so that (2.6.2) represents an AM carrier. Then (2.6.3) becomes

$$c_0(t) = \text{Real} \{a_0(t)e^{j(\omega_c t + \psi)}\} \tag{2.6.4}$$

where

$$a_0(t) = \int_{-\infty}^{\infty} a(t-x)h_c(x)e^{-j\omega_c x} \, dx \tag{2.6.5}$$

Hence the output envelope function $a_0(t)$ appears as a complex time function obtained by filtering $a(t)$ with a complex impulse response $h_c(x)e^{-j\omega_c x}$. Thus impulse response corresponds to the equivalent transfer function

$$\tilde{H}_c(\omega) = \int_{-\infty}^{\infty} h_c(x) e^{-j\omega_c x} e^{-j\omega x} \, dx$$

$$= \int_{-\infty}^{\infty} h_c(x) e^{-j(\omega_c + \omega)x} \, dx$$

$$= H(\omega + \omega_c) \qquad\qquad (2.6.6)$$

Thus the filter $\tilde{H}_c(\omega)$ is equivalent to the positive frequency filter function $H_c(\omega)$ shifted to $\omega = 0$, as shown in Figure 2.17. The filtered carrier $c_0(t)$ therefore has a complex envelope obtained by filtering $a(t)$, the amplitude modulation of the input carrier, with the equivalent filter $\tilde{H}_c(\omega)$. If $A(\omega)$ is the Fourier transform of $a(t)$, then the transform of $a_0(t)$ is

$$A_0(\omega) = A(\omega)\tilde{H}_c(\omega) \qquad\qquad (2.6.7)$$

Note that $a_0(t)$ in (2.6.5) [i.e., the transform of (2.6.7)] is, in general, a complex time function, having both an amplitude and phase variation at each t, and the output carrier $c_0(t)$ is both amplitude and phase modulated. However, $a_0(t)$ is purely real (no phase variation) if its transform $A_0(\omega)$ has a magnitude even in ω and a phase odd in ω. (Recall Problem 1.1.) Since $a(t)$ is real, this even and odd symmetry occurs in $A_0(\omega)$ if $\tilde{H}_c(\omega)$ is also even in magnitude and odd in phase about $\omega = 0$. This requires the filter $H_c(\omega)$ to have the same symmetry about ω_c. Thus if the carrier filter is symmetric about the carrier frequency, no phase modulation is induced on the output carrier, and the filtered carrier simply has an amplitude modulation obtained by filtering the input modulation with the equivalent filter $\tilde{H}_c(\omega)$. This means the effect on the modulation of bandpass filtering of the AM carrier can be determined by considering instead the effect of passing the modulation alone through the equivalent low pass filter. Thus to achieve a given distortion the bandpass filter must be as wide about $\omega = \omega_c$ to pass $c_0(t)$ as a low pass filter must be about $\omega = 0$ to pass $a(t)$ with the same distortion level. The problem of designing bandpass AM filters reduces

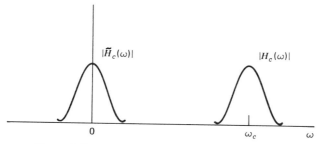

Figure 2.17. Bandpass filter and low pass equivalent.

therefore to a straightforward problem in linear low pass filter design. In fact, AM bandpass filters are often designed in exactly this manner—first designing the low pass equivalent, then using frequency translation (see Problem 1.14) to obtain the desired bandpass network.

When the amplitude modulation involves binary digital waveforms, the bandpass filtering can directly affect the bit waveforms. If the modulating waveform has the form of (2.1.1), a symmetric bandpass filter will produce an output AM carrier whose amplitude becomes

$$a_0(t) = \sum_k d_k \hat{s}(t - kT_b) \qquad (2.6.8)$$

with $\hat{s}(t)$ now a waveform whose transform is

$$F_{\hat{s}}(\omega) = F_s(\omega)\tilde{H}(\omega) \qquad (2.6.9)$$

This means that spectrum of the carrier at the filter output will contain a continuous portion in Figure 2.7 that has been filtered about the carrier with the transfer function $|\tilde{H}(\omega)|^2$. By proper selection of this low pass equivalent function, the bandpass filter can be designed to reduce out-of-band spectral energy, if desired.

The modulating waveform in (2.6.8) now has the filtered bit waveforms in (2.6.9) and in general will no longer be limited to $(0, T_b)$. Since frequency filtering tends to extend a waveform in time, the filtered AM carrier now contains bit waveforms that could spread into adjacent bit intervals. This causes a form of symbol-to-symbol waveform addition, or what is commonly referred to as *intersymbol interference*. (This will be of prime concern in considering digital decoding in Chapter 8.) Hence narrow bandpass filtering of an AM digital carrier can reduce spectral tails, but may introduce intersymbol interference in the digital modulating waveform.

Filtering Angle Modulated Carriers

In this case we reconsider (2.6.2) with $a(t) = A$, $\theta(t) = \Delta m(t)$, so that (2.6.3) becomes

$$c_0(t) = A \text{ Real} \left\{ e^{j\omega_c t} \int_{-\infty}^{\infty} h(x) e^{-j\omega_c x} e^{j\theta(t-x)} \, dx \right\} \qquad (2.6.10)$$

We recognize again the equivalent low pass filter impulse response in the integrand, but the desired modulation waveform $m(t)$ now appears in the exponential function. We therefore do not obtain the simple modulation-filtering interpretation as we did for AM. Analysis methods for (2.6.10) generally involve some type of expansion of the exponential modulation term. The most common expansion procedure is the quasi-linear method [4] in which the modulation is expanded in a Taylor series and the exponential expansion is terminated after several terms. Thus we write

$$\theta(t - x) = \theta(t) - \frac{\theta'(t)x}{1!} + \frac{\theta''(t)x^2}{2!} \cdots \qquad (2.6.11)$$

where the primes denote differentiations with respect to t, corresponding to a Taylor series expansion of $\theta(t - x)$ about $x = 0$. We then have for the integral in (2.6.10)

$$\int_{-\infty}^{\infty} h(x)e^{-j\omega_c x} \exp j\left[\theta(t) - \theta'(t)x + \frac{\theta''(t)x^2}{2} \cdots\right] dx \qquad (2.6.12)$$

If we retain the first three terms only, (2.6.10) can be written as

$$c_0(t) = \text{Real}\left\{Ae^{j\omega_c t + j\theta(t)} \int_{-\infty}^{\infty} h_c(x)e^{-j\omega(t)x}e^{j\theta''(t)x^2/2} \, dx\right\} \qquad (2.6.13)$$

where we have let $\omega(t) \overset{\Delta}{=} \omega_c + \theta'(t)$, which is the instantaneous frequency of $c(t)$. If we further expand

$$\exp\left[j\frac{\theta''(t)x^2}{2}\right] \approx 1 + j\frac{\theta''(t)x^2}{2} \qquad (2.6.14)$$

and note from Appendix A, A.1,

$$\int_{-\infty}^{\infty} x^2 h_c(x)e^{-j\omega x} \, dx = -\frac{d^2}{d\omega^2} H_c(\omega) \qquad (2.6.15)$$

then

$$c_0(t) \approx \text{Real}\left\{Ae^{j\omega_c t + j\theta(t)}\left(H_c[\omega(t)] - j\frac{\theta''(t)}{2} H_c''[\omega(t)]\right)\right\}$$

$$\approx \text{Real}\left\{Ae^{j\omega_c t + j\theta(t)}H_c[\omega(t)]\left[1 - j\frac{\omega'(t)}{2}\left(\frac{H_c''[\omega(t)]}{H_c[\omega(t)]}\right)\right]\right\} \qquad (2.6.16)$$

The preceding represents a somewhat complicated approximation to the exact filter output, but nevertheless does give some indication of the filtering effect. The first term is called the *quasi-linear* response term and has the following interpretation. Let us write the filter transfer function as $H_c(\omega) = |H_c(\omega)| \exp[j\angle H_c(\omega)]$ where $\angle H_c(\omega)$ is the phase function and $|H_c(\omega)|$ the magnitude. Then the first term in (2.6.16) yields an output component

$$A|H_c[\omega(t)]| \cos\{\omega_c t + \theta(t) + \angle H_c[\omega(t)]\} \qquad (2.6.17)$$

That is, at time t_0, the input amplitude is changed by the magnitude of the filter function at the instantaneous frequency $\omega(t_0)$, whereas the phase at that frequency, $\angle H_c[\omega(t_0)]$, is added to the input phase. Thus the quasi-linear term instantaneously evaluates the output amplitude and phase, as the frequency of the input carrier slides over the filter function. From the quasi-linear term we see that the bandpass filter should have essentially flat magnitude and linear phase over the maximum frequency excursions of $c(t)$, if negligible amplitude and phase distortion are to occur. However, the quasi-linear term adequately represents $c(t)$ only if the remaining terms in (2.6.16) can be neglected. This means it is necessary that

$$\frac{1}{2}\left|\left(\frac{d\omega(t)}{dt}\right)\left(\frac{dH_c^2[\omega(t)]/d\omega^2}{H_c[\omega(t)]}\right)\right| \ll 1 \qquad (2.6.18)$$

or equivalently,

$$\left|\frac{dH_c^2[\omega(t)]/d\omega^2}{H_c[\omega(t)]}\right| \ll \frac{2}{|d\omega(t)/dt|} \qquad (2.6.19)$$

for all t. The term on the left depends on the specific function $H_c(\omega)$, whereas the term on the right depends only on the rate of change of the input carrier frequency. In essence, (2.6.19) requires that the frequency cannot change too rapidly relative to the filter function slope characteristics. Unfortunately, it is difficult to evaluate the left side of (2.6.19), although tabulations of its maximum values for several common types of filter have been presented [5].

Other expansion methods [6, 7] have been proposed for (2.6.10), but all require the computation of fairly complicated terms that invariably must be done numerically. These expansions are applicable primarily when the carrier modulation index is small and the total distortion is low, so that only a few terms of the expansion are necessary to characterize the filter distortion.

When the frequency modulation involves a digital baseband waveform (FSK), the effect of the bandpass filter is extremely difficult to assess from the previous equations. One generally resorts to simulation and real time spectral analysis to estimate bit waveform distortion caused by the filter.

For a PSK carrier, (2.5.7) shows that the filtering effect is identical to (2.6.4) for the filtered AM case. Filtered PSK carriers can therefore be analyzed exactly as AM carriers with a pulse digital modulation. This means that the combination of spectrum reduction and intersymbol interference in PSK can be determined by a similar low pass equivalent as in (2.6.9).

REFERENCES

1. Abromowitz, M. and Stegun, I. *Handbook of Mathematical Functions*, National Bureau of Standards, Washington, D.C., 1965, Chap. 9.
2. Watson, G.A. *Treatise on Beseel Functions*, Cambridge University Press, Cambridge, England, 1958.
3. Proakis, J. *Digital Communications*, McGraw-Hill, New York, 1986, Chap. 3.
4. Panter, P. *Modulation, Noise, and Spectral Analyses*, McGraw-Hill, New York, 1965.
5. Baghdady, E. (ed.) *Lecture on Communication Systems*, McGraw-Hill, New York, 1960.
6. Bedrosian, E. and Rice, S. "Distortion and Crosstalk of Filtered Angle Modulated Signals," *Proc. IEEE*, vol. 56, January 1968, pp. 2–13.
7. Williams, O. "A Comparison of the Carson-Fry and Rice-Bedrosian Methods of FM Analysis, *IEEE Trans. Comm.*, vol. COM-52, August 1973, pp. 972–980.

PROBLEMS

2.1 (2.1) How much bandwidth is needed to transmit a 600 line television picture having the same line resolution and line scanning time as standard TV?

2.2 (2.1) How long will it take to print out a newspaper page (2 ft long and 1 ft wide) if a 4 kHz facsimile system is used with a line scan rate of 40 lines per inch? How much must the bandwidth increase if the printout is desired in 1 min?

2.3 (2.1) Prove that the Fourier transform of $f_s(\tau)$ in (2.1.8) is given by (2.1.14a), where $F_s(\omega)$ is the Fourier transform of $s(t)$. Prove (2.1.14b) by using the sampling identity in Appendix A.

2.4 (2.1) Show that the Fourier transform of the raised cosine pulse

$$s(t) = \frac{1}{2}\left[1 - \cos\left(\frac{2\pi t}{T_b}\right)\right], \qquad 0 \le t \le T_b$$

is that given in Figure 2.3.

2.5 (2.1) Given the waveform $m(t)$ in (2.1.25), let $\{a_k\}$ be a sequence of independent binary $(0, 1)$ random variables with probability p that $a_k = 1$. (a) Show that the autocorrelation function of $m(t)$ is

$$R_m(\tau) = \mathscr{E}[m(t)m(t+\tau)]$$

$$= \frac{p}{T_b} r_{s_1}(\tau) + \frac{1-p}{T_b} r_{s_0}(\tau) - r_s(\tau) + \sum_j r_s(\tau + jT_b)$$

where $r_{s_i}(\tau) = \int_{-\infty}^{\infty} s_i(t) s_i(t + \tau) \, dt$ and $s(t) = [p s_1(t) + (1 - p) s_0(t)]$.
(b) Determine the Fourier transform of part a to obtain (2.1.26).

2.6 (2.2) Determine the frequency spectrum of the waveform $a(t)p(t)$, where $a(t)$ is a positive waveform with transform $A(\omega)$ and $p(t)$ is a periodic, $(0, 1)$ square wave with period T sec.

2.7 (2.3) Let $a(t)$, $b(t)$ be random processes with known correlations $R_a(\tau)$, $R_b(\tau)$ and $R_{ab}(\tau)$. (a) Determine the correlation and spectral density of the general quadrature carrier:

$$c(t) = A a(t) \cos (\omega_c t + \psi) + A b(t) \sin (\omega_c t + \psi)$$

Assume ψ uniform over $(0, 2\pi)$ and independent of $a(t)$ and $b(t)$. (b) Determine the power of $c(t)$ in part (a). (c) What are the conditions required for $c(t)$ to have a constant amplitude for all t?

2.8 (2.3) Determine $c(t)$ and its corresponding transform for the special case of an SSB-AM carrier with modulation $a(t) = b \sin \omega_a t$. Show a plot of the transform magnitude. Show that the required $H_a(\omega)$ for SSB-AM has an impulse response

$$h(t) = \frac{1}{\pi t}$$

2.9 (2.4) A voice tone at frequency 4 kHz is sent by FM over a system with bandwidth of 24 kHz. What is the maximum permitted value of frequency deviation Δ_ω assuming a Carson rule bandwidth?

2.10 (2.4) A baseband waveform of 10 kHz is frequency modulated onto a carrier. Find the Carson rule bandwidth when the following frequency deviation coefficients are used: (a) $\Delta_f = 1 \, \text{kHz/V}$, (b) $\Delta_f = 100 \, \text{kHz/V}$, (c) $\Delta_f = 1 \, \text{MHz/V}$.

2.11 (2.4) An oscillator can be frequency deviated only 10% of its resonant frequency. Assume sine wave modulation and determine the maximum modulating frequency that can be used to frequency modulate the oscillator when operated at a resonant frequency of 2 GHz and with $\beta \geq 5$. What is the Carson rule bandwidth that is required with this FM signal?

2.12 (2.4) A carrier is frequency modulated by two sine waves

$$C_1 \cos (\omega_1 t + \theta_1) + C_2 \cos (\omega_2 t + \theta_2)$$

Derive the result in (2.4.16) for the case $K = 2$.

2.13 (2.4) An FM carrier is frequency modulated with a random process $m(t)$. The frequency variation at any t is given by $\Delta_f m(t)$. The probability that its frequency lies in the interval $(f_1, f_1 + df)$ is therefore the probability that $m(t)$ lies in the interval $(f_1/\Delta_f, f_1/\Delta_f + df/\Delta_f)$. The power spectral value at f_1 is therefore given by that of a sine wave at f_1 times the probability that $f = f_1$. Use these facts to show that the FM spectrum must approximate in shape the probability density of $m(t)$ at any t.

2.14 (2.5) We wish to determine the rms bandwidth (Problem 1.13) of a PM carrier phase modulated with a Gaussian noise process with zero mean and autocorrelation $R_\theta(\tau)$. (a) Show that the spectral spread is due only to the term $\exp[R_\theta(\tau)]$. (b) Write the rms bandwidth B_{rms} in terms of $S_{\text{ex}}(\omega)$, the transform of $\exp[R_\theta(\tau)]$. (c) Using transform theory, show that part b is

$$B_{\text{rms}} = \frac{2[-e^{R_\theta(0)}R_\theta''(0)]^{1/2}}{e^{R_\theta(0)/2}}$$

$$= 2[-R_\theta''(0)]^{1/2}$$

where the primes denote derivatives [recall $R_\theta'(0) = 0$]. (d) Use Problem 1.29 to show

$$B_{\text{rms}} = 2[R_{\theta'}(0)]^{1/2}$$

Explain what this means in terms of carrier frequency.

2.15 (2.5) A carrier at ω_c is phase modulated by the baseband waveform in (2.5.4). How much power is in the frequency components at (a) $\omega_c + 2\omega_1$, (b) $\omega_c - 3\omega_1 + 2\omega_2$, (c) $\omega_c - 5\omega_4$, (d) $\omega_c + \omega_1 - 3\omega_2 + 4\omega_4$?

2.16 (2.5) An RF carrier is phase modulated with the sum of K sine waves $\sum_{i=1}^{K} \Delta_i \sin(\omega_i t + \theta_i)$ plus a binary waveform $\Delta_r r(t)$, where $r(t) = \pm 1$ and has transform $F_r(\omega)$. (a) Derive the resulting PM spectrum and roughly sketch its shape. (b) Write the amplitude of the carrier component.

2.17 (2.5) Generalize the condition on the carrier phase angle ψ in (2.5.10) for the autocorrelation function to be stationary (not dependent on t).

2.18 (2.5) A Gaussian white noise process of spectral level N_0 one-sided, is filtered in an ideal filter of bandwidth B_1 Hz and is used to phase modulate a pure carrier at frequency ω_c and amplitude A. What is the power in the carrier component of the modulated carrier?

2.19 (2.5) Given the system in Figure P2.19, what is the power in the spectral line at frequency $\omega_c + 2\omega_1$ of the PM modulated carrier?

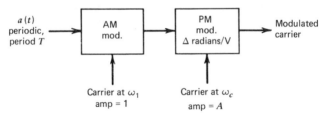

Figure P2.19.

2.20 (2.6) The front end of a receiver has the filter function

$$H_c(\omega) = \frac{1}{1 + j[(\omega - \omega_c)/2\pi B_3]^2}$$

(a) Determine the power in the output signal when receiving the AM signal

$$c(t) = a(t) \sin (\omega_c t + \psi)$$

with

$$a(t) = a_0 + b_0 \sin (\omega_m t)$$

(b) Repeat when the filter input signal is a random waveform having a bandlimited white noise power spectrum of one-sided level S_0 and bandlimited to $2B_n$ Hz about the carrier frequency. (c) Determine how wide B_3 should be in part b in order to pass 96% of the input waveform power.

2.21 (2.6) Given the carrier filter $H_c(\omega) = \exp[-(\omega - \omega_c)^2/2\sigma^2]$. Find the maximum rate of change of carrier frequency that can be tolerated for quasi-linear analysis.

3

CARRIER TRANSMISSION

After the modulated carrier is produced, it must be prepared for transmission to the receiver. Since the modulated carrier is usually formatted at a carrier frequency much lower than the desired transmission frequency, a typical RF transmitter generally involves stages of frequency conversion and power amplification, as shown in Figure 3.1. Frequency conversion (translating the modulated carrier at one frequency to a band at another frequency) requires additional carrier waveforms, produced from subsystems referred to as *frequency generators*. In this chapter these transmitter blocks are reviewed, with the primary objective of modeling their operation and understanding their inherent limitations. Later in the chapter, the effect of the propagation channel on the transmitted field is examined.

3.1 OSCILLATORS, CARRIER WAVEFORMS, AND FREQUENCY GENERATORS

A key requirement in any communication system is the generation of a pure unmodulated carrier at a fixed frequency. These sine wave tones are needed for modulation, as we saw in Chapter 2, and may also be needed for the frequency conversions shown in Figure 3.1.

Carrier tones are generated from oscillator circuits, the latter composed of active devices coupled to tuning mechanisms via some type of feedback [1], as shown in Figure 3.2. Resonances of the tuning circuits allows a sustained feedback oscillation to occur, producing an output tone at the resonant frequency. The oscillator tuning circuits commonly used are the resistance, inductance, capacitance (RLC) electronic or microwave circuit, the crystal quartz resonator, and the atomic resonators. RLC circuits are the simplest and easiest to construct and therefore are the more frequently used oscillator. However component imperfections and aging often makes it difficult to set and maintain a precise tone frequency over long time intervals. Crystal oscillators use the crystal structure itself as a resonant

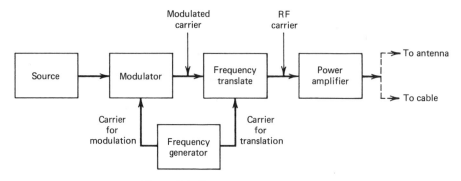

Figure 3.1. Transmitter block diagram.

circuit (or a component of a resonant circuit) to produce sharply tuned resonances and relatively stable output tones. A cut quartz crystal with properly attached electrodes acts like a high Q series—parallel electronic resonant circuit [2–5]. The frequency of the oscillation can be set by either the series resonance or the parallel resonances of the crystal. These resonant frequencies depend on the way the crystal is cut and formed and the crystal thickness. The parallel resonant frequency is usually used for oscillator frequencies below about 20 MHz, since high frequency oscillation with parallel resonance require extremely thin crystal cuts. Hence the series resonance of the crystal is usually used for the higher frequencies. Quartz crystal has a sharply tuned resonance, producing highly stable oscillators. Crystal oscillators at 5 and 10 MHz are common communication elements in modern RF systems.

The common atomic resonator is the cesium beam, which uses a stream of cesium atoms to interact with a magnetic field so as to produce an almost perfect oscillator at the specific frequency 9.152 GHZ. Rubidium resonators, using light beams interacting with rubidium vapor, produce a fixed oscillation at 6.8 GHZ. Atomic oscillators are often inserted as frequency mea-

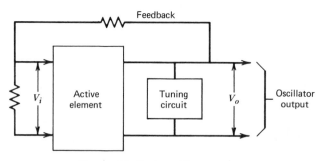

Figure 3.2. Basic oscillator system.

surement standards and are used primarily as reference tones for systems requiring extreme frequency accuracy, such as time measurement and ranging systems.

Often there is need to vary the oscillator output frequency according to an external control signal. This is achieved by using the control voltage to vary or adjust the tuning circuit parameter, resulting in an output frequency directly proportional to the control. Such devices are referred to as *voltage controlled oscillators* (VCOs). *RLC* oscillators are most often used as a VCO, since electronic elements such as resistors and capacitors are easily adjusted, and voltage control over a relatively wide frequency range is usually feasible. Voltage controlled crystal oscillators (VCXOs) can also be used, but frequency range is often limited, because of the sharp tuning of the crystal structures. Hence a VCXO is only used if the "pulling range" of the oscillator is relatively small.

An ideal oscillator produces a pure sinusoidal carrier with fixed amplitude, frequency, and phase, as we assumed in Chapters 1 and 2. Practical oscillators, however, produce carrier waveforms with parameters that may vary in time, due to temperature changes, component aging, and inherent tuning circuit noise. Amplitude variations are somewhat tolerable since they can be easily controlled with electronic clipper circuits and limiting amplifiers. More important to a communication system are the variations in frequency and phase that may appear on an oscillator output. Although preliminary system design may be based on the supposition of ideal carriers, the possibility of imperfect oscillators and the degradation they may produce must eventually be considered.

Frequency offsets in oscillators are usually specified as a fraction of the oscillator design frequency. This fraction is generally normalized by a 10^{-6} factor and stated in units of parts per million (ppm). An offset of Δf Hz in an oscillator designed for f_0 Hz output frequency will therefore be stated as having an offset of $(\Delta f/f_0)10^6$ ppm. Thus, for example, a 5 MHz oscillator, specified as having a stability of ± 2 ppm, will be expected to produce an output frequency that is within $\pm 2 \times 10^{-6} \times 5 \times 10^6 = \pm 10$ Hz of the desired 5 MHz output.

Oscillator frequency offsets are contributed primarily by frequency uncertainty (inability to exactly set the desired frequency), frequency drift (long term variations due to component changes, stated in ppm/sec), and short term, random frequency variations. The latter appears as a phase jitter on the oscillator, which effectively converts the fixed carrier phase ψ of an ideal oscillator to a randomly varying phase noise process $\psi(t)$. This phase noise typically has a spectrum that is predominantly low frequency, extending out to about several kilohertz [6–9]. This means that the phase noise process generally varies slowly relative to phase modulation waveforms, justifying the assumption that $\psi(t)$ is approximately constant over the time variations of the modulation. Since internal oscillator phase noise will add directly to any phase or frequency modulation placed on that oscillator carrier, phase noise will always be of primary concern in angle modulated systems. The specific phase noise spectrum becomes important in phase tracking analysis and is considered again in Chapter 10. In general, *RLC* oscillators will tend to have higher phase noise than crystal oscillators, while atomic resonators have almost negligible phase noise.

Frequency generators are constructed from oscillators by multiplying a reference oscillator carrier to the desired frequencies needed throughout the transmitter. Multiplication of an oscillator carrier frequency can be achieved by straightforward generation of its harmonics via an electronic nonlinearity. The most common is a simple squaring device followed by a bandpass filter, as shown in Figure 3.3a. If the oscillator produces the reference tone $A \cos(\omega_0 t + \psi)$, its square is

$$[A \cos(\omega_0 t + \psi)]^2 = \frac{A^2}{2} + \frac{A^2}{2} \cos(2\omega_0 t + 2\psi) \qquad (3.1.1)$$

This corresponds to a constant (dc) signal plus the second harmonic tone at $2\omega_0$. The latter is extracted by the bandpass filter centered at $2\omega_0$, producing a direct frequency doubling at the output. Higher order nonlinearities can be used, but the higher harmonics tend to be progressively reduced in amplitude. Additional amplification is then necessary to restore the desired carrier amplitude. Note that the phase is also multiplied by the same frequency multiplication factor. This means that any phase noise on the

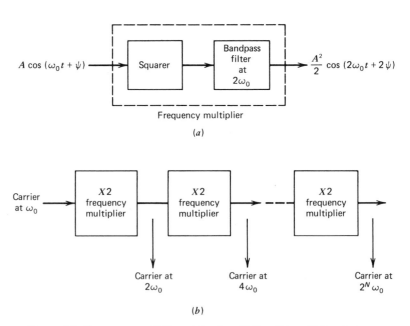

Figure 3.3. Frequency multipliers: (a) ×2 multiplier, (b) multiplier chain.

input carrier will have its spectrum scaled up during frequency multiplication.

By cascading N sequential stages of a "times two" multipler, the reference can be multiplied to $2^N\omega_0$ (Figure 3.3b). Multiplier chains of this type are commonly used in modern systems to generate multiple carrier frequencies from a single reference oscillator, as shown. Subsystems of this type, producing all the required carriers from a common source, are referred to as *master frequency generators* (MFGs).

Frequency multiplication can also be achieved by frequency synthesizers [10], as shown in Figure 3.4. These are feedback systems in which the reference oscillator is used to control the frequency of a separate VCO, via a frequency divider circuit. Divide-down counters are obtained by simple cycle counting subsystems that produce an output pulse every N input cycle zero crossings. A low pass filter then passes the first harmonic of the pulse sequence, having a frequency of $1/N$th of the divider input frequency. If the VCO output frequency in Figure 3.4 is at f_0 Hz, then the divide-down frequency is at f_0/N Hz. A frequency comparison circuit produces an output error voltage proportional to the frequency difference between the reference carrier frequency f_r and the divider output carrier. This error voltage is then used to control the VCO frequency so as to null out the frequency difference. The loop therefore stabilizes such that $f_0/N - f_r = 0$, or at the VCO frequency

$$f_0 = Nf_r \tag{3.1.2}$$

Hence the synthesizer (VCO) output frequency is maintained at a frequency that is a direct multiplication of the input reference oscillator frequency by the divided factor N. Since divider circuits can be easily constructed with relatively high (2000–4000) factors, synthesizers can be achieved with similar multiplication factors. Furthermore, since the reference oscillator is generally selected as an extremely stable reference, the output frequency of the VCO will be maintained with a similar stability. Thus the feedback operation effectively shifts the frequency stability of the reference to the higher frequency VCO. Frequency synthesizers are becoming increasingly

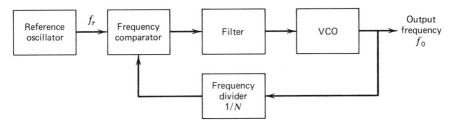

Figure 3.4. Frequency synthesizer.

important in modern high frequency systems, and entire frequency generators can be constructed by using a single reference to control a multiple set of VCO frequencies via parallel divide-down feedback loops.

There has also been rapid advancement in the development of digital frequency generators (DFG). These devices use digital software in conjunction with a stable reference oscillator to produce a subharmonic output carrier. A DFG operates by storing in memory a sequence of numerical values of one period of a sine wave. These numbers are then continually read out in sequence at a prescribed clock rate driven by the reference

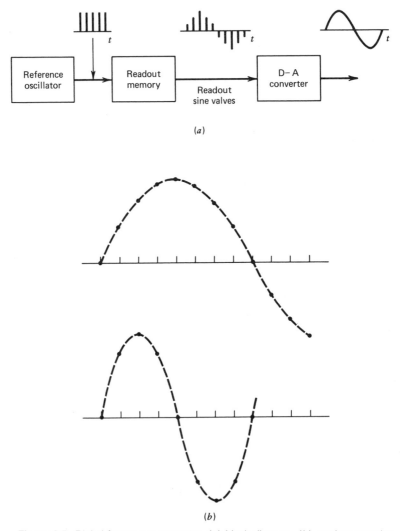

Figure 3.5. Digital frequency generator: (a) block diagram, (b) readout samples.

oscillator. This produces in series the discrete numerical values of a sine wave, which are then converted to an analog waveform by digital–analog (D-A) conversion, forming the output carrier, as shown in Figure 3.5a.

Since the sine wave values (including the zero crossings) are clocked by the reference oscillator, the output frequency contains the periodic stability of the reference, with the voltage accuracy of the stored samples. If there are M stored sine words in memory, the output frequency in Figure 3.5a will be $f_1 = f_0/M$, where f_0 is the reference frequency.

Suppose that we now alter the readout from memory by reading out every other value at the same clock rate (Figure 3.5b). We now produce a higher output frequency at $f_2 = 2f_1 = 2f_0/M$. By reading out every fourth sine wave value, the output frequency is again doubled. Continuing in this manner, the output frequency can increase to $f_0/4$ (at least four values, including zero crossings, are needed to define a sine wave). Hence a DFG is primarily used for generating low frequency, fine-resolution harmonics, with the lower limit depending only on the storage capability of the memory.

3.2 CARRIER MIXING AND UPCONVERSION

In Section 3.1 it was shown how an oscillator carrier can be frequency multipled to form a frequency generator. Often, however, the modulated carrier itself must be frequency shifted to place it at the desired RF transmission frequency. This frequency shifting to a higher frequency band (called *upconversion*) is accomplished by a mixer circuit.

An ideal mixer (Figure 3.6a) corresponds to an electronic multiplier followed by a bandpass filter. The inputs to the multiplier are the modulated carrier and a pure carrier mixing tone provided by the frequency generator. Denote the general modulated carrier again by

$$c(t) = a(t) \cos (\omega_c t + \theta(t) + \psi) \tag{3.2.1}$$

and denote the mixing carrier by

$$l(t) = A_l \cos (\omega_l t + \psi_l) \tag{3.2.2}$$

By trigonometrically expanding the product, we see that the result of multiplying produces the sum and difference frequency terms

$$c(t)l(t) = G_m \left\{ \frac{A_l a(t)}{2} \cos \left[(\omega_c + \omega_l)t + \theta(t) + \psi + \psi_l \right] \right.$$
$$\left. + \frac{A_l a(t)}{2} \cos \left[(\omega_c - \omega_l)t + \theta(t) + \psi - \psi_l \right] \right\} \tag{3.2.3}$$

where G_m is the multiplier gain constant. The bandpass filter eliminates the

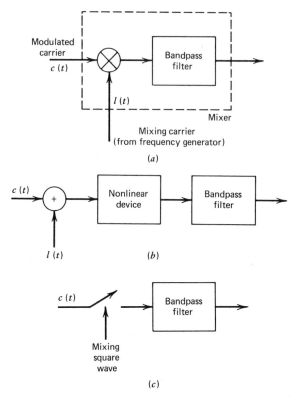

Figure 3.6. Mixer block diagrams: (a) multiplier, (b) sum-nonlinear model, (c) switching model.

difference frequency for upconversion, assuming $(\omega_c + \omega_l) \gg |(\omega_c - \omega_l)|$. The output of the mixer is then

$$c_o(t) = K_m a(t) \cos\left[(\omega_c + \omega_l)t + \theta(t) + \psi + \psi_l\right] \qquad (3.2.4)$$

where $K_m = G_m A/2$. The mixer output is itself a modulated carrier, having the same amplitude and phase variation of the input carrier, but at the sum frequency $(\omega_c + \omega_l)$. The input modulated carrier therefore has been frequency translated from ω_c to $(\omega_c + \omega_l)$ while preserving its modulation. The amplitude modulation is scaled by the overall mixer gain K_m, while the angle modulation is unaffected if the mixing carrier phase ψ_l is constant. By properly selecting ω_l, the modulated carrier can be upconverted to any desired RF frequency. This is why the frequency generator subsystem, which produces the mixing carriers, plays an important role in RF transmitters. Note that the amplitude A_l of the mixing carrier in (3.2.2) is important only in setting the overall mixer gain K_m.

In practice the carrier multiplication in Figure 3.6a is accomplished by either nonlinear or switching devices. In the nonlinear method a squaring device, as in Figure 3.3a, or a general amplifier nonlinear gain characteristic is typically used to generate the required cross product. The two carriers that are to be mixed are first summed, then the nonlinearity generates the required product term that is filtered off by the bandpass mixer filter, as shown in Figure 3.6b. A second mixing method is by switching circuitry, in which the input modulated carrier is "chopped," or gated, in time by a square wave at the carrier mixing frequency, as in Figure 3.6c. The switching or gating operation is generally accomplished with diodes, transistors, or transformers. The gating operation performs an effective multiplication of the input carrier by the gating square wave, of which the fundamental component generates the desired mixer product.

In some RF systems upconversion is accomplished via several stages of mixing. The modulated carrier is first upconverted to an intermediate frequency, then upconverted again to the desired RF frequency. This simplifies the design of the frequency generator by reducing the frequency multiplication factors it must provide.

3.3 CARRIER AMPLIFICATION

After the modulated RF carrier is generated and frequency converted, the transmitter power amplifier then must amplify the carrier signal to the desired power value for transmission, over either the antenna or cable propagation system, as was shown in Figure 3.1 We shall find that system performance invariably depends on the amount of carrier amplification that can be generated at the transmitter. Therefore, there always exists a requirement for extremely high power amplification at carrier frequencies. Furthermore, this amplification must extend over the complete bandwidth of the modulated carrier, so as not to cause carrier amplitude or phase distortion. In some transmitters there may be multiple carriers to be amplified by a single power amplifier. The power amplifier must have sufficient gain and bandwidth to properly amplify all carriers simultaneously.

Conventional electronic amplifiers are limited by parasitic capacitance and electron transit times, both tending to reduce gain at the higher carrier frequencies. For this reason, cavity amplifiers are necessary for producing higher power levels over wide, high frequency bandpass bandwidths. Such devices are physically large and usually have relatively poor amplifier *efficiency*. Efficiency is the ratio of the carrier power output to the prime power (batteries, generators, solar cells, etc.) that must be provided to operate the amplifier. The lower the efficiency, the less the carrier power for a given amount of prime power, or, conversely, the more prime power to achieve a desired carrier power. Also important is the fact that the unused prime power (one minus the efficiency times the prime power) represents power that must be dissipated as heat in some way at the transmitter. In systems where prime power is constrained by weight or cost limitations, the efficiency of the power amplifier becomes a critical design parameter, often dictating the ultimate hardware components.

The most common types of high gain cavity amplifiers are the *klystron* and the *traveling wave tube* (TWT) [11–17]. Both devices can produce amplification factors on the order of 50–60 dB with efficiencies of about 40–60%. However, the klystron is generally quite large in physical size (3–6 ft long), which limits its usefulness to Earth-based operation. The TWT, because it is more compact, has emerged as the basic spaceborne transmitter amplifier for satellites and space vehicles. However, TWT amplifiers exhibit saturation effects when operated at high input (*drive*) power levels. Figure 3.7 shows a typical TWT curve for available output power versus input drive power and indicates the pronounced saturation of the amplification. For low values of input power, the output power is linearly related to the input, and the device operates as a conventional linear amplifier. As the input waveform amplitude is increased, the amplifier becomes nonlinear, with the output amplitude eventually becoming saturated as the attainable power from the amplifier levels off. Achievement of the maximum power in the amplifier is therefore accompanied by a non-linear amplification of the input signal, as exhibited by the saturation effect. When only a single RF carrier is being amplified, this nonlinearity poses no serious problem, since the amplifier harmonics produced will be well outside the RF bandwidth of the carrier. Thus single carrier power amplifiers are usually operated with as large an input amplitude as possible so as to

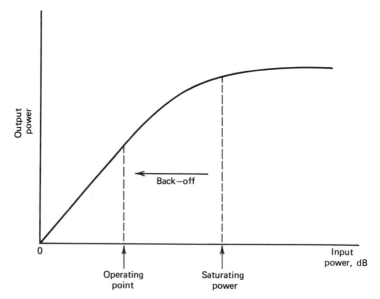

Figure 3.7. Power amplifier characteristic.

achieve the maximum available output power. The amplifier is said to be *saturated*, and the drive power at which the saturation occurs is called its *saturating power*. When multiple carriers are involved, the nonlinearity of the amplifier causes carrier cross-products, or *intermodulation* interference, to occur. By reducing the total input drive power from its saturating power, a more linear amplification is produced (with less intermodulation) but with less total output power to be divided among the carriers. The ratio of saturating drive power to desired operating drive power is called the amplifier *backoff*. Increasing backoff produces less output power but less intermodulation interference, and a tradeoff study must be made to assess the effect of each in selecting power amplifier operating points.

Solid state amplifiers are also used for power amplification. The two most common solid state devices are the *tunnel diode* [15] amplifier and the *maser* [14]. Tunnel diodes are compact, have fairly good efficiencies and low noise levels, and can provide 20–40 dB of amplification, but they are usually limited in their peak power output capability. Thus tunnel diodes are mostly used in low level transmitters or as drive amplifiers in cascade with more powerful cavity amplifiers. Masers provide much larger gain values but usually must be cooled to operate well below room temperatures for best performance. This requires additional cooling apparatus that must also be incorporated at the transmitter.

Two relatively new solid state amplifiers that have been developed for high frequency power amplification are the field-effect transistor (FET) and the impact avalanche and transit time (IMPATT) amplifier [17]. These devices have been shown to provide moderate gains with up to 1–5 W outputs at frequencies of 10–100 gHz. Techniques for combining multiple devices into a single higher power source are being developed.

All power amplifiers exhibit decreasing available peak power levels with increasing frequency. This inherent falloff in transmitter power capability is a major impediment to the design of communication systems at the higher RF frequencies (X-band and above). In later studies we will see how transmitter power limitations can impact the overall capability of a communication link.

3.4 COUPLING LOSSES

In any transmitter the parameter of particular importance is the actual amount of amplified carrier power coupled into the antenna or guide system from the amplifier. This depends not only on the amplifier size but also on the connecting circuitry from the amplifier output to the antenna or cable input terminals. Power levels quoted for carrier amplifiers are usually for matched impedance loading (i.e., the input impedance of the cable or antenna is equal to the output impedance of the amplifier). When the impedances are not matched, a power loss occurs in the coupling. This can be accounted for as shown in Figure 3.8a. If v_0 is the open circuit output voltage of the amplifier, Z_a its output resistance, and Z_t the input resistance of the transmission antenna or cable terminals, the transmitter power (voltage squared divided by resistance) at the antenna or cable input is then

$$P_t = v_0^2 \left[\frac{Z_t}{Z_t + Z_a} \right]^2 \left(\frac{1}{Z_t} \right) \tag{3.4.1}$$

Under matched load conditions $Z_a = Z_t$ and $P_t = v_0^2/4Z_a \stackrel{\Delta}{=} P_{amp}$, which is the stated amplifier output power. For the mismatched case, however, (3.4.1) becomes

$$P_t = P_{amp} \left[\frac{4Z_a Z_t}{(Z_a + Z_t)^2} \right] \tag{3.4.2}$$

The bracket accounts for the power coupling loss into the transmission terminals. If the antenna is remote from the amplifier, connecting lines or waveguides are necessary, which inject additional loss. Now coupling losses may occur at both the waveguide and antenna input, in addition to losses in the guide itself (Figure 3.8b). If L_g is the ratio of the guide output power to its input power when fed and terminated by a matched impedance, (3.4.2) becomes instead

$$P_t = P_{amp} L_g \left[\frac{4Z_t Z_g}{(Z_t + Z_g)^2} \right] \left[\frac{4Z_a Z_g}{(Z_a + Z_g)^2} \right] \tag{3.4.3}$$

where Z_g is the ohmic impedance of the guide. When properly matched, $Z_t = Z_g = Z_a$ and $P_t = L_g P_{amp}$. Thus the brackets in (3.4.3) account for the

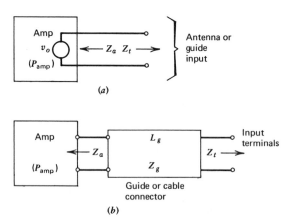

(a)

(b)

Figure 3.8. Amplifier coupling circuity: (a) direct amplitifier coupling, (b) cable connected amplifier coupling.

additional coupling losses. Waveguide loss parameters L_g are usually specified in decibel loss per length, and therefore the power loss in the guide is directly dependent on its length. This can become a rather serious effect when the antenna is remotely located from the power amplifier. It should also be pointed out that waveguide impedance values Z_g are functions of frequency, and the impedance matching condition is frequency dependent. Ideally, Z_g is designed to be matched at the carrier band center, that is, the carrier center frequency.

When the coupling involves a waveguide, cable, or transmission line, the coupling loss can be obtained directly from the *voltage standing wave ratio* (VSWR). The latter is the ratio of the maximum to minimum voltage inside the guide. If a guide has impedance values that cause a complex reflection coefficient Γ to exist, the power transfer through the guide is known to be [12]

$$\frac{P_t}{P_{amp}} = 1 - |\Gamma|^2 \qquad (3.4.4)$$

The reflection coefficients is related to the VSWR by

$$|\Gamma| = \frac{(VSWR) - 1}{(VSWR) + 1} \qquad (3.4.5)$$

The coupling loss in (3.4.4) is then

$$\frac{P_t}{P_{amp}} = \frac{4(VSWR)}{(1 + VSWR)^2} \qquad (3.4.6)$$

Since VSWR values are relatively easy to measure in a waveguide (by simply a voltmeter probe inserted in the line), coupling losses can be accurately determined from (3.4.6), without knowledge of the actual internal impedances.

3.5 TRANSMITTING ANTENNAS

In communication systems using unguided space transmission, the carrier waveform is propagated from the transmitter by use of a transmitting antenna. An antenna is simply a transducer that converts electronic signals into electromagnetic fields, or vice versa. A transmitting antenna converts the amplified carrier signal into a propagating electromagnetic field. Any transmitting antenna is composed of a feed assembly that illuminates an aperture or reflecting surface, from which the field then radiates. This field is transmitted by the antenna as a propagating plane wave with a prescribed polarization and spatial distribution of its field power density. The spatial

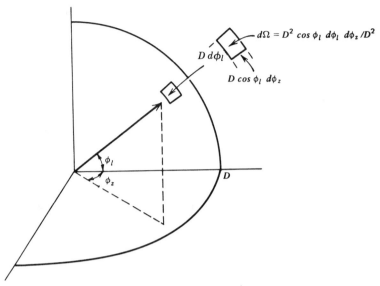

Figure 3.9. Antenna geometry.

power distribution of a transmitting antenna is described by its *antenna pattern*, the latter defined as

$$w(\phi_l, \phi_z) = \text{power transmitted per unit solid angle in the direction } (\phi_l, \phi_z)$$

$$(3.5.1)$$

where (ϕ_l, ϕ_z) are the elevation and azimuth angles defining a ray line direction from the geometric center of the antenna, as shown in Figure 3.9. Thus the antenna pattern indicates the amount of field power that will pass through a unit solid angle[†] in a given direction out from the antenna center.

In addition to its power content, an electromagnetic field also has a designated *polarization* (orientation in space). This polarization is determined by the manner in which the electromagnetic field is excited at the antenna feeds prior to propagation. The common polarizations in communication links are *linear* and *circular*. In linear polarization the electromagnetic field is aligned in one planar direction throughout the entire propagation, as shown in Figure 3.10a. These directions are usually designated as horizontal or vertical polarizations (relative to the antenna coordinates). In circular polarization (CP) the field is excited and transmitted with components in two orthogonal coordinates (one horizontal and one vertical) that are

---

[†]Recall that the solid angle of a cone enscribed on a sphere of radius D is given by the surface area of its spherical cap, divided by D^2.

phased so that the combination of the two produces a resultant field polarization that appears to rotate circularly as the wave propagates (Figure 3.10*b*). Depending on the sign of the phasing, the resultant polarization vector can be made to rotate clockwise [called *right-hand* circular polarization (RHCP)] or rotate counterclockwise [*left-hand* circular polarization (LHCP)].

The antenna pattern is often normalized to form the *antenna gain function*

$$g(\phi_l, \phi_z) = \frac{w(\phi_l, \phi_z)}{\text{power per unit solid angle of an isotropic radiator with same total transmitted power}} \quad (3.5.2)$$

An *isotropic* radiator is an antenna that radiates equally in all directions. The total power transmitted by an antenna P_T is the integral of $w(\phi_l, \phi_z)$ over the unit sphere,

$$P_T = \int_{\text{unit sphere}} w(\phi_l, \phi_z)\, d\Omega = \int_{-\pi}^{\pi} \int_{-\pi/2}^{\pi/2} w(\phi_l, \phi_z) \cos\phi_l\, d\phi_l\, d\phi_z \quad (3.5.3)$$

where $d\Omega = (\cos\phi_l)\, d\phi_l\, d\phi_z$ is the differential solid angle (Figure 3.9). The power per unit solid angle of an isotropic antenna with the same total power is then $P_T/4\pi$. Therefore

$$g(\phi_l, \phi_z) = \frac{w(\phi_l, \phi_z)}{P_T/4\pi} \quad (3.5.4)$$

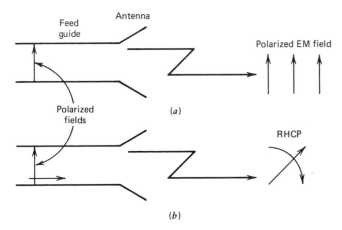

Figure 3.10. Generation of polarized EM fields: (*a*) linear polarization, (*b*) circular polarization.

Thus the gain function of a transmitting antenna is simply a normalized form of the antenna pattern. Note that the area under the gain function is always 4π. This means that if the gain function is increased in one direction, it must be suitably decreased in other directions so as to maintain constant area. In general, a transmitting antenna is designed to transmit its radiated field in one specific ray line, referred to as the antenna *main lobe*. Transmissions in other directions are called *sidelobes* of the antenna and represent power transmission in unwanted directions. Although a function of two dimensional angles, gain patterns are often displayed in terms of a planar angle, as shown in Figure 3.11. This plots the relative gain in angular directions from the antenna origin, showing clearly the mainlobe and sidelobes.

Rather than deal with the actual antenna gain function, we often simply specify its *gain* and *field of view*. The gain g of a transmitting antenna is the maximum value of its gain function:

$$g \stackrel{\Delta}{=} \max_{\phi_l, \phi_z} g(\phi_l, \phi_z) \tag{3.5.5}$$

and is often stated in decibels. The antenna solid angle field of view is a measure of the solid angle into which most of the transmitted field power is concentrated; that is, the field of view is a measure of the directional properties of the antenna. Field of view can be defined in several ways from the $g(\phi_l, \phi_z)$ pattern, such as the 3 dB field of view (the solid angle over which the gain function falls to one-half its peak value, often called the *half-power field of view*), the *6 dB field of view*, *mainlobe field of view*, and so on.

Usually field of view is defined in terms of planar angles, using mainlobe plots similar to Figure 3.11 for both the azimuth and elevation planes. For symmetric gain pattern, the planar beamwidth Φ_b, in radians in any plane, is related to the solid angle field of view in steradians by (Problem 3.15)

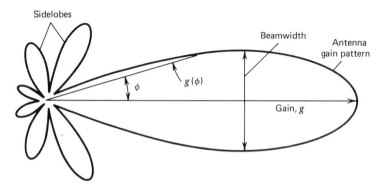

Figure 3.11. Antenna planar gain function $g(\phi)$.

$$\Omega_{\text{fv}} = 2\pi \left(1 - \cos \frac{\Phi_b}{2}\right)$$

$$\approx \frac{\pi}{4} \Phi_b^2, \qquad \Phi_b \ll 1 \tag{3.5.6}$$

Thus beamwidth measurements can be converted to field of view, using (3.5.6).

An intrinsic property of any antenna is that its maximum gain g and half-power beamwidth ϕ_b in radians at wavelength λ are related to the physical area of the antenna $\mathscr{A}$ by [11, 18–21]

$$g = \left(\frac{4\pi}{\lambda^2}\right) \rho_a \mathscr{A} \tag{3.5.7a}$$

$$\phi_b = c \frac{\lambda}{d\sqrt{\rho_a}} \tag{3.5.7b}$$

where $\mathscr{A}$ is the antenna aperture cross-sectional area, d the cross-sectional diameter, c is a proportionality coefficient dependent on the type of antenna, and ρ_a the antenna *aperture loss* factor. This loss factor accounts for antenna diffraction losses and also depends on the type of antenna and on how the antenna aperture is electromagnetically illuminated. Note that the antenna gain is always proportional to the square of the carrier frequency and to the antenna area, while the beamwidth varies inversely with frequency and size. The larger the antenna or the higher the frequency, the larger is the gain and the narrower the beamwidth. Thus a given antenna has an increasingly more directional pattern as higher frequencies are used. At a fixed frequency, the pattern becomes more directional as the antenna is made larger. Equations (3.5.7) are the key transmitting antenna equations that are of most use to communication engineers, since they relate gain and beamwidth to antenna size and carrier frequency.

The common antenna types are the linear dipole, the helix, the horn, the antenna array, and the parabolic reflector. These are shown in Figure 3.12, along with a sketch of their gain patterns and parameter values. The dipole has a pattern that is hemispherical and produces a propagating field polarized in the direction of the dipole. The automobile whip antenna is an example of a dipole antenna. Helix and horns are smaller antennas with reasonable directivity but higher sidelobes than parabolic reflectors. A helix antenna produces a circularly polarized field, while a horn is generally used to produce linearly polarized fields. An array is a group of small antennas (dipoles, horns, or helices) properly separated so that if a carrier is phase shifted and individually radiated from each element, the propagating fields will reinforce in some directions and interfere in others. The net result will be a combined beam with directivity. By properly selecting the phase shifts between array elements, the directivity of the beam can be oriented in a

Antenna type	Pattern	Gain g	Half-power beamwidth
Short dipole $l \ll \lambda$	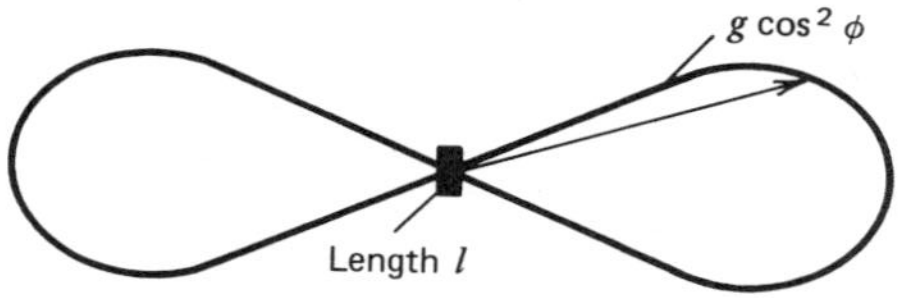 $g\cos^2\phi$, Length l	1.5	$90°$
Long dipole $l \gg \lambda$, $l = \lambda/2$		1.5, 1.64	$47°$, $78°$
Helix	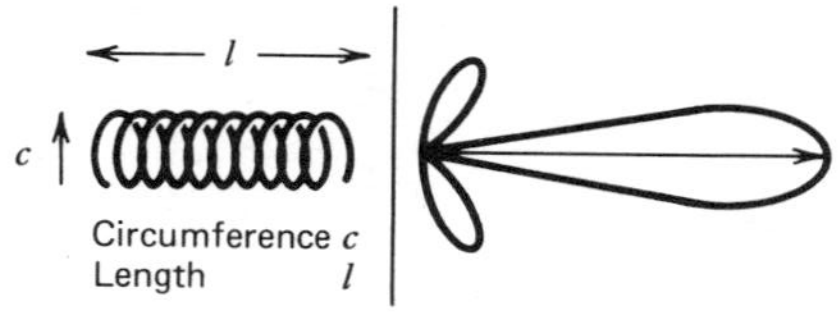 Circumference c, Length l	$15\left[\dfrac{cl}{\lambda^2}\right]$	$52°\left[\dfrac{\lambda}{c\sqrt{l}}\right]$
Square horn dimension d 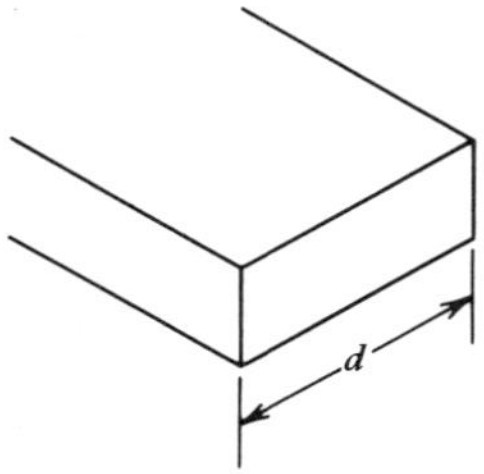	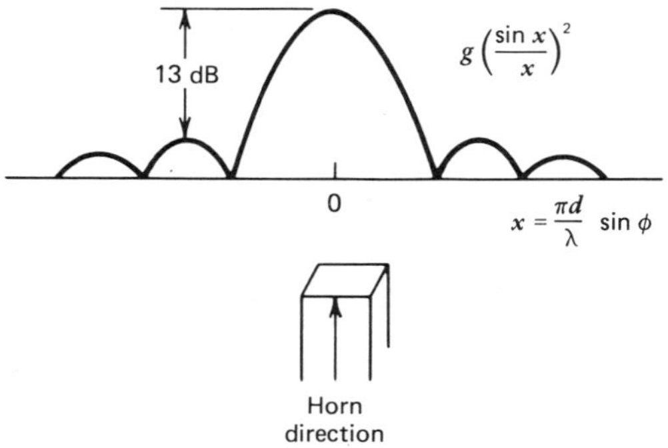 $g\left(\dfrac{\sin x}{x}\right)^2$, $x = \dfrac{\pi d}{\lambda}\sin\phi$, 13 dB, Horn direction	$\dfrac{4\pi d^2}{\lambda^2}$	$\dfrac{0.88\lambda}{d}$ rad

Figure 3.12. Common antenna characteristics.

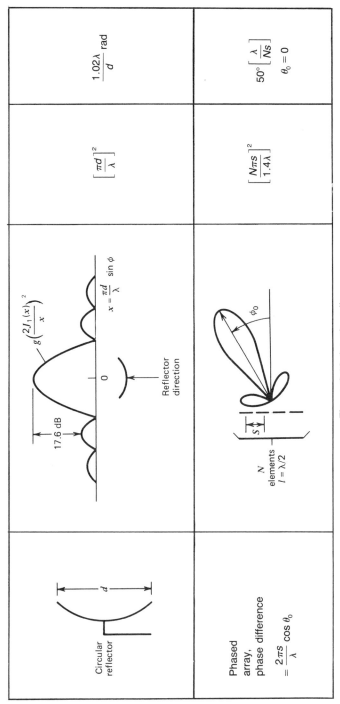

Circular reflector	$g\left(\dfrac{2J_1(x)}{x}\right)^2$ $x = \dfrac{\pi d}{\lambda}\sin\phi$	$\left[\dfrac{\pi d}{\lambda}\right]^2$	$\dfrac{1.02\lambda}{d}$ rad
Phased array, phase difference $= \dfrac{2\pi s}{\lambda}\cos\theta_0$		$\left[\dfrac{N\pi s}{1.4\lambda}\right]^2$	$50°\left[\dfrac{\lambda}{Ns}\right]$ $\theta_0 = 0$

Figure 3.12 (*continued*).

given direction. Hence a phase array can theoretically produce beams in arbitrary off-axis directions. As we see from Figure 3.12, gain increases with the number of array elements, and thus high gain is achieved with large arrays. The beamwidth also changes with the beam direction, being narrowest for broadside beams and becoming increasingly larger for off-axis beams. Phased-array antennas represent a rapidly advancing technology that will continue to undergo substantial development. Phase arrays allow beam steering without physically moving an antenna; they simply readjust phase shifters in the array feed. This electronic steering can be used to advantage to point toward moving receivers, as in an orbiting satellite system.

The most popular antenna is the parabolic reflector, or *dish*. Figure 3.13 shows two different feed diagrams for a parabolic dish. The field to be transmitted is excited in the feed waveguide with the desired polarization and is then radiated to the reflector. The feed may be located in front at the focus of the parabolic dish, or it may be fed from behind using a subreflector. If the dish is uniformly illuminated by the feed, the reflected transmitted gain pattern is circular symmetric, as given in Figure 3.12. The gain, half-power beamwidth ϕ_b, and the sidelobe gain depend on the manner in which the feed radially distributes the field intensity over the dish. Table 3.1 lists the relation between type of illumination, the resulting aperture ef-

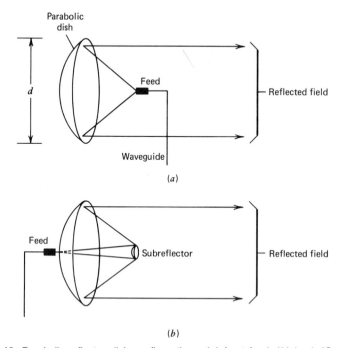

Figure 3.13. Parabolic reflector dish configurations: (*a*) front feed, (*b*) back (Cassagranian) feed.

TABLE 3.1 Parameter Values of Parabolic Reflectors[a]

γ	ρ_a, %	ϕ_b, rad	Sidelobe, dB	Gain g
0	100	$1.02 \, \lambda/d$	-17.6	$9.86 \, (d/\lambda)^2$
1	75	$1.27 \, \lambda/d$	-24.6	$7.39 \, (d/\lambda)^2$
2	55	$1.47 \, \lambda/d$	-30.6	$5.4 \,\,\, (d/\lambda)^2$
3	44	$1.65 \, \lambda/d$	-40.6	$4.83 \, (d/\lambda)^2$

[a]Antenna feed distribution over dish $= [1 - (2r/d)^2]^\gamma$, $r =$ radial distance, $r \leq d/2$.

ficiency ρ_a, beamwidth ϕ_b, and the sidelobe peak values. Note that reduction of sidelobes is accompanied by a spreading of the radiated beam.

For a parabolic antenna having a diameter of d meters and $\rho_a = 1$, the gain and planar half-power beamwidth equations [the latter obtained by using (3.5.7)] at frequency f_c, reduce to

$$g = \left(\frac{\pi}{3}\right)^2 \left(\frac{df_c}{10^8}\right)^2 \qquad (3.5.8a)$$

$$\phi_b = 3.06 \left(\frac{10^8}{f_c d}\right) \text{ radians} \qquad (3.5.8b)$$

For $d = 1$ m, we need $f_c = 10^9 = 1$ GHz for a 20 dB ($g = 100$) gain, whereas a frequency of 100 MHz requires $d = 10$ m for the same gain. Thus, for practical antenna sizes, we must transmit fairly high waveform frequencies in order to obtain sufficient antenna gain. This now demonstrates a basic reason why modulation and upconversion are used prior to transmission. It allows low baseband frequencies to be shifted up to high carrier frequency ranges where sufficient antenna gains and reasonable antenna sizes are available. If modulation were not used, we would have to deal with the antenna parameters based on baseband frequencies instead of carrier frequencies. We obtain the same conclusion if we instead designed an antenna using beamwidth instead of specifying a desired gain. This would occur, for example, if our prime objective was to achieve a specified narrow beamwidth in order to confine power transmission properly. Equation (3.5.8b) also dictates large antennas and high carrier frequencies. For example, a 2 m antenna would have to be operated at 1 GHz to have a beamwidth of only 18°.

A nomogram relating gain and beamwidth to parabolic antenna diameter d and carrier frequency f_c in (3.5.8) is shown in Figure 3.14. (A straight line connecting two of the parameters will read off the corresponding value of the third parameter.) It should be remembered that the listed value of d is for a unit value of aperture loss, $\rho_a = 1$. The true physical antenna diameter needed is related to the diameter value listed by $d_{\text{phy}} = d/\sqrt{\rho_a}$.

In addition to efficiency loss, the roughness of an antenna dish surface

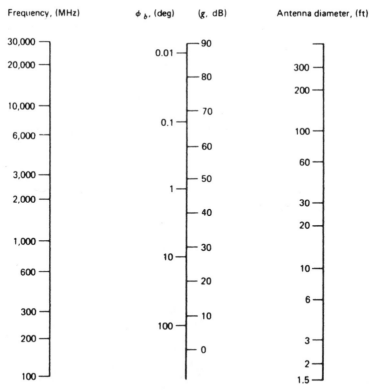

Figure 3.14. Nomogram of reflector dish antenna: g = gain (dB), ϕ_B = beamwidth (degrees).

can cause radiation scattering and loss of gain in the desired direction. This surface roughness loss is typically given as

$$L_r = e^{-(4\pi\sigma_r/\lambda_c)^2} \tag{3.5.9}$$

where σ_r is the rms roughness in wavelength dimensions [22]. Figure 3.15 plots L_r for several ranges of values as a function of frequency. Note that as higher carrier frequencies are used, the loss due to the roughness of the antenna surface eventually becomes important and overcomes the theoretical increase in gain with frequency. Thus, while (3.5.8) predicts that an arbitrarily large gain is theoretically feasible by continuing to increase frequency, Figure 3.15 shows that surface scatter effects eventually will begin to degrade the gain. Surface roughness often becomes critical in reflecting antennas that are exposed to weather, extreme temperature variations, or radiation effects in space. Both the cost and weight of improving surface reflectivity generally increase as higher tolerance is required.

The value of an antenna gain function in communication analysis is that it

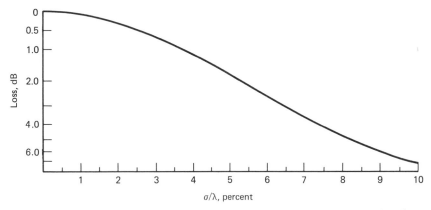

Figure 3.15. Antenna roughness loss: σ_r = rms roughness, λ = carrier wavelength.

allows immediate calculation of the amount of transmitted field power that will impinge on a normal receiving surface area of size $\mathscr{A}$ located a distance D in direction (ϕ_{l_0}, ϕ_{z_0}) in free space from the transmitting antenna (Figure 3.16). The power over $\mathscr{A}$, $P_{\mathscr{A}}$, is from (3.5.1)

$$P_{\mathscr{A}} = w(\phi_{l_0}, \phi_{z_0})\Omega_a \qquad (3.5.10)$$

where Ω_a is the solid angle subtended by $\mathscr{A}$, when viewed from the antenna. If $D \gg \sqrt{\mathscr{A}}$, then

$$\Omega_a \cong \frac{\mathscr{A}}{D^2} \qquad (3.5.11)$$

and from (3.5.2), (3.5.10), and (3.5.11) we have

$$\begin{aligned} P_{\mathscr{A}} &= w(\phi_{l_0}, \phi_{z_0})\frac{\mathscr{A}}{D^2} \\ &= \left[\frac{P_t g(\phi_{l_0}, \phi_{z_0})}{4\pi D^2}\right]\mathscr{A} \end{aligned} \qquad (3.5.12)$$

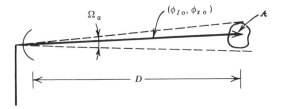

Figure 3.16. Antenna power flow diagram.

The quantity in brackets has units of power/area, and therefore represents the power density (called the field *intensity*, or *flux density*) of the propagating electromagnetic field at the distance D. The term in the numerator is often referred to as the *effective isotropic radiated power* (EIRP) of the transmitter in the direction (ϕ_{l_0}, ϕ_{z_0}). That is,

$$\text{EIRP} = P_t g(\phi_{l_0}, \phi_{z_0}) \tag{3.5.13}$$

The EIRP represents the field power that the transmitter can send in the specific direction. It is called an *effective isotropic power* since it represents the equivalent power that an isotropic ($g = 1$) antenna must have in order to send the same power in the same direction. Note that EIRP depends on the total power amplification of the transmitter, including circuit and coupling losses, and on the antenna gain in the direction of the receiving area. Also note that electromagnetic field intensity decreases as distance squared, even when propagating in an undistorting (free space) medium.

While the gain g represents the peak gain of the antenna, the actual gain used in the EIRP calculation depends on the angular direction, generally from transmitter to receiver; that is, it depends on how accurately the antenna is pointed at the receiver. This accuracy depends, in turn, on the pointing error, which arises from both the inability to aim an antenna in exactly the right direction and the inaccurate knowledge of the target location. The presence of a pointing error means that the power to the receiver is determined by the antenna gain on the skirts of the pattern, rather than by the peak gain g. For a parabolic antenna of diameter d with a pointing error ϕ_e, this will be given by

$$g(\phi_e) = g \left[\frac{2J_1(\pi d\phi_e/\lambda)}{(\pi d\phi_e/\lambda)} \right]^2 \tag{3.5.14}$$

where g is the peak gain. Using the Bessel function approximation for small argument,

$$J_1(x) \approx \frac{x}{2}\left(1 - \frac{x^2}{8}\right) \approx \frac{x}{2} e^{-x^2/8} \tag{3.5.15}$$

the mispointed gain in (3.5.14) is approximately

$$g(\phi_e) \approx g e^{-(\pi d\phi_e/2\lambda)^2}$$
$$= g e^{-2.6(\phi_e/\phi_b)^2} \tag{3.5.16}$$

where ϕ_b is the half-power beamwidth ($\phi_b = 1.02\lambda/d$). Equation (3.5.16) indicates that an exponential decrease in gain occurs for small pointing errors, as the pointing error increases relative to beamwidth. When narrow

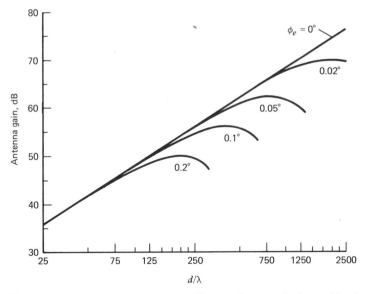

Figure 3.17. Antenna gain versus d/λ. The pointing error is denoted by ϕ_e.

beams are used, a given pointing error therefore becomes more critical. Since beamwidth depends directly on the antenna d/λ (diameter/wavelength) ratio, pointing-error losses increase exponentially with this product. Figure 3.17 plots $g(\phi_e)$ as a function of d/λ for several values of pointing errors. This pointing-error dB loss is added directly to surface roughness dB loss, L_r in (3.5.9), in lowering EIRP. Note that while roughness losses prevent use of high frequencies, pointing errors prevent use of narrow beams, and therefore constrain both frequency and antenna size.

3.6 CARRIER PROPAGATION CHANNELS

When designing a communication system, the properties of the propagation path from transmitter to receiver must be taken into account. Proper characterization of the path is equivalent to defining the communication channel shown in Figure 1.1. Electromagnetic propagating channels can be roughly separated into guided and unguided transmissions. Again, a thorough discussion of electromagnetic propagation in all types of medium is beyond our scope here. In this section, however, we receive the salient features of each of the channels and attempt to summarize much of the extensive work dealing with channel measurements and models. Our basic objective is to indicate the principal characteristics of the channel and to make the system engineer aware of the manner in which they affect the overall communication operation.

Guided Channels

In a guided channel the electromagnetic field is confined to a closed path or "pipe" from transmitter to receiver. Such channels occur whenever hardwire, cabling, or fibers are used, such as for telephone lines, cable transmissions, or short range internal communications. The guided channel is used primarily for convenience. It has the advantage of being completely shielded from external interference and is therefore extremely suitable for establishing a link in areas dense with electromagnetic fields. Also, the guided channel can be inserted to avoid propagation obstacles and to establish privacy links, as in home cable television. It also has the advantage of allowing complete control of the electromagnetic field during propagation by the insertion of internal amplifiers (*repeaters*) spaced along the line. The field can be amplified and regenerated at these points to restore it to proper power levels. This becomes significant when other forms of field amplification are not possible, such as with transoceanic telephone cabling.

The basic disadvantages of guided channels are the cost of implementing such channels and the excessive attenuation of the field in the guide. Guide attenuation values are typically stated in terms of decibel loss per unit length, so that the total attenuation applied to the field from transmitter to receiver depends on the guide length. Although such attenuation factors depend on the type, size, and material of the cable, in general, lower attenuation is achieved only with larger cable sizes and therefore increased cost. Low cost, lossy lines are limited in their usable length and confine the available link to relatively short distances (several tens of miles). As stated, cable attenuation can be overcome either by amplifier insertion or by resorting to larger cables at increased cost. One must therefore trade off the increased cost of less lossy cables against the additive cost of amplifier insertions and their accompanying amplifier noise.

An added complexity is that attenuation factors also depend on propagating frequency. Figure 3.18a shows a typical plot of metallic coaxial cable attenuation versus propagating frequency for two representative coaxial sizes. The line attenuation increases with frequency and decreases with cable size. This rapid increase with frequency generally confines operation to the 10–400 MHz range. The variation with frequency also means that different frequencies of the transmitted waveform may undergo different attenuation and therefore produce an effective filtering (amplitude distortion) of the signal. This is important when low frequency baseband waveforms are transmitted directly, such as with telephone lines. This effect requires line *equalization* to even out the frequency response over the waveform bandwidth, adding to the cost of the line. By modulating onto a carrier and operating at higher frequencies, where the bandwidth is a smaller percentage of the propagating frequencies, this effect becomes negligible, although the overall attenuation is greater. In addition, guides must be properly terminated to avoid reflections and allow most efficient coupling out of the propagating energy. This termination requires impedance matching to the

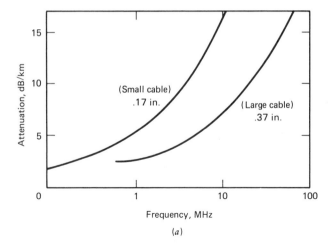

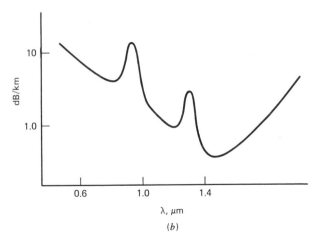

Figure 3.18. Cable attenuation loss: (*a*) metallic cable decibel loss per length versus frequency, (*b*) glass fiber decibel loss per length versus optical wavelength.

receiver input circuitry. Unfortunately, cable impedances are quite low (50–1000 ohms) while the input impedance of receiver amplifiers is much larger. This means that impedance matching transformers must be inserted, again adding to the system cost.

An important example of a guided channel is the cable TV system that dispenses multiple-channel TV to our homes from central feeder locations. In cable TV the VHF and UHF channels are placed on carriers side by side on the frequency axis, generally covering bands from 50 MHz to about

400 MHz. These are than sent over 75 ohm dielectric coaxial cables to the homes. Because of the high losses, as evidenced by Figure 3.18a, transmission distances are limited to about a mile, and upper frequency bands must often be amplified (usually with repeaters placed on telephone poles) to maintain signal levels for TV reception. At the home, the channels from the feeder lines are then converted to the proper RF frequency for insertion into the TV receiver.

Figure 3.18b shows the corresponding attenuation loss for glass fibers at optical wavelengths. The significant loss reduction for the optical link vis-à-vis its electronic counterpart in Figure 3.18a is the primary reason why fiber links are becoming so important in point-to-point guided systems. Several kilometers of fiber will only produce a few dB of loss, while metallic cables over the same distances may have losses as high as 50–100 dB. This loss reduction will more than compensate for the lower power light sources and increased optical noise of the fiber link (see Chapter 12.) This optical electronic cable comparison will be further examined in Section 4.5.

Unguided Channels

An unguided channel occurs when an antenna is used and the transmitted electromagnetic field allowed to propagate freely from the antenna. When the field propagates in the Earth's atmosphere, power losses may occur because of the nature of the atmosphere itself. In addition, further losses may be imposed by the Earth itself. The magnitude of these additional losses will depend on the manner in which the wave is transmitted and the particular properties of the atmosphere itself.

Two basic propagation channels can be defined for the unguided electromagnetic wave (Figure 3.19): (1) the ground wave channel and (2) the space wave channel. Each is appropriate in a specific application, and each defines a slightly different propagation channel model. A communication

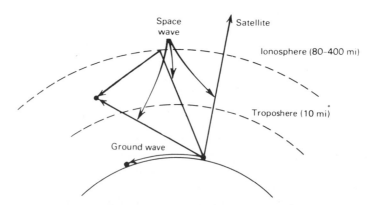

Figure 3.19. Channels for unguided carrier propagation.

engineer must carefully distinguish the most accurate channel model before proper assessment of propagation losses can be made. In the following discussion, we summarize the important characteristics of each of these channels.

Ground Wave Channel

When both transmitter and receiver are located within a few meters of the Earth, radio waves propagate as guided waves with the Earth itself serving as a waveguide wall for the propagating field. Such models are applicable in radio, land based mobile, and hand-held communication systems. Ground wave channels are characterized by strong attenuation with distance, and attenuation values that increase with frequency. These effects are generated identically to waveguide attenuation in guided channels. For this reason ground wave transmission is restricted to frequencies HF and below and over relatively short (10–50 km) distances even when no obstructions (mountains, buildings, trees, etc.) are present. Figure 3.20 shows a typical attenuation characteristic associated with ground wave propagation, showing the rapid deterioration of field strength with both frequency and distance.

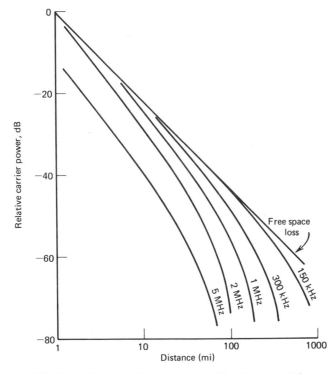

Figure 3.20. Ground wave carrier power loss with distance and frequency.

Space Wave Channel

In space wave channels the wave propagates directly from transmitter to receiver through space without influence from the Earth. Such systems characterize Earth–airplane, airplane–airplane links, and satellite links, as shown in Figure 3.19. For land based systems, the space model is appropriate only if the antennas are raised high enough (using towers, rooftops, or mountain tops) to avoid the ground attenuation. This raising of the antennas to achieve a space channel generally restricts their size, and higher frequencies (UHF, VHF) must be used to achieve suitable gains.

Space wave propagation is susceptible to the effects of the atmosphere. Recall that the Earth is surrounded by a collection of gases, atoms, water

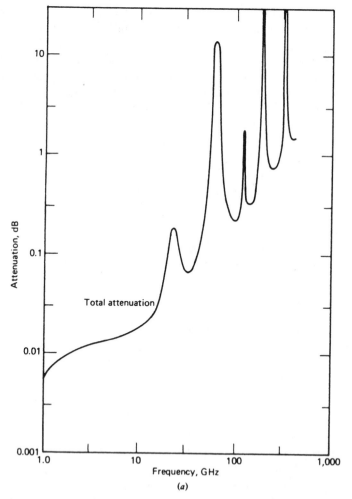

Figure 3.21. Atmospheric attenuation due to absorption: (a) horizontal path at sea level.

droplets, pollutants, and so on captured by the Earth's gravity field and extending to an altitude of about 400 miles, constituting the Earth's atmosphere. The heaviest concentration of these particulates is near the Earth, with particle density decreasing with altitude. The particles in the upper part of the atmosphere (*ionosphere*) absorb and reflect large quantities of radiated energy from the sun. The absorbed energy is then reradiated in all directions by the ionosphere. In addition, the absorbed energy ionizes the ionospheric atoms, producing bands of upper atmospheric free electrons that surround the Earth. These electrons directly interact with any electromagnetic field passing through them. As a result, a field propagating in the Earth's atmosphere undergoes a power loss.

Losses are caused by absorption and scattering of the field by the atmospheric particulates. These effects become more severe as the field carrier frequency is increased to a point where the wavelength begins to approach the size of the particulates. Figure 3.21a shows the average decibel loss that can be expected as a function of frequency for the clear atmosphere. Below about 10 GHz, total atmospheric attenuation is nominal (<2 dB); at frequencies above that, however, it increases rapidly. Higher amounts of attenuation occur at those particular frequencies having

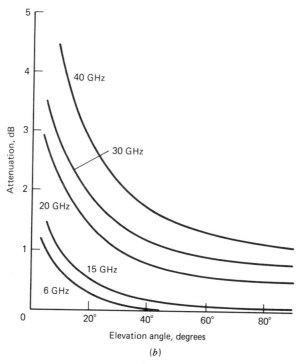

Figure 3.21. Atmospheric attenuation due to absorption: (*b*) as a function of elevation angle.

wavelengths corresponding to specific gas molecules in the atmosphere. For example, severe absorption due to water vapor molecules occur at about 22 and 118 GHz. This attenuation increase at the higher frequencies is the primary disadvantage of shifting operation to the higher frequency bands in space systems.

Since the density of particulates in the atmosphere decreases with distance above Earth, atmospheric attenuation is lower at the higher altitudes. This means that the total attenuation expected in a space link will depend on both the link elevation angle and the slant range. Most severe attenuation occurs at angles close to the horizon and decreases for vertical paths directly overhead. Figure 3.21*b* plots the atmospheric attenuation at various frequency bands as a function of elevation angle measured from ground. From these curves, it is clear that it is preferable that links are established above a minimum elevation angle. For example, domestic satellite systems serving particular areas are usually designed to ensure that the satellite will be higher than some acceptable viewing angle.

The most serious atmospheric effect in a space link is rainfall. Water droplets scatter and absorb impinging radiation, causing attenuation many times the clear air losses shown in Figure 3.21. Rain effects become most severe at wavelengths approaching the water drop size, which is dependent on the type of rainfall. In heavy rains raindrop size may approach a centimeter, and severe absorption may occur at frequencies as low as 10 GHz. Thus rainfall effects can become extremely severe at frequencies at X-band and above.

If a space link is to be maintained during a rainfall, it is necessary that enough extra power (called *power margin*) be transmitted to overcome the maximum additional attenuation induced by the rain. Hence it is necessary to have an accurate assessment of expected rain loss when evaluating link parameters. The expected additional rain loss depends on the operating frequency, the amount of rainfall, and the path length of the propagation through the rain.

To evaluate rain loss, we first obtain the expected rainfall rate in millimeters per hour for the region of the space link. Table 3.2 lists the designations relating type of rain and rainfall rate. We then use the rainfall rate to estimate the decibel loss per path length at the operating frequency.

TABLE 3.2 Rainfall in Space Link Region

Rainfall rate, mm/hr	Designation
0.25	Drizzle
1.25– 12.5	Light rain
12.5 – 25.0	Medium rain
25.0 –100.0	Heavy rain
>100	Tropical downpour

This estimate is generated from combinations of empirical data and mathematical models that fit the data. The rainfall attenuation model commonly used [23] is of the form

$$\frac{\text{Decibel loss}}{\text{Length}} = ar^b \tag{3.6.1}$$

where r is the rainfall rate and a and b are frequency dependent coefficients obtained via curve fitting and extrapolation [24–34]. Figure 3.22 plots (3.6.1) as a function of frequency and rainfall rate. Note the increase in loss as both the operating carrier frequency and the rainfall rate increase.

The mean path length of the rain is then determined for the given elevation angle. This length also depends on rainfall rate, as shown in Figure

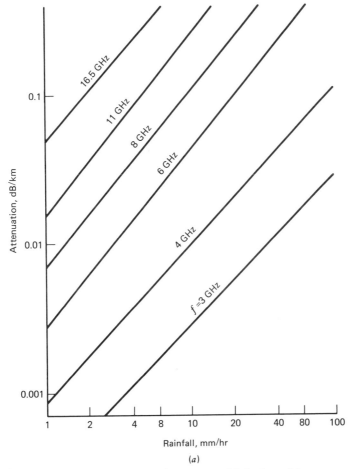

Figure 3.22. (*a*) Rainfall attenuation versus rainfall rate and frequency.

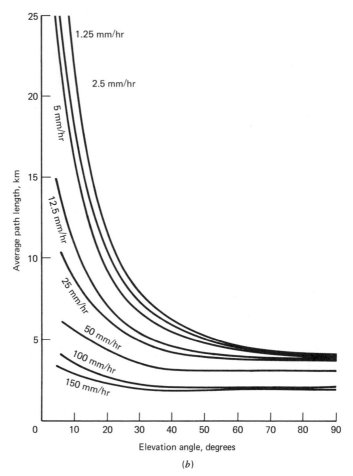

Figure 3.22. (*b*) Average rainfall path length versus elevation angle and rainfall rate.

3.22*b*, since rainclouds with heavy rain are, in general, closer to the ground. These path length curves are also obtained by fitting mathematical models to measured data [23]. With the mean path distance estimated, expected rainfall attenuation is obtained by multiplying the rain decibel/length ratio by the mean path length.

Rainfall attenuation computed as just described is actually a conditional attenuation dependent on the rainfall rate selected. Since the latter is itself a statistical phenomenon, it is often more meaningful to take into account the probability that a given rainfall rate will be exceeded. Data on such probabilities are usually available for most regions of the world. When these probabilities are taken into account we compute rain attenuation statistics similar to that shown in Figure 3.23. This shows, for a particular region, the expected percent of time that a given rainfall attenuation level will be

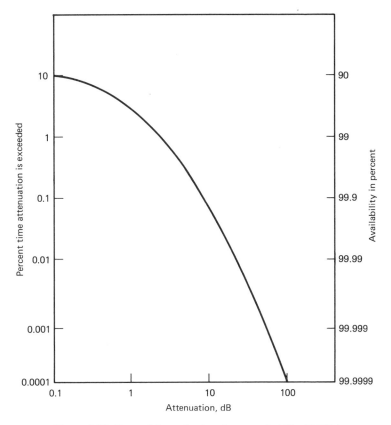

Figure 3.23. Percent time abscissa is exceeded ($f = 10$ GHz).

exceeded. Since Figure 3.22 allows us to convert rainfall rate to expected attenuation at specific frequencies, these rainfall probabilities allow us to associate a probability with a given attenuation being exceeded at that frequency and elevation angle.

Curves of this type are much more meaningful for allowing safe design margins and preventing an inordinate amount of link overdesign. If $P(\alpha)$ is the percentage of time an attenuation α is exceeded, then $1 - P(\alpha)$ is the percent *availability* of the link with an attenuation less than α. The value $P(\alpha)$ is often called the percent *outage* time of the link, with an attenuation exceeding α. This shows, for example, that if a given space link can withstand a rain attenuation of 7 dB, it will be available 99.5% of the time, or equivalently, it will have a rain outage 0.5% of the time (about 45 hr per year).

Some space wave channels have been designed to take advantage of the atmospheric properties by making use of the ionosphere and troposphere boundaries (Figure 3.19). The ionosphere is a dense collection of free electrons trapped in a belt around the Earth. To the longer wavelength carriers (HF and below), this ionosphere appears as a conducting surface that reflects carrier energy. This reflected energy may be used as a primary path in a long range link. Such reflections often allow communications over distances where line-of-sight space wave propagation is no longer possible. Unfortunately, the attenuation of the field caused by boundary reflection is generally quite high. In addition, the movement of the belt from hour to hour and its dependence on time of the year and galactic activity (such as solar flares) produces a variable channel that often varies from a strong to a weak link in short time periods. For this reason reliable channels can be defined only for limited time periods. Nevertheless, reflecting wave links have been successfully implemented and used for long range, over-the-horizon systems. In addition, these waves are often responsible for commerical radio stations being heard far outside their transmission range on certain clear evenings.

Using reflecting channels at frequencies below HF is basically disadvantageous because of required antenna sizes. On the other hand, as frequency is increased into the VHF and UHF range, these smaller wavelengths penetrate the ionosphere and are absorbed or scattered rather than reflected. Hence reflecting channels cannot be maintained at the higher frequency bands. The *critical frequency* of the atmosphere is that frequency below which a reflected wave can be satisfactorily maintained. This critical frequency depends on the electron density of the ionosphere, which, in turn, depends on the time of day, season of the year, and so on. Published data, obtained from years of experimental measurements, are available for specifying this frequency. Critical frequencies generally fall in the range 5–20 MHz. This is why long range commercial radio is possible via reflected waves but commercial television is not.

Satellite and deep space systems involve transmissions out through the Earth's atmosphere to orbiting stations. Hence electromagnetic energy must purposely pass through the ionosphere belt rather than be reflected. Hence these outer space channels must use frequencies well above the critical frequency (see Chapter 11). As the carrier frequency is increased into the VHF and SHF bands a larger portion of the field energy passes through the ionosphere, rather than being scattered or reflected. Thus outer space channels are almost always operated in the SHF band or above (see Table 1.3). The links are still susceptible to the atmospheric absorption, of course, but the effect is minimized somewhat when vertical paths are used. In addition, typical deep space links to stationary satellites ($\approx 20{,}000$ mi), the moon ($\approx 232{,}000$ mi), or nearly planets ($\approx 10^8$ mi) clearly involve distances only a small part of which includes the Earth's atmosphere. Thus the added decibel loss introduced by the atmospheric path is a relatively small part of the total electromagnetic loss caused by the propagation over such long distances. Of course, transmission outside the Earth's atmosphere (e.g., between two space satellites) would have no atmospheric losses.

In summary, we have qualitatively investigated the carrier propagation channel and its effect on carrier transmission. We have presented some representative curves and figures to aid in understanding these effects. Unfortunately, most of these propagation phenomena are functions of many external and immeasurable parameters, specialized to each specific situation, such as weather conditions, geographic location, and season of the year. For this reason a single curve is inadequate to describe all cases. In addition, most documented results report average data, and in any given channel, variations from the average may be quite significant. We point out that other types of channel can be defined that have not been considered here but arise in special applications. Underwater communications (sonar systems) and seismic propagation channels represent two such special examples. Such systems require more complicated electromagnetic analyses for proper modeling.

Throughout our channel discussion we have repeatedly emphasized the dependence of the channel effect on frequency, sometimes favoring higher carrier frequencies and sometimes lower. This again illustrates the importance of transmitter modulation and frequency translation to select the transmission frequencies that best interface with channel. This then serves as another important consideration in the carrier frequency designation.

3.7 OTHER CHANNEL EFFECTS

The discussion in the previous section was concerned with the power losses introduced by the channel during field propagation. However, in addition to these power losses the channel may introduce other effects on the carrier waveform. These additional effects may take the form of channel filtering, polarization distortion, Doppler shifting, fading, or multipaths. Although these effects are usually considered secondary, and are often ignored in first order analyses, the system designer should be aware of their possibility, since in some situations they may become the limiting factor on performance. In this section we review their causes and characteristics.

Filtering

When fields propagate in a cable or in the atmosphere, different frequencies may actually flow at different velocities. Thus different frequencies within the modulating waveform may experience slightly different propagation times, referred to as *dispersion*. The overall effect is to produce inherent nonuniform frequency delays, which causes a phase distortion of the carrier waveform. This distortion can be interpreted as being caused by an effective channel filtering on the carrier waveform. This filtering tends to limit the available carrier modulation bandwidth. For relatively quite atmospheric conditions, the space channel dispersion bandwidth is generally quite large, on the order of hundreds of gigahertz, and is usually not of concern in the RF carrier range, where bandwidths are never this large. Turbulent atmospheric conditions, however, may reduce this channel bandwidth to several gigahertz, and therefore may be significant factor in optical and millimeter carrier systems. Channel filtering is usually measured by observing pulse spreading. An RF carrier is amplitude modulated by a pulse and sent over the space channel. The pulses are then made continually narrower, and the received pulse is monitored until a fixed percentage of pulse spreading has been observed. Channel bandwidth is then related to the reciprocal of the pulse width at which the spreading was noted.

Polarization Loss

An electromagnetic field is transmitted with a specific spatial polarization, as was shown in Figure 3.10. During propagation the atmosphere may cause a rotation of this polarization angle. For example, a vertically polarized transmitted field may have its field orientation shifted to some angle off the vertical during propagation. This field rotation is caused by the collective effects of the atmospheric particulates that spatially shift the impinging radiation field during forward scattering. The principal contribution is from water droplets and water vapor that act as nonspherical lenses that reorient the polarization direction. As a result, field rotation is most severe during rain, clouds, or misty conditions and tends to be wavelength dependent. Polarization angle shift is only about 2–6° at C-band and K-band frequencies during clear air transmission, but may increase to 20–40° during heavy turbulence and rainfall.

Field rotation produces a polarization shift that causes a linearly polarized field to no longer be aligned with a linearly polarized matched receiver. Only the component of the received field in the same spatial alignment as the receiving antenna will produce detected field power. Since the component will decrease as $\cos^2(\theta_p)$, where θ_p is the polarization angle shift, field rotation with linearly polarized fields leads directly to field power loss. This is commonly referred to as *polarization loss* and must be treated as an additional atmospheric space loss.

For a circularly polarized field, the rotation caused by the atmosphere will only cause an advancement or retardation of the rotating field, which will be indistinguishable from the field rotation itself when observed at the receiver. Hence circularly polarized fields with matching receiver antennas are not affected by polarization rotations.

Doppler

Whenever relative motion exists between transmitter and receiver, the received carrier frequency is not the same as the transmitted carrier frequency. This frequency shifting during transmission is due to the Doppler effect of propagating waves between nonstationary points. The effect is to cause an overall shift of the carrier frequency. A system designer must allow for, or compensate for, this Doppler shift whenever it becomes significant. In certain cases the Doppler may be completely predictable, as with satellites with fixed orbits. Doppler effects due to unintentional transmitter or receiver motion may have to be estimated and properly compensated in system design.

If the relative velocity of the transmitter toward (away from) the receiver is $+v$ $(-v)$ m/sec, the receiver frequency of the RF carrier at f_c Hz is given by

$$f_r = \frac{(1 + v/c)f_c}{[1 - (v/c)^2]^{1/2}} \tag{3.7.1}$$

where c is the speed of light. Since $v/c \ll 1$, an accurate approximation to the Doppler shift from f_c in (3.7.1) is given by

$$f_{\text{Dop}} = f_r - f_c \approx \left(\frac{v}{c}\right) f_c \tag{3.7.2}$$

and the Doppler frequency offset is directly proportional to the relative velocity. Note that v is the velocity component along the line-of-sight vector between transmitters and receiver. For velocity directions at other angles, only the velocity component along this line will produce the Doppler effect. It should also be pointed out that if there is a constant acceleration—linearly changing velocity—between transmitter and receiver, the Doppler frequency will also appear to be linearly changing when observed at the receiver.

Fading

The channel losses discussed in Section 3.6 represented an average type of power loss. However, over a short period of time a communication signal may have severely reduced power values corresponding to deep "fades" of the transmitted signal. This fading can be considered as an instantaneous attenuation of the carrier amplitude. Analysis or design of a system in the presence of these fades requires one to describe these attenuation factors statistically.

In general, fading channels are separated into *flat* fading (all frequencies over the transmission band fade simultaneously) and *bandwidth-selective* fading (portions of the band fade more than others). In a flat fading channel the carrier waveform can be considered to be undistorted but effectively multiplied by a fade variable. Such behavior is characteristic of space links passing through a turbulent atmosphere. Flat fading can be slow (rate of fade much slower than the reciprocal of the highest modulating frequency), and the carrier waveform can be considered to be multiplied by a random amplitude factor. Probability statistics of this random amplitude therefore describe the properties of the fade. With fast fading, the rate of fade is considered to be significant relative to the carrier amplitude time variation. The overall result is an effective multiplication of the carrier by a rapidly varying amplitude function that represents the fading phenomena. In this case the carrier is in fact "modulated" by the fading amplitude, which appears as a carrier amplitude distortion at the time of reception. The rapidly varying fading effect is generally random and described by a suitable stochastic process model. Fast fading can generally be corrected only by "dividing out" the fading effect by some type of automatic gain control or fading estimation, the latter via channel probing.

Bandwidth selective fading is effectively equivalent to the insertion of a bandpass RF channel filter into the carrier transmission path when the fade occurs. Typically, the filter effect is to cause the band edge frequencies to have a fade either greater or weaker than the bandcenter frequencies. The result is a sudden loss of high frequency content (loss of modulation waveform edges) in the former case, or a loss of low frequency content (waveform droop) in the latter case. The type and severity of the resulting modulation distortion therefore depends on the actual fade characteristic. Attempts to compensate via filter equalization can only be achieved by accurate channel monitoring.

Multipath

Often the receiver receives not only the direct transmission from the transmitter but also secondary transmissions caused by reflections or echoes of the direct transmission (Figure 3.24). These secondary receptions generally arrive later than the primary transmission (since they propagate over longer paths) and therefore appear as interfering electromagnetic waves. These secondary paths, called *multipaths*, are predominant when reflecting surfaces appear near the receiver. Such multipath effects are particularly

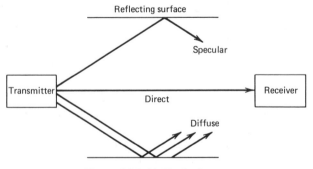

Figure 3.24. Multipath diagram.

noticeable in Earth-based mobile and maritime systems, where reflections from buildings, mountains, and oceans are unavoidable. It is precisely this multipath that produces "ghosts" in home television reception.

Although multipaths may be well reduced in amplitude, a large number can combine at the receiver input as an effective intereference. Often one component of the multipath tends to dominate, and appears as an identical, delayed replica of the transmitted signal itself. This is called the *specular* component of the multipath. The weaker multipaths tend to accumulate into an effective noise, called the *diffuse* multipath components. Thus the typical multipath carrier signals is generally modeled as

$$c_m(t) = a_c c(t) + a_s c(t - \tau_s) + \text{(diffuse noise)} \qquad (3.7.3)$$

where $c(t)$ is the desired carrier signal, the a values account for power losses, and τ_s is the transmission delay of the specular component. Specific statistics for the diffuse noise are difficult to measure, and the latter noise is often modeled as an additive noise process.

Elimination of the effect of the specular component is often of fundamental importance in multipath environments. Of particular significance is the fact that the specular interference is itself the transmitted carrier signal, and the desired carrier term cannot be made to dominate by simply increasing the transmitter carrier power. In fact, all three terms in (3.7.3) depend on transmitter power.

When the differential path length is the proper number of wavelengths, the specular component from the multipath can arrive out of phase with the direct component, causing an interference effect that subtracts from the desired waveform contribution, causing a signal cancellation or fade. This cancellation effect tends to happen most often at wavelengths around bandcenter, and multipath typically produces a center band frequency notch that appears as a form of frequency-selective fading (Figure 3.25). This

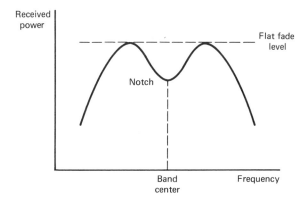

Figure 3.25. Frequency selective band filtering due to multipath fading.

characteristic "M" curve is commonly associated with mobile radio links [35–38]. As the receiving vehicle moves the notch location may, in fact, shift, or drift, over the frequency band, making equalization techniques extremely difficult.

In certain cases the effect of the specular part can be reduced by proper signal selection if the delay τ_s is known or can be estimated (Problem 3.22). Sometimes specular multipath can be eliminated by accurate antenna pointing and antenna pattern control to avoid reception of the secondary path.

Doppler, fading, and multipath are, of course, of concern only in unguided space field transmission systems and are not characteristic of guided links. Cables, although having attenuation and often dispersive filtering effects, do not exhibit these other properties, which is a primary advantage in their use.

REFERENCES

1. Smith, J. *Modern Communication Circuits*, McGraw-Hill, New York, 1986.

2. Frerking, M. *Crystal Oscillator Design*, Van Nostrand-Reinhold, New York, 1978.

3. Heising, R. *Quartz Crystals and Electronic Circuits*, Van Nostrand-Reinhold, New York, 1946.

4. Matthys, R. *Crystal Oscillator Circuits*, Wiley, New York, 1983.

5. Parzen, B. *Design of Crystal and Other Harmonic Oscillators*, Wiley, New York, 1983.

6. Gagliardi, R.M. *Introduction to Communications Engineering*, 1st ed., Wiley, New York, 1978, Appendix C.

7. Lindsey, W. and Chie, C. "Theory of Oscillator Instability Based on Structure Functions," *Proc. IEEE*, vol. 64, no. 12, p. 197.

8. Kroupa, V. "Noise Properties in PLL Systems," *IEEE Trans. Comm.*, vol. COM-30, no. 10, October 1982, pp. 2244–2253.

9. Vanier, J. and Teta, M. "Phase Locked Loops with Atomic Frequency Standards," *IEEE Trans. Comm.*, vol. COM-30, no. 10, October, 1982, pp. 2355–2362.

10. Rohde, U. *Digital PLL Frequency Synthesizers*, Prentice-Hall, Englewood Cliffs, N.J., 1983.

11. Angelakos, D. and Everhart, T. *Microwave Communications*, McGraw-Hill, New York, 1968.

12. Reich, H. *Microwave Theory and Techniques*, Van Nostrand, Princeton, N.J., 1953.

13. Singer, J. *Masers*, Wiley, New York, 1959.

14. Meyer, J. "Systems Applications of Solid State Masers," *Electronics*, vol. 33, no. 45, November 1960, pp. 58–63.

15. Howes, M.J. and Morgan, D.V. *Microwave Devices*, Wiley, New York, 1979.

16. Strauss, R., Bretting, J. and Metivier, R. "TWT for Communication Satellites," *Proc. IEEE*, vol. 65, March 1977, pp. 387–400.

17. Hoefer, W.J. "Microwave Amplifiers," in *Digital Communications—Microwave Applications*, K. Feher, ed., Prentice-Hall, Englewood Cliffs, N.J., 1981, Chap. 4.

18. Kraus, J. *Antennas*, McGraw-Hill, New York, 1950.

19. Silver, S. *Microwave Antenna Theory and Design*, MIT Radiation Laboratory Series, vol. 12, McGraw-Hill, New York, 1949.

20. Terman, F.E. *Electronic and Radio Engineering*, 4th ed., McGraw-Hill, New York, 1956.

21. Jordan, E.C. *Electromagnetic Waves and Radiating Systems*, Prentice-Hall, Englewood Cliffs, N.J., 1950.

22. Ruze, R. "Effect of Surface Error Tolerance on Efficiency and Gain of Parabolic Antennas," *Proc. IEEE*, vol. 54, April 1966.

23. Ippolita, L. *Radiowave Propagation in Satellite Communications*, Van Nostrand-Reinhold, New York, 1986.

24. Engelbrecht, R. "The Effect of Rain on Satellite Communications," *RCA Rev.*, vol. 40, June 1979.

25. Hogg, D. and Chu, T. "The Role of Rain in Satellite Communications," *Proc. IEEE*, September 1975.

26. Calo, S., Schiff, L. and Staras, H. "Effects of Rain on Multiple Access Transmission via Satellite," *Proc. Int. Conf. Comm.*, Toronto, 1978, pp. 30.1.1–30.1.6.

27. Crane, R.K. "Prediction of the Effect of Rain on Satellite Communication Systems," *Proc. IEEE*, vol. 65, March 1977, pp. 456–474.

28. Medhurst, R.G. "Rainfall Attenuation of Centimeter Waves," *IEEE Trans. Antennas Propag.*, vol. AP-13, July 1965, pp. 550–563.

29. Bullington, K. "Radio Propagation at Frequencies Above 30 Megacycles," *Proc. IRE*, vol. 35, October 1947, p. 1122.

30. Burrows, C.R. and Attwood, S.S. *Radio Wave Propagation*, Academic, New York, 1949.

31. Reed, H.R. and Russell, C.M. *Ultra High Frequency Propagation*, Wiley, New York, 1953.

32. Millman, G.H. "Atmospheric Effects on VHF and UHF Propagation," *Proc. IRE*, vol. 46, August 1958, pp. 1492–1501.

33. Van Vleck, J.H. "The Absorption of Microwaves by Oxygen," *Phys. Rev.*, vol. 71, April 1, 1947, pp. 413–424.

34. Van Vleck, J.H. "The Absorption of Microwaves by Uncondensed Water Vapor," *Phys. Rev.*, vol. 71, April 1, 1947, pp. 425–433.

35. Rummler, W.D. "A New Selective Fading Model: Application to Propagation Data," *Bell Syst. Tech. J.*, vol. 58, May–June 1979, pp. 1037–1071.

36. "Special Issue on Digital Radio," *IEEE Trans. Comm.*, vol. COM-27, no. 12, December 1979.

37. Feher, K. *Digital Communications*: *Microwave Applications*, Prentice-Hall, Englewood Cliffs, N.J., 1981.

38. Anderson, C.W., Barber, S.G. and Patel, R.N. "The Effect of Selective Fading on Digital Radio," *IEEE Trans. Comm.*, vol. COM-27, December 1979, pp. 1870–1876.

PROBLEMS

3.1 (3.1) (a) Using the feedback loop equations in Figure 1.16, derive the required condition for the system shown in Figure 3.2 to oscillate. Assume the active device has a gain G from input voltage to output current, and the tuned circuit has impedance $Z_t(\omega)$. (b) Use the result of part a to determine the frequency of oscillation if the active device is a transistor ($G = -gm$) and if the tuning circuit has the form of a parallel LC circuit.

3.2 (3.1) *Frequency noise* is defined as the derivative of phase noise. Show that the power spectral density of frequency noise of an oscillator is quadratic in frequency when the phase noise has a flat spectrum.

3.3 (3.1) Show that the *rms offset frequency* of an oscillator is the square root of the mean squared value of its phase noise spectrum.

3.4 (3.1) Given a sequence of phase values θ_i, $i = 1, 2, 3, \ldots, N$, of an oscillator. The *Allan variance* (AV) is defined as

$$AV = \frac{1}{N-3} \sum_{n=3}^{N} (\theta_n - \theta_{n-1})^2 - (\theta_{n-1} - \theta_{n-2})^2$$

Show that the AV is an estimate of the difference frequency stability of the oscillator.

3.5 (3.1) The frequency synthesizer shown in Figure 3.4 is modified to that in Figure P3.5. (a) Determine the frequency at the output at which the system will stabilize. (b) Explain how this design can improve the multiplication factor and/or the resolution of the synthesizer. (*Hint:* Q can be an integer and/or fractional multiplier.)

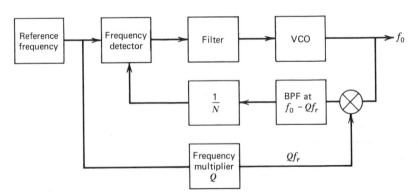

Figure P3.5.

3.6   (3.1) Let N sequential values of a sine wave (taken at $360/N$ degrees apart) be stored in memory as k-bit words. These sample words are then read out at a clock rate f_s words/sec. (a) Plot the sequence of sample values, carefully labeling the time axis, and connect the points to form a continuous sine wave. (b) Use the same readout rate, but assume that every other sample value is read out, and plot the sequence waveform on the same time axis. (c) Repeat for every fourth sample value. Show that the maximum and minimum frequency of the output sine wave is

$$\text{Minimum frequency} = \frac{f_s}{2^k}$$

$$\text{Maximum frequency} = \frac{f_s}{4}$$

3.7   (3.1) A *fractional divider* can be obtained as follows. After each N input zero crossings, produce an output crossing. Do this for $M - 1$ output crossings. For the Mth, count one extra input crossing before producing the output crossing. Show that this will fractionally divide the input frequency.

3.8   (3.1) A carrier squaring device is used to obtain frequency doubling of a unit power carrier. Determine the carrier power loss in decibels for the doubled frequency carrier. Repeat for the tripled frequency carrier when a cubic device is used.

3.9   (3.4) Consider the transmitter amplifier system in Figure 3.8b. The amplifier has a 50 W output under matched loading and an output impedance of 500 ohms. The cable is 20 ft long and has an impedance of 100 ohms from either end and a power loss of 0.1 dB/ft. The antenna has an input impedance of 1000 ohms. Determine the amplifier power coupled into the antenna.

3.10   (3.4) A 100 W amplifier with 1000 ohm output impedance feeds into a 1 m antenna. The antenna has its input impedance approximated by $Z_g \cong 800\mathscr{A}/\lambda^2$. Determine the power coupled into the cable at each of the frequencies: (a) 30 MHz, (b) 300 MHz, (c) 3 GHz.

3.11   (3.4) A waveguide has a measured maximum field voltage in the guide of 1.035 V and minimum of 0.9 V. Determine the waveguide power loss in decibels.

3.12   (3.5) (a) A rectangular antenna of width a and length b produces at wavelength λ the gain function

$$g(\phi_l, \phi_z) = K \left\{ \frac{\sin\left[(\pi a \sin \phi_l)/\lambda\right] \sin\left[(\pi b \sin \phi_z)/\lambda\right]}{\left[(\pi a \sin \phi_l)/\lambda\right]\left[(\pi b \sin \phi_z)/\lambda\right]} \right\}^2$$

where K is a proportionality constant. Determine the half-power beamwidth in degrees for the pattern in the $\phi_l = 0$ and in the $\phi_z = 0$ plane. (b) Repeat for the symmetrical pattern

$$g(\phi_l, \phi_z) = K \left[\frac{2J_1[\pi a \sin \phi)/\lambda]}{(\pi a \sin \phi)/\lambda} \right]^2$$

where $J_1(x)$ is the Bessel function.

3.13   (3.5)   A stationary satellite, placed in orbit at 20,000 mi above the Earth, is to transmit to the Earth at the frequency of 4 GHz. A parabolic antenna is to be used having an aperture efficiency of 50%. (a) Determine the size of the antenna that will transmit a field of view essentially covering the Earth. (b) What is the resulting effective transmitting gain? (The Earth has a diameter of approximately 6876 mi.)

3.14   (3.5)   A communication station on Earth is to transmit to the moon at 2 GHz. Design an antenna pair (transmitting plus receiving) so that the transmission from Earth essentially covers the moon, and the total effetive antenna gain (product of each gain) is 70 dB. Use the same efficiencies as in Problem 3.13. Assume the distance to the moon is 200,000 mi and its diameter is 0.27 of the Earth's.

3.15   (3.5)   Derive the relation between antenna solid angle Ω_{fv} and planar angle beamwidth Φ_b for a symmetric gain function. (*Hint:* Relate the area of the spherical cap of a cone to its planar angle.) Determine the approximate behavior as $\Phi_b^2 \ll 1$.

3.16   (3.5)   A 2 m circular antenna at 1 GHz has a circular gain pattern shown in Figure P3.16 as a function of azimuth angle. It is fed by 30 dBW of power and has an aperture efficiency of 0.95. What is the transmitted field power density at a distance of 1000 m located at 15° from the main antenna lobe direction?

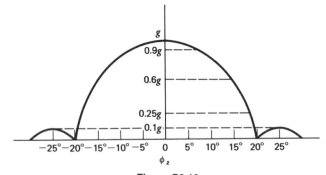

Figure P3.16.

3.17   (3.5)   Electromagnetic power decreases with distance propagated. A student claims "If we use a receiving area equal to the wavefront area, no power is lost during transmission." Is the student correct? Explain.

3.18   (3.5)   Two identical antennas [gain pattern $= g \exp(-\phi^2/2\phi_b^2)$], have diameters D_1 and D_2 and are separated by distance x, and each observes the same plane wave transmitter (Figure P3.18). Assume that $2\phi_b$ is the half-power beamwidth of the antennas. The transmitter is located on a line L a distance z away ($z \gg x$). Determine the point on L where the transmitter will deliver the same power to each antenna.

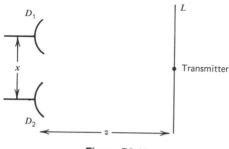

Figure P3.18.

3.19   (3.6)   The attenuation distribution curve in Figure 3.23 is convex. In some models the curve is shown concave. Explain how the shape of the curve will influence conclusions concerning link availability–link margin tradeoffs.

3.20   (3.7)   An Earth station communicates at S band with a satellite at 22,000 mi. Determine the maximum Doppler shift that can be expected because of the rotation of the Earth. Assume that the satellite is stationary while the Earth rotates under it. (The diameter of Earth is 6876 mi.)

3.21   (3.7)   A pulse 10 nsec wide is spread to a 50 nsec width during propagation through a medium. Using Problem 1.3, estimate the filtering bandwidth of the medium.

3.22   (3.7)   A T sec burst of a sine wave of amplitude A and frequency ω_c is transmitted to a receiver. Determine the shortest multipath delay that can be tolerated to ensure that the cross-correlation of direct and multipath signal is less than 10% of the direct signal power. Assume $\omega_c T \gg 1$, and the multipath signal has the same amplitude as the direct signal when received. The cross-correlation of two signals is defined as

$$\gamma_{12} = \frac{1}{T} \int_0^T s_1(t)s_2(t) \, dt$$

3.23 (3.7)   Show that a differential delay between parallel paths of the same propagating waveform will effectively "filter" the waveform. Estimate the filter bandwidth.

4

CARRIER RECEPTION

In Chapter 3 we examined carrier transmission and discussed the propagation effects that occur when a modulated carrier is sent over a communication channel as an electromagnetic field. To the communication engineer, the parameter of primary interest is the actual amount of transmitted power that can be collected at the receiving station. This depends not only on the transmission effects but also on the structure of the receiving subsystem and its ability to intercept the electromagnetic field. Following field reception, the system designer is faced with the task of properly processing the resultant waveforms to achieve the most efficient communication link for useful information transfer. In this chapter we investigate the reception of the electromagnetic field and the filtering and other types of carrier processing that may be applied to the recovered waveform. Primary emphasis is on subsystem modeling and design, and on performance evaluation.

4.1 RECEIVING SUBSYSTEMS AND ANTENNAS

The receiver subsystem for collecting and processing the electromagnetic field is shown in Figure 4.1. If the transmitted field is sent as an unguided wave, a receiving antenna must be used to convert the impinging field to an electronic waveform. The resulting signal is then passed to the receiver *front end* electronics. In a guided system no receiving antenna is needed, and the field in the transmission line or cable need only be coupled directly into the front end. In the front end the carrier waveform is generally filtered to remove as much intereference as possible without distorting the desired carrier waveform; it is then amplified to obtain a desired power level. During this front end processing other types of electronic interference may be imposed, and the effect of any front end filtering on the modulated carrier must be accounted for.

Since modulation was introduced at the transmitter to allow the baseband information signal to be carried to the receiver, a corresponding demodula-

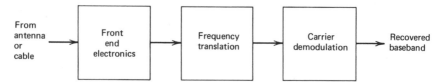

Figure 4.1. Receiver carrier processing system.

tion operation must be performed at the receiver to extract the baseband. This carrier demodulation is perhaps the key operation of the receiver, and the overall success of the link depends to a large extent on the ability to demodulate the baseband waveform properly. Usually other forms of waveform processing, such as frequency translation, may be applied following front end filtering in order to prepare for the carrier demodulation.

Receiving antennas convert the impinging electromagnetic field to an electronic waveform. In order to do this efficiently, the antenna must be properly aligned with the polarization of the field being received. If the field is linear polarized, and a dipole antenna (or an array of dipoles) is used, the dipoles must be aligned with the direction of field polarization. This is characteristic of automobile whip antennas (aligned with the vertically polarized radio waves) and the rooftop TV antennas (aligned with the horizontally polarized TV waves). If a reflector or horn antenna is used, the polarization matching is done in the waveguide or feedline after the antenna. The received field is converted to an identical polarized field in the guide, as shown in Figure 4.2. A probe is then inserted in the guide to detect the field. For a linear polarized field, the probe must be aligned with the field direction. For a circular polarized field, a phase shifter is placed in the

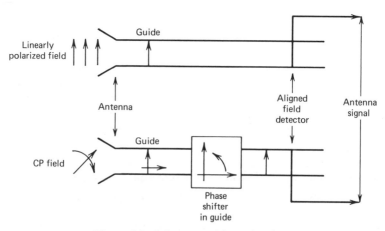

Figure 4.2. Antenna receiving subsystem.

guide to shift the time varying horizontal component onto the vertical component, combining the two into a linear polarized field, as shown. The detector probe is then aligned with the linear combination. The phase shifter must spatially shift in the proper direction, depending on whether the field is right-hand or left-hand circularly polarized. It is clear from Figure 4.2 why unintentional field rotation of a linearly polarized field during propagation could produce a power loss (polarization mismatch) at the receiving antenna, while CP fields will not be affected by additional field rotation.

An antenna aligned to receive a vertically polarized field will not receive a horizontally polarized field and vice versa. Likewise, an antenna for RHCP will not receive an LHCP field. This permits two independently modulated electromagnetic fields, each with their separate orthogonal polarizations, to be transmitted over the same channel, and to be separately received by two different receiving antennas, the latter oriented for each of the polarization states. This is the basis for *dual-polarization* communications, in which a common carrier bandwidth can be used simultaneously over the same link without mutual interference. Many modern satellite links are designed in this way.

Receiving antennas can be described in terms of an effective cross-sectional area $\mathscr{A}(\phi_l, \phi_z)$ associated with an arriving ray direction (ϕ_l, ϕ_z) defined with respect to the receiving antenna coordinates (Figure 4.3). This function indicates the effective collecting area that the antenna presents to an arriving plane wave from this direction. The effective area function therefore describes a receiving pattern over all directions from the antenna coordinates. The maximum area

$$\mathscr{A}_m = \max_{\phi_l, \phi_z} \mathscr{A}(\phi_l, \phi_z) \qquad (4.1.1)$$

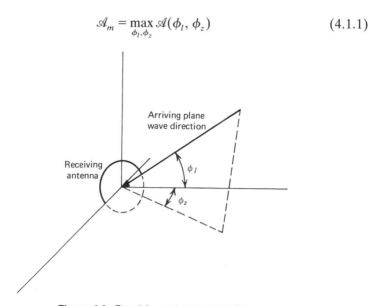

Figure 4.3. Receiving antenna geometry.

is the largest area presented to arriving fields. This maximum area is less than the physical antenna area $\mathscr{A}_p$ by the receiving antenna aperture loss factor ρ_a in (3.5.7),

$$\mathscr{A}_m = \mathscr{A}_p \rho_a \tag{4.1.2}$$

However, from the reciprocity condition (Reference 1, Chapter 4) of transmitting and receiving plane waves, it is known that

$$\frac{\mathscr{A}(\phi_l, \phi_z)}{\mathscr{A}_m} = \frac{g(\phi_l, \phi_z)}{g} \tag{4.1.3}$$

where $g(\phi_l, \phi_z)$ is the gain function of Section 3.5 if the same antenna had been used for transmission. This means that, except for scaling factors, an antenna's receiving and transmitting patterns are the same, and the ability of an antenna to collect field power from a given direction is identical to its ability to transmit. Thus the main transmission direction of a gain pattern while transmitting becomes the primary direction of field reception while receiving, and the sidelobe values represent reduced receiving capability in off-axis directions. If we relate (4.1.3) to (3.5.7), we see that for receiving antennas,

$$\frac{\mathscr{A}_m}{g} = \frac{\lambda^2}{4\pi} \tag{4.1.4}$$

Combining (4.1.4) and (4.1.3) shows

$$\mathscr{A}(\phi_l, \phi_z) = \frac{\lambda^2}{4\pi} g(\phi_l, \phi_z) \tag{4.1.5}$$

in any direction (ϕ_l, ϕ_z). Equation (4.1.5) therefore relates antenna receiving area patterns directly to the gain pattern and the electromagnetic field frequency. Note that the receiving field of view is now

$$\Omega_{fv} \cong \frac{\lambda^2}{\rho_{ap} \mathscr{A}_p} \text{ steradians} \tag{4.1.6}$$

associated with the gain pattern. This now becomes the solid angle associated with maximal power reception of the receiving antenna. This means that high gain antennas have large effective receiving areas, as indicated by (4.1.5), and small fields of view for collecting power in only certain directions. This latter property is important for eliminating unwanted electromagnetic interference arriving from off-axis directions. This condition of high receiver selectivity therefore dictates high carrier frequencies in order to achieve reasonable antenna sizes. Thus the modulation of the baseband up to carrier frequencies aids us not only in carrier transmission but in carrier reception as well.

Note that (4.1.5) and (4.1.6) show that the design of receiving antennas can be equivalently stated in terms of designing transmitting gain patterns, and the use of transmitting design charts, as in Figure 3.14, can be applied here. Therefore, if we desire a receiving antenna with a given receiving field of view, we can design the antenna as if it were to be a transmitting antenna with the same field of view.

4.2 RECEIVER CARRIER POWER

In analyzing or designing a communication system it is always necessary to know the actual amount of carrier power that can be transmitted to the receiver. As we shall see, this parameter plays a key role in assessing ultimate system performance. Based on our analysis to this point, we can determine the amount of carrier power theoretically to be recovered at the receiver antenna output terminals because of the power sent from the transmitter.

Consider first the carrier transmission channel shown in Figure 4.4 operating with unguided field propagation at a fixed carrier frequency f_c. We assume that the transmitter power amplifier produces the modulated carrier signal $c(t)$ with carrier power P_{amp}. That is, P_{amp} is the value of P_c at the amplifier output, where P_c has one of the forms in Sections 2.2–2.4 depending on the type of modulation. The carrier power coupled into the antenna terminals is then given by

$$P_t = P_{amp} L_t \qquad (4.2.1)$$

where L_t accounts for any guide or coupling losses at the transmitter [bracketed terms in (3.4.2) and (3.4.3)]. The EIRP of the transmitter in the direction of the receiver is obtained from (3.5.13) as

$$\text{EIRP} = P_t g_t(\phi_l, \phi_z) \qquad (4.2.2)$$

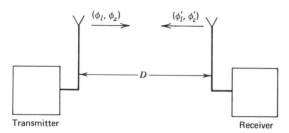

Figure 4.4. Transmitting and receiving channel.

where $g_t(\phi_l, \phi_z)$ is the transmitting gain function in the direction of the receiver, the latter considered a point located a distance D from the transmitter. During propagation, this EIRP will be reduced by any added channel losses L_a (absorption, rainfall, etc.) due to the atmosphere. The field power density (field intensity) of the propagated plane wave arriving at the receiver antenna after propagating a distance D is obtained from (3.5.12) as

$$P_{\text{den}} = \frac{(\text{EIRP})L_a}{4\pi D^2} \qquad (4.2.3)$$

Note that L_a accounts for all the additional propagation losses due to the atmospheric propagation channel ($L_a = 1$ for free space) and may also be a function of distance D as well as carrier frequency, as discussed in Section 3.6. The field plane wave power collected at the output of a receiving antenna presenting an effective area $\mathcal{A}(\phi_l', \phi_z')$ to the transmitter plane wave arriving from the transmitter direction (ϕ_l', ϕ_z') is then

$$P_r = P_{\text{den}} \mathcal{A}(\phi_l', \phi_z') \qquad (4.2.4)$$

This effective receiving area can be related to its gain function $g_r(\phi_l', \phi_z')$ using (4.1.5), so that we can write instead

$$P_r = P_{\text{den}} \left(\frac{\lambda^2}{4\pi}\right) g_r(\phi_l', \phi_z') \qquad (4.2.5)$$

where λ is the carrier wavelength. The collected antenna power actually fed into the receiver front end must be reduced by any polarization mismatches and receiving antenna coupling losses. Thus the total collected carrier power at the receiver front end terminals is

$$P_r = \left(\frac{\text{EIRP}}{4\pi D^2}\right) L_a \left(\frac{\lambda^2}{4\pi}\right) g_r(\phi_l', \phi_z') \qquad (4.2.6)$$

It is convenient to interpret

$$L_p = \left(\frac{\lambda}{4\pi D}\right)^2 \qquad (4.2.7)$$

as an effective *free space propagation loss*, because of the manner in which it appears in (4.2.6). If the antennas are properly aligned, the transmitter and receiver directions are located at the peak of the gain functions, and (4.2.6) reduces to

$$P_r = (\text{EIRP})L_a L_p g_r$$
$$= P_{\text{amp}} L_t L_a L_p g_t g_r \qquad (4.2.8)$$

where g_t and g_r are now the gains of the transmitting and receiving antennas at the carrier frequency. Equation (4.2.8) summarizes the effect of the complete carrier transmission subsystem as the power flows from transmitter amplifier to receiving antenna terminals and involves simply a product of the gains and losses along the way. It is often convenient to express this in decibels as

$$(P_r)_{dB} = (P_{amp})_{dB} + (L_t)_{dB} + (L_a)_{dB} + (L_p)_{dB} + (g_t)_{dB} + (g_r)_{dB} \quad (4.2.9)$$

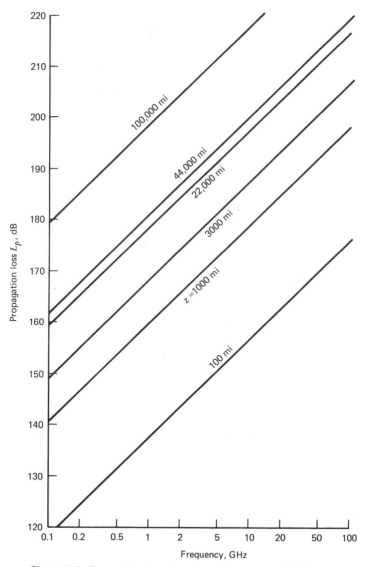

Figure 4.5. Propagation loss L_p versus frequency and distance.

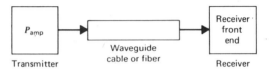

Figure 4.6. Cable transmission system.

where loss factors will appear as negative decibel values. The propagation loss factor in (4.2.7), when converted to decibels, reduces to the specific form

$$(L_p)_{dB} = -36.6 - 20 \log [D(\text{mi}) f_c(\text{MHz})] \qquad (4.2.10)$$

A plot of $(L_p)_{dB}$ is given in Figure 4.5 and aids in evaluating $(L_p)_{dB}$ from the preceding equation for typical carrier frequencies and transmission distances D. We see that computation of receiver power in decibels requires simply the addition of appropriate gain and loss decibels. This computation is often called a *power budget* and is a basic part of link analysis.

When the propagation is by guided channels (Figure 4.6) the analysis is similar. The amplifier power coupled into the transmission cable or fiber is again given by (4.2.1), where L_t accounts for any transmitter coupling losses. Cable attenuation losses are typically stated in decibels per unit length of cable and include both the electromagnetic propagation loss and added losses due to cable walls, radiation, and internal reflections. If the cable has a length D and an attenuation loss factor of α dB/length, the input decibel power is reduce by αD dB to give the cable output decibel power. If an amplifier or repeater of gain G_a is inserted a distance D_1 down the cable, the power after propagating a distance D_1 is now reduced by $\alpha D_1 - (G_a)$ dB. If a sequence of amplifiers is spaced along the cable, the propagating power is reduced by the loss of each section of cable and increased by the total inserted power gain. The final power coupled out of the cable and into the receiver front end is then further reduced by the coupling losses L_r. Thus for guided channels, the power flow equation becomes

$$(P_r)_{dB} = (P_{\text{amp}})_{dB} + (L_t)_{dB} - \alpha D + \sum_i (G_i)_{dB} + (L_r)_{dB} \qquad (4.2.11)$$

where D is the total cable length and the sum is over the number of cable amplifiers inserted. Again a power budget can be used for conveniently tabulating and summing the required terms.

4.3 BACKGROUND NOISE

In addition to collecting the desired carrier field from the transmitting antenna, a receiving antenna also collects noise energy from background sources in its field of view. This background energy is due primarily to random noise emissions from galactic, solar, and terrestrial sources, constituting the sky background. The amount of noise collected by the antenna is the ultimate limitation to the sensitivity of the receiving system, since it determines the weakest carrier signal that can be distinguished. The amount of pure background noise received depends on both the amount radiated from the sky background and the portion of that collected by the receiving antenna.

The background noise source is generally described by its *radiance* function, defined as

$$\mathcal{H}(\phi_l, \phi_z) = \begin{array}{l} \text{power received per Hz from the background} \\ \text{direction } (\phi_l, \phi_z) \text{ per unit area of receiving} \\ \text{antenna per unit solid angle of the source} \end{array} \quad (4.3.1)$$

Basically $\mathcal{H}(\phi_l, \phi_z)$ is the noise power spectral density level of the background one would receive from a unit solid angle source if a unit area receiving antenna is pointed at the sky at angle (ϕ_l, ϕ_z) from the antenna coordinates. The total background noise power in watts received from the sky background over a bandwidth B Hz is then

$$P_b = B \int_{\text{sphere}} \mathcal{H}(\phi_l, \phi_z) \mathcal{A}(\phi_l, \phi_z) \, d\Omega \quad (4.3.2)$$

where $\mathcal{A}(\phi_l, \phi_z)$ is again the effective area the antenna presents to the direction (ϕ_l, ϕ_z). By substitution from (4.1.5), this can be rewritten in terms of the receiving antenna gain function $g_r(\phi_l, \phi_z)$ as

$$P_b = \frac{B\lambda^2}{4\pi} \int_{\text{sphere}} \mathcal{H}(\phi_l, \phi_z) g_r(\phi_l, \phi_z) \, d\Omega \quad (4.3.3)$$

Thus by knowing the background radiance function and the antenna gain pattern, one can compute the receiver background noise power directly from (4.3.3). Radiance functions for the sky are obtained by employing a blackbody radiation model to describe the noise emissions. From Planck's law, we know that a body (in this case, the sky) at temperature $T°(\phi_l, \phi_z)$ Kelvin in the direction (ϕ_l, ϕ_z) produces a radiance function at frequency f of

$$\mathcal{H}(\phi_l, \phi_z) = \frac{hf^3}{c^2} \left[\frac{1}{\exp\left[hf/\kappa T°(\phi_l, \phi_z)\right] - 1} \right] \quad (4.3.4)$$

where c is the speed of light, $h = $ Planck's constant $= 6.624 \times 10^{-34}$ W-sec/Hz, and $\kappa = $ Boltzmann's constant $= 1.379 \times 10^{-23}$ W/degree Kelvin-Hz.

If $hf/\kappa T° \ll 1$, then (4.3.4) is well approximated by

$$\mathcal{H}(\phi_l, \phi_z) \cong \kappa \left(\frac{f}{c}\right)^2 T^\circ(\phi_l, \phi_z) = \left(\frac{\kappa}{\lambda^2}\right) T^\circ(\phi_l, \phi_z) \qquad (4.3.5)$$

Note that when $T^\circ = 300$ K, $hf/\kappa T^\circ = 1$ at 6000 GHz, so that the preceding approximation is quite accurate for RF and microwave wavelengths and for reasonable temperatures. Thus, for blackbody radiation, the radiance function $\mathcal{H}(\phi_l, \phi_z)$ is directly related to the temperature functions $T^\circ(\phi_l, \phi_z)$ describing the sky background. Such temperature functions are available as *radio temperature maps* [2], which indicate temperature contours of the sky when viewed from Earth. When (4.3.5) is used in (4.3.3), we have

$$P_b = \frac{B\kappa}{4\pi} \int_{\text{sphere}} T^\circ(\phi_l, \phi_z) g_r(\phi_l, \phi_z) \, d\Omega \qquad (4.3.6)$$

and background noise power can be determined by integrating the known sky temperature function directly. Note that the collected power P_b depends on the antenna orientation through the gain pattern $g_r(\phi_l, \phi_z)$. This equation suggests that an overall effective background temperature of the sky for a particular receiving antenna can be defined as

$$T_b^\circ = \frac{1}{4\pi} \int_{\text{sphere}} T^\circ(\phi_l, \phi_z) g_r(\phi_l, \phi_z) \, d\Omega \qquad (4.3.7)$$

This allows (4.3.6) to be written as simply

$$P_b = \kappa T_b^\circ B \qquad (4.3.8)$$

When expressed in this way, the received background noise appears to have a flat, one-sided power spectral density level of κT_b° W/Hz over the bandwidth B. The effective temperature T_b° is determined by weighting the temperature function of the sky by the actual antenna gain pattern and integrating in (4.3.7). The integration is often simplified by using temperature function approximations. In general, the sky is composed of localized galactic sources (sun, moon, planets, stars, etc.), each having a temperature T_i° approximately constant over a spatial solid angle Ω_i when viewed from the antenna site, and a relatively constant sky background of temperature T_s°. The effective temperature of the overall background in (4.3.7) with this model simplifies to

$$T_b^\circ = T_s g_s + \sum_i T_i^\circ g_i \qquad (4.3.9)$$

where

$$g_i = \frac{1}{4\pi} \int_{\Omega_i} g_r(\phi_l, \phi_z) \, d\Omega$$

$$\qquad (4.3.10)$$

$$g_s = \frac{1}{4\pi} \int_{\text{rest of sky}} g_r(\phi_l, \phi_z) \, d\Omega$$

Since the antenna function has a constant area of 4π, $(\Sigma_i\, g_i) + g_s = 1$. Thus the background temperature is collected as a weighted sum of individual galactic source temperatures and a constant sky background. Note that each gain coefficient g_i in (4.3.9) depends on the location of the particular source within the gain pattern. Thus a particular galactic source does not contribute much noise to the antenna if it lies on the fringes of the gain pattern. Some discrete sources (called *radio stars*) appear as discrete points in the sky $(\Omega_i \approx 0)$, and their effect is accounted for by treating them as spatial "delta functions" in (4.3.7). Hence a point source of temperature T_1° in direction (ϕ_{l_1}, ϕ_{z_1}) contributes a term $T_1^\circ g(\phi_{l_1}, \phi_{z_1})$ to the sum in (4.3.9).

Table 4.1 summarizes some common sources of galactic noise in the sky, listing their average temperature [2–9]. The sun is obviously the basic noise generator, having a beamwidth of about 0.5° as viewed from the Earth. Because of its relatively high temperature, the sun can contribute quite significantly to receiver noise if it appears in the antenna field of view. Most starts are relatively weak noise radiators and can be neglected. Some of the larger stars are more active noise sources with the temperatures listed, but since they occupy such small beamwidths relative to typical antenna patterns, their contribution is generally insignificant. The planets and the moon are passive reflectors of the sun's energy, with the moon being most dominant, since it has the large beamwidth ($\approx 0.5°$). The effect of the outer planets is lessened because they occupy extremely small fields of view. For passive radiators, the amount of noise energy contributed to the Earth is accounted for by modifying its incident sun energy by a constant called the *albedo* (ratio of reflected to incident energy). This albedo is a function of frequency, and received energy from these sources generally decreases with increasing frequency (i.e., at higher frequencies more sun energy is absorbed by them than is reflected). When the albedo is taken into account,

TABLE 4.1   Typical Background Source Temperatures and Beamwidths

Source	Temperature, °K	Beamwidth, degrees
Sun	6000	0.5
Moon	200	0.5
Stars (at 300 MHz)		
Cassiopeia	3700	$<10^{-3}$
Cygnus	2650	$<10^{-3}$
Taurius	710	$<10^{-3}$
Centaurus	460	$<10^{-3}$
Planets (10 MHz–10 GHz)		
Mercury	613	2×10^{-3}
Venus	235	6×10^{-3}
Mars	217	4.3×10^{-3}
Jupiter	138	1.3×10^{-3}
Saturn	123	5.7×10^{-3}
Earth (from moon)	300	2

the planets appear as localized sources with the approximate temperatures listed.

The temperature T_s° in (4.3.9) is due to the clear sky background and appears as an accumulation of a continuum of cosmic and galactic noise sources, influenced by the frequency effects of the radiation albedo. As a result, the clear sky appears as an almost uniform noise source in all directions, with an equivalent temperature that descreases with frequency. Typical background clear sky temperature T_s° is shown in Figure 4.7 as a function of frequency.

When we view from the Earth, sky temperature is also affected by the atmosphere. The gases and vapors tend to absorb sun energy and reradiate a portion of this energy isotropically as noise. This atmospheric reradiation predominates at frequencies above 10 GHz and tends to increase beyond this point as the wavelengths approach particulate size. To a receiving antenna this reradiation appears as an extraneous source of background noise that tends to increase the sky temperature T_s° at higher frequencies. Thus the sky background appears as a combination of galactic effects that fall off with frequencies, as in Figure 4.7, and atmospheric effects that begin for short wavelengths and increase with frequency. When these two predominate effects are combined, one obtains an effective sky temperature behavior, as a function of frequency, shown in Figure 4.8. An important result of these dependencies is a region between 1 and 10 GHz, where an

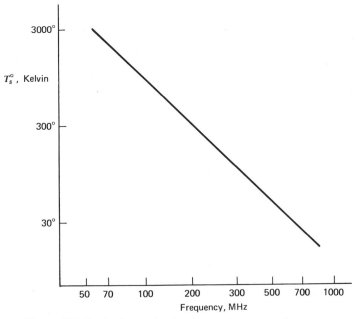

Figure 4.7. Sky background noise temperature versus frequency.

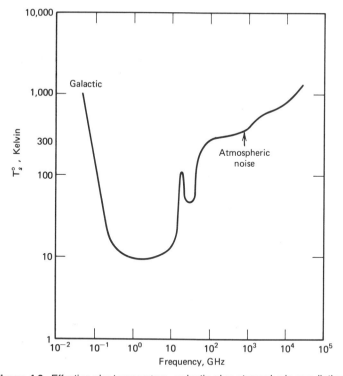

Figure 4.8. Effective sky temperature, galactic plus atmospheric reradiation.

antenna aimed toward the sky receives less noise power than at a higher or lower frequency. This noise "window" is particularly advantageous for low power deep space and satellite communication systems and is the principal reason why such systems utilize primarily carrier frequencies in the SHF band. Note that background temperatures as low as 10–100° can be obtained by operating in the window. This background temperature can increase slightly (by about 50°) when the reradiation effect is augmented by water droplets (i.e., during rain, clouds, or misty conditions).

Although the sky background is the primary source of receiver noise, there are other types of interference that may appear during reception. One important type is *radio frequency interference* (RFI) due to other transmitting sources. This noise becomes important when the desired transmitter is located in an area dense with other operating transmitters, as, for example, with a satellite receiver aimed back toward the Earth. The amount of RFI can only be assessed with knowledge of these secondary source locations and their EIRP capability. An evaluation of the degree of interference then requires the computation of a link budget for each such source relative to the receiving antenna. RFI is particularly bothersome since it often is concentrated in frequency bands rather than being spectrally spread, as with pure background noise interference.

Another type of noise that appears with certain Earth based receivers is urban noise. Such noise is due to various electronic emissions associated with a typical urban environment, such as vehicles, communications, equipment, and so on. Such noise is usually impulsive, or burstlike, in nature and is generally modeled as a *shot noise* [10] process at the receiver front end. Shot noise interference takes the form of independently occurring random impulses obeying some prescribed distribution in time. When the rate of occurrence is high and fairly uniform in time, the shot noise pulses overlap into a continuous noise process with a well-defined power spectral density. When the rate of occurrence is low, however, the impulsive nature of the process is emphasized, preventing it from being treated mathematically as a continuous noise process, and care must be used in analytical treatments. Lastly, diffuse multipath noise, as in (3.7.3), may also contribute to the total receiver interference.

4.4   RECEIVER FRONT END ANALYSIS

At the receiver the total antenna signal is coupled into the receiver front end circuitry, as shown in Figure 4.9a. Since the antenna may be located away from the actual front end electronics, a waveguide or cable may be used to provide this coupling. As stated in Section 3.4, such coupling devices inherently introduce attenuation into the antenna signal. The antenna produces at the front end input the carrier signal having the power level P_r, as computed in (4.2.8). In addition, any background noise and interference received by the antenna will produce a noise voltage that is superimposed on the receiver carrier signal. The combined carrier plus noise waveform is coupled through the cable to the receiver front end electronics.

The receiver front end circuitry has the prime objective of amplifying and filtering the relatively weak antenna signal. We model this front end filter-amplifier as having a filter transfer $\sqrt{G_c}H_c(\omega)$, Figure 4.9b, where G_c is the front end power gain, and $H_c(\omega)$ represents a unit height carrier filter function. In reality the system may involve a cascade of several filters and amplifiers, so that $\sqrt{G_c}H_c(\omega)$ represents the combined effect of the cascade. The filter characteristic $H_c(\omega)$ represents the filtering that the front end provides for the received carrier signal. As such, the filter is bandpass in nature, typically having a flat, unit level gain function over the bandwidth of the received carrier signal. Outside this band the filter should fall off rather quickly in frequency to remove as much noise and interference as possible. Thus the carrier filter function is a bandpass filter tuned to the receiver carrier signal, and therefore has the basic form sketched in Figure 4.9c. In many cases the carrier frequency (i.e., the actual location of the carrier spectrum) is not known exactly (e.g., because of Doppler), and the carrier

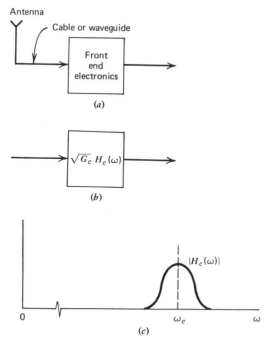

Figure 4.9. Receiver front end: (*a*) block diagram, (*b*) filter model, (*c*) filter frequency function.

filter may have to be several times larger in bandwidth to accommodate this uncertainty. The bandwidth actually needed in the front end filter also depends on the amount of distortion we are willing to accept in the modulated carrier. We might, however, point out that the construction of a suitable carrier filter for the receiver front end is not a trivial task. Difficulties are due to the following factors:

1. Filter bandwidths less than 0.1% of the center frequency often lead to unsatisfactory parameter values for practical realization. This means it is difficult to build arbitrarily narrow bandpass filters at the high RF carrier frequencies usually used.

2. The requirement of flat inband response and fast rolloff tends to be conflicting. In order to get sharp rolloff, one must often settle for nonflat inband response, as with Chebychev bandpass filters. On the other hand, flat inband behavior is generally accompanied by rounded-off band edges, as with Butterworth filters.

3. High order bandpass filters must generally be constructed as cascaded filter stages. Each state, however, tends to introduce signal attenuation (*insertion loss*) that may severely reduce the overall front end gain.

4. Sharp cornered filters are invariably accompanied by severe phase distortion (nonlinear phase response) at the band edges. Such phase effects can seriously distort carrier waveforms and generally must be corrected by additional front end equalizing circuitry.

Even after the carrier filter is properly designed for the front end, the carrier waveform must be amplified without distortion to provide the front end signal for the receiver processing that follows. For the model in Figure 4.9, the carrier power produced at the output of the front end is then

$$P_{FE} = P_r L_g G_c \qquad (4.4.1)$$

where G_c accounts for the front end amplifier power gain and L_g for any guide losses from antenna to front end input. Of course, all antenna noise in the front end bandwidth is amplified as well.

When providing the amplification for the antenna signal, the receiver front end electronics invariably adds additional circuit noise to that already present as received background noise. This circuit is due both to thermal noise generated in passive resistors and semiconductors and to electronic noise resulting from flicker, shot, and tube noises in active amplifiers. Noise properties of any electronic device, whether passive or active, are accounted for by specification of its *noise figure*. Let us digress temporarily to discuss this parameter.

Consider an electronic device (filter, amplifier, etc.), as shown in Figure 4.10. We assume it is being fed by a source with output resistance Z_s that is matched to the input resistance Z_{in} of the device. We assume the source is at temperature T_0° Kelvin and the power gain of the device from input to output is G. It is well known [11–13] that a resistor Z_s at temperature T_0° generates a thermal noise voltage whose one-sided voltage spectral density is $4\kappa T_0^\circ Z_s$ V^2/Hz, where κ is Boltzmann's constant. The noise power from the source in a bandwidth B that will be coupled into the electronic device under matched impedance conditions is then

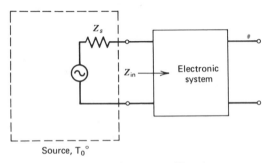

Figure 4.10. Electronic system with noisy source.

$$\text{Input noise power} = (4\kappa T_0^{\circ} Z_s)\, B \left(\frac{Z_{\text{in}}}{Z_s + Z_{\text{in}}}\right)^2 \frac{1}{Z_{\text{in}}}$$

$$= \kappa T_0^{\circ} B \qquad (4.4.2)$$

since $Z_s = Z_{\text{in}}$. Thus the power coupled in under matched conditions depends only on the source temperature and the prescribed bandwidth, and not on the value of the source resistance. Equation (4.4.2) implies that the input thermal noise power has a flat spectral density of level κT_0° W/Hz, one-sided, over the bandwidth B. The output noise power of the device due to this input thermal noise is given by the product of the device input power and the device power gain. Hence

$$\begin{array}{c}\text{Noise power output of device in } B \text{ Hz} \\ \text{due to source thermal noise at } T_0^{\circ}\end{array} = \kappa T_0^{\circ} BG \qquad (4.4.3)$$

The noise figure F of the device is defined as

$$F \overset{\Delta}{=} \frac{\text{total output noise in } B \text{ when input source temp } T_0^{\circ} = 290^{\circ}}{\text{output noise due to only the source at } T_0^{\circ} = 290^{\circ}}$$

$$(4.4.4)$$

Noise figure is therefore the ratio of total output noise to that due to the input noise alone, when the input source is at a specified temperature of 290°K. The total output noise is that due to the input source noise plus that due to internally generated noise within the device itself. Denoting the output noise power from internal sources as P_{int}, (4.4.4) becomes

$$F = \frac{\kappa(290^{\circ})BG + P_{\text{int}}}{\kappa(290^{\circ})BG}$$

$$= 1 + \frac{P_{\text{int}}}{\kappa(290^{\circ})BG} \qquad (4.4.5)$$

We see that $F \geq 1$, and $F = 1$ only for a device with no internal noise (i.e., a noiseless device). Thus the noise figure of an electronic device indicates the amount by which the output noise will be increased over that of a noiseless device. Standard methods are known for measuring device noise figure (Problem 4.18). Typical noise figure values for receiver amplifiers are generally in the range 2–12 dB. From (4.4.5) we can solve for P_{int} as

$$P_{\text{int}} = \kappa[(F-1)290^{\circ}]BG \qquad (4.4.6)$$

which relates the output noise power due to internal sources to the device noise figure. When we compare it to (4.4.2) we see that internal noise is produced at the device output as if it were caused by an input source at

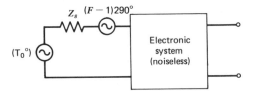

Figure 4.11. Equivalent noise input.

temperature $(F-1)290°$. Thus we can account for internal device noise by simply adding to the input a matched source of temperature $(F-1)290°$ and treating the device as if it were noiseless. This equivalent source noise is added directly to any source noise already present, as shown in Figure 4.11. The temperature $(F-1)290°$ is called the *equivalent device temperature*. Note that it is not simply the physical temperature of the device, although the latter certainly affects the value of F. Low noise receiver front ends using maser amplifiers or cryogenically cooled amplifiers can have noise figures as low as 1.1 (0.4 dB), and equivalent temperatures of about 30°K.

In many cases the electronic device may actually be composed of a cascade of devices, each with its own noise figure and power gain. One must therefore evaluate the noise figure of the cascade in order to assess the total output noise power that will occur. Consider the cascade of two devices as shown in Figure 4.12, each with noise figure F_1 and F_2 and power gain G_1 and G_2, respectively. We assume the devices are impedance matched at their input and output. The noise power at the output of the first device when a 290° matched source is placed at its input is $\kappa(290°)BG_1 + \kappa(F_1 - 1)290°BG_1$. This noise has the equivalent noise of the second stage, $\kappa(F_2 - 1)290°B$, added to it. The noise figure of the overall cascade is then

$$F = \frac{\kappa(290)°BG_1G_2 + \kappa(F_1 - 1)290°BG_1G_2 + \kappa(F_2 - 1)290°BG_2}{\kappa(290°)BG_1G_2}$$

$$= F_1 + \frac{F_2 - 1}{G_1} \tag{4.4.7}$$

We see that the noise figure of the cascade of the two devices is related to the individual noise figures by (4.4.7). Note that if the first stage gain G_1 is large, the contribution of the second stage is reduced, and the noise figure of the cascade is approximately equal to that of the first stage alone.

Figure 4.12. Cascade of two noisy systems.

The result can be easily extended to more than two devices, and the overall noise figure for a cascade of N devices, each with gain and noise figure (G_i, F_i), is (Problem 4.12)

$$F = F_1 + \frac{F_2 - 1}{G_1} + \frac{F_3 - 1}{G_1 G_2} + \cdots + \frac{F_N - 1}{G_1 G_2 \cdots G_{N-1}} \qquad (4.4.8)$$

Thus the noise figure contribution of each stage is reduced by the product of gains up to that stage. For this reason, it is always advantageous to place noisier devices further back in the cascade, while using high gain, low noise devices at the beginning.

Any electronic device, having an input and output, must necessarily have a noise figure. Consider any passive, lossy device, such as a transmission line, cable, waveguide, attenuator, or power splitter. Let T_g° be its phsycial temperature and L_g be the power loss factor of the line (ratio of output to input power). The output noise power due to a matched input source at T_g° is then $\kappa T_g^\circ B L_g$. However, the total noise at the output terminals under matched output conditions must be $\kappa T_g^\circ B$, since the output (i.e., looking back into the guide) appears as a pure resistance at temperature T_g°. This means that $\kappa T_g^\circ B L_g + P_{\text{int}} = \kappa T_g^\circ B$, or

$$P_{\text{int}} = \kappa (1 - L_g) T_g^\circ B$$

$$= \kappa \left(\frac{1 - L_g}{L_g} \right) T_g^\circ B L_g \qquad (4.4.9)$$

This immediately identifies the lossy waveguide as having an equivalent input noise temperature of $(1 - L_g) T_g^\circ / L_g$. Equating this to (4.4.6) shows that the lossy guide noise figure is

$$F_g = 1 + \left(\frac{1 - L_g}{L_g} \right) \frac{T_g^\circ}{290^\circ} \qquad (4.4.10)$$

when operated at temperature T_g°. When $T_g^\circ = 290^\circ$, $F_g = 1/L_g$.

We can now use the receiver noise figure definitions to determine the amount of noise contributed by the actual receiver front end. Consider again the receiver model in Figure 4.9. The antenna signal produces the carrier waveform along with the background noise having the equivalent temperature T_b°, as determined from (4.3.7). The combined waveform is then coupled into the receiver front end, the latter having the overall noise figure F computed by the previous analysis. By using equivalent noise sources, as in Figure 4.11, the equivalent one-sided noise spectral level at the antenna terminals is then

$$N_0 = \kappa T_{\text{eq}}^\circ \; W/Hz \qquad (4.4.11)$$

where

$$T^{\circ}_{eq} = T^{\circ}_b + (F - 1)290^{\circ} \tag{4.4.12}$$

After the front end filtering, the one-sided noise spectral density of the front end output is

$$\hat{S}_n(\omega) = N_0 L_g G_c |H_c(\omega)|^2 \tag{4.4.13}$$

The resulting output noise power, caused by background, thermal, and internal sources, is then

$$P_n = N_0 L_g G_c B_c \tag{4.4.14}$$

where

$$B_c = \frac{1}{2\pi} \int_0^{\infty} |H_c(\omega)|^2 \, d\omega \tag{4.4.15}$$

Here B_c represents the one-sided noise bandwidth of the carrier front end filter shown in Figure 4.9. If we again assume that the front end subsystem has a bandwidth wide enough to pass the carrier waveform without distortion, the carrier power at the filter output is that given by (4.4.1). The resulting ratio of carrier power to total noise powers at the front end filter output is then

$$\begin{aligned} \text{CNR} &= \frac{P_r G_c L_g}{\kappa T^{\circ}_{eq} G_c L_g B_c} \\ &= \frac{P_r}{\kappa T^{\circ}_{eq} B_c} \end{aligned} \tag{4.4.16}$$

The preceding CNR parameter therefore serves as an indication of how well the front end subsystem has received the transmitted carrier. We see that this CNR depends only on the received carrier power P_r in (4.2.8), the carrier filter noise bandwidth in (4.4.15), and the equivalent temperature defined in (4.4.12). This latter parameter can be interpreted as the overall effective noise temperature of our receiver, taking into account all noise sources during reception and filtering. It is interesting to note that CNR does not depend at all on the power gain G_c of the front end subsystem. This is due to the fact both signal and noise are being power ampified during front end processing. Of course, the signal and noise powers individually do depend on the system gain, and the latter will therefore be important in determining the dynamic range of the actual voltage variations in the front end.

We also see that CNR is related directly to B_c in (4.4.15), and obviously this should be as small as possible for optimum performance. Since $|H_c(\omega)|$ is normalized to one at bandcenter, B_c can be reduced only by careful

TABLE 4.2 Bandpass Filter Noise Bandwidths Relative to 3 dB Frequency for Butterworth and Chebychev Filters

Filter	Order	$\dfrac{B_c}{B_{3\,dB}}$	Filter	Order	$\dfrac{B_c}{B_{3\,dB}}$	Filter	Order	$\dfrac{B_c}{B_{3\,dB}}$
Butterworth	1	1.570	Chebychev	1	1.570	Chebychev	1	1.57
	2	1.220	($\epsilon = 0.1$)	2	1.15	($\epsilon = 0.158$)	2	1.33
	3	1.045		3	0.99		3	0.86
	4	1.025		4	1.07		4	1.27
	5	1.015		5	0.96		5	0.81
	6	1.010		6	1.06		6	1.26

control of the filter charcteristic, and using a filter that is no wider than necessary. Practical filter construction, however, often makes this a difficult task, as was pointed out earlier. In order to evaluate B_c we must perform the integration in (4.4.15), which requires specific knowledge of the function $H_c(\omega)$. For many such functions, B_c can be directly related to the 3 dB bandwidth of the filter. Table 4.2 lists some common front end filter types and their corresponding $B_c/B_{3\,dB}$ ratios.

The equivalent temperature T°_{eq} in (4.4.1) depends on both the background and the receiver noise figure. Background temperature T°_b can be reduced only by controlling the antenna pattern. The front end noise figure depends on the electronic elements and the way in which the receiver is constructed. Consider again the receiver model in Figure 4.13a, showing the antenna signal coupled through a lossy cable to a front end amplifier having noise figure F_a. The cable is at temperature T°_g and has power loss L_g. Using (4.4.7), the front end noise figure is then

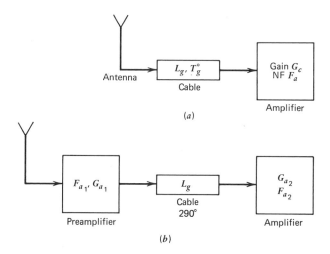

Figure 4.13. Antenna coupling models: (a) cable–amplifier, (b) amplifier–cable–amplifier.

$$F = F_g + \frac{F_a - 1}{L_g}$$

$$= 1 + \left(\frac{1 - L_g}{L_g}\right)\left(\frac{T_g^\circ}{290^\circ}\right) + \frac{F_a - 1}{L_g} \qquad (4.4.17)$$

The noise figure depends on both temperature and loss of the cable. Cooling, or insulating the front end, lowers T_g° and thereby reduces the noise figure. If the cable is at $T_g^\circ = 290^\circ$, then

$$F = 1 + \left(\frac{1 - L_g}{L_g}\right) + \frac{F_a - 1}{L_g}$$

$$= \frac{F_a}{L_g} \qquad (4.4.18)$$

We see that the cable loss acts to effectively increase the noise figure of the amplifier in determining the overall noise figure. Hence excessive cabling between antennas and front end electronics tends to produce noisy receivers.

Often the noise figure can be improved by redesigning the front end. Consider instead the receiver in Figure 4.13b showing the insertion of an RF preamplifier at the antenna terminals, prior to the cable connection. The noise figure is now

$$F = F_{a_1} + \frac{(F_{a_2}/L_g - 1)}{G_{a_1} L_g} \qquad (4.4.19)$$

Note that now the front end noise figure can be controlled by using a preamplifier with high gain ($G_{a_1} \gg 1$) and low noise figure F_{a_1}. As a result, when the front end uses an RF preamplifier, its noise figure determines the entire front end noise figure if its gain is suitably high. Thus a basic requirement in preamplifier design is for high gain, *low noise amplifiers* (*LNAs*) for this purpose. Figure 4.14 plots some preamplifier noise figures and gain values as a function of carrier frequency. Solid state microwave amplifiers, such as *tunnel diode amplifiers* (TDA), with their inherent low noise contributions and moderately high power gains, are almost univerally used for front ends. Unfortunately, TDAs have decreasing gain values at the higher microwave frequencies (>10 GHz), and it is expected that TDAs will be rapidly replaced by newer technological advances in LNA design, such as field-effect transistors (FET).

Earlier we showed that front end elements further down the cascade contribute less to overall noise figure if high gain precedes it, and lossy elements tend to increase noise figure. Hence an LNA should be placed as close to the antenna terminals as possible, prior to any significant coupling

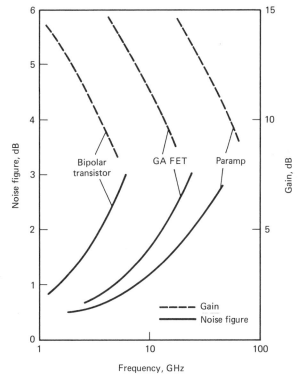

Figure 4.14. Low noise amplifiers: gain and noise figure versus frequency.

losses. Cable losses leading from antenna to amplifier increase the effective receiver noise and should be minimized. Typically, front end noise figures are about 5–10 dB, producing receiver noise temperatures of about 1000–3000°K. An Earth-based station can generally use extensive cooling to lower its front end noise figure to the 2–5 dB range. Hence Earth-station noise temperatures are more typically about 50–800°K.

It should be pointed out that when front end noise figures are significantly reduced, the front end noise is dominated by the background noise in (4.4.12). However, the background noise may be dominated by reradiation effects that tend to be weather dependent. As a result, low noise figure receivers tend to have significant variations in noise power, making its design performance somewhat less stable.

4.5 LINK BUDGETS

The carrier to noise ratio (CNR) in (4.4.16) is an important parameter in determining the receiver performance, since it represents the ratio of carrier power to the total noise power. Typically, communication links require P_r to be about 10 times greater than the noise power (i.e., $\text{CNR} \approx 10 \, \text{dB}$) for adequate receiver processing.

When the transmission parameters for P_r in (4.2.6) are inserted, we have

$$\text{CNR} = \frac{(\text{EIRP}) L_p L_a g_r}{\kappa T^\circ_{eq} B_c} \tag{4.5.1}$$

When stated in decibels, this becomes

$$(\text{CNR})_{dB} = (\text{EIRP})_{dB} + (L_p)_{dB} + (L_a)_{dB} + (g_r)_{dB}$$
$$- (\kappa)_{dB} - (T^\circ_{eq})_{dB} - (B_c)_{dB} \tag{4.5.2}$$

The resulting CNR in decibels is therefore obtained by adding and subtracting the decibel values given above. Note that Boltzmann's constant in (4.5.1), when converted to decibels, has the value $(\kappa)_{dB} = -228.6$, which enters (4.5.2) as a positive number in the link budget. Table 4.3 evaluates an example link budget over satellite ranges, showing the way in which the entries of (4.5.2) are individually determined and combined.

Since the receiver bandwidth B_c is often dependent on the modulation format, we often isolate the RF link power parameters by normalizing out bandwidth dependence. We therefore define instead RF carrier to noise level ratio,

TABLE 4.3 Link Budget Example

Transmitted power P_T (=30 W)	15 dBW	
Transmitter antenna gain g_t	55 dB	
EIRP		70 dBW
Path length (23,000 m)		
Frequency (14 GHz)		
L_p		−207 dB
Atmospheric loss (rain)		−5 dB
Polarization loss		−1 dB
Pointing loss		−0.6 dB
Receiver antenna (diameter = 3 ft, efficiency = 55%) gain g_r	40 dB	
Background temperature $T_b = 100°$		
Receiver noise figure $F = 7.86$ dB		
Receiver noise temperature $T_{eq} = 1584°$	32 dB	
Receiver $g/T°$		8 dB
Boltzmann's constant		228.6 dB
Bandwidth (10 MHz)		−70 dB
CNR		23 dB

$$\frac{C}{N_0} = \frac{P_r}{\kappa T^\circ_{eq}} \tag{4.5.3}$$

which does not depend on B_c. In later analysis with digital systems, we shall find that an important parameter is the ratio of the carrier energy E_b in a bit time to the noise spectral level N_0 at the receiver. This can be obtained directly from C/N_0 as

$$\frac{E_b}{N_0} = \left(\frac{C}{N_0}\right) T_b \tag{4.5.4}$$

where T_b is the bit time. Thus, from knowledge of the link C/N_0, which depends only on the RF link parameters, we can directly compute either analog CNR by dividing in bandwidth, or digital E_b/N_0 by multiplying in bit times.

It is often convenient to factor (4.5.3) as

$$\frac{C}{N_0} = \left[\frac{\text{EIRP}}{\kappa}\right] [L_a L_p] \left[\frac{g_r}{T^\circ_{eq}}\right] \tag{4.5.5}$$

The first bracket contains only transmitter parameters, the second bracket contains propagation parameters, and the last bracket contains receiver parameters. Thus (4.5.5) separates out the contribution of each subsystem to the total C/N_0. This interpretation is interesting in that it shows, for example, that in terms of C/N_0, the only effect of the receiving system is through the ratio g_r/T°_{eq} (i.e., the ratio of the receiver antenna gain to its equivalent noise temperature). In fact, it shows that as far as the receiver is concerned, performance can be maintained with a lower receiver gain (smaller antenna) if T°_{eq} can be suitably reduced. Hence there is a direct tradeoff of receiver antenna size and receiver noise temperature in achieving the desired performance. Receiver temperature can be controlled by careful control of the antenna background and receiver electronics (lower noise figure). Since antenna size directly impacts overall receiver cost and construction, this type of receiver noise quality-antenna size tradeoff is often desirable. Figure 4.15 plots the value of g/T°_{eq} for a given frequency–antenna size product at various values of noise temperature T°_{eq}.

Conversely, use of a high gain antenna allows a noisier receiver to be employed. Such a receiver will generally dominate the noise contribution from the background. This makes the performance of the system less sensitive to parameters influencing the background contribution, such as field of view and rainfall; that is, small changes in T°_b will not affect T°_{eq} and therefore C/N_0.

One must be careful, however, not to overemphasize the g/T° parameter in attempting to improve satellite performance, since it often leads to serious misconceptions. For example, at a glance Figure 4.15 implies that

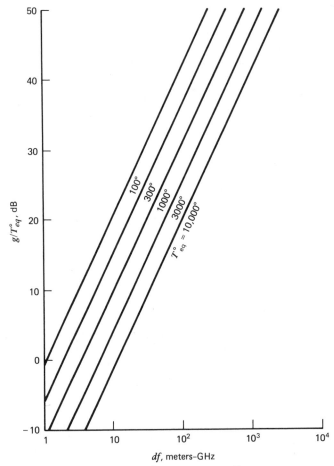

Figure 4.15. Receiver g/T°_{eq} factor versus T°_{eq} amd *df* product.

g/T° can be improved by using a higher carrier frequency with a given antenna size, since this directly increases receiving antenna gain. However, we recall from (4.2.7) that propagation loss also increases as f^2, so that with a fixed-transmitter EIRP in (4.5.5), C/N_0 does not directly depend on carrier frequency. (In fact, it may be reduced since the atmospheric losses L_a generally increase with f.) This can be made more obvious by reinserting (4.1.6) and (4.2.1) to rewrite (4.5.3):

$$\frac{C}{N_0} = \left[\frac{(\text{EIRP})L_a}{4\pi\kappa T^\circ_{eq}} \right] \left[\frac{\mathscr{A}_r}{D^2} \right] \tag{4.5.6}$$

or a fixed transmitting EIRP and T°_{eq}, C/N_0 is independent of frequency nd depends only on propagation path length and receiving area $\mathscr{A}_r$. The apparent dependence on frequency appears in L_p only if we choose to write L_p in terms of receiving gain g_r. Thus selecting a higher frequency band will *not* directly aid the receiving system. Of course, increasing frequency does aid the transmitter when a given EIRP can be achieved with a smaller transmitting antenna.

When the field is propagated from transmitter to receiver by guided transmission, as with long distance cables, for example, no background noise is involved. Instead, noise generated from the transmitter power amplifier and from any amplifiers inserted in the cable appears at the front end input and replaces the background noise power. This cable noise can be determined by computing the cable noise figure and inserting an equivalent noise temperature at the cable input. If the cable is simply a lossy line, its noise figure follows from (4.4.10). If the line contains inserted amplifiers, the noise figure must be calculated. This is most conveniently done by considering the line as a cascade of individual sections, each composed of a lossy section of line feeding an amplifier having a specified noise figure. For example, consider the cable system in Figure 4.16, showing a source at 290°K feeding a cable containing M identical amplifiers spaced at equal distances. We can consider the cable to have M identical sections, each composed of a lossy line of loss L_1 and an amplifier of gain G_a and noise figure F_a. We assume the cable and source are at 290°K, and the cable amplifier gain G_a exactly compensates for the line loss L_1; that is, $L_1 G_a = 1$ for each section. Using the result of (4.4.8), the overall noise figure of the cable is

$$F_c = \frac{F_a}{L_1} + \frac{(F_a/L_a) - 1}{L_1 G_a} + \frac{(F_a/L_1) - 1}{(L_1 G_a)^2} \cdots$$

$$= M\left(\frac{F_a}{L_1}\right) - M + 1 \tag{4.5.7}$$

The noise temperature of the combined cable and source, referred to the cable input terminals, is then $T^\circ_{eq} = 290° + (F_c - 1)290° = F_c(290°)$. This now

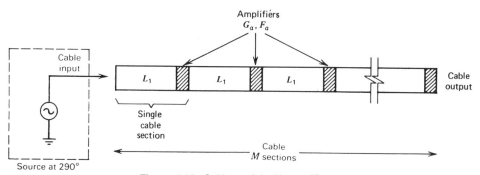

Figure 4.16. Cable model with amplifiers.

represents the effective temperature of the noise that the cable and source insert into the cable input. The carrier power at the cable output was given in (4.2.11). The resulting CNR at the cable output is then

$$\mathrm{CNR} = \frac{P_r (L_1 G_c)^M}{\kappa F_c 290° B_c (L_1 G_c)^M}$$

$$\approx \frac{P_{\mathrm{amp}} L_t L_r}{\kappa M (F_a / L_1) 290° B_c} \qquad (4.5.8)$$

We see that the CNR is inversely proportional to the amplifier noise figure and the number of amplifiers inserted. In fact, (4.5.8) indicates that the source power P_{amp} must increase with the number of amplifiers if CNR is to remain constant. At first glance this might appear to indicate a disadvantage in using amplifiers at all. However, it must be remembered that the gain of the amplifier is canceling out the line loss. If no amplifiers were used, F_c would be equal to the reciprocal of the total cable loss; that is, $F_c = 1/L_1^M$. This means that the effective noise would increase exponentially with M instead of only proportionally. Even if the amplifiers were quite noisy, the former effect would be much more serious than the latter (see Problem 4.25).

An important consideration in any guided cable system is the length of the link that can be used before an amplifier or repeater must be inserted. It is in this regard that optical fibers became important. Consider the optical fiber link in Figure 4.17. The design objective is to have as long a link as possible while achieving a specified CNR at the output receiver. The optical fiber link is governed by the same basic equations as the metallic cable. The analysis details are given in Chapter 12. To summarize briefly, the light source transmits a light wave with power P_t. It arrives at the fiber output (receiver input) after a fiber attenuation loss of αD dB. The effective CNR at the output is therefore

$$\mathrm{CNR} = \frac{P_t [10^{-\alpha D/10}]}{(hf/\eta) B_c} \qquad (4.5.9)$$

where h is Planck's constant [see (4.3.4)], f is the optical frequency, and η is the efficiency factor of the optical detector. If this is compared to (4.5.8),

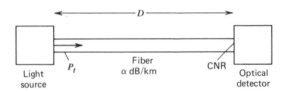

Figure 4.17. Optical fiber model.

the fiber channel appears to have an effective receiver quantum noise spectral level of (hf/η), as contrasted with the RF noise level $N_0 = kF290°$. At optical frequencies $(f \approx 3 \times 10^{14})$, hf/η is approximately 20 dB greater than the RF noise level. Fiber links therefore have higher receiver noise levels than RF cable links.

However, the difference lies in the numerator, where the fiber loss factor α is significantly lower than the loss for the metallic cable (recall Figure 3.18). The fiber attenuation over several kilometers may only be a few decibels, while the cable loss may be 60 dB or higher. Even with a 20 dB higher noise level, the fiber link still achieves about a 40 dB numerator advantage in achieving the same CNR. This means, for example, a fiber link with a 1 mW light source can use a link 10 times longer than a metallic cable link with a 1 W transmitter amplifier. This is a primary reason why fiber systems are competitive in modern guided link communications.

4.6 RECEIVER NOISE PROCESSES

In Section 4.4 we concentrated solely on the power values of the receiver noise. However, in subsequent discussions we must have a more definitive analytical model for the noise processes generated in our receivers. Recall that our primary sources of receiver noise were the blackbody radiation noise and the thermal circuit noise. In the past each of these noises has been rigorously analyzed, both theoretically and experimentally, and each has been related to the accumulated effect of a large number of independent random movements of charged particles. When such models are developed, these noise sources invariably reduce to zero mean noise processes obeying Gaussian statistics, the latter obtained by application of some form of the central limit theorem (Section B.3). Thus noise processes in communication receivers are almost universally accepted as Gaussian random processes.

Earlier in this chapter we developed a receiver noise model in which the power spectral density of the receiver noise, referred to the receiving antenna output terminals, is given by $\kappa T_{eq}^{°}$ W/Hz, where $T_{eq}^{°}$ was given in (4.4.12). Thus the receiver antenna noise has a spectrum that is flat over all microwave frequencies and has a two-sided spectral level (Figure 4.18a):

$$S_{ant}(\omega) = \frac{\kappa T_{eq}^{°}}{2} = \frac{N_0}{2} \qquad (4.6.1)$$

Thus receiver noise is inherently a Gaussian white noise process when referred to the antenna terminals. After front end filtering, this spectrum is shaped by the carrier filter function, producing the two-sided front end output noise spectrum in (4.4.13), which we write here as

$$S_n(\omega) = \tfrac{1}{2}N_0 G_1 |H_c(\omega)|^2 \qquad (4.6.2)$$

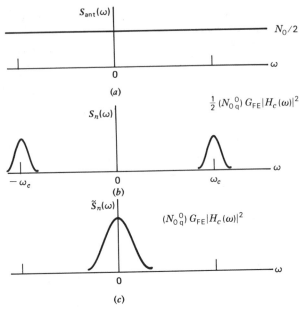

Figure 4.18. Receiver noise spectral densities: (a) antenna white noise, (b) bandpass front end noise, (c) noise quadrature component spectrum.

where G_1 is the total front end gain (guide attentuation and amplifier gain). Note that this noise spectrum has been shaped by the filter function (Figure 4.18b) and limited to the spectral extent of the filter $H_c(\omega)$. Since this filter typically occupies a narrow symmetrical band about the filter center frequency, the filtered front end noise is a narrowband Gaussian process with this spectrum. Bandpass Gaussian noise processes of this type can be analytically written as (see Section B.6)

$$n(t) = n_c(t) \cos (\omega_c t + \psi) - n_s(t) \sin (\omega_c t + \psi) \qquad (4.6.3)$$

where ψ is an arbitrary phase angle, ω_c is the filter frequency, and $n_c(t)$ and $n_s(t)$ are independent Gaussian noise processes, each having the two-sided spectrum

$$\tilde{S}_n(\omega) = N_0 G_1 |\tilde{H}_c(\omega)|^2 \qquad (4.6.4)$$

where $\tilde{H}_c(\omega)$ is again the low frequency, shifted version of $H_c(\omega)$, as given in (2.6.6). This spectrum is shown in Figure 4.18c. We see that the two-sided low frequency spectrum $\tilde{S}_n(\omega)$ is simply a low frequency version of the bandpass noise spectrum about ω_c. That is, $\tilde{S}_n(\omega)$ is obtained by simply shifting the one-sided version of the spectrum $S_n(\omega)$ in (4.6.2) to the origin. The phase angle ψ is purely arbitrary and can be selected to be any desired (nonrandom) value without altering the statistical property of $n(t)$.

The representation in (4.6.3) is called a *quadrature expansion* of the bandpass noise $n(t)$, and $n_c(t)$ and $n_c(t)$ are its *quadrature components*. Note that the components themselves each have the same power as the noise $n(t)$ itself. That is,

$$\text{Power of } n_c(t) = \text{power of } n_s(t)$$

$$= \frac{1}{2\pi} \int_{-\infty}^{\infty} \tilde{S}_n(\omega) \, d\omega$$

$$= \frac{1}{2\pi} \int_{-\infty}^{\infty} S_n(\omega) \, d\omega$$

$$= \text{power of } n(t) \tag{4.6.5}$$

We emphasize that each quadrature component does not have half the total power, as one might expect. This becomes a basic point in later analyses.

In summary, then, we see that receiver noise can be modeled in either of two ways. We can consider the noise as a Gaussian white noise process of spectral level N_0 appearing at the input to the receiver front end, as in Figure 4.19a, or we can consider the noise as a bandpass noise process, having the spectrum $S_n(\omega)$ in (4.6.2), appearing at the front end output, as in Figure 4.19b. Note that the latter model is exactly equivalent to adding the white noise at the front end input and taking into account the filtering within. Of course, both models produce exactly the CNR in (4.4.16) at the front end output.

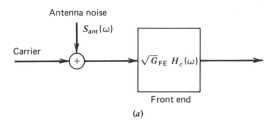

(a)

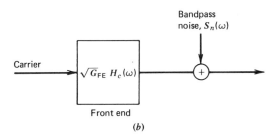

(b)

Figure 4.19. Receiver noise models: (a) with antenna noise, (b) with equivalent bandpass front end noise.

4.7   CARRIER PLUS NOISE WAVEFORMS

We wish to examine in detail the front end output waveform. Recall that this output is composed of the desired carrier waveform $c(t)$ superimposed with the bandpass noise process $n(t)$ of the receiver front end. We denote this combined waveform as

$$x_c(t) = c(t) + n(t) \qquad (4.7.1)$$

The carrier waveforms we have been discussing are of the general form

$$c(t) = a(t) \cos [\omega_c t + \theta(t) + \psi] \qquad (4.7.2)$$

where we remember that $a(t)$ or $\theta(t)$ may be constant, depending on the type of modulation. The noise $n(t)$ is the zero mean, narrowband, Gaussian process discussed in Section 4.6 and having the quadrature expansion in (4.6.3). The combined waveform in (4.7.1) is then

$$x_c(t) = a(t) \cos [\omega_c t + \theta(t) + \psi] + n_c(t) \cos (\omega_c t + \psi) - n_s(t) \sin (\omega_c t + \psi)$$
$$(4.7.3)$$

where we have taken the arbitrary phase angle of the noise process to be identical to that of the carrier. If we substitute the identities

$$\cos (\omega_c t + \psi) = \cos [\theta(t)] \cos [\omega_c t + \theta(t) + \psi] + \sin [\theta(t)] \sin [\omega_c t + \theta(t) + \psi]$$
$$\sin (\omega_c t + \psi) = \cos [\theta(t)] \sin [\omega_c t + \theta(t) + \psi] - \sin [\theta(t)] \cos [\omega_c t + \theta(t) + \psi]$$
$$(4.7.4)$$

then the total signal can be combined instead as

$$x_c(t) = [A + \hat{n}_c(t)] \cos [\omega_c t + \theta(t) + \psi] - \hat{n}_s(t) \sin [\omega_c t + \theta(t) + \psi]$$
$$(4.7.5)$$

where we have introduced

$$\hat{n}_c(t) \overset{\Delta}{=} n_c(t) \cos [\theta(t)] + n_s(t) \sin [\theta(t)]$$
$$\hat{n}_s(t) \overset{\Delta}{=} n_s(t) \cos [\theta(t)] - n_c(t) \sin [\theta(t)] \qquad (4.7.6)$$

The noise components $\hat{n}_c(t)$ and $\hat{n}_s(t)$ are modified forms of the quadrature components of the bandpass noise and involve the carrier phase modulation $\theta(t)$. Equation (4.7.5) can now be written as

$$x_c(t) = [a(t) + \hat{n}_c(t)] \cos(\omega_c t + \theta(t) + \psi) - \hat{n}_s(t) \sin(\omega_c t + \theta(t) + \psi)$$

$$= \text{Real}\,\{\alpha(t)e^{jv(t)}e^{j(\omega_c t + \theta(t) + \psi)}\}$$

$$= \alpha(t) \cos[\omega_c t + \theta(t) + \psi + v(t)] \tag{4.7.7}$$

where now

$$\alpha(t) = \{[a(t) + \hat{n}_c(t)]^2 + \hat{n}_s^2(t)\}^{1/2} \tag{4.7.8a}$$

$$v(t) = \tan^{-1}\left[\frac{\hat{n}_s(t)}{a(t) + \hat{n}_c(t)}\right] \tag{4.7.8b}$$

Here $\alpha(t)$ is called the *envelope* of the combined signal $x_c(t)$, and $v(t)$ is referred to as the *phase noise*. Thus the effect of adding bandpass receiver noise to the carrier in (4.7.2) is to change the amplitude $a(t)$ to the envelope process $\alpha(t)$ and to introduce the additive phase noise $v(t)$ to the carrier phase modulation. Roughly speaking, the receiver interference has added noise variations to the amplitude and phase of the original transmitted carrier. Note that the phase noise depends not only on the input noise but also on the carrier amplitude and phase.

The statistical properties of the modified noise components in (4.7.6) can be derived from those of the quadrature components in Section 4.6. First we note that if there is no phase modulation [$\theta(t) = 0$], as in the case of standard AM, then $\hat{n}_c(t) = n_c(t)$, $\hat{n}_s(t) = n_s(t)$, and the components are identical to the bandpass noise components. If $\theta(t) \neq 0$, then $\hat{n}_c(t)$, $\hat{n}_s(t)$ have zero mean value [since $n_c(t)$ and $n_s(t)$ have zero mean] and their autocorrelation functions are

$$R_{\hat{n}_c}(\tau) = \mathcal{E}[\hat{n}_c(t)\hat{n}_c(t + \tau)]$$

$$= R_{n_c}(\tau)\{\cos[\theta(t)] \cos[\theta(t + \tau)]\}$$

$$+ R_{n_s}(\tau)\{\sin[\theta(t)] \sin[\theta(t + \tau)]\} \tag{4.7.9}$$

where $R_{n_c}(\tau)$ and $R_{n_s}(\tau)$ are the autocorrelation functions of $n_c(t)$ and $n_s(t)$, respectively. However, these are identical, and (4.7.9) reduces to

$$R_{\hat{n}_c}(\tau) = R_{n_c}(\tau)\{\cos[\theta(t) - \theta(t + \tau)]\} \tag{4.7.10}$$

If $\theta(t)$ is a random process, the bracket must be further averaged over the joint statistics of $\theta(t)$ and $\theta(t + \tau)$. Similarly, the autocorrelation of $\hat{n}_s(t)$ and the cross-correlation of $\hat{n}_c(t)$ and $\hat{n}_s(t)$ are (Problem 4.29)

$$R_{\hat{n}_s}(\tau) = R_{n_s}(\tau)\{\cos[\theta(t) - \theta(t + \tau)]\}$$

$$R_{\hat{n}_c\hat{n}_s} = R_{n_c}(\tau)\{\sin[\theta(t) - \theta(t + \tau)]\} \tag{4.7.11}$$

Thus the modified components $\hat{n}_c(t)$ and $\hat{n}_s(t)$ have their autocorrelation as that of $n_c(t)$ and $n_s(t)$, modified by the effect of the carrier phase function $\theta(t)$. We note this autocorrelation is, in general, not stationary. However, in most applications we can simplify (4.7.10) and (4.7.11). If the carrier phase modulation does not change significantly over the correlation time of the $n_c(t)$ process, we may state

$$\theta(t) \approx \theta(t + \tau) \qquad \begin{array}{l} \text{over values of } \tau \text{ for which } R_{n_c}(\tau) \\ \text{is essentially nonzero} \end{array} \qquad (4.7.12)$$

Then we may argue that

$$R_{\hat{n}_c}(\tau) = R_{\hat{n}_s}(\tau) \approx R_{n_c}(\tau)$$

$$R_{\hat{n}_c \hat{n}_s}(\tau) \approx 0 \qquad (4.7.13)$$

That is, the noise processes in (4.7.6) become approximately stationary and uncorrelated with autocorrelation given by $R_{n_c}(\tau)$. The condition in (4.7.12) is basically equivalent to the statement that the bandwidth of the spectrum of the bandpass noise is much larger than that of the carrier phase modulation $\theta(t)$. This means that the front end bandwidth must exceed the phase modulation bandwidth. Since angle modulated systems are usually operated as wideband systems (i.e., given large phase or frequency deviations), this bandwidth condition is usually satisfied. When (4.7.13) is valid, the power spectral density of $\hat{n}_c(t)$ and $\hat{n}_s(t)$ are identical to those of $n_c(t)$ and $n_s(t)$. Thus, from (4.6.4),

$$S_{\hat{n}_c}(\omega) = S_{\hat{n}_s}(\omega) = \tilde{S}_n(\omega) \qquad (4.7.14)$$

where again $\tilde{S}_n(\omega)$ is the low frequency equivalent of the bandpass noise spectrum.

Although the autocorrelation and spectrum of the processes $\hat{n}_c(t)$ and $\hat{n}_s(t)$ can be determined in this way, we often need to determine their actual probability densities for later analysis. If the carrier phase process $\theta(t)$ is a deterministic time function, then (4.7.6) corresponds to a sum of Gaussian variables at each t, and these modified components are therefore also Gaussian. If, however, $\theta(t)$ is itself a random process, then their individual probability density at any t is no longer obvious. To determine this, we must investigate formally the transformation of the random variables $[n_c(t), n_s(t), \theta(t)]$ at any t over to the random variables $\hat{n}_c(t)$, $\hat{n}_s(t)$ at the same t, using (4.7.6). At a given t denote the random variable $n_c(t)$ by n_c, $\theta(t)$ by θ, and so on, and consider the conditional probability density $p(\hat{n}_c, \hat{n}_s|\theta)$. We write this as the formal transformation of densities, conditioned on θ (see Section B.2)

$$p(\hat{n}_c, \hat{n}_s|\theta) = \frac{1}{|J|} \, p(n_c, n_s|\theta)\Big|_{[n_c, n_s]} \tag{4.7.15}$$

where J is the Jacobian of the transformation in (4.7.6)

$$J = \det \begin{bmatrix} \cos\theta & \sin\theta \\ -\sin\theta & \cos\theta \end{bmatrix} = 1 \tag{4.7.16}$$

and $[n_c, n_s]$ represents the substitution for n_c and n_s in terms of $\hat{n}_c$ and $\hat{n}_s$ obtained by inverse solving (4.7.6). Therefore

$$p(\hat{n}_c, \hat{n}_s|\theta) = p(n_c, n_s|\theta)\Big|_{[n_c, n_s]} \tag{4.7.17}$$

Now since $n_c(t)$ and $n_s(t)$ are independent of the carrier phase process $\theta(t)$, the density on the right is jointly Gaussian in the variables n_c and n_s. After substituting $[n_c, n_s]$, (4.7.17) becomes

$$\begin{aligned} p(\hat{n}_c, \hat{n}_s|\theta) &= \frac{1}{2\pi R_{n_c}(0)} \exp\left[-\frac{(n_c^2 + n_s^2)}{2R_{n_c}(0)} \right]\Bigg|_{\substack{n_c = \hat{n}_c \cos\theta - \hat{n}_s \sin\theta \\ n_s = \hat{n}_s \cos\theta + \hat{n}_c \sin\theta}} \\ &= \frac{1}{2\pi R_{n_c}(0)} \exp\left[-\frac{(\hat{n}_c^2 + \hat{n}_s^2)}{2R_{n_c}(0)} \right] \end{aligned} \tag{4.7.18}$$

The resulting density is that of a pair of independent Gaussian random variables. This establishes that $\hat{n}_c(t)$ and $\hat{n}_s(t)$ are themselves jointly Gaussian at any t and independent of the probability density on $\theta(t)$. Therefore, each process in (4.7.6) is itself Gaussian, no matter what the density on $\theta(t)$, and the correlation functions of these processes are given in (4.7.10) and (4.7.11). Under the condition of (4.7.12), these processes therefore become independent Gaussian processes each with spectral density given in (4.7.14). As such, they are then statistically identical to the quadrature components $n_c(t)$ and $n_c(t)$ and cannot be distinguished from them in any way. This means that without any loss of rigor we can replace the modified components $\hat{n}_c(t)$ and $\hat{n}_s(t)$ by their quadrature counterparts in subsequent analysis, so long as (4.7.12) is a fairly accurate approximation.

4.8 FREQUENCY TRANSLATION AND IF PROCESSING

It is common in communication receivers to translate the front end waveform to a lower frequency range for further processing prior to demodulation. This lower frequency range is called the *intermediate frequency* (IF) of the receiver. The basic advantage of translating to a lower frequency (downconversion) is that electronic devices, such as filters, amplifiers, and demodulators, are easier to construct at the lower frequencies. For example, a carrier bandwidth of 100 kHz at a carrier frequency of 1 GHz would require a front end filter whose width is 0.01% of center frequency. Such a filter would place severe constraints on parameter values. However, if this waveform could be shifted to a carrier frequency of 10 MHz, only a 1% filter bandwidth would be required.

Frequency translation at the receiver is accomplished using mixers, similar to those at the transmitter in Section 3.2, except frequency downconversion is used. A downconverting mixer is shown in Figure 4.20. The local carrier is again used to mix (multiply) with the RF waveform to form sum and difference carrier frequencies, of which the difference is now selected by the mixer bandpass filter. By proper selection of the mixing carrier frequency, the difference frequency can be adjusted to any desired IF band. Since the RF carrier is received with the additive receiver noise, the combination of modulated carrier and RF noise waveforms is processed in the mixer.

Let us write the signal $x_c(t)$ in the same form as in (4.7.1) and write the local carrier signal as $A_l \cos(\omega_l t + \psi_l)$, where ψ_l represents any phase variation that may exist. We again assume that the mixer multiplier has a gain (or attenuation) G_m and contains a filter tuned to the difference frequency $(\omega_c - \omega_l)$ [or $(\omega_l - \omega_c)$, whichever is positive]. The multiplier performs the multiplication of $x_c(t)$ with the local carrier signal, while multiplying by the gain G_m, and the resulting signal is then filtered to provide the mixer output. The multiplier output is therefore

$$G_m x_c(t) A_l \cos(\omega_l t + \psi_l) = G_m[c(t) + n(t)][A_l \cos(\omega_l t + \psi_l)] \quad (4.8.1)$$

We now substitute from (4.7.2), expanding the product of cosine terms into sum and difference frequencies. Noting that the filter eliminates sum

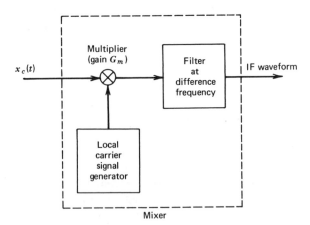

Figure 4.20. Frequency mixer.

frequency terms [assuming $(\omega_c + \omega_l) \gg |\omega_c - \omega_l|$] while passing only the difference frequency terms, the mixer output signal is then

$$x_{IF}(t) = K_m a(t) \cos[\omega_{cl}(t) + \theta(t) + \psi - \psi_l] + n_m(t) \qquad (4.8.2)$$

where again $K_m = G_m A_l/2$, $\omega_{cl} = |\omega_c - \omega_l|$, and $n_m(t)$ is the mixer output noise. The first term corresponds to the carrier mixing, in which the RF carrier is shifted to IF with the same amplitude and phase functions as discussed in Section 3.2. Note that if the local carrier phse ψ_l is constant, phase modulation on the carrier is not affected by the mixer.

The second term in (4.8.2) is due to the noise. Mixer output noise can be obtained by inserting the quadrature expansion of $n(t)$ in (4.8.1) and retaining the difference frequency terms, yielding

$$n_m(t) = K_m[n_c(t) \cos(\omega_{cl}t - \psi_l + \psi) - n_s(t) \sin(\omega_{cl}t - \psi_l + \psi)] \quad (4.8.3)$$

If ψ_l is a constant, then (4.8.3) is merely a shifted version of the front end noise; that is, $n_m(t)$ has a spectrum identical in shape to $S_n(\omega)$, except it is shifted to ω_{cl}. If ψ_l is a function of t (due to phase variations on the mixing carrier), then (4.8.3) must be expanded by using (4.7.4), resulting in

$$n_m(t) = K_m[\hat{n}_c(t) \cos(\omega_{cl}t + \psi) - \hat{n}_s(t) \sin(\omega_{cl}t + \psi)] \qquad (4.8.4)$$

where $\hat{n}_c(t)$ and $\hat{n}_s(t)$ are again the modified components in (4.7.6), only with the local carrier phase $\psi_l(t)$ involved instead of the phase modulation $\theta(t)$. If, however, $\psi_l(t)$ varies slowly with respect to the front end noise variations, the argument following (4.7.12) can be used to insert $n_c(t)$ and $n_s(t)$ in (4.8.4). In this case, $n_m(t)$ is again merely a shifted version of the front end noise. Thus we can conclude that as long as the local carrier phase variation is negligible relative to that of the noise variations, mixer output noise is simply a frequency shifted version of the bandpass noise appearing at its input, filtered by the IF bandpass filter. Since the latter is generally narrower in bandwidth than the RF filters, it determines the IF noise bandwidth. Hence the mixer output noise $n_m(t)$ has the spectrum determined from (4.6.2) as

$$S_{n_m}(\omega) = \tfrac{1}{2} N_0 G_1 K_m^2 |H_{IF}(\omega)|^2 \qquad (4.8.5)$$

where $H_{IF}(\omega)$ is the transform of the mixer output filter. It can also be shown that the mixer noise in (4.8.5) is also a Gaussian noise process for *any* random phase process $\psi_l(t)$, by the same development leading to (4.7.18) (see Problem 4.31).

We can compute the resulting CNR after IF mixing. From (4.8.2) we see that the carrier signal is scaled by the mixer gain K_m, while the noise is produced with the spectrum in (4.8.5). The IF CNR is therefore

$$\text{CNR}_{\text{IF}} = \frac{K_m^2 P_c}{K_m^2 N_0 G_1 B_{\text{IF}}}$$

$$= \frac{P_r}{N_0 B_{\text{IF}}}$$

$$= (\text{CNR})_{\text{RF}} \left(\frac{B_c}{B_{\text{IF}}} \right) \tag{4.8.6}$$

where $(\text{CNR})_{\text{RF}}$ is the RF input CNR in (4.4.16), and B_{IF} is the noise bandwidth of the IF bandpass filter. The mixing has identically scaled the carrier and noise powers and reduced the noise power by the narrower IF filtering. This is why the IF CNR is generally increased over that at the RF stage and why a receiver can often operate with extremely low front end CNR. For example, if the RF filter band is 200 MHz while the IF band is 10 MHz, a CNR_{IF} of 10 dB can be obtained with an RF CNR of -3 dB.

In actual system design, some consideration must be given to the selection of the IF frequency ω_{cl}. If further filtering is to be accomplished in the IF stage, we generally desire ω_{cl} to be about 1000 times larger than the bandwidth of the modulated carrier. This makes it easier to construct IF bandpass filters that sufficiently reject out-of-band noise. On the other hand, the IF stage must interface with subsequent signal processors, such as demodulators, which may be preselected to operate at specific frequencies, thereby removing some freedom in IF selection. In addition, consideration must be given to the actual effect of the local carrier frequency on the other elements of the receiver. Since multipler attenuation G_m may be significant, the amplitude of the local carrier A_l is generally quite large to produce an acceptable mixer gain K_m. This means a relatively high powered local carrier is being produced in the IF stage of the receiver, and its coupling into nearby elements may produce undesirable frequency interference. Of special concern is the possibility of the local carrier "leaking" back into the receiver front end. For this reason, IF is often selected such that the local frequency ω_l is not within the bandwidth of the front end filter. This requires $\omega_l < \omega_c - (2\pi B_c/2)$, or

$$\omega_{cl} > 2\pi \frac{B_c}{2} \tag{4.8.7}$$

Mixers can also be used to translate a received carrier at one RF frequency to another RF frequency, instead of an IF frequency. This, in fact, represents the basic operation of the communication satellite relay. Here an RF carrier is received from the ground, frequency translated at the satellite to another RF carrier frequency, power amplified, and retransmitted back to Earth. Such a system is the basic for an RF *transponder* discussed earlier in Section 1.5. Satellite transponding systems are discussed in more detail in Chapter 11. Conditions similar to those in (4.8.7) are necessary to prevent the retransmitted carrier frequencies from coupling back into the transponder front end.

4.9 AUTOMATIC GAIN CONTROL

The total power level of the RF waveform may vary significantly during reception, as a result of changes in received carrier power and noise levels. This alters the peak voltage values that must be handled by the electronics. For example, a power increase of 40 dB means that the electronics must process waveforms whose peak voltages may suddenly be 100 times larger.

To combat these power variations, we often insert some type of receiver automatic gain control (AGC) to limit the extent of the power variation. The objective of an AGC circuit is simply to reduce the receiver gain levels when the input power is high and conversely increase gain when the input power falls. This limits the peak voltage variation that occurs in subsequent processing stages.

AGC can be applied at either the RF or IF stages. The simplest and most common system [14, 15] is one that simply monitors (measures) the power and feeds back a control signal to increase or reduce amplifier gain levels. Such a system is shown in Figure 4.21. Power measurement is accomplished by a squaring circuit followed by a low pass filter. The filter time averages (integrates) the squared waveform, producing a voltage proportional to the waveform power over the past filter integration time. This instantaneously measured power is compared to a preselected bias power level, with the feedback voltage generated from the difference. The voltage is then used to adjust the gain of the control amplifier. Most control amplifiers are selected

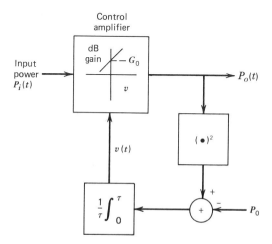

Figure 4.21. Automatic gain control (AGC) subsystem.

to have a decibel gain variation that is linear with respect to the input control voltage. Thus any power variation of the RF (or IF) waveform from the preselected bias power will automatically adjust the decibel gain of the control amplifier.

If we denote $P_i(t)$ as the amplifier time varying input power, and $P_o(t)$ as its output, we can relate them through the amplifier gain variation

$$P_o(t) = P_i(t)[G_0 10^{-c_1 v(t)/10}]$$
$$= G_0 P_i(t) e^{-0.23 c_1 v(t)} \tag{4.9.1}$$

where $v(t)$ is the control voltage, G_0 the amplifier static gain, and c_1 is a proportionality constant in decibels per volt. Note that the control amplifier has an effective instantaneous gain

$$\frac{P_o(t)}{P_i(t)} = \frac{G_0}{Z(t)} \tag{4.9.2}$$

with $Z(t)$ defined as

$$Z(t) = e^{0.23 c_1 v(t)} \tag{4.9.3}$$

The amplifier therefore compensates for power variations by dividing its static gain by the function $Z(t)$.

The control voltage $v(t)$ in Figure 4.21 is obtained by integrating the power difference, or equivalently

$$\frac{dv}{dt} = \frac{K}{\tau}[P_o(t) - P_0] \tag{4.9.4}$$

where K is the integration gain factor and τ is the integration time constant. This assumes that an accurate measurement of amplifier output power is made. The amplifier gain adjustment factor in (4.9.3) satisfies the equation

$$\frac{dZ}{dt} = e^{0.23 c_1 v(t)} \left[0.23 c_1 \frac{dv}{dt} \right]$$
$$= 0.23 c_1 Z(t) \left[\frac{K}{\tau}(P_o(t) - P_0) \right] \tag{4.9.5}$$

Substituting from (4.9.2) yields

$$\frac{dZ}{dt} + K_L Z(t) = K_L G_0 \frac{P_i(t)}{P_0} \tag{4.9.6}$$

with $K_L = 0.23 c_1 P_0 K / \tau$. This shows that $Z(t)$ is the output of a first order filter having time constant $\tau / 0.23 c_1 P_0 K$ and input $P_i(t)/P_0$. Thus the AGC operates by filtering the normalized input power and using the output to divide the amplifier gain. For the loop to be "fast enough" it is necessary that the filtering time constant be shorter than the expected time fluctuations in the input power. If this occurs, $Z(t) \approx G_0 P_i(t)/P_0$, and in (4.9.1), $P_o(t) = P_0$. That is, a fast AGC will hold the amplifier power output at the fixed power level P_0 in spite of input power variations.

A problem with basic AGC systems of this type is that the amplifier gain is controlled by the total input power $P_i(t)$. Since the input is generally composed of carrier plus noise waveform, the control gain may be determined by the noise alone if the noise power exceeds the carrier power. High noise levels could saturate the receiver by reducing the AGC gain so as to reduce the carrier power levels, thus prohibiting subsequent carrier processing. This so-called AGCing on noise is always of concern in receivers that operate at low CNR. Some methods for developing AGC control signals that depend only on the carrier power level are discussed in Chapter 10.

4.10 NONLINEAR PROCESSING AND CARRIER LIMITING

Often a communication receiver involves some type of nonlinearity during the processing of either the front end or the IF carrier signal. This nonlinearity may be undesired, such as an amplifier having a nonlinear gain or saturation effects, or may be intentionally placed in the system, such as a limiter or a rectifier. It is therefore important in system design to understand the effects of such nonlinearities on the waveforms in the receiver and to account for them properly in system analysis. To consider this, we present some general results concerning nonlinear analysis of carrier waveforms. We concentrate only on the "memoryless" type of nonlinear device, in which the present value of the device output depends only on the present value of its input and not on any of its past.

Assume that the nonlinearity of the system is described by the function $g(x)$. Thus if $x(t)$ is the time process at the nonlinearity input, the output is

$$y(t) = g[x(t)] \qquad (4.10.1)$$

as shown in Figure 4.22. A common analytical procedure [16, 17] is to expand $y(t)$ in terms of the transform of the function $g(x)$. That is, if we denote $G(\omega)$ as Fourier transform of $g(x)$, then $y(t)$ can be rewritten as

Figure 4.22. Nonlinear device.

$$y(t) = \frac{1}{2\pi} \int_{-\infty}^{\infty} G(\omega) e^{j\omega x(t)} \, d\omega \qquad (4.10.2)$$

We point out that since typical nonlinearities $g(x)$ may have different characteristics for $x > 0$ and $x < 0$, $G(\omega)$ is actually a two-sided transform that often must be computed separately for $x > 0$ and $x < 0$ (Problem 4.33). This means that the inverse transform indicated in (4.10.2) may require separate contour integrations for the cases where $x(t) > 0$ and $x(t) < 0$. This point is discussed in more detail in References 16 and 17.

The use of (4.10.2) for analyzing nonlinearities in terms of their transforms is particularly convenient for the type of carrier waveforms we have been considering. Let us represent the general noisy carrier in the receiver (at either the front end or IF) as

$$x_c(t) = \alpha(t) \cos [\omega_c t + \theta(t)] \qquad (4.10.3)$$

where $\theta(t)$ represents the combined phase variation due to the carrier modulation and any phase noise. We can then use the Jacobi–Anger identity (Section A.2)

$$e^{j\omega\alpha \cos \beta} = \sum_{i=0}^{\infty} \varepsilon_i I_i(j\omega\alpha) \cos (i\beta) \qquad (4.10.4)$$

where $I_i(z)$ is the modified Bessel function of the first kind and order i [18], $\varepsilon_0 = 1$, and $\varepsilon_i = 2$, $i \neq 0$. Equation (4.10.2) becomes

$$y(t) = \sum_{i=0}^{\infty} c_i [\alpha(t)] \cos [i\omega_c t + i\theta(t)] \qquad (4.10.5)$$

where

$$c_i(\alpha) = \frac{\varepsilon_i}{2\pi} \int_{-\infty}^{\infty} G(\omega) I_i[j\omega\alpha] \, d\omega \qquad (4.10.6)$$

Equation (4.10.5) is a general expression for the output of any nonlinearity in (4.10.1) when the input is given by (4.10.3). Note that the output always appears as the sum of harmonically related, modulated carriers with amplitude variations depending on the type of nonlinearity transform $G(\omega)$. We therefore see that harmonic generation is inherent in nonlinear processing. The effect of these harmonics must be carefully examined for their interference on later system components. Some common types of receiver nonlinearity and the resulting forms for $c_i(\alpha)$, evaluated from (4.10.6), are summarized in Table 4.4. Note that in certain cases only a finite number of harmonics appear, whereas in other cases only even (or odd) harmonics are generated. Clearly, the frequency multipliers described in the preceding section can be easily constructed from such nonlinear elements.

TABLE 4.4   Common Nonlinearities and Parameters

Device	Nonlinear Function $g(x)$	$c_i(\alpha)$
Half-wave νth power law		$\alpha^\nu C(\nu, i)$
Full wave (even) νth power law		$2\alpha^\nu C(\nu, i),$ i even $0,$ i odd
Full wave (odd) νth power law		$2\alpha^\nu C(\nu, i),$ i odd $0,$ i even
Half-wave linear bandpass		$\dfrac{\alpha}{\pi}, i = 1$
Full wave square law bandpass		$\dfrac{\alpha^2}{2}, i = 2$
Fullwave odd, $\nu = 0$ bandpass		$\left(\dfrac{4}{\pi}\right), i = 1$

$$C(\nu, m) = \frac{\varepsilon_m (G_{am}[\nu + 1])}{2^{\nu+1}\left(G_{am}\left[1 - \dfrac{m - \nu}{2}\right]\right)\left(G_{am}\left[1 + \dfrac{m + \nu}{2}\right]\right)}, \qquad G_{am}[z] = \int_0^\infty e^{-t} t^{z-1}\, dt$$

Since nonlinearities tend to produce harmonics, the nonlinear device is generally followed by a bandpass filter tuned to the carrier frequency. The combination of the nonlinear element followed by the bandpass filter is often called a *bandpass nonlinearity*. For such devices the only harmonic component appearing at the output is that corresponding to the index $i = 1$ in (4.10.5). For example, the output of the bandpass full wave odd νth power law devices in Table 4.4 would be

$$y(t) = C(\nu, 1)\alpha^{\nu}(t) \cos [\omega_c t + \theta(t)] \qquad (4.10.7)$$

The output is therefore a carrier with the same phase variation as the input, but with a modified amplitude variation.

One type of nonlinear operation that often occurs in receiver models is limiting. Limiting can be produced unintentionally by saturation effects in amplifiers, but in many systems limiters are purposely inserted in the IF channel to improve performance. The limiting prevents high peak voltages caused by noise spikes from exceeding allowable dynamical ranges in the receiver processing. In addition, limiters are useful for maintaining power and amplitude control during the subsequent demodulation operation.

An ideal *hard limiter* is a nonlinear device whose input–output function $g(x)$ is shown in Figure 4.23. Mathematically, the hard limiter is simply a $\nu = 0$, full wave, odd device listed in Table 4.4. A *bandpass hard limiter* (BPL) is a hard limiter followed by a bandpass filter tuned to the input carrier frequency, the latter allowing only the frequency components within its bandwidth to pass. If the input to a bandpass limiter is a noisy carrier waveform $c(t) = \alpha(t) \cos [\omega_c t + \theta(t)]$, the bandpass limiter output is given by (4.10.7), with $\nu = 0$. Since $C(0, 1) = 4V_L/\pi$, the BPL output is

$$y(t) = \left(\frac{4V_L}{\pi}\right) \cos [\omega_c t + \theta(t)] \qquad (4.10.8)$$

Note that the output produces a RF carrier with the same phase and frequency modulation as the input but with a constant amplitude level. In

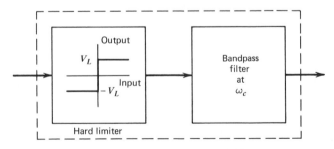

Figure 4.23. Bandpass hard limiter.

effect, the BPL eliminates amplitude modulation, while preserving angle modulation. Obviously, limiters should not be used when AM carriers are involved. We also see that the power at the output of a bandpass limiter [power in $y(t)$ in (4.10.8)] is always given by

$$P_L = \frac{1}{2}\left(\frac{4V_L}{\pi}\right)^2 = \frac{8V_L^2}{\pi^2} \tag{4.10.9}$$

for any input process. Thus by proper adjustment of the limiting level V_L, the amplitude levels and power output of the BPL can be accurately adjusted. This allows control of the peak amplitude variations of the resulting carrier waveform during the subsequent receiver processing.

When the input to the BPL is composed of the sum of an angle modulated carrier plus additive receiver noise, it is often desirable to known the extent by which the carrier waveform has been preserved in passing through the BPL. This is difficult to determine from (4.10.8), since the limiter output noise is incorporated entirely into the phase noise of the carrier. However, it can be shown [16, 17] that the carrier power in the first harmonic term of the limiter output is given by

$$P_{co} = \frac{2V_L^2}{\pi} \rho_i e^{-\rho_i}\left[I_0\left(\frac{\rho_i}{2}\right) + I_1\left(\frac{\rho_i}{2}\right)\right]^2 \tag{4.10.10}$$

where ρ_i is the limiter input CNR,

$$\rho_i \overset{\Delta}{=} \frac{A^2/2}{N_0 B_c} \tag{4.10.11}$$

Since the total power at the output of the BPL is P_L in (4.10.9), it follows that the noise and cross-product power terms must constitute the difference. Hence the BPL output noise power is

$$P_{no} = P_L - (P_{co}) \tag{4.10.12}$$

The resulting bandpass limiter output CNR is then

$$\text{CNR}_{\text{BPL}} = \frac{P_{co}}{P_{no}}$$

$$= \frac{P_{co}/P_L}{1 - (P_{co}/P_L)} \tag{4.10.13}$$

where

$$\frac{P_{co}}{P_L} = \left(\frac{\pi}{4}\right) \rho_i e^{-\rho_i}\left[I_0\left(\frac{\rho_i}{2}\right) + I_1\left(\frac{\rho_i}{2}\right)\right]^2 \tag{4.10.14}$$

A plot of the normalized ratio

$$\Gamma \overset{\Delta}{=} \frac{CNR_{BPL}}{\rho_i} \tag{4.10.15}$$

is shown in Figure 4.24 as a function of ρ_i. The result shows the way in which the CNR is altered in passing through the BPL. Note that CNR_{BPL} has the asymptotic behavior

$$CNR_{BPL} \cong \begin{cases} \left(\dfrac{\pi}{4}\right)\rho_i, & \text{for} \quad \rho_i \ll 1 \\ 2\rho_i, & \text{for} \quad \rho_i \gg 1 \end{cases} \tag{4.10.16}$$

Thus the effect of the BPL is to cause an increase in the CNR if the ratio is large, but to cause a slight degradation (by about 2 dB) if the input CNR is low. We point out that CNR_{BPL} is the output ratio of the carrier power to the total output interference power in the limiter bandpass filter bandwidth.

Since the CNR is altered in passing through the limiter, we can interpret limiting as effectively modifying the input carrier and noise powers at the output. Hence we denote α_s^2 and α_n^2 as the limiter carrier and noise *suppression* factors, defined by

$$P_{co} = \alpha_s^2 P_c$$
$$P_{no} = \alpha_n^2 (N_0 B_c) \tag{4.10.17}$$

Since $CNR_{BPL} = P_{co}/P_{no}$, it follows that

$$CNR_{BPL} = \frac{\alpha_s^2 P_c}{\alpha_n^2 (N_0 B_c)} = \left(\frac{\alpha_s^2}{\alpha_n^2}\right)\rho_i \tag{4.10.18}$$

This therefore defines the suppression factor ratio as

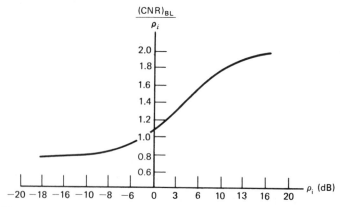

Figure 4.24. Bandpass limiter CNR suppression ratio (ρ_i = input CNR).

$$\frac{\alpha_s^2}{\alpha_n^2} = \Gamma \qquad (4.10.19)$$

Thus, as far as CNR is concerned, the BPL alters the carrier and noise amplitudes by the suppression factors in (4.10.17), where the ratio of these factors is given by the parameter Γ in Figure 4.24. Note that the suppression parameters themselves depend on ρ_i and the manner in which the carrier and noise amplitudes are effectively modified by the BPL depends on their relative strength. We carefully emphasize that (4.10.17) is a model for the BPL output only as far as CNR is concerned. Note that it implies that the output noise is merely an amplitude adjusted version of the input noise, when in fact the limiter output interference is composed of both noise and carrier cross products. In "linearizing" the BPL in this way, we adjust the carrier and interference ratio so that the output CNR is correct, but we must use care in subsequent processing analysis when we model the output in this way.

To determine the individual suppression values, we use (4.10.9) and (4.10.17) to write

$$\alpha_s^2 P_c + \alpha_n^2 (N_0 B_c) = P_L \qquad (4.10.20)$$

Substituting from (4.10.19) and solving for the limiter output carrier power $\alpha_s^2 P_c$ defines the suppression factor as

$$\alpha_s^2 P_c = P_L \left[\frac{\text{CNR}_{\text{BPL}}}{1 + \text{CNR}_{\text{BPL}}} \right] \qquad (4.10.21a)$$

Similarly, we obtain the noise suppression factor with

$$\alpha_n^2 (N_0 B_c) = P_L \left[\frac{1}{1 + \text{CNR}_{\text{BPL}}} \right] \qquad (4.10.21b)$$

Thus the α_s and α_n suppression factors are such that they divide the available bandpass limiter output power P_L in accordance with the ratios in (4.10.21). As CNR_{BPL} increases, a greater portion of the limiter output power is associated with the carrier.

REFERENCES

1. Angelakos, D. and Everhart, T. *Microwave Communications*, McGraw-Hill, New York, 1968.
2. Stephenson, R.G. "External Noise," in *Space Communications*, A. Balikrishnan, ed., McGraw-Hill, New York, 1970.
3. Ippolito, L. *Radiowave Propagadion in Satellite Communications*, Van Nostrand-Reinhold, New York, 1986.

4. Freeman, R.L. *Telecommunication Transmission Handbook*, Wiley, New York, 1975.

5. Ko, H. "The Distribution of Cosmic Radio Background Radiation," *Proc. IRE*, vol. 46, no. 1, January 1958, pp. 218–225.

6. Cottony, H. and Johler, J. "Cosmic Radio Noise Intensity in the UHF Band," *Proc. IRE*, vol. 40, 1946, pp. 1487–1489.

7. Smith, A. "Extraterrestrial Noise in Space Communications," *Proc. IRE*, vol. 48, no. 4, April 1960, pp. 594–603.

8. Mayer, C. "Thermal Radio Radiation from the Moon and Planets," *IEEE Trans. Antennas Propag.*, vol. 12, no. 7, December 1964, pp. 902–913.

9. Pratt, W. *Laser Communication Systems*, Wiley, New York, 1969, Chap. 6.

10. Davenport, W. and Root, W. *Theory of Random Signals and Noise*, McGraw-Hill, New York, 1958.

11. Pierce, J. "Physical Sources of Noise," *Proc. IRE*, vol. 44, May 1956, pp. 601–608.

12. Bennett, W. *Electrical Noise*, McGraw-Hill, New York, 1960.

13. Van der Ziel, A. *Noise*, Prentice-Hall, Englewood Cliffs, N.J., 1954.

14. Ohlson, J. "Exact Dynamics of AGC," *IEEE Trans. Comm.*, vol. COM-22, no. 1, January 1974, pp. 72–78.

15. Gagliardi, R. "Coupled AGC–Costas Loops with AM/PM Conversion," *IEEE Trans. Comm.*, vol. COM-28, January 1980, pp. 122–127.

16. Davenport and Root, loc. cit, Chap. 12.

17. Thomas, J. *Statistical Communication Theory*, Wiley, New York, 1969, Chap. 6.

18. Abramowitz, A. and Stegun, I. *Handbook of Mathematical Functions*, National Bureau of Standards, Washington, D.C., 1965, Chap. 9.

PROBLEMS

4.1   (4.1)   A 2 m circular antenna operates at 1 GHz with an aperture loss factor of 0.5. (a) What is the maximum area it will present to an arriving plane wave? (b) Repeat if the carrier frequency was 100 MHz.

4.2   (4.2)   Suppose the transmitted electromagnetic field is not an infinite plane wave but instead has a finite field cross-sectional area as it propagates. After propagating a distance D its wavefront has area $\mathcal{A}_f$. Assume it is transmitted with a given EIRP to a receiver of area $\mathcal{A}$ at D. Show that the receiver power collected in free space is

$$\left(\frac{\text{EIRP}}{4\pi D^2}\right)\mathcal{A}, \quad \text{if} \quad \mathcal{A} \le \mathcal{A}_f$$

$$\left(\frac{\text{EIRP}}{4\pi D^2}\right)\mathcal{A}_f, \quad \text{if} \quad \mathcal{A}_f \le \mathcal{A}$$

4.3 (4.2) A 100 W transmitter amplifier feeds a 1 GHz carrier into matched antenna terminals. The antenna has a 20 dB gain in the direction of the receiver. The transmission channel is a 400 km link, of which half of the path involves atmospheric propagation with a 0.01 dB/km loss. The receiving antenna is 1 m with 0.5 aperture loss factor and no coupling loss. How much power in watts will be collected at the receiver terminals?

4.4 (4.2) A transmitter radiates 10 w at carrier frequency 1.5 GHz with the following gain pattern:

$$g(\phi_l, \phi_z) = (100)e^{-(\phi_l^2 + \phi_z^2)/2(15°)^2}$$

A receiver is located 30° off antenna boresight, a distance 100 km away, and has a 0.1 rad receiving beamwidth aimed at the transmitter. Neglect aperture, coupling, and atmospheric losses, and determine the carrier power at the receiving antenna.

4.5 (4.2) Two parabolic antennas are pointed as shown in Figure P4.5. If the diameter of each is increased (separately), will the received power increase or decrease? Explain.

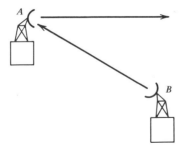

Figure P4.5.

4.6 (4.2) Given the system shown in Figure P4.6, answer true or false and explain or show why: (a) for maximum received carrier power, doubling the carrier frequency provides the same advantage as doubling either antenna diameter, (b) repeat part a, with the constraint that the transmit beamwidth is fixed at ϕ_{b_1}, (c) if it is a two-way link (transmit and receive at both ends) with same P_T and a beamwidth constraint on Trans. 1 it is better to use higher frequencies with the 2-1 link.

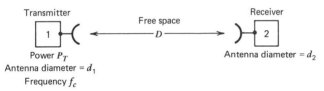

Figure P4.6.

4.7   (4.2)   A source couples 100 W into a 100 km cable having 2.0 dB/km loss. Ten amplifiers are to be inserted into the cable for carrier amplification. How much gain should each have in order to achieve 2 W of carrier power at the output? Neglect output coupling losses.

4.8   (4.3)   An Earth-based antenna is pointed at the moon with gain function

$$g(\phi_l, \phi_z) = \left(\frac{2}{\Phi_b^2}\right) e^{-(\phi_l^2 + \phi_z^2)/2\Phi_b^2}$$

where $2\Phi_b$ is its beamwidth. Assume that the temperature of the moon is 200 K over its surface area and the surrounding sky is 50°K. Neglect other background sources. Use the approximation $d\Omega = d\phi_l \, d\phi_z$. (a) Determine the effective background temperature of the antenna when $2\Phi_b$ is set equal to the beamwidth subtended by the moon. (b) Sketch the behavior of the effective temperature as a function of Φ_b.

4.9   (4.3)   An antenna observes a constant temperature background sky. A student claims "If we increase the antenna area, its gain increases and we receive more background noise." Is this assumption correct? Explain.

4.10   (4.3)   A passive planet occupies a 1° field of view with temperature $T_0°$ K. A "black hole" is a discrete source (delta function in space), and assume that it has the same temperature. Explain qualitatively what will happen in each case if we point an antenna directly at the source and begin to increase the antenna area.

4.11   (4.3)   Often a source is described by its *immittance function* $\mathcal{N}$, which is the product of its radiance and subtended field of view, $\mathcal{N} = \mathcal{H}\Omega_s$. Assume $\mathcal{N}$ is a constant and the antenna receiving area is $\mathcal{A}$ over its field of view, Ω_{fv}. Show that

$$P_b = \begin{cases} B\mathcal{N}\mathcal{A} & \text{if} \quad \Omega_s \le \Omega_{\mathrm{fv}} \\ B\mathcal{N}\mathcal{A}\,\dfrac{\Omega_{\mathrm{fv}}}{\Omega_s} & \text{if} \quad \Omega_{\mathrm{fv}} < \Omega_s \end{cases}$$

4.12   (4.4)   (a) Show that a cascade of three devices, having gain and noise figure G_i, F_i, respectively, will have an overall noise figure of

$$F = F_1 + \frac{F_2 - 1}{G_1} + \frac{F_3 - 1}{G_1 G_2}$$

(b) Generalize the result to N stages.

4.13 (4.4) Given the parallel combination of two devices with noise figures F_1, F_2 and power gains G_1, G_2, determine the overall noise figure. Assume a one-half power split at the input.

4.14 (4.4) Determine the noise figure of the resistor network shown in Figure P4.14. Assume $R_s = R_1 + R_2$. (*Hint:* Insert equivalent noise sources and compute total output power.)

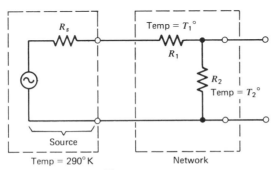

Figure P4.14.

4.15 (4.4) Given two lossy coupling lines in cascade, with power gain L_1 and L_2, and temperature T_1° and T_2°, respectively. Determine the overall noise figure and equivalent input temperature of the cascade. Assume matched impedances.

4.16 (4.4) Determine the noise figure of the system shown in Figure P4.16. Assume a one-half power split at the input.

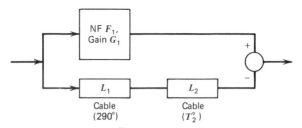

Figure P4.16.

4.17 (4.4) Determine the noise figure of the system shown in Figure P4.17.

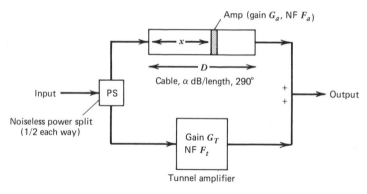

Figure P4.17.

4.18 (4.4)   Noise figure of a device can be measured in the following way. A resistor at 290°K is connected at the input and the output noise power is measured. The resistor is heated until the output noise power exactly doubles, and the resistor temperature T_2° is noted. Show that

$$F = \frac{T_2^\circ - 290^\circ}{290^\circ}$$

4.19 (4.4)   The front end of an RF receiver is shown in Figure P4.19 where all elements are perfectly matched. How much power gain must the tunnel diode amplifer have to produce a front end noise figure of 10 dB?

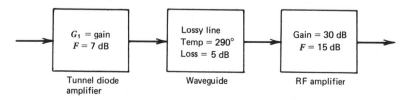

Figure P4.19.

4.20 (4.4)   Find the necessary G_{amp} for the system in Figure P4.20 to obtain a NF of F_0.

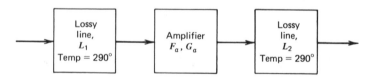

Figure P4.20.

4.21 (4.4) An antenna feeds a 20-ft external cable (0.5 dB loss/ft) that couples into an amplifier (noise figure = 3 dB). The antenna background temperature is $4 \times 290° = 1160°$. What is the maximum temperature of the cable in order to ensure the amplifier output has no more than four times the noise due to the background? Assume matched loading.

4.22 (4.5) (a) A deep space link is to transmit an RF carrier with 10 kHz bandwidth at 5 GHz from Earth to a spacecraft near Venus over a free space channel (10^8 m). A 200 ft transmitting antenna at the ground terminal is available. The spacecraft employs an isotropic antenna coupled to its front end with a line having 3 dB loss and operates at room temperature. The front end amplifier noise figure is 10 dB. The background sky presents an effective temperature of 10 dB below room temperature. How much transmitter power must be available to ensure transmission with a 10 dB SNR at the spacecraft in the carrier bandwidth? (b) For downlink transmission (spacecraft to Earth), a 10 W transmitter is available. The ground receiver can be maintained at 7 dB below room temperature with a 2 dB amplifier noise figure and a 3 dB coupling loss. The same sky background as in part a is also present. The spacecraft antenna can be aligned to provide some directivity over isotropic transmission. Show a plot of the available RF bandwidth that can be theoretically provided at a 10 dB RF SNR in the downlink, as a function of the spacecraft antenna gain.

4.23 (4.5) The CNR in a free space antenna communication link is being discussed. (a) Student 1 claims: "If we increase carrier frequency, we increase the gain of the receiving antenna, thereby increasing ($g/T°$), and we increase CNR." Is this correct? Explain. (b) Student 2 claims: "Increasing frequency also increases transmitting antenna gain, which will also increase CNR." Is this correct? Explain. (c) Student 3 claims: "If we fix the transmitter beamwidth, then the CNR is the same no matter what frequency we use." Is this correct? Explain.

4.24 (4.5) A low level, noiseless power source of -103 dBW feeds into a 100 km cable at 290°K. The cable has a loss of 0.2 dB/km. An amplifier of gain 70 dB at noise figure of 4 dB is to be inserted into the cable. An output SNR of 30 dB is desired in a 1 MHz bandwidth. How far down the cable should the amplifier be inserted (or does it make any difference)? Neglect coupling losses.

4.25 (4.5) A long cable transmission line is composed of M sections, each section having a line of loss L, followed by an internal amplifier. The gain of the amplifier is selected so that it just cancels the line loss. The amplifier has a noise figure of F_a. The line is fed by a power of P_1 W from a source of temperature 290°K. The cable is also at 290°K. (a) Determine the SNR of output if there was just one section. (b)

Determine the output SNR for M sections. (c) Determine the output SNR if no amplifiers were used. (d) Let $F_a = 10$ dB, $M = 10$, and $L = 3$ dB, and comment on whether long lines should have noisy amplifiers inserted.

4.26 (4.6) A random process has the two-sided spectrum shown in Figure P4.26a. A quadrature expansion is made at 1 GHz. (a) Sketch the spectrum of the quadrature components. (b) Are they uncorrelated processes? (c) If the process is passed through the system shown in Figure P4.26b, what is the output power?

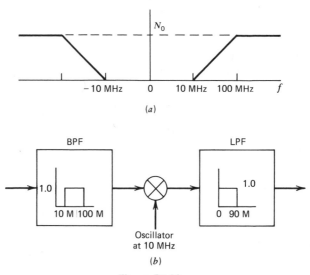

(a)

(b)

Figure P4.26.

4.27 (4.6) Given the quadrative noise expansion in (4.6.3). Assume that $n_c(t)$ and $n_s(t)$ are uncorrelated and each has the same spectral density $\tilde{S}_n(\omega)$. Determine the correlation function of $n(t)$ and compute its power spectrum. (*Hint:* See Section B.6.)

4.28 (4.6) Show that if white noise (level $N_0/2$) is to be represented as in (4.6.3), then $n_c(t)$ and $n_s(t)$ will not be uncorrelated and each will have spectrum N_0, $|\omega| \le \omega_c$ and $N_0/2$, $|\omega| > \omega_c$.

4.29 (4.7) Derive (4.7.10) and (4.7.11) from (4.7.6), for the modified noise components.

4.30 (4.7) Verify the Jacobian in (4.7.16) for the transformation defined in (4.7.6).

4.31 (4.8) Prove that the mixer noise $n_m(t)$ in (4.8.4) is always Gaussian, even if ψ_l is random. [*Hint:* Define an auxiliary process $n'_m(t) =$

$n_s(t) \cos \psi_l - n_c(t) \sin \psi_l$ and determine the joint density of n_m and n'_m.]

4.32 (4.9)   A carrier occupies a 1 MHz RF bandwidth at 1 GHz. What local oscillator frequency is required to generate an IF with a 1% bandwidth?

4.33 (4.10)   Let $G_+(\omega)$ be the Fourier transform of $g(x)$ over $x > 0$. Show that for an even nonlinear device ($g(-x) = g(x)$) we have $G(\omega) = G_+(\omega) + G_+(-\omega)$, whereas for an odd device ($g(-x) = -g(x)$) we have $G(\omega) = G_+(\omega) - G_+(-\omega)$.

4.34 (4.10)   A receiver uses an RF band at three times the IF frequency band. There is the possibility of IF signals leaking back into the RF band. Explain the events, in terms of RF interference, that will occur if the IF contains an amplifier that is (a) a full wave odd device, (b) a full wave even device, (c) a squaring device. (*Hint:* Refer to Table 4.4.)

4.35 (4.10)   The carrier-plus-noise waveform into a hard-limiter amplifier has CNR = 3 dB. The amplifier output power is 10 W. Estimate how much power is in the output carrier and in the output noise at the amplifier output. Repeat for CNR = 10 dB and −3 dB.

5

CARRIER
DEMODULATION

Following RF and IF processing, the communication receiver must demodulate the carrier to recover the baseband information signal. In this chapter we consider this demodulation operation when the latter is performed in the presence of receiver noise. We assume that all front end processing of the carrier waveform has been accounted for in the description of the carrier and the interfering IF noise is the narrowband Gaussian process discussed in Section 4.7. We concentrate only on the cases where the carrier may be of the AM, FM, or PM type. In each case we are primarily interested in the properties of the demodulated signal and its relation to the corresponding transmitter baseband signal. An outgrowth of our study is an attempt to model the entire carrier subsystem from transmitter baseband to demodulator output. A result of this type allows us to concentrate separately on baseband and carrier subsystem design.

5.1 THE DEMODULATING SUBSYSTEM

The demodulation portion of the carrier subsystem at the receiver is shown in Figure 5.1. The antenna signal is processed in the RF and IF stages, as discussed. The output of the IF filter stage serves as the input to the receiver carrier demodulator. This input is the sum of the IF carrier and the IF noise waveform. We again write this combined waveform as

$$x_{\mathrm{IF}}(t) = c(t) + n(t) \tag{5.1.1}$$

where the carrier and noise are referred to the demodulator input. The carrier $c(t)$ is at the IF frequency ω_{IF}, and the narrowband IF noise $n(t)$ again has the quadrature expansion

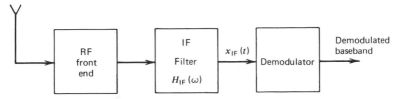

Figure 5.1. The receiver demodulating subsystem.

$$n(t) = n_c(t) \cos (\omega_{IF} t + \psi) - n_s(t) \sin (\omega_{IF} t + \psi) \qquad (5.1.2)$$

relative to this frequency. The quadrature components $n_c(t)$ and $n_s(t)$ are independent Gaussian processes, each having spectral density

$$S_{n_c}(\omega) = S_{n_s}(\omega) = \tilde{S}_n(\omega)$$

$$= N_0 |\tilde{H}_{IF}(\omega)|^2 \qquad (5.1.3)$$

where

$$N_0 = \kappa T^{\circ}_{eq} G_1 \qquad (5.1.4)$$

and $\tilde{H}_{IF}(\omega)$ is the low frequency equivalent of the IF bandpass filter function $H_{IF}(\omega)$. The gain G_1 accounts for the total receiver gain from antenna to demodulator input. Since the power of the receiver carrier signal at the antenna output terminals is P_r, the power in the carrier at the demodulator input is

$$P_c = G_1 P_r \qquad (5.1.5)$$

where P_r was computed in Section 4.2. We are now interested in determining the performance of various demodulating systems that operate on the IF waveform in (5.1.1). The structure and performance of these demodulators depend on the type of modulation involved.

5.2 AMPLITUDE DEMODULATION

We first consider demodulation of an AM carrier. We write the AM carrier at the demodulator in the standard form

$$c(t) = C[1 + m(t)] \cos (\omega_{IF} t + \psi) \qquad (5.2.1)$$

where $m(t)$ is the baseband modulation, ψ is the arbitrary carrier phase shift, and the amplitude coefficient C satisfies (5.1.5); that is,

$$P_c = \frac{C^2}{2}(1 + P_m) = G_1 P_r \qquad (5.2.2)$$

where P_m is the power of the modulation $m(t)$. We assume all filtering effects of the RF and IF stages have been incorporated into the description of $m(t)$. The input to the demodulator in Figure 5.1 is then the sum of the carrier in (5.2.1) and the noise process in (5.1.2). This combined waveform can be expanded as in (4.7.7),

$$x_{\mathrm{IF}}(t) = \alpha(t) \cos [\omega_{\mathrm{IF}} t + \psi + v(t)] \qquad (5.2.3)$$

where

$$\alpha(t) = (\{C[1 + m(t)] + n_c(t)\}^2 + n_s^2(t))^{1/2} \qquad (5.2.4)$$

and $v(t)$ is the phase noise. Let us first consider an AM demodulator that consists of an ideal envelope detector followed by a low pass filter $H_m(\omega)$, as shown in Figure 5.2. This filter is selected to have a bandwidth large enough to pass the modulation $m(t)$ without distortion, and therefore has a bandwidth of approximately B_m Hz, the latter the bandwidth of $m(t)$. The ideal envelope detector is considered to be a device that produces an output voltage directly proportional to the envelope of the input. In practice, envelope detectors are constructed from rectifier circuits with long time constants [1–4], so as to follow the envelope function but be immune to the high frequency carrier variations. When the IF signal $x_{\mathrm{IF}}(t)$ in (5.2.3) is the input, the envelope detector yields the output signal

$$\begin{aligned} x_1(t) &= K_d[\text{envelope of } x_{\mathrm{IF}}(t)] \\ &= K_d \alpha(t) \\ &= K_d(\{C[1 + m(t)] + n_c(t)\}^2 + n_s^2(t))^{1/2} \qquad (5.2.5) \end{aligned}$$

where K_d is the proportionality (gain) constant of the detector. Thus the

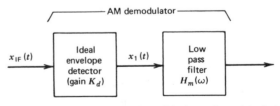

Figure 5.2. The AM demodulator (ideal envelope detector).

envelope detector produces an output that contains the desired baseband $m(t)$ in combination with the noise components. In particular, we note that it does not yield the baseband as a separate, distinct output unless the noise processes are zero. However, if we can assume that[†]

$$C[1 + m(t)] \gg |n_c(t)|, |n_s(t)| \qquad (5.2.6)$$

we can approximate

$$
\begin{aligned}
x_1(t) &\cong K_d C[1 + m(t)] + K_d n_c(t) \\
&= K_d C + K_d C m(t) + K_d n_c(t) \qquad (5.2.7)
\end{aligned}
$$

The envelope detector output now appears as the sum of the baseband signal $K_d C m(t)$ plus an additive noise $K_d n_c(t)$. (The constant $K_d C$ can be theoretically removed by subtraction or ac coupling.) The output filter $H_{\hat{m}}(\omega)$ is designed to pass the baseband modulation without distortion while filtering the noise process $K_d n_c(t)$. This latter noise has the spectrum $K_d^2 \tilde{S}_n(\omega)$, where $\tilde{S}_n(\omega)$ is given in (5.1.3). The total noise power appearing at the demodulator filter output is therefore

$$
\begin{aligned}
P_n &= \frac{K_d^2}{2\pi} \int_{-\infty}^{\infty} \tilde{S}_n(\omega) |H_m(\omega)|^2 \, d\omega \\
&= \frac{K_d^2 N_0}{2\pi} \int_{-\infty}^{\infty} |\tilde{H}_{\mathrm{IF}}(\omega)|^2 |H_m(\omega)|^2 \, d\omega \qquad (5.2.8)
\end{aligned}
$$

Assuming that the filter does not distort the baseband signal $m(t)$, the SNR at the demodulator output is then

$$
\begin{aligned}
\mathrm{SNR}_d &= \frac{K_d^2 C^2 P_m}{K_d^2 N_0 (2\tilde{B}_m)} \\
&= \frac{C^2 P_m}{2 N_0 \tilde{B}_m} \qquad (5.2.9)
\end{aligned}
$$

where $\tilde{B}_m$ is the one-sided noise bandwidth

$$\tilde{B}_m = \frac{1}{2\pi} \int_0^{\infty} |\tilde{H}_{\mathrm{IF}}(\omega)|^2 |H_m(\omega)|^2 \, d\omega \qquad (5.2.10)$$

Thus the ideal envelope detector shown in Figure 5.2 demodulates with the output SNR in (5.2.9), provided that (5.2.6) is valid. The latter corresponds

---

[†]This condition requires that the signal envelope exceed the noise processes at every t. Since the noise is random, one can only interpret this statement in a statistical sense. Hence we mean "for most of the time" the envelope will exceed the noise.

to a condition that the carrier envelope be strong enough relative to the noise. This occurs with a high probability if the power in the carrier is sufficiently greater than that of the noise components in the IF stage. That is,

$$\text{CNR}_{\text{IF}} \triangleq \frac{G_1 P_r}{N_0 B_{\text{IF}}} = \frac{P_r}{\kappa T_{\text{eq}}^{\circ} B_{\text{IF}}} \gg Y \qquad (5.2.11)$$

where Y is a suitably selected threshold value (see Problem 5.1) and B_{IF} is the noise bandwidth of the IF filter stage. In practical systems Y is generally taken as a value in the range 10–20 dB. We therefore conclude that an ideal envelope detector will demodulate the baseband with the SNR in (5.2.9), provided the demodulator input has a carrier to noise power ratio in (5.2.11) that satisfies an appropriate threshold. If the IF and low pass filter are taken to be ideal filters over the bandwidths $2B_m$ and B_m, respectively, corresponding to the modulated and baseband bandwidths, the noise bandwidth in (5.2.10) then becomes

$$\tilde{B}_m = \frac{1}{2\pi} \int_0^{2\pi B_m} d\omega = B_m \qquad (5.2.12)$$

and the output SNR is identical to (5.2.9) with $\tilde{B}_m$ replaced by B_m. In this case the output SNR can be written in terms of the demodulator input CNR by using $B_{\text{IF}} = 2B_m$ and substituting (5.2.11) into (5.2.9), yielding

$$\text{SNR}_d = \frac{C^2 P_m}{2N_0 B_m} = 2\left(\frac{P_m}{1 + P_m}\right) \frac{C^2(1 + P_m)/2}{2N_0 B_m}$$

$$= 2\left(\frac{P_m}{1 + P_m}\right) \text{CNR}_{\text{IF}} \qquad (5.2.13)$$

Thus the demodulator output SNR is directly related to the input CNR and depends on the power that can be put into the modulation. In standard AM we required $|m(t)| \leq 1$ to avoid overmodulation, which forces $P_m \leq 1$. If $P_m = 1$, its largest possible value, then $(\text{SNR})_d = (\text{CNR})_{\text{IF}}$ and the demodulated SNR is equal to that of its carrier input. Therefore, to obtain an increase in SNR_d with ideal envelope AM demodulation, we can only attempt to increase the IF CNR, which requires transmitting higher power P_r, or reducing the receiver noise temperature T_{eq}°.

If overmodulation occurs with envelope detection, the envelope demodulator will produce a distorted baseband during these time periods, since it does not respond to the carrier phase that reflects these sign changes. Thus envelope detection of nonstandard AM carriers, such as suppressed carrier waveforms, is not advisable in AM demodulation. There are, however, other methods for demodulation in these cases, as is considered in Section 5.3. SSB-AM carriers with a vestigial or reinserted carrier can be envelope detected, as we showed in (2.3.6).

Note from (5.2.4) that if the threshold condition is satisfied, so that (5.2.6) is valid, the demodulator output appears as the sum of the desired baseband $m(t)$ and the filtered version of the quadrature noise. Thus, above threshold, AM demodulation with ideal envelope detection effectively recovers the carrier modulation without distortion while simple adding low pass noise to the waveform. However, below threshold, where (5.2.6) has reversed inequalities, (5.2.5) instead becomes

$$x_1(t) \approx \{2C[1 + m(t)]n_c(t) + n_c^2(t) + n_s^2(t)\}^{1/2} \qquad (5.2.14)$$

and the baseband $m(t)$ appears only with a multiplication by the noise. That is, the signal disappears and the noise totally dominates. Thus, as the IF CNR decreases below threshold, the noise effect increases and the usable signal is suppressed, causing a rapid deterioration in the demodulation process. The overall effect is to produce a typical AM SNR_d characteristic, as shown in Figure 5.3. Above threshold Y, (5.2.13) is valid, producing a linear increase with $(CNR)_{IF}$. Below threshold, SNR degrades quickly, as the IF noise dominates the demodulation operation. As stated, there is always some question regarding the actual value of the threshold Y, as measured from empirical data used to generate such curves, since the "knee" of the SNR_d curves tends to be somewhat broad.

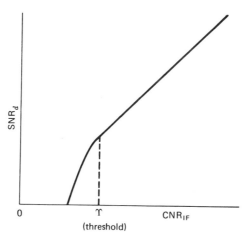

Figure 5.3. Typical AM SNR_d curve.

5.3 AM COHERENT DEMODULATION

A second way to demodulate an AM carrier is by the use of the system in Figure 5.4, where frequency mixing is used for demodulation. We replace the ideal envelope detector by a carrier mixer as shown and use the low pass filter $H_m(\omega)$ as the mixer filter. The mixer uses a local carrier generated at the receiver having the same frquency ω_{IF} as the IF carrier. We write this local oscillator signal as the pure carrier

$$A_l \cos(\omega_{IF}t + \psi_l) \qquad (5.3.1)$$

where ψ_l represents its phase angle. The mixer filter output is that discussed in Section (4.8) and has the form

$$K_m C[1 + m(t)] \cos(\psi - \psi_l) + n_m(t) \qquad (5.3.2)$$

where K_m is the mixer gain and $n_m(t)$ is the mixer noise in (4.8.4) with $\omega_{cl} = 0$. Thus the AM product demodulator produces the baseband signal term

$$K_m C \cos(\psi - \psi_l)m(t) \qquad (5.3.3)$$

which appears as the desired baseband modulation multiplied by the cosine of the phase error between input carrier and local oscillator. If $\psi_l = \psi$, the local oscillator is said to be *phase synchronized* or *phase coherent* with the input carrier, and the demodulated output contains the undistorted baseband signal component. Since the mixer noise has the same spectral shape as $\tilde{S}_n(\omega)$ in (5.1.3), the mixer noise power at the output filter is again

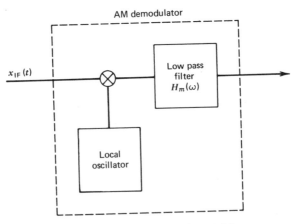

Figure 5.4. AM coherent demodulator.

given by (5.2.8). The resulting demodulator output SNR is then the ratio of the signal power to the noise power. Hence

$$
\begin{aligned}
\text{SNR}_d &= \frac{K_m^2 C^2 P_m}{K_m^2 N_0 (2\tilde{B}_m)} \\
&= \frac{C^2 P_m}{2 N_0 \tilde{B}_m}
\end{aligned}
\tag{5.3.4}
$$

Note that this is identical to that of the ideal envelope detector system operating above threshold. Hence an AM product demodulator operated in phase synchronism produces the same output SNR_d as an ideal envelope detector. When a phase error exists between the carrier signals, the output SNR_d instead becomes

$$
\text{SNR}_d = \frac{C^2 P_m}{2 N_0 \tilde{B}_m} (\cos^2 \psi_e)
\tag{5.3.5}
$$

where ψ_e is the phase error $(\psi - \psi_l)$. In this case the output SNR_d is suppressed from the synchronized case by the squared cosine error. Since $\cos \psi_e \to 0$ as $\psi_e \to 90°$, (5.3.5) shows the importance of maintaining phase coherence during AM mixer demodulation. If ψ_e is considered a random variable (because either ψ or ψ_l may be random), $\cos^2 \psi_e$ must be replaced by $\mathscr{E}[\cos^2 (\psi_e)]$, where $\mathscr{E}$ is the expectation operator over the density of the phase error. It is important to note that (5.3.5) was obtained without requiring any explicit threshold condition, as in (5.2.11). In essence, the CNR condition is replaced by a condition of maintaining small phase error between IF and demodulator oscillator carriers. Later we see that this phase synchronism can be achieved with tracking loops, and the small phase error condition does in fact imply an inherent threshold condition on the IF carrier power.

If the phase error ψ_e is a function of time, then (5.3.3) corresponds to a multiplication of the desired baseband signal by the cosine error time function. This latter effect can cause complete baseband distortion. For example, if the local oscillator differs by frequency ω_d from the IF carrier, then $\psi_e(t) = \omega_d t$ and (5.3.3) becomes instead $K_m Cm(t) \cos \omega_d t$. This will appear as a shift of the spectrum of $m(t)$ to the carrier offset frequency ω_d. This resulting signal corresponds to an offset baseband spectrum that may be distorted by the baseband output filter if ω_d is large. That is, the frequency error may shift the baseband spectrum outside the bandpass of the output filter.

The notion of maintaining phase coherency between a received and local carrier signal is fundamental throughout much of our subsequent study. Such coherence requires the local carrier to learn and "keep up with" the incoming carrier phase, an operation that can be achieved with a tracking system that forces the local carrier phase to follow that of the received carrier. This operation is discussed later in this chapter and in more detail in Chapter 10. Maintaining phase coherence, which for many years was considered a difficult electronic task, has been shown over the past decade to be both feasible and in fact relatively simple to achieve. For this reason phase coherent systems have become an integral part of modern communication systems.

Since (5.3.4) follows from (5.3.2) even if overmodulation occurred, we see that coherent AM demodulation can be used to obtain undistorted demodulation of suppressed carrier AM. However, the local oscillator must remain phase coherent during the overmodulation. That is, the local oscillator must track the carrier phase ψ even though the total phase of the carrier changes by π radians each time overmodulation occurs. Subsystems that achieve phase referencing in spite of extraneous carrier phase shifts are considered in Chapter 10. In the suppressed carrier AM case, P_c in (5.2.2) is replaced by $C^2 P_m/2$ and (5.2.13) becomes

$$\text{SNR}_d = 2\text{CNR}_{\text{IF}} \tag{5.3.6}$$

Thus a factor of 2 is gained in output SNR by using suppressed carrier AM with coherent demodulation. However, the task of deriving the necessary phase coherence is now made more difficult without the presence of the carrier component.

SSB-AM carriers can also be demodulated with coherent demodulation methods. When the SSB-AM carrier in (2.3.1) is multiplied with the phase coherent reference in (5.3.1), we obtain

$$c(t)[\cos (\omega_{\text{IF}}t + \psi_l)] = K_m a(t)[1 + \cos (2\omega_{\text{IF}}t + 2\psi_l)]$$
$$+ K_m \hat{a}(t)[\sin (2\omega_{\text{IF}}t + 2\psi_l] \tag{5.3.7}$$

The low pass filter then removes the double IF frequency terms, producing the desired amplitude modulation $a(t)$. Note that the quadrature term involving $\hat{a}(t)$ is used to eliminate the sideband in the transmitted waveform but does not contribute to the demodulated waveform.

Coherent AM demodulation can be achieved with other implementations as well. The phase coherent reference carrier can be added to the received IF carrier, and the combined waveform squared and filtered, as shown in Figure 5.5. The squaring generates the required cross-product multiplication term, with the low pass filter recovering the modulation term. In vestigial carrier systems, a portion of the carrier is already present, and only the squaring and filtering is needed for demodulation. In effect, with the coherent carrier present (either by insertion or as a vestigial carrier), the squaring and filtering is performing as an ideal envelope detector. This is, in fact, one common technique for constructing AM envelope detectors.

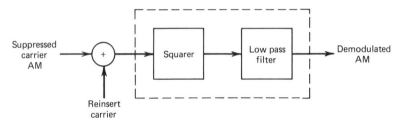

Figure 5.5. AM squarer–filter demodulator with carrier insertion.

5.4 FREQUENCY DEMODULATION

In this section we consider the demodulation of an FM carrier. The demodulator input waveform is assumed again to have the form in (5.1.1) when $c(t)$ now corresponds to a general FM carrier signal having the form

$$c(t) = A \cos [\omega_{IF} t + \theta(t)] \qquad (5.4.1)$$

where now

$$\theta(t) = \Delta_\omega \int m(t) \, dt$$

$$\frac{A^2}{2} = P_r G_1 \qquad (5.4.2)$$

and Δ_ω is the frequency deviation in rad/sec. The baseband signal $m(t)$ occupies a bandwidth of B_m Hz, and we assume that $m(t)$ is normalized so that $P_m = 1$. The carrier signal $c(t)$ occupies an IF bandwidth given approximately by the Carson rule bandwidth

$$B_{IF} = 2(\beta + 1) B_m \ \text{Hz} \qquad (5.4.3)$$

where $\beta = \Delta_\omega / 2\pi B_m$ is the carrier modulation index. We are interested primarily in the case where $\beta \gg 1$, so that the B_{IF} is several times the modulating bandwidth B_m, corresponding to wideband FM. The waveform to be demodulated is the sum of (5.4.1) and the narrowband Gaussian noise process of (5.1.2). As discussed in Section 4.7, the IF signal can then be written as in (4.7.7):

$$x_{IF}(t) = \alpha(t) \cos [\omega_{IF} t + \theta(t) + v(t)] \qquad (5.4.4)$$

where the phase noise is

$$v(t) = \tan^{-1} \left[\frac{n_s(t)}{A + n_c(t)} \right] \qquad (5.4.5)$$

Here we have replaced the modified noise components $\hat{n}_c(t)$ and $\hat{n}_s(t)$ with $n_c(t)$ and $n_s(t)$ by using the wideband condition of (4.7.12). Consider the processing of the preceding IF signal by the FM demodulator in Figure 5.6, composed of the cascade connection of an ideal frequency detector followed by the low pass filter $H_m(\omega)$. An ideal frequency detector is a device whose output voltage is proportional to the instantaneous frequency variation of the input. Such devices are often called *frequency discriminators* and can be constructed by combining tuned circuits and using the slope of the gain function, or can be implemented by zero crossing counters and integrators (Problem 5.7). In the absence of noise, the ideal FM demodulator responds to the instantaneous frequency of $c(t)$ about ω_{IF}, and therefore yields an output directly proportional to the baseband signal $m(t)$. When noise is present, however, the frequency detector responds to the frequency of the IF signal in (5.4.4). Its output is therefore

$$x_1(t) = K_d[\omega(t) - \omega_{IF}]$$

$$= K_d \frac{d}{dt}[\theta(t) + v(t)] \qquad (5.4.6)$$

where K_d is the detector gain. Expanding the derivative of the phase noise process yields

$$x_1(t) = K_d \left[\Delta_\omega m(t) + \frac{[A + n_c(t)]n_s'(t) - n_s(t)n_c'(t)}{[A + n_c(t)]^2 + n_s^2(t)} \right] \qquad (5.4.7)$$

where the primes denote differentiation with respect to t. The first term is the desired baseband, and the second term is the effect of the phase noise, the latter causing undesired deviations of the detector output. An understanding of this noise interference is hindered by the rather complex manner in which the noise components appear. If, however, we apply an assumption similar to (5.2.6)

$$A \gg |n_c(t)|, |n_s(t)| \qquad (5.4.8)$$

we can then expand (5.4.7) as

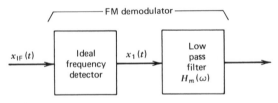

Figure 5.6. FM demodulator (ideal frequency detector).

$$x_1(t) \cong K_d \left[\Delta_\omega m(t) + \frac{n_s'(t)}{A} + \frac{n_c(t)n_s(t)}{A^2} - \frac{n_s(t)n_c'(t)}{A^2} + O\left(\frac{1}{A^3}\right) \right]$$

$$(5.4.9)$$

where $O(1/A)^3$ represent terms of order $(1/A)^3$ or higher. If we neglect all terms of second order or higher in $1/A$, we simplify to

$$x_1(t) = K_d \Delta_\omega m(t) + K_d \left(\frac{n_s'(t)}{A} \right) \qquad (5.4.10)$$

Thus, under the condition of (5.4.8), the FM detector output noise is

$$n_1(t) = \frac{K_d n_s'(t)}{A} = \left(\frac{K_d}{A} \right) \frac{dn_s(t)}{dt} \qquad (5.4.11)$$

The quadrature noise component $n_s(t)$ has the power spectrum $\tilde{S}_n(\omega)$. Hence the noise in (5.4.11) has the corresponding spectrum (recall Problem 1.29)

$$S_{n_1}(\omega) = \left(\frac{K_d}{A} \right)^2 \omega^2 \tilde{S}_n(\omega) = \left(\frac{K_d}{A} \right)^2 \omega^2 N_0 |\tilde{H}_{\text{IF}}(\omega)|^2 \qquad (5.4.12)$$

Thus the output of the ideal frequency detector is the sum of the desired modulating signal plus a noise process whose spectral density is given by (5.4.12). This spectrum is sketched in Figure 5.7 assuming an ideal IF filter over the bandwidth B_{IF}. Since the output filter has a bandwidth B_m much less than B_{IF} when wideband FM is used, the demodulator output noise is considerably reduced. If the output filter has transfer function $H_m(\omega)$ and passes the baseband signal $m(t)$ without distortion, the FM demodulator output has baseband signal power

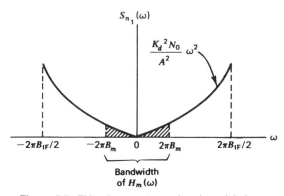

Figure 5.7. FM noise spectrum after demodulation.

$$P_s = K_d^2 \Delta_\omega^2 P_m = K_d^2 \Delta_\omega^2 \tag{5.4.13}$$

and noise power

$$P_n = \left(\frac{K_d}{A}\right)^2 \frac{N_0}{2\pi} \int_{-\infty}^{\infty} \omega^2 |\tilde{H}_{\mathrm{IF}}(\omega)|^2 |H_m(\omega)|^2 \, d\omega \tag{5.4.14}$$

If the output filter has a rectangular filter function over B_m Hz, then

$$P_n = \left(\frac{K_d}{A}\right)^2 \frac{N_0}{2\pi} \int_{-2\pi B_m}^{2\pi B_m} \omega^2 \, d\omega$$

$$= \frac{2K_d^2 N_0 (2\pi)^2 B_m^3}{3A^2} \tag{5.4.15}$$

The FM demodulator output SNR is then

$$\mathrm{SNR}_d = \frac{P_s}{P_n} = \frac{3A^2 K_d^2 \Delta_\omega^2}{2K_d^2 N_0 (2\pi)^2 B_m^3}$$

$$= 3\beta^2 \left[\frac{A^2/2}{N_0 B_m}\right] \tag{5.4.16}$$

where again $\beta = \Delta_\omega / 2\pi B_m$. The preceding is called the *FM improvement formula* and reveals the basic advantage of FM communications. It relates the output SNR to the parameter β and $(A^2/2N_0 B_m)$. The latter parameter is the ratio of the IF carrier power $(A^2/2)$ to input IF noise power in the bandwidth of the baseband $m(t)$. Note that if the parameters A, B_m, and N_0 are considered fixed, the output SNR_d in (5.4.16) increases with the parameter β, which, in turn, depends on the transmitted deviation Δ_ω. This improvement in demodulated SNR appears as a "quieting" effect on observations of the output. For large output SNR, FM systems should therefore operate with a large frequency deviation relative to their base-bandwidth. However, recall that the IF carrier (Carson rule) bandwidth is given by $2(\beta + 1)B_m$, and therefore also increases with β. Hence FM improvement in output SNR is achieved at the expense of required FM bandwidth. This latter effect illustrates the advantage of using wideband FM systems. By increasing the transmitted carrier bandwidth, we obtain an increase in the demodulated output SNR. Recall that no such tradeoff was possible in AM systems. Of course, physical limitations on the amount by which a carrier can be deviated, or on the available carrier bandwidth, will ultimately constrain the achievable SNR_d predicted by (5.4.16).

The preceding results are of course predicated on (5.4.8). We can again relate this to a condition on the probability of the noise components having values less than the carrier amplitude, as in Problem 5.1. Thus the condition that this probability be sufficiently high is equivalent to a condition that

$$\mathrm{CNR_{IF}} = \frac{A^2/2}{N_0 B_{IF}} \geq Y \qquad (5.4.17)$$

for some Y. Equation (5.4.17) plays the role of an FM threshold condition on the demodulator input (IF) carrier power, just as in the AM case in (5.2.11). Note that as the index β increases, the carrier power needed to satisfy the preceding threshold also must increase, since B_{IF} increases also with β. In many communication systems, transmitter power is at a premium, whereas bandwidth is expendable (e.g., satellite links). For these systems FM is extremely suitable, since it allows this tradeoff of power for bandwidth. That is, if we desire a given SNR_d, then (5.4.16) indicates that bandwidth (β) can be traded off for carrier power ($A^2/2$) by increasing β and decreasing A^2, while keeping SNR_d fixed. This tradeoff can be made until A^2 reaches a lower value such that (5.4.17) is no longer satisfied. Equation (5.4.16) can be rewritten directly in terms of $\mathrm{CNR_{IF}}$ by substitution from (5.4.3),

$$\mathrm{SNR}_d = 6\beta^2(\beta + 1)\mathrm{CNR_{IF}} \qquad (5.4.18)$$

This allows a direct relation between demodulator input and output SNR in FM receivers.

Operation below threshold renders the approximation in (5.4.8) no longer valid, and the effect of the higher order noise terms in (5.4.9) must be considered. These remaining terms cause an increase in the output noise and a rapid deterioration of the output SNR occurs, just as in AM. Note that the first second order term [third term in (5.4.9)] has a power spectrum given by the convolution of the spectra of the quadrature components, and

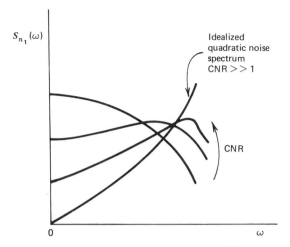

Figure 5.8. Variation in output noise spectrum as $\mathrm{CNR_{IF}}$ decreases.

therefore has a spectrum concentrated about zero frequency. The effect of including this term is to add more low frequency noise, converting the output noise spectrum from the quadratic form in Figure 5.6 to a more flattened spectrum, as shown in Figure 5.8. Thus as FM carrier power is decreased, the addition of the higher order terms in (5.4.9) causes an increase in the output inband noise and a reduction of the output SNR from that predicted by the FM improvement formula in (5.4.16). Figure 5.9

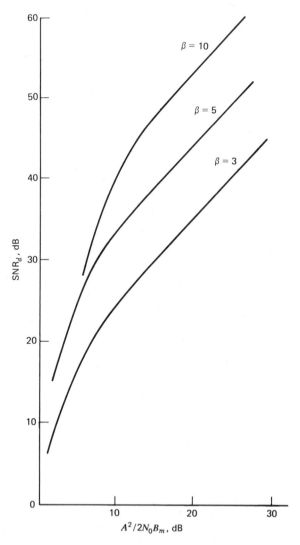

Figure 5.9. FM improvement curves. Output SNR_d versus CNR in the modulation bandwidth B_m (β = modulation index).

shows typical FM output SNR_d curves, exhibiting the manner in which the FM improvement formula in (5.4.16) is no longer valid below threshold. The "knee" of the curves involved is typically on the order of $CNR_{IF} \approx$ 12–16 dB. Below threshold operation of FM detectors has been studied rigorously [5–8] from a spectral analysis point of view, and more recently by modeling the higher order effects in terms of phase "jumps" and applying an appropriate "click" analysis [9]. Each of these approaches adequately predicts the changing demodulator noise spectrum as the input carrier power is reduced, as depicted in Figure 5.8.

5.5 PHASE DEMODULATION

When the carrier is phase modulated with the baseband signal $m(t)$, demodulation is accomplished by recovering the phase of the received IF signal. In PM systems the IF carrier is given by

$$c(t) = A \sin[\omega_{IF}t + \theta(t)]$$
$$= A \sin[\omega_{IF}t + \Delta m(t)] \qquad (5.5.1)$$

where again $A^2/2$ is the IF carrier power and Δ is the phase modulation index. When the preceding carrier is received in the presence of additive Gaussian IF noise, the resulting signal can be combined into the IF signal, as in (5.1.1):

$$x_{IF}(t) = c(t) + n(t)$$
$$= \alpha(t) \sin[\omega_{IF}t + \Delta m(t) + v(t)] \qquad (5.5.2)$$

where $v(t)$ is the phase noise in (5.4.5). Let us consider the behavior of the preceding phase modulated IF signal when acted on by an ideal phase detector (i.e., a device whose output is proportional to the phase deviation of the input IF signal). The overall PM demodulator is shown in Figure 5.10. The output low pass filter has a bandwidth corresponding to the bandwidth B_m of the baseband $m(t)$. From (5.5.2) the phase detector output will be

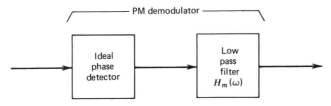

Figure 5.10. The PM demodulator (ideal phase detector).

$$x_1(t) = \Delta m(t) + v(t) \qquad (5.5.3a)$$

where

$$v(t) = \tan^{-1}\left[\frac{n_s(t)}{A + n_c(t)}\right] = \frac{n_s(t)}{A} + O\left(\frac{1}{A^2}\right) \qquad (5.5.3b)$$

If we again require

$$\frac{A^2/2}{N_0 B_{\text{IF}}} \geq \Upsilon \qquad (5.5.4)$$

we can approximate $v(t)$ by dropping the second order terms in $1/A$ in the expansion of the inverse tangent function. We then have

$$x_1(t) = \Delta m(t) + \frac{n_s(t)}{A} \qquad (5.5.5)$$

The output of the demodulator is then the modulation component with power $\Delta^2 P_m$ and a Gaussian noise component of power $2\tilde{B}_m N_0/A^2$, where $\tilde{B}_m$ is the noise bandwidth in (5.2.10). Assuming $P_m = 1$, we can write the idealized output SNR for PM demodulation as

$$\text{SNR}_d = \frac{\Delta^2}{2N_0 \tilde{B}_m/A^2} = \Delta^2 \left(\frac{A^2/2}{N_0 \tilde{B}_m}\right) \qquad (5.5.6)$$

which is similar to the FM result with $3\beta^2$ replaced by Δ^2. The threshold condition in (5.5.4) again requires a suitable CNR in the IF bandwidth, the latter given by $B_{\text{IF}} = 2(\Delta + 1)B_m$. Thus the PM system also trades off bandwidth for carrier power, and the discussion for FM improvement is appropriate here for PM as well. However, the problem of obtaining wideband PM improvements in practice is hindered by the fact that Δ, the transmitter phase deviation, cannot be arbitrarily increased. This would require then that the ideal phase detector assumed in Figure 5.10 be able to detect absolute phase deviations rather than modulate 2π detection (i.e., distinguish between θ degrees and $(n2\pi + \theta)$ degrees). Since practical phase detectors [1] involve sine wave phase comparisons, this 2π multiple is often lost, and the absolute value of the phase deviation cannot be detected. This limits the useful phase deviations to $\Delta \leq \pm \pi$, and the use of wideband PM is therefore not as popular as wideband FM.

5.6 FM AND PM DEMODULATION WITH FEEDBACK TRACKING

The demodulation of FM and PM considered in Sections 5.4 and 5.5 was accomplished by detection circuits that recovered the carrier modulation.

Angle modulated carriers can also be demodulated by the use of feedback tracking. In the latter schemes, an output waveform is fed back to control an oscillator phase that is mixed with the modulated carrier to perform the demodulation. The entire feedback subsystem now plays the role of a carrier demodulator, with the output of the feedback system producing a demodulated waveform. Because of the feedback operation, such forms of demodulators are less susceptible to circuit parameter variations, and have the basic advantage that the demodulation can be achieved with less received carrier power than with the standard discriminator circuits of the previous sections.

An angle demodulator using feedback tracking is depicted in Figure 5.11. The input is the IF carrier to be demodulated. The mixer beats the IF carrier with the output of a local oscillator. The local oscillator is at the same frequency as the IF carrier, and the mixing operation produces a low pass baseband waveform at the mixer output. This waveform is low pass filtered to produce the loop output and to reduce the higher frequency output noise. The output is also fed back to control the frequency variation of the local oscillator. Thus the local oscillator is itself a form of voltage controlled oscillator (VCO), having its output frequency varied proportionally to its input waveform.

The tracking loop output is also filtered by the output demodulator filter to provide the demodulated baseband. This output filter is used to achieve the desired spectrum shaping and final noise reduction for the demodulation. The loop in Figure 5.11, composed of the phase detector mixer, low pass loop filter, and VCO, is referred to as a *phase tracking loop*, or *phase lock loop* (PLL).

To understand how feedback causes demodulation to occur, let us write the IF carrier at the loop input in Figure 5.11 as in (5.4.1)

$$c(t) = A \cos \left[\omega_{IF} t + \theta(t) \right] + n_{IF}(t) \qquad (5.6.1)$$

where ω_{IF} is the IF frequency, $n_{IF}(t)$ is the IF noise process, and again $\theta(t)$ is the angle modulation, corresponding to either FM or PM. We write the local oscillator waveform as

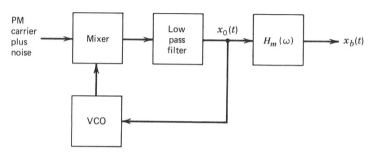

Figure 5.11. Phase tracking demodulator.

$$x_l(t) = \sin\left[\omega_{\mathrm{IF}}t + \theta_l(t)\right] \qquad (5.6.2)$$

where $\theta_l(t)$ accounts for the phase variation inserted by the feedback. Note that while (5.6.1) involves a cosinusoidal waveform, $x_l(t)$ uses a sinusoid at that same frequency, and thereby assumes a fixed $90°$ phase shift (aside from the modulation) between these carriers. The output of the mixer is then

$$x_1(t) = K_m A \sin \theta_e(t) + n_m(t) \qquad (5.6.3)$$

where K_m is the mixer gain, $n_m(t)$ the mixer output noise, and

$$\theta_e(t) = \theta(t) - \theta_l(t) \qquad (5.6.4)$$

The mixer baseband signal therefore involves the sine of the instantaneous phase error between the IF and the local carrier waveform. If we assume this phase error $\theta_e(t)$ is small enough so that

$$\sin \theta_e(t) \approx \theta_e(t) \qquad (5.6.5)$$

then (5.6.3) simplifies to

$$x_1(t) = AK_m[\theta(t) - \theta_l(t)] + n_m(t) \qquad (5.6.6)$$

The waveform $x_1(t)$ is low pass filtered by $K_1 F(\omega)$ to generate the loop output $x_o(t)$. The constant K_1 represents the gain level of the filter. The Fourier transform of the loop output waveform is then

$$X_o(\omega) = K_1 F(\omega)[AK_m \Phi_e(\omega) + N_m(\omega)] \qquad (5.6.7)$$

where

$$\Phi_e(\omega) = \Phi(\omega) - \Phi_l(\omega) \qquad (5.6.8)$$

and $\Phi(\omega)$ is the transform of the angle modulation $\theta(t)$. We have used capitals to denote Fourier transforms of the corresponding lowercase time functions. Also, we have symbolically denoted $N_m(\omega)$ as the transform of the noise process $n_m(t)$ in order to determine the effect of the loop filtering, without being concerned with its specific meaning. The signal $x_o(t)$ serves as the loop output and is also fed back to control the frequency of the local oscillator. Since the frequency of the VCO is controlled by its input waveform, its phase $\theta_l(t)$ is proportional to the integral of $x_o(t)$. Hence we can write

$$\Phi_l(\omega) = K_2 \frac{X_o(\omega)}{j\omega} \qquad (5.6.9)$$

where K_2 is proportionality constant associated with the oscillator and indicates the radian phase shift per volt of control signal. Combining (5.6.8) and (5.6.9) with (5.6.7) yields

$$X_o(\omega) = K_1 F(\omega) A K_m \left[\Phi(\omega) - K_2 \frac{X_o(\omega)}{j\omega} \right] + K_1 F(\omega) N_m(\omega)$$

$$(5.6.10)$$

Solving for $X_o(\omega)$ shows

$$X_o(\omega) = \left[\frac{A K_m K_1 F(\omega)}{1 + A K F(\omega)/j\omega} \right] \Phi(\omega) + \left[\frac{K_1 F(\omega)}{1 + A K F(\omega)/j\omega} \right] N_m(\omega)$$

$$= \left[\frac{A K_m K_1 F(\omega) j\omega}{j\omega + A K F(\omega)} \right] \left[\Phi(\omega) + \frac{N_m(\omega)}{A K_m} \right] \qquad (5.6.11)$$

where

$$K = K_1 K_2 K_m \qquad (5.6.12)$$

is the total component gain around the loop. Thus the output of the phase tracking loop appears as a filtered version of the carrier phase modulation in (5.6.1), plus an added filtered noise term.

The loop output $x_o(t)$ is then filtered by the output filter function $H_m(\omega)$ to produce the final demodulator output $x_b(t)$, having transform

$$X_b(\omega) = H_m(\omega) X_o(\omega) \qquad (5.6.13)$$

The output filter can now be designed in conjunction with the type of modulation. If the IF carrier is frequency modulated,

$$\Phi(\omega) = \Delta_\omega \frac{M(\omega)}{j\omega} \qquad (5.6.14)$$

in (5.6.11) and $H_m(\omega)$ can be designed as

$$H_m(\omega) = \begin{cases} \dfrac{j\omega + A K F(\omega)}{A K_m K_1 F(\omega)}, & |\omega| \le 2\pi B_m \\ 0, & \text{elsewhere} \end{cases} \qquad (5.6.15)$$

Using (5.6.14) and (5.6.15), the demodulator output in (5.6.13) becomes

$$X_b(\omega) = \Delta_\omega M(\omega) + j\omega N_m(\omega)/A K_m \qquad (5.6.16)$$

The demodulated output has therefore recovered the frequency modulating waveform $m(t)$ plus additive noise. Since the mixer noise has the power spectrum $K_m^2 \tilde{S}_n(\omega)$ given in (4.6.4), the demodulated output noise in (5.6.16) will have the spectrum

$$S_{n_b}(\omega) = \frac{\omega^2 N_0}{A^2}, \qquad |\omega| \le 2\pi B_m \qquad (5.6.17)$$

similar to (5.4.12). Hence the demodulated SNR_d is identical to (5.4.16), and therefore performs identically to an ideal FM detector.

A similar result holds for a PM carrier. In this case,

$$\Phi(\omega) = \Delta M(\omega) \qquad (5.6.18)$$

and the output filter can now be designed as

$$H_m(\omega) = \begin{cases} \dfrac{j\omega + AKF(\omega)}{AK_m K_1 F(\omega) j\omega}, & |\omega| \le 2\pi B_m \\[2mm] 0, & \text{elsewhere} \end{cases} \qquad (5.6.19)$$

The demodulator output will have the transform

$$X_b(\omega) = \Delta M(\omega) + N_m(\omega)/AK_m \qquad (5.6.20)$$

Again the demodulator has recovered the phase modulating waveform with additive noise having power $P_n = (K_m^2 N_0/A^2 K_m^2) 2B_m = 2B_m (N_0/A^2)$ as in (5.5.5). The resulting demodulator output SNR_d is identical to (5.5.6) corresponding to that of the ideal phase detector.

Hence the tracking demodulator in Figure 5.11, with a properly designed output filter, can be made to perform identically to an ideal FM or PM detector. This occurs without having to apply any threshold conditions as we did in (5.4.17) and (5.4.4). Instead we had to apply the small phase error condition in (5.6.5). This is reminiscent of the coherent AM demodulation operation and suggests again the notion of phase synchronization between the local and IF carrier. However, here the IF carrier phase is varying according to the angle modulation, and the local carrier phase must continually maintain the synchronism throughout the variations. We say that the two carriers are "locked" in phase, and the loop is "in lock," if the phase error continually remains small. When the phase error becomes so large that (5.6.5) is no longer valid, we say the loop is "out of lock," and the preceding linear analysis is no longer applicable. (We must then resort to nonlinear loop theory for further analysis, as we shall do in Chapter 10.) Thus the feedback demodulating system in Figure 5.11 will demodulate the carrier as long as the VCO phase "tracks" (remains close to) the input carrier phase variation. For this reason feedback demodulators of this type are often referred to as *phase tracking* demodulators.

The PLL in Figure 5.11 can be redrawn as shown in Figure 5.12a. Its

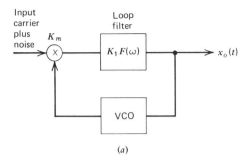

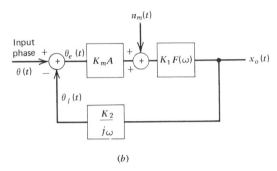

Figure 5.12. The phase lock loop: (a) The basic loop, (b) the equivalent linear model.

output $x_o(t)$ is identical to that which would be produced by the linear feedback system shown in Figure 5.12b. (Recall Figure 1.16c.) The latter is called the *baseband equivalent* system to the feedback tracking loop and is equivalent in the sense that it generates the same output waveform $x_o(t)$. Note that the input to the equivalent baseband loop is the phase modulation $\theta(t)$ of the IF carrier, while the mixer noise is effectively added at the point shown. In the loop, the mixer becomes a phase subtractor, the loop VCO represents a phase integrator, while the loop filtering $K_1 F(\omega)$ is identical to the filtering in the actual system. We emphasize that the baseband equivalent loop is only to aid analysis and does not indicate the actual system implementation, the latter shown instead in Figure 5.11. However, any design of the loop filter using the equivalent loop model is of course directly applicable to the actual feedback loop filter.

Note also in Figure 5.12 that the IF carrier amplitude A appears as an effective loop gain parameter in the baseband equivalent system. This means that loop performance will necessarily depend on the value of this amplitude. Furthermore, it means that any loop design based on the equivalent system will require a particular value of A, and performance will vary if the IF carrier has a different amplitude. For this reason amplitude control devices, such as limiters or automatic gain control systems, are often inserted prior to phase tracking demodulators to stabilize the carrier amplitude.

The output filter $H_m(\omega)$ acts to both limit the output noise to the modulation bandwidth and compensate for the loop filtering on the modulation itself. For example, the filter in (5.6.15) for the FM case behaves as a form of differentiating network. Since the loop tracks the carrier phase, the external differentiator is needed to produce a baseband output related to carrier frequency. For the PM case, $H_m(\omega)$ is a form of a sharp cutoff, low pass filter. If the actual spectrum of the modulating waveform $m(t)$ is known, these demodulating filters can in fact be further optimized, as we shall see in Section 6.2.

The advantage of using a PLL in cascade with an output filter as a phase tracking PM demodulator is that the loop can be designed for best phase tracking performance, while the output filter can be designed to produce the desired demodulation transfer function. Since the output filter is not part of the feedback loop, it can be designed to have any desired shape without affecting loop tracking or stability. Because of its simplicity of design, the PLL has emerged as a basic component in modern communication system design [9–12]. As we shall see in later discussions, its basic operation (i.e., causing an oscillator to phase track a carrier waveform) makes it advantageous in several other important communication applications as well.

5.7 LINEAR PLL MODULATION TRACKING

The PLL shown in Figure 5.12 must operate as a linear system, which means that its error in tracking the input phase variation must be relatively small for all t. Let us examine the requirements for constructing a PLL to achieve the small error condition and determine the parameters that will influence the magnitude of the phase error.

The PLL block diagram was shown in Figure 5.12a, and its equivalent baseband model was shown in Figure 5.12b. It is first convenient to define

$$H(\omega) \overset{\Delta}{=} \frac{AK[F(\omega)/j\omega]}{1+[AKF(\omega)/j\omega]} \tag{5.7.1}$$

as the loop gain function of the PLL, where $F(\omega)$ is the loop filter and $K = K_1 K_2 K_m$ is the total component gain of the loop. From Figure 5.12b we see that $H(\omega)$ is actually the transfer function from the baseband loop input to VCO output of the baseband equivalent system. The error in tracking, $\theta_e(t) = \theta(t) - \theta_l(t)$, is then the signal appearing at the subtractor output. Using linear feedback analysis, its transform is then

$$\Phi_e(\omega) = \Phi(\omega) - \Phi_l(\omega)$$

$$= \left[\frac{1}{1 + AK[F(\omega)/j\omega]} \right] \Phi(\omega) - \left[\frac{AK[F(\omega)/j\omega]}{1 + AK[F(\omega)/j\omega]} \right] \frac{N_m(\omega)}{K_m A}$$

$$(5.7.2)$$

Using (5.7.1), this can be written as

$$\Phi_e(\omega) = [1 - H(\omega)]\Phi(\omega) - H(\omega) \left(\frac{N_m(\omega)}{K_m A} \right) \qquad (5.7.3)$$

The first term is the contribution to the phase error due to the carrier phase modulation $\theta(t)$. The second is due to the mixer noise entering the loop. The problem of designing modulation tracking loops therefore reduces to proper design of the loop gain function $H(\omega)$, which, in turn, depends on the loop filter function $F(\omega)$ and the loop gain K. The filtering operation indicated by (5.7.3) represents a filtering of $\theta(t)$ with the transfer function $[(1 - H(\omega)]$ and a filtering of $n_m(t)/K_m A$ with $H(\omega)$. Since the modulation $\theta(t)$ is expected to have a low pass spectrum, increasing the bandwidth of $H(\omega)$ [letting $H(\omega) \approx 1$ over a wider frequency range] reduces the phase error due to modulation but increases the mean square phase error due to noise. Thus the design of modulation tracking loops basically involves a tradeoff of modulation error versus noise error. Note that if no extraneous phase shifts or noise were involved, then adequate signal error reduction requires that $|H(\omega)| = 1$ over the bandwidth of the modulation process $m(t)$. That is, the loop transfer function $H(\omega)$ should have a bandwidth roughly equivalent to the modulation bandwidth. When extraneous phase modulation and noise is present, this bandwidth may have to be larger or smaller, depending on the dominant contributor to the total error. This tradeoff is examined in detail in Section 5.8.

Specific forms of the loop gain function $H(\omega)$ for several common types of loop filter functions $F(\omega)$ are listed in Table 5.1. The *order* of the loop is an indication of the degree of filtering provided by the loop, and therefore specifies loop complexity. Most PLL are designed to be of the first or second order to avoid stability problems prevalent with more complex loops. Third order loops are used only in special situations where severe phase modulation dynamics are involved. First order loops involve no loop filtering and are the easiest to design and analyze. Second order loops have loop filters that can be constructed as relatively simple *RC* networks (Problem 5.25). The second order, high gain PLL has the general loop gain function

$$H(\omega) = \frac{1 + 2\zeta(j\omega/\omega_n)}{-(\omega/\omega_n)^2 + 2\zeta(j\omega/\omega_n) + 1} \qquad (5.7.4)$$

The parameter ζ is called the loop *damping factor*, and ω_n is called the loop

TABLE 5.1   Tabulation of Loop Filters and Loop Bandwidths

Loop Filter	Loop Order	$H(\omega)$	B_L
$F(\omega) = 1$ (no loop filter)	1	$\dfrac{AK}{j\omega + AK}$	$\dfrac{AK}{4}$
$F(\omega) = \dfrac{j\omega\tau_2 + 1}{j\omega\tau_1 + 1}$ $\omega_n^2 = \dfrac{AK}{\tau_1}$ $2\zeta\omega_n = \dfrac{1 + AK\tau_2}{\tau_1}$	2	$\dfrac{1 + j[(2\zeta/\omega_n) - (AK)^{-1}]\omega}{-(\omega/\omega_n)^2 + j(2\zeta/\omega_n)\omega + 1}$	$\dfrac{\omega_n}{8\zeta}\left[1 + \left(2\zeta - \dfrac{\omega_n}{AK}\right)^2\right]$
$F(\omega) = \dfrac{j\omega\tau_2 + 1}{j\omega\tau_1}$ $\omega_n^2 = \dfrac{AK}{\tau_1}$ $2\zeta\omega_n = \dfrac{AK\tau_2}{\tau_1}$	2	$\dfrac{1 + j(2\zeta/\omega_n)\omega}{-(\omega/\omega_n)^2 + j(2\zeta/\omega_n)\omega + 1}$	$\dfrac{\omega_n}{8\zeta}(1 + 4\zeta^2)$
$F(\omega) = \dfrac{(j\omega)^2 + aj\omega + b}{(j\omega)^2}$	3	$\dfrac{AK[(j\omega)^2 + j\omega a + b]}{(j\omega)^3 + AK(j\omega)^2 + AKaj\omega + bAK}$	$\dfrac{AK(aAK + a^2 + b^2)}{4(aAK - b)}$

natural frequency. The damping factor is tyically in the range $0.5 \le \zeta \le 1.5$ and is an indication of the stability of the loop. As the damping factor approaches zero, the loop becomes less stable, and a much tighter tolerance must be placed on the mixer, oscillator, and filter in terms of their time delay contribution within the loop. The loop natural frequency ω_n is an indication of the loop bandwidth and its reciprocal specifies loop response time. The relationship between these loop parameters and the parameters of the loop components and filter coefficients is also listed in Table 5.1. Figure 5.13 shows plots of $|H(\omega)|$ in (5.7.4) as a function of ω/ω_n for several values of ζ. Most loops are operated with $\zeta = 0.707$, which is referred to as a *critically damped loop.*

From (5.7.3) we see that for any PLL the corresponding phase error in the time domain is given by

$$\theta_e(t) = \theta_{em}(t) + \theta_{en}(t) \qquad (5.7.5)$$

where $\theta_{em}(t)$ is the error time function associated with the inverse transform of the first term and $\theta_{en}(t)$ is that due to the noise. Since the mixer noise is a Gaussian noise process, $\theta_{en}(t)$ itself evolves as a Gaussian process, and the first term depends on the type of modulation model used. For any $H(\omega)$ the mean squared value of the noise process $\theta_{en}(t)$ at any t can be obtained by

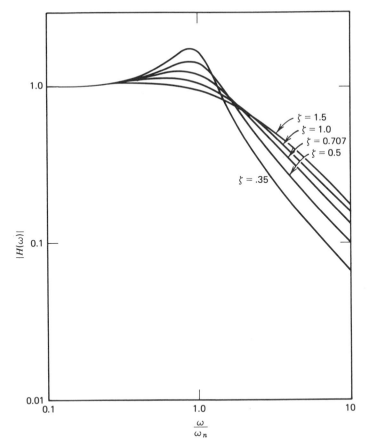

Figure 5.13. Loop gain function for second order PLL (ζ = damping factor, ω_n = loop natural frequency).

integrating over the spectral density of the last term in (5.7.3). Since the mixer noise has spectral density $K_m^2 \tilde{S}_n(\omega)$, this is given by

$$\sigma_n^2 = \frac{1}{2\pi} \int_{-\infty}^{\infty} |H(\omega)|^2 \frac{\tilde{S}_n(\omega)}{A^2} \, d\omega$$

$$= \frac{N_0}{A^2} \left[\frac{1}{2\pi} \int_{-\infty}^{\infty} |H(\omega)|^2 \, d\omega \right] \qquad (5.7.6)$$

where we have assumed $\tilde{H}_{IF}(\omega) = 1$ over the bandwidth of $H(\omega)$. If we denote

$$B_L \stackrel{\Delta}{=} \frac{1}{2\pi} \int_0^{\infty} |H(\omega)|^2 \, d\omega \qquad (5.7.7)$$

as the one-sided *loop noise bandwidth*, we can write simply

$$\sigma_n^2 = \frac{N_0 2 B_L}{A^2} \qquad (5.7.8)$$

Noting that $A^2/2$ is the IF carrier power and that $N_0 B_L$ is the noise power in the tracking loop noise bandwidth in (5.7.7), we can denote

$$\text{CNR}_L \overset{\Delta}{=} \frac{A^2/2}{N_0 B_L} \qquad (5.7.9)$$

as the loop CNR in the loop noise bandwidth. Hence

$$\sigma_n^2 = \frac{1}{\text{CNR}_L} \qquad (5.7.10)$$

Thus the mean square tracking error contributed by noise is given by the reciprocal of this CNR. It is again important to note that $H(\omega)$, and therefore B_L, is itself a function of A, as is evident from (5.7.1), and the loop noise bandwidth actually depends on the received carrier power. Therefore, in specifying a particular loop bandwidth, indication of the carrier power level at which it must be measured must also be given.

Table 5.1 also lists the relations for one-sided loop noise bandwidths B_L obtained by evaluating (5.7.7) using (5.7.1). Note that increasing the order of the loop always widens the loop noise bandwidth, while also increasing its tracking capability (shorter response time).

The modulation term $\theta_{em}(t)$ of the loop phase error has a transform given by the first term in (5.7.3). Its effect therefore depends on the form of the carrier modulation $\theta(t)$. If $\theta(t)$ is considered to be a deterministic time function, then $\theta_{em}(t)$ can be computed by inverse transforming and will contribute a deterministic time response to the loop phase error. For example, if we consider the carrier phase modulation to be a sinusoid of frequency ω_m and amplitude Δ, then $\theta_{em}(t)$ in (5.7.4) is also a sinusoid with amplitude $|1 - H(\omega_m)|\Delta$. Note that this modulation error term is approximately zero if $H(\omega_m) \approx 1$ or if ω_m is well within the bandwidth of the loop transfer function $H(\omega)$. For other types of deterministic modulation, $\theta_{em}(t)$ can be determined by straightforward linear system theory.

Alternatively, the carrier phase modulation $\theta(t)$ can be modeled itself as a Gaussian modulation process with spectrum $\Delta^2 S_m(\omega)$. The time process $\theta_{em}(t)$ is now a Gaussian response process having a zero mean value and a mean square value at any t given by

$$\sigma_m^2 = \frac{\Delta^2}{2\pi} \int_{-\infty}^{\infty} |1 - H(\omega)|^2 S_m(\omega)\, d\omega \qquad (5.7.11)$$

The error process $\theta_e(t)$ in (5.7.5) is then the sum of two uncorrelated Gaussian processes, which is itself a Gaussian process, having a total mean square value at any t of

$$\sigma_e^2 = \sigma_n^2 + \sigma_m^2 \tag{5.7.12}$$

with σ_n^2 given in (5.7.10). We see that the phase error in the modulation tracking PLL always appears as a random Gaussian time process, the latter a direct consequence of our assumption that the noise processes are Gaussian and the loop is linear. If the carrier modulation is considered a deterministic time function, then $\theta_{em}(t)$ is a known time function that serves as the mean of the process $\theta_e(t)$, the latter having mean square value σ_n^2 about this mean at each value of t.

With these models in mind, let us investigate our principal assumption that the loop is linear, which means that $\theta_e(t)$ must be small (<1) for almost all t. At any instant of time, $\theta_e(t)$ is a Gaussian random variable with zero mean (assuming $\theta_{em}(t) = 0$) and variance σ_e^2. Therefore, the probability that this error process is less than any desired phase error value θ_0 at any particular time t is

$$\mathrm{Prob}\,[|\theta_e(t)| < \theta_0] = \int_{-\theta_0}^{\theta_0} \frac{1}{\sqrt{2\pi}\sigma_e} \exp\left[\frac{-x^2}{2\sigma_e^2}\right] dx$$

$$= \mathrm{Erf}\left(\frac{\theta_0}{\sqrt{2}\sigma_e}\right) \tag{5.7.13}$$

where $\mathrm{Erf}\,(a)$ is tabulated in Appendix B. In general, we can consider $\sin\theta_e \approx \theta_e$ if $|\theta_e| \le \pi/6\,\mathrm{rad}\,(=30°)$. Figure 5.14 shows a plot of (5.7.13) for $\theta_0 = 30°$ as a function of σ_e. One can consider this probability as equivalent to the percentage of time the loop remains in linear operation. Thus

$$\mathrm{Prob}\,[|\theta_e(t)| \le 30°] = \mathrm{Erf}\left(\frac{\pi/6}{\sqrt{2}\sigma_e}\right) \tag{5.7.14}$$

For the loop to remain linear 95% of the time, we require

$$\frac{\pi/6}{\sigma_e} \ge 2$$

or

$$\sigma_e^2 \le \left(\frac{\pi}{12}\right)^2 \tag{5.7.15}$$

Hence the mean square value of the phase error at any t must be sufficiently small to guarantee linear operation. For deterministic modulation and zero modulation tracking error, $\sigma_e^2 = \sigma_n^2$, and (5.7.10) and (5.7.15) imply

$$\mathrm{CNR}_L \ge \left(\frac{12}{\pi}\right)^2 = 14.59 = 11.6\,\mathrm{dB} \tag{5.7.16}$$

For random modulation, σ_e^2 is given in (5.7.12), and we have instead

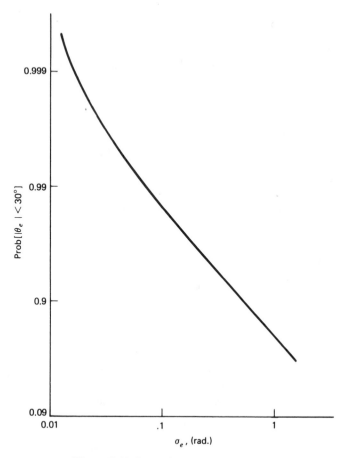

Figure 5.14. Loop phase error probability.

$$\text{CNR}_L \geq \frac{1}{(\pi/12)^2 - \sigma_m^2} \qquad (5.7.17)$$

which requires a higher value of CNR_L. We therefore conclude that a PLL phase tracking demodulator is linear only if the CNR in its loop noise bandwidth is sufficiently large. The right-hand sides of (5.7.16) and (5.7.17) serve as a threshold for determining just how large this CNR_L must be, depending on the assumptions made concerning the modulation function $m(t)$. Thus the condition that the PLL be linear is equivalent to a threshold condition on the CNR in the B_L bandwidth. If the threshold condition is violated, then the probability of the loop operating in the linear region is decreased, as is evident from (5.7.14). If the loop does not track the modulation in a linear manner, then the equivalent baseband filtering model developed earlier is no longer valid, and the demodulated baseband waveform is further distorted.

It should be emphasized here that the threshold depends on the noise in the loop bandwidth, which must be only wide enough to track the carrier phase modulation. In general, this modulation bandwidth is on the order of B_m Hz, and the loop noise bandwidths are much less than typical IF noise bandwidths. This fact demonstrates the advantage of using feedback tracking demodulation instead of standard detector-type demodulation, discussed earlier. The latter system must transmit enough carrier power to combat the noise in the entire IF bandwidths to satisfy their threshold condition [recall (5.5.4) and (5.4.17)]. However, the feedback systems must only overcome the noise in the loop bandwidth $B_L \approx B_m$ to satisfy its threshold in (5.7.16). Hence tracking demodulators can be operated with less carrier power than detector demodulators. It is precisely this fact that has fostered an increasing interest in phase tracking demodulation in modern systems.

5.8 DESIGN OF MODULATION TRACKING LOOPS

In the previous section we have shown how we can begin with a particular phase demodulating loop and, knowing the properties of the IF signal to be demodulated (i.e., knowing the modulation waveform and noise spectrum), we can determine the ability of the loop to remain in lock and act as a phase demodulator. In particular, we found that this condition was related to the loop tacking error. We can, instead, approach the problem from the opposite point of view. Given the waveform properties, we can determine the best loop design for maintaining a locked condition (i.e., minimizing the mean squared tracking error). In this section we consider this approach to the design of a modulation tracking PLL. We concentrate on two particular cases of interest.

Sinusoidal Modulation, Second Order PLL

Consider first the case where

$$\theta(t) = \Delta \sin(\omega_m t + \psi_m) \qquad (5.8.1)$$

and ψ_m is a uniformly distributed random phase angle over $(0, 2\pi)$. Since the modulation process has a power spectrum composed of delta functions at $\pm \omega_m$, the mean squared loop tracking error, from (5.7.11) and (5.7.12), is then

$$\sigma_e^2 = \tfrac{1}{2}|1 - H(\omega_m)|^2 \Delta^2 + \sigma_n^2 \qquad (5.8.2)$$

We assume a second order PLL is to be used for the modulation tracking, represented by the general loop gain function in (5.7.4). Hence

$$|1 - H(\omega_m)|^2 = \left| \frac{(\omega_m/\omega_n)^2}{(\omega_m/\omega_n)^2 + (2\zeta\omega_m/\omega_n) + 1} \right|^2 \tag{5.8.3}$$

and

$$\sigma_n^2 = \frac{N_0 B_L}{A^2/2} = \frac{N_0}{A^2} \left[\frac{\omega_n}{4\zeta} (1 + 4\zeta^2) \right] \tag{5.8.4}$$

Thus the loop mean squared error depends explicitly on the loop parameters ω_n and loop damping factor ζ. If we assume that the modulation frequency ω_m lies well within the loop natural frequency so that $\omega_m \ll \omega_n$, we can approximate

$$\sigma_e^2 \cong \left(\frac{\omega_m}{\omega_n} \right)^4 \frac{\Delta^2}{2} + \left(\frac{N_0}{A^2} \right) \left[\frac{\omega_n}{4\zeta} (1 + 4\zeta^2) \right] \tag{5.8.5}$$

The mean squared error σ_e^2 can now be minimized with respect to ω_n and ζ by solving the simultaneous equations:

$$\frac{d\sigma_e^2}{d\zeta} = 0 = \frac{N_0 \omega_n}{4} \left(\frac{-1}{\zeta^2} + 4 \right)$$

$$\frac{d\sigma_e^2}{d\omega_n} = 0 = \frac{\omega_m^4 \Delta^2}{2} \left(\frac{-4}{\omega_n^5} \right) + \frac{N_0}{A^2 4\zeta} (1 + 4\zeta^2) \tag{5.8.6}$$

The solution follows immediately as

$$\zeta = \tfrac{1}{2}$$

$$\omega_n = \left[\frac{2\Delta^2 \omega_m^4 A^2}{N_0} \right]^{1/5} \tag{5.8.7}$$

Note that a loop damping factor of $\tfrac{1}{2}$ is specified, whereas the required loop natural frequency ω_n is directly related to the index and frequency of the modulation and to the parameters of the IF noise. This, unfortunately, is a property of optimal design procedures associated with a particular type of waveform—the resulting design is directly related to the specific waveform parameters. This means that these parameters values must be known precisely before the optimal loop can be constructed. The parameters in (5.8.7) can be used in conjunction with Table 5.1 to determine the loop gain and design the appropriate loop filter function.

Random Modulation

Now consider the case where $\theta(t)$ is a Gaussian random modulating process with zero mean and power spectrum $S_m(\omega)$. The mean square error is now given by

$$\sigma_e^2 = \frac{1}{2\pi} \int_{-\infty}^{\infty} \left[|1 - H(\omega)|^2 S_m(\omega) + |H(\omega)|^2 \left(\frac{N_0}{A^2} \right) \right] d\omega \qquad (5.8.8)$$

The first term does not integrate in closed form as it did in the previous case. Hence (5.8.8) must be minimized directly. By straightforward application of the calculus of variations (Section A.3) we define

$$f[H(\omega)] = |1 - H(\omega)|^2 S_m(\omega) + |H(\omega)|^2 \left(\frac{N_0}{A^2} \right) \qquad (5.8.9)$$

We then replace $H(\omega)$ by $H_0(\omega) + \epsilon\eta(\omega)$ and compute the solution function $H_0(\omega)$ that satisfies

$$\frac{\partial f[H_0(\omega) + \epsilon\eta(\omega)]}{\partial \epsilon} \bigg|_{\epsilon=0} = 0 \qquad (5.8.10)$$

for any arbitrary function $\eta(\omega)$. This yields

$$0 = -\{[1 - H_0(\omega)]\eta^*(\omega) + [1 - H_0^*(\omega)]\eta(\omega)\} S_m(\omega)$$

$$+ [H_0(\omega)\eta^*(\omega) + H_0^*(\omega)\eta(\omega)] \frac{N_0}{A^2}$$

$$= 2 \operatorname{Real}\left\{ \left[-[1 - H_0(\omega)]S_m(\omega) + H_0(\omega)\left(\frac{N_0}{A^2} \right) \right] \eta^*(\omega) \right\} \qquad (5.8.11)$$

where * denotes complex conjugates. The optimal solution $H(\omega) = H_0(\omega)$ follows as

$$H_0(\omega) = \frac{S_m(\omega)}{S_m(\omega) + (N_0/A^2)} \qquad (5.8.12)$$

The preceding equation yields the PLL loop gain function for minimizing (5.8.8). However, (5.8.12) may not always produce a realizable loop function. (Such a constraint was not part of the initial problem formulation.) A realizable form of $H(\omega)$ is obtained by utilizing only the upper ω plane roots of the function on the right, and (5.8.12) must be interpreted as a solution in this sense. We mention that (5.8.12) is, in reality, a form of *Wiener filter* [13], about which much has been written in terms of realizable and unrealizable filter solutions [14, 15]. Thus the design of modulation tracking loops can be readily obtained by making use of well-known results from the theory of Wiener filtering.

As an example, let

$$S_m(\omega) = \frac{\Delta^2}{1 + (\omega/\omega_m)^2} \qquad (5.8.13)$$

Then in (5.8.12),

$$H_0(\omega) = \frac{C_1^2}{1 + C_2\omega^2} \qquad (5.8.14)$$

where $C_1^2 = \Delta^2/(\Delta^2 + N_0)$ and $C_2 = N_0/\omega_m^2\Delta^2$. The denominator in (5.8.14) must now be factored to determine the upper half ω plane roots. Since $(1 + C_2\omega^2) = (1 + j\sqrt{C_2}\,\omega)(1 - j\sqrt{C_2}\,\omega)$, we have

$$H_0(\omega) = \frac{C_1}{1 + j\sqrt{C_2}\,\omega} \qquad (5.8.15)$$

which is the realizable loop gain function. Note that the desired loop is in this case a first order loop with loop time constants dependent on the modulation and noise parameters. In particular, note that as the noise is weakened, $N_0 \rightarrow 0$, the required loop bandwidth increases, whereas the opposite is true when the noise is made stronger. Equation (5.7.1) identifies the required loop filter function as

$$AKF(\omega) = \frac{j\omega H_0(\omega)}{1 - H_0(\omega)} = C_1\frac{j\omega}{(1 - C_1) + j\sqrt{C_2}\,\omega} \qquad (5.8.16)$$

which can easily be constructed as an RC network.

We see therefore, from these examples, that there are well-defined theoretical approaches available for aiding in the design of modulation tracking phase demodulators. The relationship of loop design and Wiener filter theory has been shown, which generates a plethora of loop design procedures and interpretations for optimal demodulator construction. The interested reader may wish to pursue this point further in the work of Stiffler [16], Lindsey [17], Lindsey and Simon [10], and Van Trees [18].

5.9 CARRIER SUBSYSTEM MODELS

We have investigated carrier demodulation techniques associated with the various modulation formats presented in earlier chapters. This demodulation operation has the primary objective of extracting the baseband modulating waveform from the carrier and is hindered by the presence of receiver noise that contaminates the carrier signal at the receiver IF. We have seen, however, that if the strength of the noise is weak relative to that of the carrier signal, then, to a first order approximation, the demodulation operation recovers the carrier baseband waveform, but with an additive noise interference. The condition on the relative signal strengths was developed in the form of a threshold condition on the ratio of the IF carrier to noise power levels. The characteristics of the demodulated noise depend on the type of demodulation used. In all cases we were able to determine a demodulated output signal to noise ratio, which serves as an overall indicator of the performance of the entire carrier subsystem. Below threshold, when the receiver noise processes become significant relative to the carrier, we invariably noted a rapid degradation of this SNR, indicating a deterioration of the link performance.

These conclusions allow us to identify a basic carrier subsystem model representing the entire baseband transmission from transmitter to demodulator output. This model is shown in Figure 5.15. In Figure 5.15a the carrier subsystem is shown in block diagram form, with the baseband waveform to be modulated appearing at the transmitter modulator. This waveform is then modulated onto the carrier, transmitter over the carrier channel, received and processed in the receiver front end, and finally demodulated to the recovered baseband at the receiver. Each of these operations has been discussed at length in the past chapters. The demodulated baseband contains the equivalent filtered version of the transmitter modulation, where the filtering accounts for any baseband filtering, or effective baseband filtering imposed by the bandpass carrier filtering. To this is added the demodulation baseband noise $n_b(t)$. Thus, as far as the baseband waveform is concerned, it appears as if the entire carrier subsystem has merely filtered the modulating waveform and added in the demodulating noise. This leads to the block diagram representation shown in

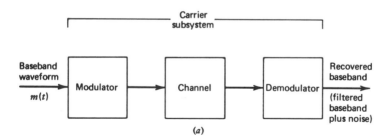

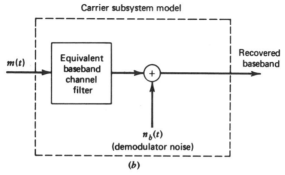

Figure 5.15. Carrier subsystem model: (a) subsystem block diagram, (b) equivalent baseband additive noise channel model.

Figure 5.15*b*, which is often referred to as the *additive noise channel model* for the carrier subsystem. Implicit in this model is the basic assumption that the transmitter carrier power is sufficient to satisfy demodulation thresholds. In addition, other carrier interference effects, such as those discussed in Section 3.7, have been neglected.

Note that in this additive noise channel model, the form of the carrier modulation enters only through the characteristics of the additive Gaussian baseband noise $n_b(t)$. This noise will have a power spectrum, $S_{nb}(\omega)$, that reflects the type of demodulation that was used. Summarizing the results of this chapter, we see that $S_{nb}(\omega)$ will have one of the forms:

$$S_{nb}(\omega) = \begin{cases} N_0|\tilde{H}_{IF}(\omega)|^2|H_m(\omega)|^2\,, & \text{for AM} \\[2mm] \left(\dfrac{N_0}{A^2}\right)\omega^2|\tilde{H}_{IF}(\omega)|^2|H_m(\omega)|^2\,, & \text{for FM} \\[2mm] \left(\dfrac{N_0}{A^2}\right)|\tilde{H}_{IF}(\omega)|^2|H_m(\omega)|^2\,, & \text{for PM} \end{cases} \qquad (5.9.1)$$

where $N_0 = \kappa T^{\circ}_{eq}G_1$, $A^2/2$ is the IF carrier power, $\tilde{H}_{IF}(\omega)$ is the equivalent IF filtering referred to the baseband, and $H_m(\omega)$ is the demodulator low pass filtering. When the filters involved are ideal filters, (5.9.1) simplifies to give the demodulated SNR_d values computed earlier. In the next chapter we make use of the additive noise model, to investigate in detail the baseband portion of the transmitter and receiver subsystems.

REFERENCES

1. Smith, J. *Modern Communication Circuits*, McGraw-Hill, New York, 1986.
2. Williams, R. *Communication System Analysis and Design*, Prentice-Hall, Englewood Cliffs, N.J., 1987.
3. Carlson, B. *Communication Systems*, 3rd ed., McGraw-Hill, New York, 1986.
4. Gregg, W. *Analog and Digital Communications*, Wiley, New York, 1977.
5. Blackman, N. *Noise and Its Effect on Communications*, McGraw-Hill, New York, 1966.
6. Middleton, D. *Statistical Communication Theory*, McGraw-Hill, New York, 1960.
7. Goldman, S. *Frequency Analysis, Modulation, and Noise*, McGraw-Hill, New York, 1948.
8. Stumpers, F. "Theory of FM Noise," *Proc. IRE*, September 1948.
9. Taub, H. and Schilling, D. *Principles of Communication Systems*, 2nd ed., McGraw-Hill, New York, 1986, Chap. 10.
10. Lindsey, W. and Simon, M. *Telecommunication Systems Engineering*, Prentice-Hall, Englewood Cliffs, N.J., 1973.

11. Viterbi, A. *Principles of Coherent Communications*, McGraw-Hill, New York, 1966.

12. Gardner, F.M. *Phaselock Techniques*, Wiley, New York, 1966.

13. Wiener, N. *Extrapolation, Interpolation, and Smoothing of Stationary Time Series*, MIT Press, Cambridge, Mass., 1949.

14. Lee, Y. *Statistical Theory of Communication*, Wiley, New York, 1960, Chaps. 14–19.

15. Thomas, J. *Introduction to Statistical Communication Theory*, Wiley, New York, 1969, Chap. 5.

16. Stiffler, J. *Theory of Synchronous Communications*, Prentice-Hall, Englewood Cliffs, N.J., 1971, Chap. 5.

17. Lindsey, W. *Synchronization Theory in Control and Communications*, Prentice-Hall, Englewood Cliffs, N.J., 1971.

18. Van Trees, H.L. *Detection, Estimation and Modulation Theory*, Part 2, Wiley, New York, 1971, Chap. 2.

PROBLEMS

5.1 (5.2) Let $n_c(t)$ be a zero mean Gaussian random process with power σ^2. (a) Write the probability that $|n_c(t)| \leq A$ at any t. (b) Show that part a will be satisfied with a probability greater than P if $A^2/2\sigma^2 \geq (\mathrm{Erf}^{-1}[P])^2$.

5.2 (5.2) An AM demodulator uses the envelope detector shown in Figure 5.2. (a) Derive an expression for the output SNR_d when the AM modulation $m(t)$ has power spectrum $S_m(\omega) = [1 + [\omega/2\pi B_m)^2]^{-1}$ and $H_m(\omega) = 1$, $|\omega| \leq 2\pi B_m$. Assume preceding threshold operation, white noise, and infinite IF bandwidths. (b) Repeat when $H_m(\omega)$ is replaced by the low pass filter $H_m(\omega) = [1 + (j\omega/2\pi B_m)]^{-1}$.

5.3 (5.3) Show that a phase coherent AM detector that is phase coherent with the sinusoidal term of a SSB-AM signal will demodulate the SSB-AM signal.

5.4 (5.3) Show that a suppressed carrier AM waveform can be demodulated by adding to it a phase coherent pure carrier and squaring the resulting combined signal.

5.5 (5.3) (a) Derive the expression for the output SNR_d for an ideal suppressed carrier AM system, assuming phase coherent demodulation. Assume that the IF carrier has power P_c, modulation has unit power, and bandwidth is B_m Hz. (b) Show that $\mathrm{SNR}_d = 2\,\mathrm{CNR}_{IF}$, where CNR is the IF CNR.

5.6 (5.4) Assume that the IF noise is white over the IF bandwidth and sketch the shape of the power spectrum of the next $(1/A^2)$ terms in (5.4.9).

5.7 (5.4) A device produces a unit impulse every time a positive going zero crossing occurs. Show that placing a T sec integrator after the device produces an approximation to a FM detector. What is the required relation between T and the bandwidth of the modulation?

5.8 (5.4) An FM system has a receiver RF threshold of 20 dB. How much received carrier power and RF bandwidth is needed to transmit a 1 kHz baseband signal with a demodulated SNR of 40 dB ($N_0 = 10^{-10}$ W/Hz)?

5.9 (5.4) (a) An FM system uses a modulating sine wave at 10 kHz and a deviation of 100 kHz. If the demodulator operates at a threshold of 15 dB, what is the demodulated SNR? (b) If the deviation was limited to 1% of the carrier frequency, what is the maximum SNR_d attainable with a 3 MHz RF carrier?

5.10 (5.4) A communication link is to transmit a 6 MHz TV channel over a 36 MHz RF bandwidth. A demodulation input threshold of 12 dB is required. The noise level is -150 dBW/Hz. (a) How much received carrier power is needed, and what SNR_d will be achieved, if AM is used? (b) Repeat for FM, assuming that the maximum possible frequency deviation is used.

5.11 (5.4) A communication system has an RF bandwidth of B Hz and can provide an RF CNR of Y in the bandwidth B. We require a demodulated SNR of Y and can use either AM or FM modulation with standard demodulation. Which system will allow the largest baseband bandwidth?

5.12 (5.4) Two sine waves at frequency f_1 and f_2, each with deviation Δ_ω, are added and transmitted by FM modulation on an RF carrier. An ideal FM receiver is operated above threshold with received carrier

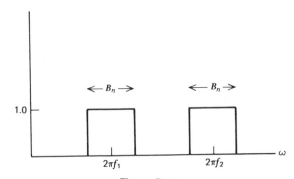

Figure P5.12.

power P_r and additive RF noise of level N_0. The demodulated output is filtered by the filters shown in Figure P5.12. (a) Determine the output SNR of each filter. (b) Repeat for an ideal PM system, with mod index Δ.

5.13 (5.4) A pure carrier of amplitude A at 1 GHz and a white noise process of level N_0 are recieved at the input to a system composed at a bandpass filter at 1 GHz with bandwidth 20 MHz, followed by an ideal FM detector and low pass filter with 20 MHz bandwidth. (a) Determine the output spectral density. (b) Determine the output power. (c) Repeat part a with the carrier shifted to 1.005 GHz (system remains the same).

5.14 (5.4) (a) Derive an expression for the output SNR_d of an FM demodulator in which the low pass output filter has transfer function $|H(\omega)|^2 = [1 + (\omega/2\pi B_m)^4]^{-1}$. Assume that $P_m = 1$, wideband FM $(\Delta_\omega \gg 2\pi B_m)$, white noise, above threshold operation, and that the filter passes $m(t)$ without distortion. (b) Compare the result to the ideal filter case.

5.15 (5.4) An FM system uses a standard demodulator with a 15 dB threshold. It is to achieve an output SNR of 50 dB when transmitting a 10 kHz modulating signal. The carrier system parameters are: range loss -165 dB, isotropic transmitting antenna, receiving antenna gain 40 dB, and receiver noise equivalent spectral level $N_0 = -191$ dBW/Hz. Neglecting pointing, coupling, and atmospheric losses, determine the required carrier bandwidth and transmitter power to maintain the system.

5.16 (5.4) A sine wave at frequency 8 MHz frequency modulates a carrier at 1 GHz with a deviation of Δ_ω. An ideal FM receiver, with IF bandwidth of 20 MHz, is operated above threshold with received carrier power P_r and additive noise of level N_0. The frequency detected output is filtered by a bandpass filter centered at frequency 8 MHz with an ideal bandwidth of 6 MHz. Determine the output SNR_d (in integral form).

5.17 (5.4) An FM system is operated with the following RF parameters:

Transmitter power = 10 kW   Rain attenuation = 0.2 dB/mi
Range 500 mi   Free space loss = 150 dB
Transmitted antenna gain = 60 dB   Sky background = 50 K
RF bandwidth = 0.2 MHz   Receiver noise figure = 3 dB
Basebandwidth = 10 kHz   Receiver antenna gain = 10 dB

What is the achievable baseband SNR_d at the receiver after FM demodulation?

5.18   (5.4)   An FM communication system transmits a baseband of B_m Hz and requires a demodulated SNR_d of 60.7 dB. Two FM receivers are available having the RF sections shown in Figure P5.18. Which receiver will require the least RF carrier power to satisfy the conditions? Neglect background noise and assume that B_{IF} is adjusted to the Carson Rule bandwidth of the carrier.

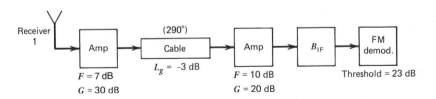

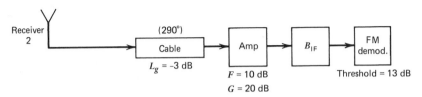

Figure P5.18.

5.19   (5.5)   Consider again the system described in Problem 5.13, but with the FM detector replaced by an ideal phase detector. Determine the output spectral density of the low pass filter output when the pure carrier and noise of Problem 5.13 is inserted at the input.

5.20   (5.6)   A PLL having a loop noise bandwidth of 1 kHz has an unmodulated carrier plus white Gaussian noise of level $N_0 = 10^{-10}$ W/Hz fed into it. Using linear theory, determine the required carrier power to guarantee a mean squared tracking error of 0.01 radian^2.

5.21   (5.6)   Verify the bandwidth B_L of the first order loop in Table 5.1.

5.22   (5.6)   A PLL with noise bandwidth of 100 Hz receives the process

$$x(t) = A \cos [\omega_0 t + m(t)] + n(t)$$

where $n(t)$ is white noise $(N_0 = 10^{-4}$ W-sec). The loop operates linearly, and the tracking of $m(t)$ produces a mean squared error of 0.46 radians. What is the necessary value of A to maintain a total rms tracking error of 0.707 radians?

5.23   (5.7)   In the frequency synthesizer loop shown in Figure 3.4 the frequency detector is replaced by a phase detector (multiplier).

Derive an equivalent baseband phase model to describe the phase on the output VCO in terms of the phase on the input reference oscillator. Explain how the reference phase noise spectrum is filtered in appearing on the output carrier.

5.24 (5.8)   A PLL tracks a carrier in additive noise (N_0) using a loop with filter $F(\omega)$ and loop gain K. Determine the loop noise bandwidth as the carrier amplitude $A \to 0$.

5.25 (5.8)   Show that the loop filter function can be constructed as either of the RC networks in Figure P5.25. Show that the filter parameters and time constants are related as shown.

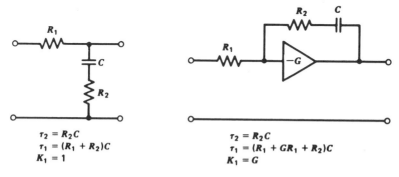

$$\tau_2 = R_2 C$$
$$\tau_1 = (R_1 + R_2)C$$
$$K_1 = 1$$

$$\tau_2 = R_2 C$$
$$\tau_1 = (R_1 + GR_1 + R_2)C$$
$$K_1 = G$$

Figure P5.25.

5.26 (5.9)   Complete the design problem in (5.8.7). That is, design the loop filter and gain using Table 5.1 and Problem 5.25.

5.27 (5.10)   A PM carrier with a phase deviation $\Delta \ll 1$ is passed through a bandpass carrier filter function $H_c(\omega)$. It is then ideally phase demodulated to produce the modulation. Show that the equivalent filter function $H_{bc}(\omega)$ shown in Figure 5.15 corresponds to $\tilde{H}_c(\omega)$, the low frequency version of $H_c(\omega)$.

6

BASEBAND WAVEFORMS, SUBCARRIERS, AND MULTIPLEXING

Our study to this point has concentrated on the carrier portion of the communication system. We have shown that under proper operating conditions, the entire carrier transmission can be represented by a basic additive noise model operating on the transmitter baseband waveform. The baseband waveform was considered to be specified by a known bandwidth B_m and power level P_m. We now wish to examine more specifically the actual characteristics of typical baseband waveforms and the manner in which the desired source information can be extracted at the receiver. Recall from Figure 1.2a that the baseband is formed from the source signal and therefore contains, in some context, the desired information. The specific form of the baseband waveform depends on the type of source, and whether a single source or a group of sources is involved. In this chapter the various baseband formats are investigated.

6.1 THE BASEBAND SUBSYSTEM

The baseband subsystem represents the portion of the communication link involving the source and baseband waveforms. As such, the baseband subsystem contains components at both the transmitter and receiver ends, which are interconnected by the carrier subsystem previously considered. A typical baseband subsystem model is shown in Figure 6.1. The source signal is baseband converted to the baseband waveform, which is then transmitted over the carrier subsystem via modulation and demodulation. The latter subsystem can be modeled by the additive noise and filtering channel shown in Figure 5.15. The baseband waveform recovered at the receiver must then be baseband deconverted to regenerate the source waveform. It is evident

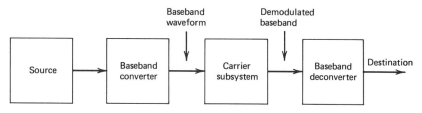

Figure 6.1. The baseband subsystem block diagram.

that a study of the baseband subsystem reduces primarily to a study of the converting and deconverting operations.

The form of the baseband waveform $m(t)$ depends on the conversion format and the type of information source. When a single source is involved, its waveform may be used directly as the baseband signal, or it can be converted by an encoding, filtering, or modulaton format. When many sources are involved, the source set must be multiplexed (combined together) to form the baseband. In all cases the resulting baseband waveform is determined by the inherent properties of the source signal itself and the processing imposed by the converter. Recall that sources are classified as either analog or digital. An analog source produces a continuous waveform in time, and its time variation represents the information sent to the receiver destination, as with audio or video sources. A digital source produces a sequence of binary symbols, and the baseband signal is formed by encoding the source symbols into a baseband waveform.

If the source waveform is used directly as the baseband signal, then clearly $m(t)$ has the same characteristics as the source itself. However, in many cases it may be advantageous to filter the source waveform prior to carrier modulation, or perhaps to modulate the source waveform onto a secondary carrier (subcarrier) to form the baseband signal. The advantage of filtering is that we can shape the baseband spectrum prior to carrier modulation, which allows us to emphasize or deemphasize specific portions of the source spectrum so as best to combat channel distortion. Subcarrier modulation allows us to control the location of the baseband spectrum along the frequency axis and thereby shift the baseband away from undesired interference. In the next sections we examine these forms of baseband conversion operations.

6.2 BASEBAND FILTERING SYSTEMS

With baseband filtering, the source signal is simply filtered prior to carrier modulation. At the receiver the demodulated baseband may then be further filtered to attempt to reconstruct the original source frequency characteristics. Using the additive noise model to represent the carrier subsystem, the overall baseband link from source to destination will have the form shown in Figure 6.2. The filter function $H_t(\omega)$ represents the premodulation filtering inserted at the transmitter for source waveform shaping. The filter function $H_{bc}(\omega)$ represents the channel filtering in Figure 5.15 effectively applied to the baseband by the channel during carrier transmission. The filter $H_b(\omega)$ represents baseband filtering inserted at the receiver following carrier demodulation. The spectrum $S_{nb}(\omega)$ is the spectral density of the added demodulator baseband noise, and $S_d(\omega)$ is the source spectrum. Typically, $H_t(\omega)$ and $H_b(\omega)$ are in the hands of the system designer, whereas $H_{bc}(\omega)$, $S_d(\omega)$, and $S_{nb}(\omega)$ are fixed by system constraints. The baseband design problem is then to select the base receiver and/or transmitting filters for the baseband link model of Figure 6.2. We assume our prime objective is to achieve as large a destination SNR, or signal distortion ratio SDR, as possible.

Consider first the case where no transmitter filtering is used [i.e., $H_t(\omega) = 1$] and the source waveform is used directly as the baseband. The receiver baseband filter $H_b(\omega)$ must then filter the baseband noise without significant distortion of the baseband signal. If we denote $d_0(t)$ as the filtered baseband signal at the destination, then a signal error term, $d(t) - d_0(t)$, can be defined, and a signal to distortion power can be determined as in (1.8.8). The total output distortion is then the sum of the error distortion power and the output noise power. Thus, as in Problem 1.32, we write

$$SDR = \frac{P_d}{\dfrac{1}{2\pi}\displaystyle\int_{-\infty}^{\infty} |1 - H_{bc}(\omega)H_b(\omega)|^2 S_d(\omega)\, d\omega + \dfrac{1}{2\pi}\displaystyle\int_{-\infty}^{\infty} |H_b(\omega)|^2 S_{nb}(\omega)\, d\omega} \tag{6.2.1}$$

where P_d is the source waveform power. With a given P_d, it is clear that SDR is maximized if the denominator in (6.2.1) is minimized. Hence we seek a baseband filter function $H_b(\omega)$ that minimizes this denominator for a

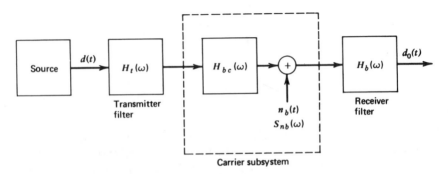

Figure 6.2. Baseband filtering model.

specified channel filter $H_{bc}(\omega)$. Equivalently, we seek the function $G(\omega) \overset{\Delta}{=} H_{bc}(\omega)H_b(\omega)$ that minimizes

$$\int_{-\infty}^{\infty} \left[|1 - G(\omega)|^2 S_d(\omega) + |G(\omega)|^2 \left(\frac{S_{nb}(\omega)}{|H_{bc}(\omega)|^2} \right) \right] d\omega \qquad (6.2.2)$$

When stated in this way, the problem is identical to the minimization carried out in Section 5.8, and by straightforward substitution we obtain the desired filter function for $H_b(\omega) = G(\omega)/H_{bc}(\omega)$, or

$$H_b(\omega) = \frac{1}{H_{bc}(\omega)} \left[\frac{S_d(\omega)}{S_d(\omega) + (S_{nb}(\omega)/|H_{bc}(\omega)|^2)} \right] \qquad (6.2.3)$$

Note that the optimal filter function inverts the channel filtering and filters according to the relative strengths of the signal spectrum and the effective noise spectrum $S_{nb}(\omega)/|H_{bc}(\omega)|^2$. This latter filtering is again a form of Wiener filter, similar to that derived in (5.8.12). At frequencies where $S_d(\omega) \gg S_{nb}(\omega)$, $H_b(\omega) \approx 1/H_{bc}(\omega)$, and the filter concentrates on compensating for the channel filtering over these frequencies. At frequencies where $H_{bc}(\omega)S_d(\omega) \ll S_{nb}(\omega)$, we have $H_b(\omega) \approx S_d(\omega)/S_{nb}(\omega) \ll 1$, and the filter effectively rejects transmission at frequencies where the noise dominates. Thus the optimal receiver baseband filter in (6.2.3) attempts to achieve a satisfactory compromise between reducing signal distortion caused by the channel filter and filtering the baseband noise from the output.

When the optimal filter function of (6.2.3) is inserted into (6.2.1), we obtain the maximum SDR that can be attained in the baseband channel. The denominator becomes

$$\frac{1}{2\pi} \int_{-\infty}^{\infty} \left\{ \left| 1 - \frac{S_d(\omega)}{S_d(\omega) + S'_{nb}(\omega)} \right|^2 S_d(\omega) + S'_{nb}(\omega) \left[\frac{S_d(\omega)}{S_d(\omega) + S'_{nb}(\omega)} \right]^2 \right\} d\omega$$

$$= \frac{1}{2\pi} \int_{-\infty}^{\infty} \frac{S'_{nb}(\omega)S_d(\omega)}{S_d(\omega) + S'_{nb}(\omega)} d\omega \qquad (6.2.4)$$

where $S'_{nb}(\omega) \overset{\Delta}{=} S_{nb}(\omega)/|H_{bc}(\omega)|^2$. That is, $S'_{nb}(\omega)$ is the baseband noise referred back to the source and represents the equivalent noise that would have to be inserted at the source to produce $S_{nb}(\omega)$. The maximum achievable baseband SDR at the receiver is then

$$\text{SDR}_{\max} = \frac{P_d}{\dfrac{1}{2\pi} \displaystyle\int_{-\infty}^{\infty} \dfrac{S'_{nb}(\omega)S_d(\omega)}{S_d(\omega) + S'_{nb}(\omega)} d\omega} \qquad (6.2.5)$$

Given the source and equivalent source noise spectra, we can therefore immediately determine the attainable SDR of the baseband subsystem. This is extremely convenient in analysis, since it allows us to assess the subsystem without having to first compute the required optimal receiver filter. Note that the SDR depends only on the integrated spectral densities in (6.2.5). Since it involves the product of the noise and source spectra at each ω, we see that we can never completely remove the baseband noise (i.e., obtain SDR$\to\infty$) as long as there is frequency overlap of the two spectra. Although we would intuitively expect this conclusion, we have here proved it to be rigorously true, and, in addition, have determined the precise manner in which the degree of overlap will influence performance.

Consider now the case where a baseband filter $H_t(\omega)$ is inserted at the transmitting end. We now have the advantage of having two separate baseband filters to adjust for best performance. We intuitively expect that the reciver filter $H_b(\omega)$ should be used primarily for noise rejection, whereas $H_t(\omega)$ can be used to compensate for signal filtering. Hence we can set

$$H_t(\omega)H_{bc}(\omega)H_b(\omega) = 1 \qquad (6.2.6)$$

while defining the transmitted baseband power

$$P_m = \frac{1}{2\pi} \int_{-\infty}^{\infty} |H_t(\omega)|^2 S_d(\omega)\, d\omega \qquad (6.2.7)$$

With (6.2.6) no source signal distortion occurs in the baseband link, and the output interference is due only to output noise. Hence the baseband output SNR is

$$\text{SNR} = \frac{P_d}{\dfrac{1}{2\pi} \displaystyle\int_{-\infty}^{\infty} |H_b(\omega)|^2 S_{nb}(\omega)\, d\omega} \qquad (6.2.8)$$

We again seek the minimization of the denominator of (6.2.8), subject to a fixed P_m in (6.2.7) and the condition of (6.2.6). From Section A.3, this requires minimization of

$$f[|H_b(\omega)|] = |H_b(\omega)|^2 S_{nb}(\omega) + \lambda \frac{S_d(\omega)/|H_{bc}(\omega)|^2}{|H_b(\omega)|^2} \qquad (6.2.9)$$

with respect to $|H_b(\omega)|$. The solution follows easily as

$$|H_b(\omega)|^4 = \frac{K^2 S_d(\omega)}{|H_{bc}(\omega)|^2 S_{nb}(\omega)} \qquad (6.2.10a)$$

where K is an arbitrary filter gain constant. The corresponding transmitter filter function, from (6.2.6), is then

$$|H_t(\omega)|^4 = \frac{S'_{nb}(\omega)}{K^2 S_d(\omega)} \tag{6.2.10b}$$

with K adjusted to satisfy (6.2.7). Thus the filter pair in (6.2.10), for the transmitter and receiver baseband subsystems, is required for maximization of SNR. Note that now only the receiver baseband filter magnitude is specified and is markedly different from the previous result of (6.2.3) when no transmitter filtering was included. Nevertheless, the filter still tends to have large gain at frequencies where the source spectrum dominates and low gain when the noise dominates. The corresponding transmitter filter essentially inverts the baseband operation, emphasizing the frequencies where the subsequent channel noise dominates and attenuating the source frequencies where the signal dominates. Such transmitter filters are often called *preemphasis filters*. Their basic role is purposely to predistort the signal spectrum so that the subsequent baseband filtering will regenerate the correct spectral shape, as indicated in (6.2.6), while producing the maximal output SNR.

6.3 SUBCARRIER MODULATED BASEBAND SYSTEMS

A baseband system using subcarrier modulation for baseband conversion is shown in Figure 6.3. The source waveform is modulated directly onto the subcarrier to form the baseband (Figure 6.3a), which is therefore itself a modulated carrier, having a baseband spectrum similar to that in Figure 6.3b. The advantage of this is that we control the location of the spectrum of the baseband signal, allowing us to shift the source spectrum away from the lower frequencies and avoid undesired baseband interference. The selection of the subcarrier frequency ω_s is arbitrary, although there may be physical constraints that limit the modulation parameters of an oscillator for a given subcarrier frequency. For example, in FM subcarriers there is a limit to the amount by which a subcarrier frequency can be deviated.

After the source signal is modulated onto the subcarrier the baseband is then modulated onto the main (RF) carrier for transmission, producing the RF carrier $c(t)$. Note there are two modulation stages—the source waveform onto the subcarrier and the subcarrier onto the RF carrier. Each modulation can be either AM, FM, or PM. For example, the source may be amplitude modulated onto the subcarrier, and the latter may be frequency modulated onto the RF carrier. This pair of modulation operations is referred to as AM|FM. One can similarly describe other types of systems as FM|FM, AM|PM, and so on.

At the receiving end, after RF and IF processing, the receiver must invert the two modulation operations and pass the IF signal through two stages of demodulation (Figure 6.3c). The first recovers the baseband subcarrier, and the second recovers the source from the subcarrier. We can analyze the

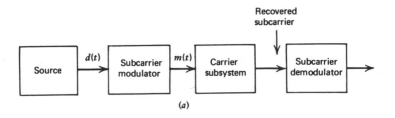

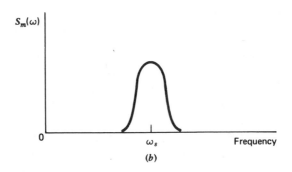

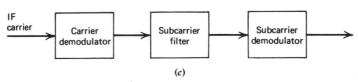

Figure 6.3. Baseband subcarrier system: (*a*) block diagram, (*b*) baseband spectrum, (*c*) receiver system.

cascade demodulation by applying successively the results of Chapter 5. In the following discussion we examine some specific examples of this type of modulation format.

AM|FM Systems

In AM|FM systems the source waveform is amplitude modulated onto the subcarrier and the baseband is frequency modulated onto the RF carrier. Denote the source waveform as $d(t)$ and assume it to occupy a bandwidth of B_d. We again assume $d(t) \leq 1$ to prevent overmodulation. The output of the carrier modulator in Figure 6.3 is the baseband signal $m(t)$ and has the normalized form

$$m(t) = [1 + d(t)] \cos(\omega_s t + \theta_s), \tag{6.3.1}$$

where ω_s and θ_s are the subcarrier frequency and phase angle, respectively. The baseband therefore has the spectrum shown in Figure 6.3b, with a subcarrier bandwidth about ω_s of $2B_d$ Hz. The upper frequency of $m(t)$ is then

$$B_m = \frac{\omega_s}{2\pi} + B_d \text{ Hz} \tag{6.3.2}$$

The baseband signal, in addition, has power

$$P_m = \tfrac{1}{2}(1 + P_d) \tag{6.3.3}$$

where P_d is the power of the source waveform $d(t)$ $(P_d \leq 1)$. The baseband frequency modulates the RF carrier producing at the receiver IF output the carrier waveform

$$c(t) = A \cos\left(\omega_c t + \Delta_\omega \int m(t)\, dt\right) \tag{6.3.4}$$

where Δ_ω is the frequency deviation of the RF carrier. This carrier occupies an IF bandwidth of approximately

$$
\begin{aligned}
B_{\text{IF}} &= 2\left[\frac{\Delta_\omega \sqrt{P_m}}{2\pi B_m} + 1\right]B_m \\
&= 2\left[\frac{\Delta_\omega}{2\pi B_m}\left(\frac{1 + P_d}{2}\right)^{1/2} + 1\right]B_m \text{ Hz}
\end{aligned}
\tag{6.3.5}
$$

where we have used the rms value of the modulation as the effective frequency deviation of the carrier. If the carrier is received with suitable power to exceed the threshold condition of the first (FM) demodulation,

$$\text{CNR}_{\text{IF}} = \frac{A^2/2}{N_0 B_{\text{IF}}} \geq Y \tag{6.3.6}$$

then the output of the demodulator has the spectrum shown in Figure 6.4. The demodulation has recovered the baseband with the addition of the quadratic frequency noise, as shown. This frequency noise has the spectrum $S_{nb}(\omega)$ in (5.9.1) for the FM case. The subcarrier bandpass filter centered at ω_s extracts the signal content only over the bandwidth occupied by the subcarrier. The signal power of the subcarrier at the filter is then

$$P_{\text{sc}} = \frac{\Delta_\omega^2}{2}(1 + P_d) \tag{6.3.7}$$

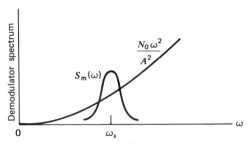

Figure 6.4. Demodulated subcarrier spectrum and FM noise.

The noise power at the filter output is obtained by integrating over the filtered spectrum of the noise. For an ideal subcarrier filter over the subcarrier bandwidth, the filter output noise power is

$$P_{ns} = \frac{2}{2\pi} \int_{\omega_s - \pi B_{sc}}^{\omega_s + \pi B_{sc}} \left(\frac{N_0 \omega^2}{A^2} \right) d\omega$$

$$= \frac{N_0}{A^2 3\pi} [(\omega_s + 2\pi B_d)^3 - (\omega_s - 2\pi B_d)^3] \tag{6.3.8}$$

Using the expansion

$$(a \pm b)^3 = a^3 \pm 3a^2 b + 3ab^2 \pm b^3$$

we reduce (6.3.8) to

$$P_{ns} = \frac{N_0}{3\pi A^2} [6\omega_s^2 (2\pi B_d) + 2(2\pi B_d)^3] \tag{6.3.9}$$

If we further assume that the subcarrier frequency is much larger than the source bandwidth, $\omega_s \gg 2\pi B_d$, then (6.3.9) is approximately

$$P_{ns} \cong 2\left(\frac{N_0 \omega_s^2}{A^2} \right) 2B_d \tag{6.3.10}$$

Since $2B_d$ is the subcarrier bandwidth, we see that (6.3.10) is equivalent to assuming that the FM demodulated noise spectrum is flat over the subcarrier bandwidth, with equivalent spectral level $N_0 \omega_s^2 / A^2$, two-sided. The input to the subcarrier demodulator therefore has a CNR, for $P_d = 1$, given by

$$\mathrm{CNR}_s = \frac{P_{sc}}{P_{ns}}$$

$$= \left(\frac{\Delta_\omega}{\omega_s} \right)^2 \left[\frac{A^2/2}{N_0 2B_d} \right] \tag{6.3.11}$$

This CNR therefore represents that available for subcarrier (AM) demodulation. Since the bracket is the IF CNR in the subcarrier bandwidth, we see that CNR_s is increased over that available in the IF. The improvement is by the square of the factor Δ_ω / ω_s. Since Δ_ω represents the RF carrier deviation by the subcarrier, Δ_ω / ω_s plays the role of a subcarrier modulation index. Again we see the availability of trading IF bandwidth for improved CNR_s, simply by increasing the deviation Δ_ω. We also note that, for a fixed deviation Δ_ω, CNR_s decreases with ω_s. This shows the disadvantage of using too large a subcarrier frequency and is simply caused by the increased frequency noise at the higher subcarrier frequencies, as seen from Figure 6.4.

The demodulated subcarrier signal must now pass into the second (AM) demodulator to recover the source signal. To achieve this, it is first necessary that the input CNR_s satisfy the AM demodulation threshold, as discussed in Sections 5.2 and 5.3. Thus we require

$$CNR_s \geq \Upsilon_{AM} \qquad (6.3.12)$$

where Υ_{AM} is the required AM threshold. This requirement allows us to determine the parameters in (6.3.11). After AM demodulation, the source signal will be recovered with a demodulated SNR given by (5.2.13),

$$SNR_d = CNR_s \qquad (6.3.13)$$

The parameter SNR_d is an indication of the overall fidelity of our communication link and is generally a parameter specified as a design constraint. That is, for a desired value of SNR_d we can use (6.3.11)–(6.3.13) to aid in the selection of the remaining system parameters.

Although the use of a subcarrier gives us control over baseband frequency location, an obvious question is whether we have paid a penalty for this advantage, as opposed to direct FM of the source waveform onto the RF carrier with no subcarrier. Recall from our FM analysis that the latter system, operating with the same peak deviation of Δ_ω, will produce a demodulated SNR of $SNR_d = 3(\Delta_\omega / 2\pi B_d)^2 (A^2 / 2N_0 B_d)$. Comparing this to (6.3.13) and (6.3.11), we see that the use of the subcarrier has produced an SNR_d that is lower by a factor proportional to $(\omega_s / B_d)^2$. We have therefore achieved a poorer source SNR_d by using the subcarrier to favorably control spectral location. As stated, this is simply due to the fact that the subcarrier forces us to operate at higher frequencies of the baseband where the demodulated baseband FM noise is greater. Tradeoffs with other system formats can also be made (Problem 6.8).

FM|FM Systems

Let us extend the discussion to FM|FM systems by repeating the analysis of

Section 6.3.1. In this case the source waveform is frequency modulated onto the subcarrier, producing

$$m(t) = \cos\left[\omega_s t + \Delta_s \int d(t)\, dt\right]$$ (6.3.14)

where Δ_s is the frequency deviation of the source onto the subcarrier. The RF carrier again appears as in (6.3.4). The baseband, subcarrier, and RF bandwidths follow as

$$B_m = \frac{\omega_s}{2\pi} + \frac{B_{sc}}{2}\ \text{Hz}$$

$$B_{sc} = 2\left(\frac{\Delta_s}{2\pi B_d} + 1\right)B_d\ \text{Hz}$$ (6.3.15)

$$B_{RF} = 2\left(\frac{\Delta_\omega}{2\pi B_m} + 1\right)B_m\ \text{Hz}$$

If (6.3.6) is satisfied, RF demodulation again produces the spectral diagram in Figure 6.4, except that the subcarrier spectrum is that of an FM signal. The filtered subcarrier power is then $\Delta_\omega^2/2$, and (6.3.10) approximates the noise power. Hence

$$\text{CNR}_s = \left(\frac{\Delta_\omega}{\omega_s}\right)^2 \left[\frac{A^2/2}{2N_0 B_{sc}}\right]$$ (6.3.16)

The preceding CNR must be large enough to satisfy the subcarrier demodulation FM threshold; that is,

$$\text{CNR}_s \geq Y_{FM}$$ (6.3.17)

Using our FM results of Section 5.4, the demodulated source signal will then be recovered with SNR of

$$\text{SNR}_d = 3\beta_s^2\, \frac{\Delta_\omega^2/2}{(N_0\omega_s^2/A^2)2B_d}$$

$$= 3\beta_s^2(\beta_s + 1)\left(\frac{\Delta_\omega}{\omega_s}\right)^2\left(\frac{A^2/2}{N_0 B_{sc}}\right)$$

$$= 6\beta_s^2(\beta_s + 1)\text{CNR}_s$$ (6.3.18)

where β_s is the subcarrier modulation index $\Delta_s/2\pi B_d$. We see now that the system receives the advantage of two consecutive stages of FM improvement. However, the RF bandwidth in (6.3.15) is also increasing faster, since both Δ_ω and B_m increase as the system attempts to take advantage of both SNR improvements. Again, a limit will exist on the amount by which β can be increased, because of the subcarrier deviation restriction.

FM|PM Systems

In FM|PM systems, phase modulation of the subcarrier onto the RF carrier is used. The baseband and carrier waveforms are then

$$m(t) = \cos \left[\omega_s t + \Delta_s \int d(t) \, dt \right] \qquad (6.3.19a)$$

$$c(t) = A \cos \left[\omega_c t + \Delta m(t) \right] \qquad (6.3.19b)$$

where Δ is now the phase modulation index of the carrier. The bandwidths are again given by (6.3.15), and the RF threshold condition is identical to (6.3.6) with the PM modulation threshold value inserted. However, the demodulated phase noise is no longer quadratic as in Figure 6.4, but rather has the two-sided flat spectrum N_0/A^2, given in (5.9.1). Thus the subcarrier CNR now becomes

$$\begin{aligned} \text{CNR}_s &= \frac{\Delta^2/2}{2(N_0/A^2)B_{sc}} \\ &= \frac{\Delta^2}{2} \left[\frac{A^2/2}{N_0 B_{sc}} \right] \end{aligned} \qquad (6.3.20)$$

The result is similar to the FM|FM case in (6.3.16), except the phase index is involved and the subcarrier frequency ω_s does not appear. Hence FM|PM systems are not directly influenced by the choice of subcarrier frequency. If CNR_s is above threshold, the source signal is finally demodulated with SNR of

$$\text{SNR}_d = 3\beta_s^2(\beta_s + 1)\text{CNR}_s \qquad (6.3.21)$$

Again the improvement caused by two stages of angle demodulation is apparent and the usual tradeoff with RF bandwidth is available. The system is constrained principally by the fact that enough carrier power $(A^2/2)$ must be provided to overcome the noise in the total IF carrier bandwidth in satisfying the PM threshold. When carrier power is limited this condition may not be easily satisfied, especially if wideband operation is used.

6.4   SOURCE MULTIPLEXING (FDM BASEBAND SYSTEMS)

When more than one sources is involved, and the information of all is to be sent simultaneously, the sources must be multiplexed (combined into a single baseband signal). This multiplexing must be such that the individual source waveforms can be separated at the receiver. A common way to achieve this is to modulate each source onto a separate subcarrier, each located at a different region of the frequency spectrum. The subcarriers can then be summed to form the multiplexed baseband signal, as shown in Figure 6.5*a*. A system of this type is referred to as a *frequency division multiplexed* (FDM) baseband system, since the source waveforms are effectively separated in frequency by the subcarriers. The modulation onto the subcarriers and the RF carrier can be done by any of the standard methods. Multiplexed systems of this type are denoted, for example, as FDM|AM|FM, indicating that FDM multiplexing is used, with AM on the subcarriers, and FM of the multiplexed baseband on the RF carrier.

A typical FDM baseband spectrum is given in Figure 6.5*b*, showing the multiplexed sources side by side, each at their respective subcarrier frequencies. The form of the spectrum about each subcarrier frequency depends on the modulation used. Often guard spacing is inserted between the spectra to obtain sufficient spectral separation. In some cases one source may be placed at zero frequency (no subcarrier used) and the remaining sources subcarrier modulated to a specific frequency range. In other cases all sources are placed on subcarriers. Theoretically, the lower end of the lower subcar-

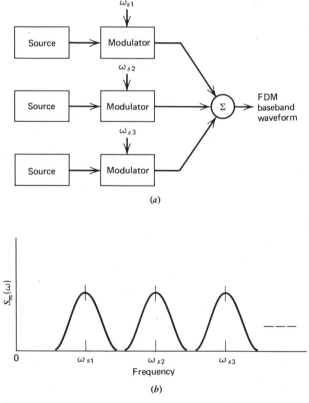

Figure 6.5. FDM subsystem: (*a*) multiplexing block diagram, (*b*) multiplexed spectrum.

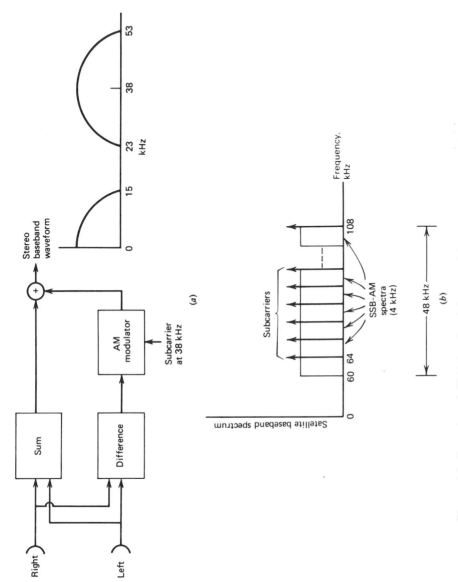

Figure 6.6. Examples of FDM baseband formats: (*a*) commercial stereo, (*b*) telephone channels.

rier channel can begin at zero frequency, but in practice it is usually located away from zero to avoid low frequency interference.

FDM baseband formats are common in our everyday commercial stereo radio, television, and telephone systems. Each television channel spectrum in Figure 2.2 is basically a form of FDM multiplexing of the video and audio information, with the voice frequency modulated onto the subcarrier at 5.25 MHz and placed above the video spectrum. Figure 6.6a shows the format for a commercial stereo radio transmission channel. The information from the left and right microphones is summed to form a source signal for direct low pass transmission. A second source signal, composed of the difference of the two microphone signals, is amplitude modulated onto a subcarrier of 38 kHz. The resulting baseband, which occupies a spectrum of about 53 kHz, is then transmitted via an FM carrier as a single stereo channel. FDM formats are also used in satellite baseband communications. Figure 6.6b shows a typical satellite repeater FDM telephone channel, containing 12 multiplexed voice channels. The baseband is part of an FDM|SSB-AM|FM system, in which single sideband AM is used to modulate each 4 kHz voice source onto the subcarrier frequencies shown. The resultant baseband spectrum is 48 kHz wide and is designated as a standardized Telephone Group [1, 2].

At the receiver, following RF demodulation, the general multiplexed spectrum Figure 6.5b is recovered. Source separation is achieved by merely filtering off the individual subcarrier channels and subcarrier demodulating (Figure 6.7). Adequate guard spacing between channels eases this subcarrier separation by the filters. If the information of several sources is desired, a parallel filter–demodulator channel must be available for each subcarrier. For example, in the stereo system in Figure 6.6a a stereo receiver demodulates both the 38 kHz subcarrier and the low pass spectra, which produces microphone sum and difference waveforms. These can then be separated

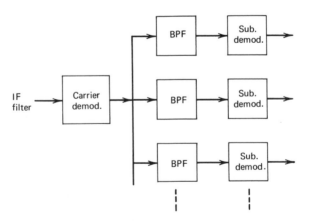

Figure 6.7. FDM receiver demultiplexing.

into left and right channels. A receiver without stereo reception capability (i.e., without a subcarrier demodulator) receives only the sum signal for monoaural performance. Thus a receiver in an FDM system demultiplexes the baseband waveform by simply bandpass filtering the desired subcarrier channel. FDM systems have this prime advantage of extreme simplicity in both multiplexing at the transmitter end and source separation at the receiving end. We emphasize that this frequency multiplexing is accomplished at the baseband level, and only a single RF carrier is being used.

FDM|AM|FM Systems

Let us examine an FDM baseband system in which AM is used on the subcarriers and FM on the RF carrier. The analysis parallels that of Section 6.3, with the added extension to more than one source. Consider N separate source waveforms $d_i(t)$, having power P_{di} and bandwidth B_{di}, respectively. The AM subcarrier multiplexing produces the baseband signal

$$m(t) = \sum_{i=1}^{N} C_i[1 + d_i(t)] \sin(\omega_{si}t + \psi_i) \qquad (6.4.1)$$

where ω_{si} are the subcarrier frequencies and ψ_i their phase angles. The baseband occupies an upper frequency bandwidth of

$$B_m = \frac{\omega_{sN}}{2\pi} + B_{dN} \qquad (6.4.2)$$

and a power level of

$$P_m = \sum_{i=1}^{N} \frac{C_i^2}{2}(1 + P_{di}) \qquad (6.4.3)$$

The RF carrier is formed again as in (6.3.4), so that C_i in (6.4.1) represents the actual frequency deviation of the ith subcarrier signal. Using an effective FM index β defined in (2.4.17), the wideband RF carrier therefore occupies a bandwidth of approximately

$$B_{RF} = 2\left[\frac{\sqrt{P_m}}{2\pi B_m} + 1\right]B_m \approx 2\left(\frac{\sqrt{P_m}}{2\pi}\right) \qquad (6.4.4)$$

The fact that B_{RF} depends on the baseband power level P_m in (6.4.3) is an important point here. Since specification of RF bandwidth is often a critical part of system design, accurate evaluation of P_m is required. Unfortunately, this may not always be an easy task. First of all, direct substitution into (6.4.3) requires knowledge of the values of all P_{di}. Often these tend to be dependent on the type of usage of the source. For example, in telephone channels P_{di} is the power content of a speaker's voice signal, which varies considerably from one person to another. A second point is that even if typical values of P_{di} can be projected, use of (6.4.3) generally leads to an overdesigned B_{RF} since, in most instances, not all sources may be active. This necessitates establishing a source activity, or *loading factor*, for the particular system involved. The problem is further complicated by the fact that activity behavior may be dependent on time of day, location of the source, and so on. In telephone analysis there has been considerable effort to collect empirical data that allow accurate evaluation of typical multiplexed telephone modulation power levels [2].

After RF demodulation the baseband spectrum is recovered in the presence of frequency demodulation noise. For a specific subcarrier channel, the analysis of Section 6.3 can be directly applied. Hence the ith subcarrier filter generates a CNR given by

$$\text{CNR}_{si} = \left(\frac{C_i}{\omega_{si}}\right)^2 \left[\frac{A^2/2}{N_0 2 B_{di}}\right] \qquad (6.4.5)$$

assuming $P_{di} = 1$. Each CNR therefore depends on the subcarrier frequency, deviation, and bandwidth. If we assume that all sources have the same bandwidth, $B_{di} = B_d$ and require the same CNR, then (6.4.5) indicates that we need

$$C_i = C\omega_{si} \qquad (6.4.6)$$

where C is a proportionality constant, in order to make CNR independent of i. Equation (6.4.6) implies that the frequency deviations of subcarriers on the RF carrier should be proportional to their subcarrier deviation. This operation of selecting transmitter frequency deviations in this way is another form of transmitter preemphasis. In effect, we purposely emphasize the higher subcarriers at the transmitter since they will be at a higher noise level at the receiver. Equation (6.4.6) corresponds to linear preemphasis, and the parameter C determines the amount of preemphasis applied. The resulting subcarrier CNR in (6.4.5) is then

$$\text{CNR}_{si} = C^2 \left[\frac{A^2/2}{N_0 2 B_d}\right] \qquad (6.4.7)$$

for all channels. Subsequent AM demodulation in each channel then produces a demodulated SNR of the same value. Thus performance is a direct function of the preemphasis parameter C. However, we also have

$$B_{RF} \cong \frac{2}{2\pi} \left[\sum_{i=1}^{N} C_i^2\right]^{1/2} = \frac{C}{\pi} \left[\sum_{i=1}^{N} \omega_{si}^2\right]^{1/2} \qquad (6.4.8)$$

and the RF bandwidth increases with C also. Equations (6.4.7) and (6.4.8) are the key design equations for FDM|AM|FM systems. The latter relates C to the allowable RF bandwidth for a given set of subcarrier frequencies, and the former determines the recovered channel SNR. For smallest RF bandwidth and largest possible SNR, the subcarrier frequency set should be selected as low as possible, which implies placing the subcarrier channels as close as possible to each other, starting at the lowest possible frequency. Unfortunately, practical filter requirements generally dictate guard spacing between channels, and low frequency interference prevents use of extremely low subcarrier frequencies.

FDM|FM|FM Systems

In FDM|FM|FM multiplexed systems the source waveforms are frequency modulated onto the subcarriers prior to FM on the RF carrier. The baseband signal is now

$$m(t) = \sum_{i=1}^{N} C_i \sin \left(\omega_{si} t + \Delta_{si} \int d_i(t) \, dt \right) \tag{6.4.9}$$

where C_i is again the carrier deviation caused by the the subcarrier and Δ_{si} is the subcarrier frequency deviation caused by the source. The power of $m(t)$ is then

$$P_m = \sum_{i=1}^{N} \frac{C_i^2}{2} \tag{6.4.10}$$

and the resultant RF, subcarrier, and basebandwidths follow as

$$B_{RF} = 2 \left[\frac{\sqrt{P_m}}{2\pi B_m} + 1 \right] B_m \approx \frac{1}{\pi} \left[\sum_{i=1}^{N} \frac{C_i^2}{2} \right]^{1/2} \tag{6.4.11a}$$

$$B_m = \frac{\omega_{sN}}{2\pi} + \frac{B_{sN}}{2} \tag{6.4.11b}$$

$$B_{si} = 2 \left[\frac{\Delta_{si}}{2\pi B_{di}} + 1 \right] B_{di} \tag{6.4.11c}$$

After RF transmission and demodulation, the ith subcarrier is filtered with the CNR obtained from (6.4.5), yielding

$$CNR_{si} = \left(\frac{C_i}{\omega_{si}} \right)^2 \left[\frac{A^2/2}{2N_0 B_{si}} \right] \tag{6.4.12}$$

So long as CNR_{si} is above the threshold, the ith FM subcarrier is demodulated with SNR,

$$\text{SNR}_{di} = 3\beta_i^2\left(\frac{B_{si}}{B_{di}}\right)\text{CNR}_{si}$$

$$= 3\beta_i^2\left(\frac{C_i}{\omega_{si}}\right)^2\left[\frac{A^2/2}{2N_0B_{di}}\right] \tag{6.4.13}$$

where $\beta_i = \Delta_{si}/2\pi B_{di}$ is the modulation index of the ith subcarrier. If we assume that all subcarrier channels use the same index and that each channel is to have the same demodulated SNR, then theoretically we can proceed as with AM subcarriers by linearly preemphasizing with $C_i = C\omega_{si}$. Under an equal subcarrier bandwidth condition, all channels would then operate with the same SNR. From a practical point of view, however, the assumption that all subcarrier channels will have the same bandwidth B_{si} may be difficult to satisfy. This is due to the deviation restriction on the Δ_{si} of the individual subcarrier oscillators. Clearly, if the subcarrier deviations are all equal in (6.4.11c), either the lower subcarriers will be overdeviated or the upper subcarriers will be underdeviated. For this reason, FDM|FM|FM are generally designed with larger deviations and bandwidths for the higher subcarrier channels to utilize always the full allowable frequency deviation. This means that B_{si} in (6.4.12) is also proportional to ω_{si} and that (6.4.12) will produce the same CNR in each channel only if

$$C_i = C\omega_{si}^{3/2} \tag{6.4.14}$$

This indicates a more rapid nonlinear preemphasis for transmitter subcarriers with FM subcarriers than with AM. Subcarrier channels selected to have bandwidths proportional to their subcarrier frequency, so as to take full advantage of the deviation limitation, are called *proportional* subcarrier bands, as opposed to the constant subcarrier bands of the AM case. On the other hand, if all channels are to have the same modulation index β_i, the source bandwidths B_{di} that can be accommodated with each Δ_{si} must also be proportional to ω_{si}. This means that small bandwidth sources should be placed on the lower subcarriers and larger bandwidth sources should be assigned to the corresponding higher subcarriers, for most efficient design.

6.5 DIGITAL SOURCES AND A-D CONVERSION

Whereas analog sources produce continuous waveforms, digital sources produce sequences of binary symbols, or bits, as their output. These symbols could be produced from command or data sources, or by the conversion of an analog source waveform into a binary sequence. The role of the communication system is then to transmit the bits as accurately as possible to the receiver. The optimal design of the communication link for accomplishing reliable bit transmission is considered in detail in Chapter 7.

However, the system designer must first understand the properties of the source itself, such as the rate at which bits are generated, and precisely what each bit represents.

By far, the most important digital source is the one that generates its source bits from an information waveform. The operation of converting a continuous waveform into a digital sequence is called *analog to digital* (A-D) conversion. A digital source employing A-D conversion has the basic block diagram in Figure 6.8. The A-D converter has the task of converting the analog waveform into an equivalent bit stream. We often say that the A-D converter "*digitizes*" the source. Basically, A-D conversion is composed of two key operations—sampling and quantizing. By sampling the source waveform periodically, we obtain a continual sequence of voltage samples. These samples correspond to the voltage values of the waveform at the sampling times. These voltage samples are then mapped, or quantized, into digital words described by binary symbols. Transmission of these symbols in sequence is then used instead of sending the original source waveform.

Both sampling and quantizing inherently introduce discrepancies into the eventual waveform reconstruction. These discrepancies are usually assessed by determining the total mean squared waveform error (MSE) that occurs after waveform reconstruction. The sampling operation produces a mean squared waveform *aliasing* error (MSAE), whereas the quantizing introduces a mean squared quantization error (MSQE). The total MSE is then the sum of the contributions from both these components. Hence we generally write

$$MSE = MSAE + MSQE \qquad (6.5.1)$$

as the total reconstructed mean squared error of the digitized waveform. We emphasize that MSE refers to an inherent source error that is due entirely to the A-D operation and is independent of the communication link itself.

Often MSE is normalized by the power of the reconstructed source waveform, P_d, and the ratio $(MSE/P_d)^{1/2}$ is often interpreted as the rms *accuracy* of the digital system. In addition, it is common practice to refer to

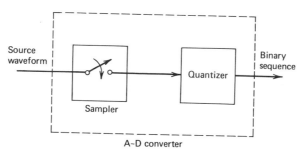

Figure 6.8. A-D conversion block diagram.

the square of its reciprocal as an effective signal to distortion ratio (SDR) for the reconstructed signal [see Equation (1.8.6)]. Thus

$$SDR = \frac{P_d}{MSE} \qquad (6.5.2)$$

and is often stated in decibels. Although it is common practice to treat the parameter SDR as if it were the same as an effective SNR, it should be noted that SDR refers to no specific noise addition. Instead, it simply accounts for the inaccuracies in the overall waveform representation and reconstruction procedures as a mean squared variation from the true waveform. In this sense SDR does retain meaning as a measure of signal fidelity. In the following paragraphs we examine the specific contributors to MSE in (6.5.1).

Sampling

The key parameter governing the waveform sampling operation in Figure 6.8 is the rate at which the samples are taken. This sampling rate is dictated by the so-called sampling theorem, which serves as a guideline for rate selection. The sampling theorem is important not only for its mathematical implication but also for its proof, which yields insight to the required construction procedure (i.e., the D-A conversion) at the receiver. Simply stated, the sampling theorem states that any time function having its transform bandlimited to B Hz can be uniquely represented by its voltage values occurring every $1/2B$ sec apart. That is, if a time function has a Fourier transform with no components outside $(-B, B)$ Hz, then we need only sample the time function every $1/2B$ sec to reconstruct completely (identically) the entire time function. When applied to source waveforms, this means that we need only accurately transmit these time samples to the receiver in order to communicate the source waveform. This extremely vital theorem is proved as follows.

Let $D(\omega)$ be the Fourier transform of the waveform $d(t)$, and let

$$D(\omega) = 0, \qquad |\omega| > 2\pi B \qquad (6.5.3)$$

Let $\bar{D}(\omega)$ be the periodic extension of $D(\omega)$, as shown in Figure 6.9. We can then write $\bar{D}(\omega)$ as

$$\bar{D}(\omega) = \sum_{k=-\infty}^{\infty} D(\omega + k4\pi B) \qquad (6.5.4)$$

Since $\bar{D}(\omega)$ is periodic, we can also expand it in a unique Fourier series in the variable ω as

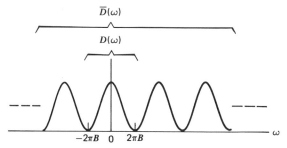

Figure 6.9. Periodic version $\bar{D}(\omega)$ of $D(\omega)$.

$$\bar{D}(\omega) = \sum_{n=-\infty}^{\infty} d_n e^{j(2\pi n/4\pi B)\omega} \tag{6.5.5}$$

Thus $\bar{D}(\omega)$ in (6.5.4) is completely specified mathematically by its Fourier coefficients $\{d_n\}$. That is, given $\{d_n\}$, $\bar{D}(\omega)$ in (6.5.5) can always be exactly constructed. Now from Fourier series we have

$$
\begin{aligned}
d_n &= \frac{1}{4\pi B} \int_{-2\pi B}^{2\pi B} \bar{D}(\omega) e^{-j(n/2B)\omega} \, d\omega \\
&= \left(\frac{1}{2B}\right) \frac{1}{2\pi} \int_{-\infty}^{\infty} D(\omega) e^{-j(n/2B)\omega} \, d\omega \\
&= \frac{1}{2B} d\left(\frac{-n}{2B}\right)
\end{aligned}
\tag{6.5.6}
$$

Thus the values of $d(t)$ at the points $t = n/2B$, $-\infty < n < \infty$, identify the sequence $\{d_n\}$, which, in turn, specifies $\bar{D}(\omega)$. From the latter, $D(\omega)$ can be determined exactly, and by use of the inverse Fourier transform, $d(t)$ is uniquely specified at all t. Hence the time samples of $d(t)$, taken at rate $2B$ samples/sec, can be used to produce $d(t)$ itself. This completes the underlying proof of the sampling theorem.

The preceding proof, in addition to establishing the sufficiency of the time samples, also suggests the manner in which the D-A reconstruction can occur. That is, $d(t)$ is simply the Fourier transform of $\bar{D}(\omega)$ limited to the range $-2\pi B \le \omega \le 2\pi B$. Therefore

$$
\begin{aligned}
d(t) &= \frac{1}{2\pi} \int_{-2\pi B}^{2\pi B} \left[\frac{1}{2B} \sum_{n=-\infty}^{\infty} d\left(\frac{-n}{2B}\right) e^{jn\omega/2B} \right] e^{j\omega t} \, d\omega \\
&= \frac{1}{2B} \sum_{n=-\infty}^{\infty} d\left(\frac{n}{2B}\right) \left[\frac{1}{2\pi} \int_{-2\pi B}^{2\pi B} e^{j[t-(n/2B)]\omega} \, d\omega \right]
\end{aligned}
\tag{6.5.7}
$$

The bracket evaluates by using

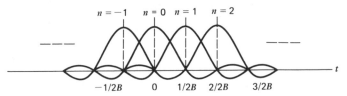

Figure 6.10. Sinc reconstruction functions: sinc $(x) = \sin{(x)}/x$.

$$\frac{1}{2\pi} \int_{-2\pi B}^{2\pi B} e^{jx\omega}\, d\omega = \frac{\sin{(x2\pi B)}}{\pi x} \tag{6.5.8}$$

Hence (6.5.7) becomes

$$d(t) = \sum_{n=-\infty}^{\infty} d\!\left(\frac{n}{2B}\right) \frac{\sin{(2\pi Bt - n\pi)}}{(2\pi Bt - n\pi)} \tag{6.5.9}$$

Thus the waveform $d(t)$ can be reconstructed by using the sample values $d(n/2B)$ as coefficients of the corresponding functions shown. These functions are simply shifted versions of the sinc $(2\pi Bt) \overset{\Delta}{=} \sin{(2\pi Bt)}/2\pi Bt$ function, the nth one centered as $t = n/2B$ (Figure 6.10). The summation of the contribution from all such functions, with the proper sample value coefficients, at any t will exactly produce $d(t)$. Note that a specific reconstruction function is required by the sampling theorem. Other techniques, such as straightline connections between sampling points, only serve as an approximation to $d(t)$, and generally produce an effective low pass filtering of the desired waveform (Problem 6.18). We also see that an infinite number of samples [terms in (6.5.9)] must be used with each one affecting $d(t)$ at all values of t. If $d(t)$ is assumed to be time limited to a T_0 sec time interval and zero elsewhere, then only $T_0/(1/2B) = 2BT_0$ time samples are needed to represent $d(t)$. However, simultaneous time limiting and band-limiting of a waveform is contradictory, although such assumptions are often made for analytical convenience, and can be somewhat justified.

The reconstruction of $d(t)$ from its time samples, using (6.5.9), although exact, requires generation of the sequence of sinc time functions. From a system design point of view, it would be convenient to have a simple interpretation to the right side of (6.5.9). To examine this, consider using the sequence of samples $d(nT)$, $T = 1/2B$, to adjust the amplitudes of a train of narrow, periodic pulses, as shown in Figure 6.11. If we denote the narrow pulses by

$$w(t) = \begin{cases} \dfrac{T}{\tau}, & -\dfrac{\tau}{2} \le t \le \dfrac{\tau}{2} \\[2mm] 0, & \text{elsewhere} \end{cases} \tag{6.5.10}$$

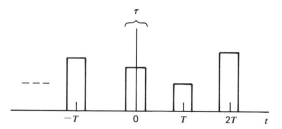

Figure 6.11. Amplitude adjusted pulse waveform.

and let its repetition period be T, then the amplitude adjusted pulse train can be written as

$$q(t) = \sum_{n=-\infty}^{\infty} d(nT)w(t-nT) \qquad (6.5.11)$$

This pulse train has Fourier transform

$$Q(\omega) = \int_{-\infty}^{\infty} q(t)e^{-j\omega t}\, dt$$

$$= \sum_{n=-\infty}^{\infty} d(nT) \int_{-\infty}^{\infty} w(t-nT)e^{-j\omega t}\, dt$$

$$= \sum_{n=-\infty}^{\infty} d(nT)e^{-jn\omega T} \int_{-\infty}^{\infty} w(t-nT)e^{-j\omega(t-nT)}\, dt \qquad (6.5.12)$$

The integral is simply the Fourier transform of $w(t)$ in (6.5.10) and does not depend on n. Hence by use of (6.5.4) and (1.4.7),

$$Q(\omega) = \left[\sum_{n=-\infty}^{\infty} D\left(\omega + \frac{2\pi n}{T}\right) \right] \left[\frac{\sin(\omega\tau/2)}{(\omega\tau/2)} \right] \qquad (6.5.13)$$

Since $T = 1/2B$, the transform $Q(\omega)$ is that of a low pass filtered version of the desired periodic spectrum $\bar{D}(\omega)$, as shown in Figure 6.12. If $\tau \to 0$ (the pulse were infinitesimally narrow and infinitely high, approaching delta functions), then $Q(\omega) \to \bar{D}(\omega)$. Thus the periodic function $\bar{D}(\omega)$, used to prove the sampling theorem, can in fact be generated by using the time samples $d(nT)$ to amplitude modulate a train of delta functions. In this case (6.5.11) becomes

$$q(t) = T \sum_{n=-\infty}^{\infty} d(nT)\delta(t-nT) \qquad (6.5.14)$$

and its transform $Q(\omega)$ is identical to $\bar{D}(\omega)$. The spectrum $D(\omega)$ is then

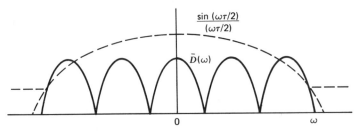

Figure 6.12. Equivalent filtering of desired waveform by pulses.

obtained by filtering off the portion of the spectrum between $(-2\pi B, 2\pi B)$, that is, by low pass filtering. The output of this low pass filter is then $d(t)$. This physically forms the basis of D-A conversion at a receiver (Figure 6.13). The sample values decoded at the receiver are used to produce amplitude modulated pulse trains that are then low pass filtered to produce the reconstructed source waveform $d(t)$. This is equivalent to assuming that the pulse train $q(t)$ in (6.5.11) or (6.5.14) was actually transmitted from the source sampler to the D-A low pass filter. (Recall that only the binary equivalent of the voltage sample values are, in fact, sent.) For this reason, $q(t)$ in (6.5.11) is often referred to as the *sampler signal* of the A-D converter. The waveform in (6.5.14) using delta functions is called the *idealized sampler* signal, and its transform is given by (6.5.13) with $\tau = 0$. To summarize, then, sampling a source waveform every T sec, sending the sample values, and reconstructing by the low pass filter method described earlier, is equivalent (as far as reconstruction is concerned) of low pass filtering the sampler signal $q(t)$ directly.

From (6.5.13) we see that the pulse train $q(t)$ need not have true delta functions, since little distortion occurs to $D(\omega)$ so long as $\tau \ll T$. On the other hand, the low pass reconstruction filter must be perfectly flat over $(-2\pi B, 2\pi B)$ to avoid distorting $D(\omega)$. Hence the low pass filter must be an ideal low pass filter. This, of course, is evident from (6.5.9), which can also be interpreted as the output produced by an input $q(t)$ of weighted delta functions for a filter whose impulse response is $\sin(2\pi Bt)/2\pi Bt$, that

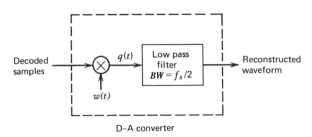

D-A converter

Figure 6.13. D-A conversion block diagram.

is, an ideal low pass filter. These ideal filter requirements can be weakened by allowing a separation between $D(\omega)$ and the nearest interfering spectrum of $\bar{D}(\omega)$. This can be achieved by interpreting the upper end of the frequency transform $D(\omega)$ in (6.5.3) as being $(B + \delta)$ instead of B. (The function will be zero over the upper δ Hz.) Sampling at the rate $2(B + \delta)$ produces a periodic frequency function $\bar{D}(\omega)$ with a spacing of 2δ between the adjacent spectra. Thus sampling faster than required simplifies the D-A filter design.

When the spectrum is truly bandlimited, sampling at the proper rate or faster allows undistorted waveform reconstruction. However, if we do not sample fast enough or if the spectrum is not truly bandlimited, an error in the reconstruction occurs as a result of sampling. This is called the aliasing error, and its mean squared value can be determined as follows. Let $d(t)$ be the waveform to be sampled and let $D(\omega)$ again be its transform, not necessarily bandlimited. Let us sample at rate f_s samples per second and reconstruct from these samples. We can again interpret the resulting reconstruction as an ideal filtering of the idealized sampler signal $q(t)$ in (6.5.14). This signal has the transform [from (6.5.13) with $\tau = 0$] given by

$$Q(\omega) = \sum_{n=-\infty}^{\infty} D(\omega + n2\pi f_s) \qquad (6.5.15)$$

This corresponds to shifts of $D(\omega)$ by all multiples of the sampling frequency f_s. The resulting spectrum is illustrated in Figure 6.14. Reconstruction by ideally filtered $Q(\omega)$ from $(-2\pi f_s/2, 2\pi f_s/2)$ no longer yields an undistorted version of the original $D(\omega)$. Instead, portions of the shifted spectrum are folded back into the desired spectrum. Stated in other words, the sampling rate f_s is capable of reconstructing frequency components of $d(t)$ only up to $f_s/2$, and all higher frequencies are lost, producing an effective error in the reconstructed waveform. The mean squared value of this signal error [difference between the true $d(t)$ and that part that will be reconstructed] is therefore given by the area under the waveform power or energy spectrum beyond $f_s/2$ Hz. This portion of the spectrum is identical to that part folded

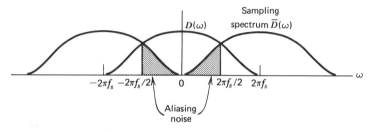

Figure 6.14. Aliasing noise diagram.

back into the range $(-2\pi f_s/2, 2\pi f_s/2)$ in $Q(\omega)$. The mean squared aliasing error (MSAE) is given by this folded integrated area. Hence[†]

$$\text{MSAE} = \frac{2}{2\pi} \int_{2\pi f_s/2}^{\infty} |D(\omega)|^2 \, d\omega \qquad (6.5.16)$$

Since the energy in the original waveform is the integral of $|D(\omega)|^2$ over all frequencies, the resulting SDR caused by aliasing becomes

$$\text{SDR} = \frac{\displaystyle\int_0^{\infty} |D(\omega)|^2 \, d\omega}{\displaystyle\int_{\pi f_s}^{\infty} |D(\omega)|^2 \, d\omega} \qquad (6.5.17)$$

The preceding can be evaluated for any specific waveform energy spectrum $|D(\omega)|^2$. As a typical example, let

$$|D(\omega)|^2 = \frac{1}{1 + (\omega/2\pi f_d)^{2\mu}} \qquad (6.5.18)$$

where f_d is the 3 dB frequency and μ determines the spectrum rolloff. Substituting into (6.5.17) yields

$$\text{SDR} = \frac{\displaystyle\int_0^{\infty} (dx/1 + x^{2\mu})}{\displaystyle\int_{f_s/2f_d}^{\infty} (dx/1 + x^{2\mu})} \qquad (6.5.19)$$

The result is plotted in Figure 6.15, as a function of the normalized sampling rate f_s/f_d, for various falloff parameter values μ. Note that sampling rates greatly in excess of twice the 3 dB bandwidth of the waveform are required to maintain acceptable SDR values, say, 30–40 dB (0.5% accuracy). This illustrates the danger in simply interpreting the 3 dB bandwidth as the bandwidth extent of $d(t)$ and in blindly applying the sampling theorem. As μ increases, the sampling rate required to produce a fixed value of SDR also decreases, approaching the theoretical minimum rate. Since spectrum falloff can be increased by additional filtering of the waveform, the results indicate the advantage of presampling filtering of the source waveform, as far as aliasing error is concerned.

In certain situations we may wish to digitize a bandpass signal. This occurs if the source produces a bandpass waveform or if the source signal is

---

[†]We use energy spectra here since the $d(t)$ we have been considering are deterministic waveforms. The resulting SDR is then a ratio of energy values. If $d(t)$ is a random process, we can equivalently deal with power spectra and compute SDR as a power ratio.

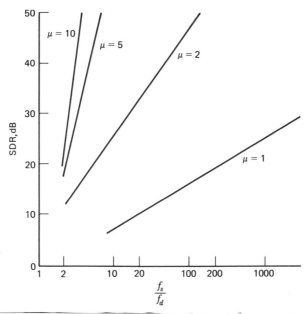

Figure 6.15. SDR due to aliasing for Butterworth shaped spectra: f_d = spectrum 3 dB bandwidth, μ = spectral falloff rate.

placed on a subcarrier prior to sampling. Reconstruction of the bandpass signal would be obtained by ideal bandpass filtering over the original bandwidth of the bandpass signal. The previous sampling discussion extends to bandpass sampling as well. Suppose that $D(\omega)$ has the bandpass spectrum shown in Figure 6.16a. If we sample the waveform $d(t)$ at rate f_s samples per second and produce the idealized sampler signal in (6.5.14) with these samples, its transform is again given by (6.5.15). This sampled spectrum is sketched in Figure 6.16b and again corresponds to the periodic shifting of $D(\omega)$ every $2\pi f_s$ rps. The actual value of $D(\omega)$ at any ω depends on the relation between the sampling frequency f_s and the original bandpass spectrum $D(\omega)$. If reconstruction is obtained by ideal bandpass filtering of $D(\omega)$ over the bandwidth of $D(\omega)$ (i.e., replacing the low pass filter in Figure 6.13 by a bandpass filter), a distortionless spectrum appears only if shifts of $D(\omega)$ do not overlap $D(\omega)$ anywhere. This requires that if the negative frequency spectrum of $D(\omega)$ is shifted just to the left of the positive frequency spectrum for some integer n, the $(n + 1)$ integer shift must move it completely to the right to avoid overlap. This requires that the shifting interval $2\pi f_s$ be at least as large as twice the positive frequency bandwidth $2\pi B$ of $D(\omega)$. Thus we require

$$f_s \geq 2B \qquad (6.5.20)$$

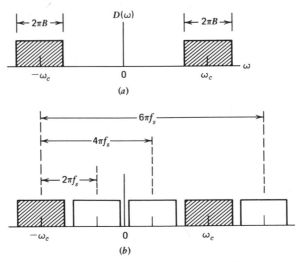

Figure 6.16. Bandpass sampling: (a) source spectrum, (b) sampled spectrum.

and the bandpass spectrum must be sampled at a rate equal to at least twice the bandpass bandwidth to avoid overlap. The actual required rate depends on the relative values of B and its spectral location (Problem 6.21).

Let us determine the condition necessary on the spectral location of $D(\omega)$ in order to allow a sampling rate of exactly $2B$ (the minimal possible in (6.5.20)). Using Figure 6.16b, it is evident that the bandpass spectrum should be located at a center frequency $\pm f_c$ such that when sampling at $f_s = 2B$ an integer number of positive shifts slides the right-hand edge of the negative frequency spectrum just up to the left-hand side of the positive spectrum. For this to occur it is necessary that $nf_s = 2[f_c - (B/2)]$, or equivalently,

$$f_c = nB + \frac{B}{2} = \left(\frac{n}{2} + \frac{1}{4}\right)f_s \tag{6.5.21}$$

for some integer n. A bandpass spectrum of bandwidth B should therefore be located at any of the center frequencies given in (6.5.21) to avoid aliasing errors (spectrum overlap) when sampling with rate $f_s = 2B$. This aids in the selection of subcarrier frequencies relative to sampling frequencies for use in bandpass sampling.

Quantization

After the waveform is sampled, the A-D subsystem must transmit the sample values of the receiver for the reconstruction. However, the sample value can take on any of a continuum of values over the voltage range, and the A-D operation can transmit only a finite number of sample values at each sampling point. It accomplishes this by mapping, or *quantizing*, the complete sample value voltage range into a finite number of selected values. This quantization operation is shown as Figure 6.17. The sample voltage range is divided into preselected intervals, and each interval is given a prescribed binary code word. All sample values occurring in a specific interval are transmitted with that interval binary word. At the receiver each binary code word is converted to a specific voltage value in the corresponding interval. This reconstructed voltage value is called the *quantization value* of the interval. Hence A-D quantization is effectively the same as converting all voltage values in an interval to the quantization value of that interval. Thus there is an inherent waveform error introduced by the quantization operation even if perfect D-A conversion followed.

The selection of the binary code for the intervals is arbitrary, but obviously a distinct code word must be available for each quantization interval. If there are L intervals, then clearly $\log_2 L$ (or the nearest largest integer) bits are needed to represent all intervals uniquely. A common technique is to use *natural* coding, in which the intervals are numbered consecutively from zero to $L - 1$, and the binary representation of each number is used as the code word for that interval. Thus $\log_2 L$ bits are produced at each sample time, and if the sampler operates as f_s samples per second, the A-D subsystem generates bits at the rate

$$R_b = f_s \log_2 L \ \text{bits/sec} \qquad (6.5.22)$$

Quantizers can take on a variety of forms, each describable by the number, size, and location of their intervals, in addition to their quantization values. Quantizers are generally constructed from arrays of diode voltage gates that

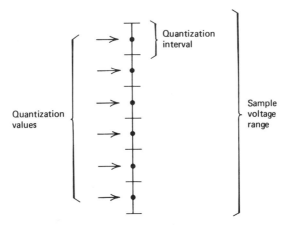

Figure 6.17. Quantization diagram.

excite binary flip-flop circuits, producing the necessary binary word for each voltage sample. The simplest type of quantizer is a *uniform* quantizer, in which all intervals are selected to have equal width over a prescribed voltage range. If a sample voltage value exceeds the quantizer range, the quantizer is said to be *saturated*. Saturating samples are generally associated with the end intervals (which is identical to assuming the end intervals extend to infinity). A quantizer with unequal intervals is called a *nonuniform* quantizer.

As stated previously, quantization introduces an intrinsic waveform error even if the remaining part of the system is error free. The mean squared quantization error (MSQE) that will be introduced at each sample value can be determined by treating the voltage sample as a random variable and averaging the mean squares sample error over all intervals. If we let MSQE_i be the mean squared quantization error when the random voltage sample is in the ith quantization interval, and if we let P_i be the probability that the sample is in the ith interval, then

$$\text{MSQE} = \sum_{i=1}^{L} (\text{MSQE}_i) P_i \qquad (6.5.23)$$

The mean squared error in an interval can be determined by computing the squared error from a sample point in the interval to the quantization value that will be reconstructed for that interval, and averaging over the probability density of the sample point. Hence if the voltage sample has the probability density $p_d(x)$, and if ξ_i is the quantization value of the ith interval, then

$$\text{MSQE}_i = \int_{I_i} (x - \xi_i)^2 p(x|i)\, dx \qquad (6.5.24)$$

where I_i represents the ith interval range and $p(x|i)$ is the probability density of the sample value when in I_i. This is given by

$$p(x|i) = \frac{p_d(x)}{P_i}, \qquad \text{for} \qquad x \text{ in } I_i \qquad (6.5.25)$$

where now

$$P_i = \int_{I_i} p_d(x)\, dx \qquad (6.5.26)$$

Equation (6.5.24) can be computed for each quantizer interval and used in (6.5.23). Thus MSQE depends on the form of the quantizer and the probability density of the sample voltage value. It is again convenient to normalize by the mean squared value of the sample voltage value

$$P_d = \int_{-\infty}^{\infty} x^2 p_d(x)\, dx \tag{6.5.27}$$

and compute an effective SDR due to quantization, just as in (6.5.17) for the aliasing error.

As an example, assume a uniform quantizer with L intervals operating over a voltage range $(-V, V)$ volts, so that each interval has width $\epsilon = 2V/L$. The center point of each interval is used as the quantization value. Let each voltage sample have a uniform distribution over $(-A, A)$ volts. Consider first the case of no saturation, $A \leq V$. Equation (6.5.24) becomes

$$\text{MSQE}_i = \int_{-\epsilon/2}^{\epsilon/2} x^2 \left(\frac{1}{\epsilon}\right) dx = \frac{\epsilon^2}{12} \tag{6.5.28}$$

for each quantization interval falling within $(-A, A)$. (We take A to be an integer multiple of ϵ). Since the result is the same for all such intervals and all intervals are equally likely to be occupied (because of the uniform density on the sample value), Equation (6.5.23) becomes

$$\text{MSQE} = \frac{\epsilon^2}{12} \tag{6.5.29}$$

Similarly, (6.5.27) evaluates to $A^2/3$. The resulting SDR due to quantization errors alone is then

$$\text{SDR} = \frac{A^2/3}{\epsilon^2/12} = \frac{L^2}{(V/A)^2} \tag{6.5.30}$$

We see that the quantization SDR varies directly with the number of intervals, but decreases by the square of ratio of the quantizer range to the signal sample range. Hence it is advantageous to insure that the quantizer "match" the expected sample range and not be any larger. Note that increasing L to improve SDR will require a subsequent increase in the A-D bit rate in (6.5.22).

Now consider $A > V$, so that saturation may occur. We assume a voltage sample outside the quantizer range will be associated with the end intervals. The mean squared interval error in (6.5.29) occurs only if the sample value falls within $(-V, V)$. Outside this range (i.e., during saturation), the mean squared saturation error (MSSE) becomes

$$\text{MSSE} \triangleq \int_{V}^{A} \left[x - \left(V - \frac{\epsilon}{2}\right)\right]^2 \left(\frac{1}{A - V}\right) dx$$

$$= \frac{(A - V + (\epsilon/2))^3}{3(A - V)} - \frac{(\epsilon/2)^3}{3(A - V)} \tag{6.5.31}$$

The total MSQE in (6.5.23) is then

$$\mathrm{MSQE} = \left(\frac{V}{A}\right)\left(\frac{\epsilon^2}{12}\right) + \left(\frac{A-V}{A}\right)\mathrm{MSSE} \qquad (6.5.32)$$

The desired signal power is again $P_d = A^2/3$. The SDR due to quantization with saturation is therefore

$$\mathrm{SDR} = \cfrac{L^2}{\left(\frac{V}{A}\right)^3 + \left(1 - \frac{V}{A}\right)^3 \left[\left(1 + \frac{V}{L(A-V)}\right)^3 - \left(\frac{V}{L(A-V)}\right)^3\right]L^2} \qquad (6.5.33)$$

Combining (6.5.30) and (6.5.33) gives the general expression for a uniform quantization with L intervals and range $(-V, V)$ operating on a uniform voltage sample of range $(-A, A)$ as

$$\mathrm{SDR} = \begin{cases} \dfrac{L^2}{u^2}, & u \ge 1 \\[3ex] \dfrac{L^2/u^3}{1 + \left[\left(\dfrac{1-u}{u} + \dfrac{1}{L}\right)^3 - \left(\dfrac{1}{L}\right)^3\right]L^2}, & u < 1 \end{cases} \qquad (6.5.34)$$

where $u = V/A$. The result is plotted in Figure 6.18 as a function of A/V for $L = 16$ and 256, corresponding to 4 and 8 bit quantizer words. The maxi-

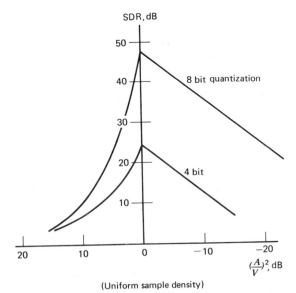

Figure 6.18. Quantizer SDR for uniform voltage sample density $(-A, A)$. Quantizer range = $-V, V$.

mum value at $A = V$ is L^2 and occurs only if the quantizer range is identical to the signal range. Note the degradation that occurs as the quantizer is mismatched, being especially severe as saturation occurs.

Once the sampling rate and quantizer size are selected for the desired SDR, the A-D source bit rate follows from (6.5.22). For example, a voice digitizer using an 8-bit (256 levels) quantizer at sampling rate $f_s = 2 \times 4\,\text{kHz} = 8000$ samples/sec will produce a source bit rate of $8 \times 8000 = 64$ kbps. If this source bit rate generates a digital pulse waveform, and this waveform is used to amplitude modulate or phase shift a carrier or subcarrier for link transmission, then a carrier main-hump bandwidth of approximately $2 \times R_b = 128\,\text{kHz}$ is needed. Note the significant bandwidth increase over the $2 \times 4 = 8\,\text{kHz}$ band needed for direct amplitude modulation by the voice waveform itself. The source rate can be reduced by decreasing the coarseness of the quantizer (lowering L) or by reducing the sampling rate, but this generally leads to lower voice quality (increased MSE). Likewise, A-D conversion of video signals with 8 bit quantization produces a bit rate of $6\,\text{MHz} \times 2 \times 8 = 96\,\text{Mbps}$, and would require a carrier bandwidth of about 200 MHz.

6.6 SOURCE RATE REDUCTION

In modern systems there is considerable interest in sending digitized waveforms over communication links. However, when the link bandwidth is limited, as is usually the case, there is a limit to the source rate that can be used with that channel. One must either use sources with reduced information bandwidths or attempt to find techniques that permit source rate reduction during digitizing without excessive sacrifice of signal fidelity. This operation of A-D conversion so as to minimize the resulting bit rate is referred to as *source encoding*.

Over the past decade, analytical approaches to generalized source encoding have been developed [3–5], and many elegant results have been derived that predict the theoretical limitations and capabilities of source encoding. Although such theoretical results are usually confined to specific system models, and unfortunately do not always yield practical design solutions, they do serve as useful guideline for system implementations. A vigorous development of source encoding is well beyond our scope here, but we may digress momentarily to indicate some ways in which standard A-D conversion may be modified to achieve some degree of source rate reduction.

One simple way to source encode is to utilize a more sophisticated quantization encoding, other than natural encoding. Consider the example illustrated in Table 6.1. A four level quantizer (or perhaps a four level command generator) can produce four possible quantization or sample values, which we simply number, as shown in column 1, but each occurs with the probability shown in column 2. Natural encoding would use the

TABLE 6.1 Source Encoding Example

Quantization or Sample Values	Probability of Occurrence	Natural Encoding	Unequal Word Length Encoding
0	$\frac{1}{8}$	00	001
1	$\frac{1}{2}$	01	1
2	$\frac{1}{4}$	10	01
3	$\frac{1}{8}$	11	000

quantization words of column 3 for representations, ignoring the relative frequency of occurrence, and would therefore continually operate at the encode rate of two bits per sample. As a modification for source encoding, we might consider using instead quantization words of different lengths, using the shortest words to code the quantization values occurring most often, and using longer words for those least used. An example of such encoding is shown in column 4. The average number of source bits used per sample will be given by the sum of the number of bits per word times the probability of each occurring. Hence, from columns 1 and 4,

$$\text{Average bits per sample} = \tfrac{1}{2}(1) + \tfrac{1}{4}(2) + \tfrac{1}{4}(3)$$

$$= 1.75 \tag{6.6.1}$$

We have therefore reduced the average rate of bit transmission over that of natural encoding by 12%. The required receiver reconstruction operation is now more complicated, however, since the words of different lengths must be recognized. This requires that the reconstruction subsystem be exactly in step with the received words, so that individual words can be separated. (Note, in this example, the appearance of either a one or three consecutive zeros signifies the end of a quantization word, provided that the examination is started at the beginning of each word.) Hence rate reduction is obtained at the expense of more stringent reconstruction processing.

The average number of bits per sample in (6.6.1) is in fact the minimum possible for the example in Table 6.1. The minimum value of the average number of bits per sample needed to encode a specific set of quantization values with known a priori probabilities is called the *entropy* of the set. If the samples are independent,[†] this entropy can be determined from the occurrence probabilities P_i of each quantization value by

$$\text{Entropy} = -\sum_{i=1}^{L} P_i \log_2 P_i \tag{6.6.2}$$

[†]If the samples are not independent (i.e., correlation exists from one sample in time to the next) then average entropy over a sequence of samples must be determined, using joint sample probabilities.

where the probabilities P_i are obtained from (6.5.26). For a given quantizer, the entropy is therefore directly dependent on the source density $p_d(x)$. In some cases (e.g., Table 6.1) we can encode each sample so as to operate at the average bits per sample given by the quantizer entropy. A coding procedure, called *Huffman* encoding [6], generates the required binary word that should be used for each quantization value so as to achieve the smallest average number of bits per sample for the given a priori probabilities. This coding technique is diagrammed in Figure 6.19 and is explained as follows. The original quantization values to be encoded are first ranked in descending order of their probabilities. The last two values are combined into a single sample value and labeled with the sum probability of the two. This combined sample is then grouped with the remaining $L - 2$ samples to form a new set of $L - 1$ sample values that are then ranked again according to their probabilities. The process is repeated for the new sample set, again adjoining the two lowest into a combined sample to be reranked with the remainders, and forming a new set of $L - 2$ samples. The process is continually repeated, keeping track of the original samples and the way they are combined (Figure 6.19), until a final set of only two samples remain. Encoding is then achieved by starting with the final two samples and assigning zeros and ones as we traverse back through the diagram. Initially we assign a zero and one to each of the last two samples. As we move back one step we append bits to our earlier bit assignments, adding a one and zero to each member of a combined sample. Whenever we label an original sample, it is removed from the procedure with the code sequence up to that

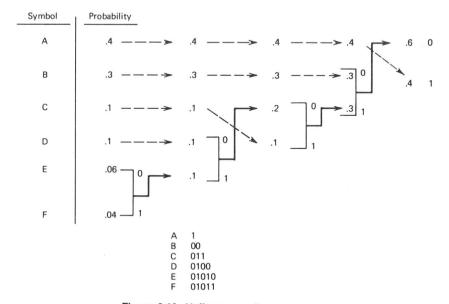

Figure 6.19. Huffman encoding example.

point. Continuing back in this manner, we eventually code all original samples with recognizable codes. It has been shown that this method achieves the minimum average number of bits per sample [7]. Note that in Figure 6.19 we do not achieve the entropy of the sample set, even though we have minimized the average number of bits to be used per sample.

Another way to achieve bit rate reduction is by the generalized A-D system shown in Figure 6.20. Rather than quantize directly the message samples as they occur, we perform a transformation $\mathcal{T}$ on the sample values prior to quantizing. Thus we instead quantize a function of a sequence of samples rather than the samples themselves. Such generalized A-D conversion systems form the basis for *redundancy removal* and *data compression* systems [5].

In redundancy removal systems, the basic role of the transformation $\mathcal{T}$ is to take advantage of correlation that may exist from sample to sample. This means that knowledge of n samples provides some information about the $(n+1)$ sample, and the latter can be quantized differently from the way it would if we were quantizing it alone. Since some information about the $(n+1)$ voltage value can be obtained from the prevous n samples, we require less bits to quantize it with the same accuracy, and the overall source bit rate can be reduced. At the receiver each sample value is reconstructed from a combination of its quantization word and its prediction from values of the previous samples. Roughly speaking, the previous samples are used to project an estimate of the present sample, and the quantization word is used to refine that estimate. We shall not derive here the internal structure of the transformation estimator $\mathcal{T}$, except to comment that its optimal design requires exact knowledge of the source sample statistics. It is common to model sources as being some form of nth order *Markov sources*, defined as a source whose present voltage sample value is statistically dependent on n of its past sample values. Obviously, the complexity of the transformation depends on the order n of the Markovian model. Thus redundancy removal systems, constructed as in Figure 6.20 with a fixed preselected transformation, are extremely model dependent in their design and performance.

When source sample statistics are not known and cannot be adequately modeled, the transformation operation can be used to perform sample to sample comparison, so that adjacent samples need not be sent unless

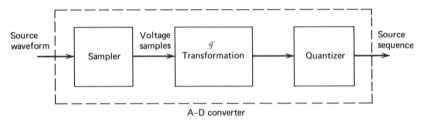

Figure 6.20. Generalized A-D convecter.

sufficiently different. In essence, the transformation $\mathcal{T}$ is being used to perform an "instantaneous correlation" measurement. Such approaches have led to the implementation of various types of *differential quantizers*, in which only sample differences are quantized instead of actual sample values. Hence if a long sequence of almost identical samples occurred, a significant reduction in the bit transmission rate can be made during their occurrence. This allows a form of adaptive adjustment of the instantaneous transmission rate, according to the actual time variations of the sample differences. However, such variations of the A-D output rate require that some type of storage (*buffers*) be provided if the transmission over the link is desired at a constant bit rate.

There has been particular interest in considering generalized A-D conversion where the transformation $\mathcal{T}$ corresponds to a class of linear orthogonal transforms [8–10]. Such transformations can be regarded as a matrix multiplication of the source samples into a set of new coordinates to represent the source message. These new coordinates are then quantized and transmitted in lieu of the original samples. At the receiver, the inverse orthogonal transform is used to convert the recovered coordinates back to the original source samples for reconstruction.

Orthogonal transformations cause the coordinate energy to be redistributed, and a transmission advantage occurs if the new coordinates can be sent with less bits (or more accuracy). For example, if N successive voltage samples $\{d_i\}$ had equal variance, each sample would have to be quantized with the same number of bits. If an orthogonal transformation of the N samples could be found that will produce N new coordinates with an unequal variance distribution, a smaller MSQE will be produced at the same bit rate by variable coordinate quantization. That is, the same number of total bits that were available for quantizing the original N samples can now be distributed to each new coordinate in proportion to its variance. The coordinates with the higher variance (energy) are therefore quantized more accurately, resulting in a smaller MSQE. (The proof of this can be found in Reference 10.) Conversely, we can transmit the coordinates with the same accuracy at a smaller bit rate. The larger the differences in the coordinate variance distribution, the greater will be the degree of improvement. Hence orthogonal transformations that redistribute the sample energy in the most nonuniform manner are the most advantageous. We point out, however, that physical implementation of the quantizer requires unequal numbers of quantization intervals for each coordinate, varying according to the variance distribution.

In general, performance improves with N, the number of samples being transformed at one time. However, the orthogonal transform requires an N by N matrix multiplication and is therefore limited by computational complexity. (In general, on the order of N^2 computations must be performed on the original source samples.) For this reason there has been interest in using orthogonal transforms that allow simplicity and ease in computation.

TABLE 6.2 Bit Rates For Digital Sources

Teletype (low rate)	75 bps
Teletype (high rate)	1.2 K
Digitized voice, PCM	
8000 samples/sec: 7-bit samples	56 K
8-bit samples	64 K
Digitized voice, differential quantizers	
High quality	40 K
Good quality	32 K
Low quality	16 K
Vocoder (synthetic voice)	2.4 K
High quality music	
15 K BW, 35 K samples/sec, 8-bit	450 K
Digital TV, high-quality color	96 M
Reduced resolution (good–excellent)	32 M
Reduced resolution (fair–good)	1.5–4.8 M
Transform encoded (good resolution)	6–11 M
Picture phone (video telephone)	6.3 M
Imagery	1–6 M

Classes of Haar [8], Walsh [9], Hadamard [11], and discrete Fourier transforms [12] have been found useful in this regard. Computer algorithms (called *fast transforms*) allow versions of these transforms to be performed with considerable reduction in computation. These fast transforms require on the order of only $(N \log N)$ computations to carry out the matrix transformation. It must also be remembered that for any transform, its inverse must also be determined for use at the reconstruction end for coordinate inversion. Thus practical source encoding by means of orthogonal coordinate transforms reduces to primarily use of orthogonal matrixes that are computational simplified, are easy to invert, and yield acceptable variance distributions for quantizing. The previously stated orthogonal transforms have been successively applied to both voice and image transmission [13–15].

Table 6.2 list some common data, voice, and video digitizing systems, and the resulting bit rates obtained by various source encoding methods. Note that source encoding can permit digitized voice to be transmitted over bandwidths well below the 100 kHz needed with standard A-D conversion. These systems, of course, require more complicated processing hardware and may reduce slightly the voice quality. Also included in the listing are the bit rates for digitizing video and imagery.

6.7 TIME DIVISION MULTIPLEXING OF DIGITAL SOURCES

In Section 6.4 we considered the simultaneous transmission of waveforms from a set of analog sources by multiplexing the source waveforms into a single waveform. With digital sources there is often a similar requirement for simultaneous transmission of the output bit sequences of a set of digital sources in the same way. For example, there may be several digital sources operating simultaneously in a particular system. The sources produce bits at various rates, and it is desired to transmit these symbols simultaneously to the receiver. One method is again to use a FDM format in which specific parts of the frequency spectrum are assigned to particular source in forming the baseband. The source bits are converted to a binary digital waveform, which is then modulated on to an individual subcarrier. The sum of all such subcarriers, one for each digital source, then forms the multiplexed baseband.

A second method is to attempt to interleave the bits from each source to produce a single bit sequence for transmission. This represents a form of *time division multiplexing* (TDM) of the sources. The operation of combining the digital sequences is sometimes referred to as *digital multiplexing*, or *parallel–serial conversion*. After reception of the multiplexed bits sequence at the receiver, a demultiplexing, or deinterleaving, operation must be performed to separate the individual bits streams. After separation these bit streams can be used simultaneously in separate channels for D-A conversion or command execution.

Interleaving bits is fairly straightforward when all sources produce bits at the same rate. We need only commutate over the source bit set with a high speed rotating switch, reading out one bit at a time from each source in sequence and continually recycling (Figure 6.21*a*). When the source rates are unequal, however, a slightly more complicated procedure must be used involving a *parallel-serial converter* (PSC) constructed from digital *shift registers*. A shift register is a cascade of binary flip-flop circuits. At prescribed time intervals (called the register *clock rate*) each stage of the register can take on the binary value of the previous stage, with the first stage taking on a new input value and the last stage dropping its bit. Thus input bit sequences can be made to shift through the register at a prescribed clock rate.

A PSC is composed of unequal shift registers placed in parallel with a single communication switch (Figure 6.21*b*). A register must be available for each different source. The different bit streams are simultaneously loaded into the appropriate registers and the entire register contents read out in sequence at fixed time intervals to accomplish the multiplexing.

Assume that there are N digital sources, having rates R_i bits/sec, $i = 1, 2, \ldots, N$, fed into a specific register of a PSC at its own rate. The register lengths (number of register stages) must be selected such that after some interval in time, all the registers will fill simultaneously. If h_i is the register length of the ith such register of a particular PSC, this loading condition requires that a time τ exists such that the number of bits of rate R_i occurring in τ sec is exactly h_i. Hence τ must be such that

$$R_i \tau = h_i \tag{6.7.1}$$

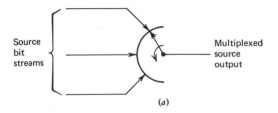

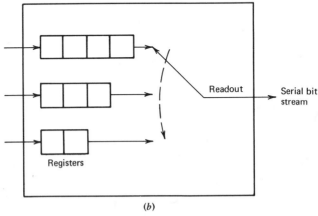

Figure 6.21. Parallel–serial converter: (*a*) equal input bit rates, (*b*) unequal input bit rates with shift registers.

for all inputs to the PSC. This immediately implies that for any two input rates, say, R_i and R_j, the corresponding register lengths must be related by

$$\frac{h_i}{h_j} = \frac{R_i}{R_j} \tag{6.7.2}$$

For smallest register lengths (simplest PSC) it is necessary that the h_i have the smallest possible integer value. Thus each h_i should be the smallest factor of the set of input rates (R_i). Equivalently, the set of input rates should be related to the set of register lengths by

$$R_i = \alpha h_i \tag{6.7.3}$$

where α is the largest common multiple of all rates. We see that a set of PSC input rates specifies the required register lengths, whereas a set of registers specifies the ratios of input rates. The minimal register size will occur if the h_i are the smallest prime factors, and the rates R_i are selected to be multiples of these prime numbers. As an example, suppose that we wish to multiplex three digital sources having the rates $R_1 = 72$ bps, $R_2 = 48$ bps, and $R_3 = 60$ bps. The smallest integer factors of these rates are then $h_1 = 6$, $h_2 = 4$, and $h_3 = 5$, respectively, and the common factor is 12 (i.e., $R_i = 12h_i$). A PSC as in Figure 6.21b would therefore require three parallel registers with six, four, and five stages, respectively.

After the registers of the PSC are filled, the bits can be read out from each register in sequence at a constant readout rate. That is, the commutating switch in Figure 6.21b moves to the first register, reads out all stored bits in sequence, then moves to the second register and reads out all bits, and so on. One complete cycle of register readout must occur in τ sec, the time it takes to fill each, with each register refilling immediately after emptying. Hence the output serial bit rate of the PSC is

$$R_o = \frac{1}{\tau} \sum_{i=1}^{N} h_i = \sum_{i=1}^{N} R_i \text{ bits/sec} \qquad (6.7.4)$$

The multiplexed output rate is therefore always equal to the sum of the individual source rates.

Whenever several data sources are multiplexed, it is generally necessary to insert synchronization bits periodically to aid in the receiver demultiplexing. If K_s synchronization bits are inserted after every K_b data bits at the output of each PSC, then the data rate of the PSC output bit stream is increased by the factor

$$R_s = \frac{K_s}{K_b/R_o} = R_o\left(\frac{K_s}{K_b}\right) \qquad (6.7.5)$$

The total effective PSC multiplexed output bit rate is then

$$R_o + R_s = R_o\left(1 + \frac{K_b}{K_s}\right) \qquad (6.7.6)$$

Thus the total multiplexed source rate is not increased significantly if the synchronization bits are a small fraction of K_b frame bits. The required number of synchronization bits depend on the accuracy to which this synchronization word must be recovered at the receiver. The exact time of arrival of the synchronization word at the receiver is needed to properly demultiplex the bit stream and synchronize the data frame. This is referred to as *frame* synchronization.

Frame synchronization is obtained at the receiver by having the receiver "look" in each data frame for the synchronization word. It does this by having the receiver baseband processor use a digital word correlator that stores the known synchronization word (Figure 6.22). As the recovered source bits are shifted through the register bit by bit, a word correlation is made with the stored word. When the transmitted word is received, its

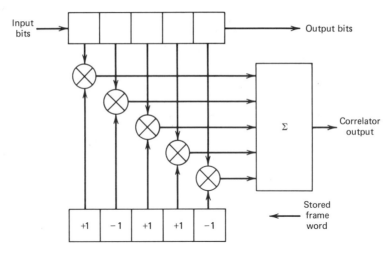

Figure 6.22. Frame word correlator.

correlation will produce a large signal output that can be noted as a threshold crossing. This crossing can be used to generate a time marker that marks all subsequent data words of the frame. During the next frame, the synchronization word is reinserted, the word is detected, and the word timing is again established for the new frame.

Deinterleaving of the multiplexed bit stream is accomplished by a commutator operating in synchronism with the arriving bit stream. The synchronized commutator taps off the group of bits of each source in sequence, feeding them to the proper destination. The synchronization bits, inserted at the PSC, maintain the required commutation synchronism for this deinterleaving.

The output bit stream of a single PSC can be combined with other PSC outputs to produce an entire multiplex hierarchy, as shown in Figure 6.23. The PSC outputs due to multiplexing source bit streams at level 1 are similarly combined to form a bit stream at level 2, which, in turn, can be further combined at level 3, and so on. The result is a structured mutiplexing format in which a large number of digital sources can be combined into a single bit stream.

Standardized PSC hardware has been established for this purpose. For example, a PSC has been developed for operating on 24 digital channels, each operating at 64 kbps (voice A-D systems). This produces a multiplexed bit rate of $24 \times 64\,\text{kbps} = 1.536\,\text{Mbps}$. With 8 synchronization bits inserted every millisecond, a resulting multiplexed bit rate of $1.536\,\text{Mbps} + 8\,\text{kbps} = 1.544\,\text{Mbps}$ is created. This is referred to as a standard T_1 digital multiplexer and is commonly used in digital subscriber service. Higher order multiplexers have been similarly established, and are listed in Table 6.3, along with the specialized source rates that can be handled.

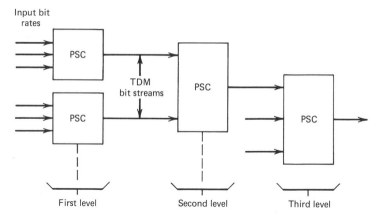

Figure 6.23. Higher level multiplexing with PSC.

Digital multiplexing hardware is generally more complicated than the simple diagrams shown. This is due to the added circuitry that is often needed to maintain clock and commutator control. Ideally, bits fed into a PSC should occur at a fixed data rate, timed by a perfect source clock. However, the timing of the source bit clock may jitter or drift, and correction circuitry must be used to reestablish a stabilized bit stream.

One way is by simple *reclocking*, in which an input bit stream is reconverted to a more stable clock. The circuitry is shown in Figure 6.24a. Midbit sampling is timed by a local bit clock, which also generates a clock pulse each bit time to represent the new bit waveforms. The input digital waveform is sampled, and the bit polarity is used to multiply the clock pulse, resulting in a pulse digital waveform at the output having the bits sequence of the input. If the input digital waveform contains clock jitter as shown in Figure 6.24b (i.e., the bit periods are not all exactly T_b sec long) the reclocking transfers the bits to the more stable local clock. Hence reclocking "cleans up" bit jitter, as long as the jitter does not exceed half a bit period.

Reclocking, however, cannot compensate for continual clock drifting. If the input bit clock tends to drift in frequency relative to the local bit clock, the input bit periods will continually decrease or increase relative to the local clock bit periods. The midbit sampling may eventually produce two samples from the same bit, or may skip over a bit. To compensate for this,

TABLE 6.3 Digital Multiplexing Hierarchy

T_1 (24 voice channels)	1.544 Mbps
T_{1c}	3.152 M
T_2 (video telephone)	6.312 M
T_3 (600 voice channels)	44.736 M
T_4	274.176 M

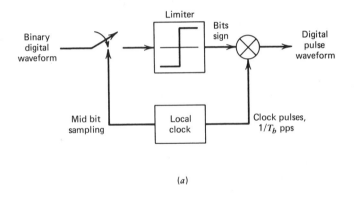

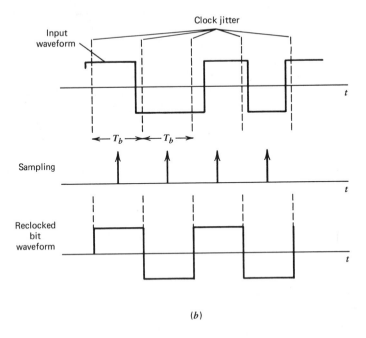

Figure 6.24. Reclocking circuit: (*a*) block diagram, (*b*) waveform diagram.

an *elastic buffer* [16] circuit is inserted, in addition to the reclocking, to stabilize the frequency drift. An elastic buffer operates as shown in Figure 6.25a. The input bits at rate r_{in} are sampled and clocked out at rate r_o by the local clock. The clock rates are monitored via a set/reset switch in which the input bit edge starts a pulse generator and the local clock samples stop the pulse. If perfect timing is maintained ($r_{in} = r_o$) the generated pulses are always of the same width. As the source clock drifts relative to the buffer output clock, the pulse width is altered accordingly (Figure 6.25b). A short term integrating filter (effectively measuring the short term dc value of the pulses) will indicate the changing pulse width. When the pulse width becomes too narrow or too wide, the input clock can be corrected. If the source clock is not accessible, the buffer can insert extra bits (*bit stuffing*) or delete bits to maintain the proper timing.

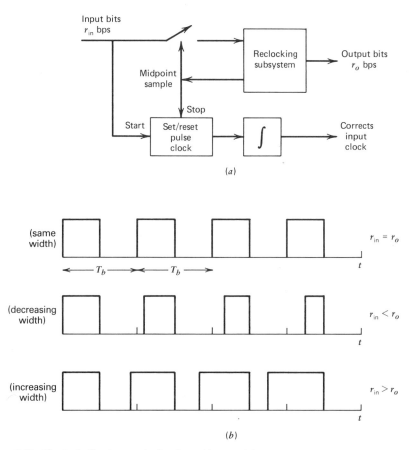

Figure 6.25. Elastic buffer for monitoring input bit rate: (*a*) system diagram, (*b*) pulse width waveforms.

REFERENCES

1. Freeman, R.L. *Telecommunication Transmission Handbook*, Wiley, New York, 1975.
2. *Transmission Systems for Communications*, 4th ed., Bell Telephone Laboratories, 1970.
3. Kurtenback, A. and Wintz, P. "Quantizing for Noisy Channels," *IEEE Trans. Comm.*, vol. COM-17, April 1969, pp. 291–302.
4. Bodycomb, J. and Haddad, A. "Some Properties of Predictive Quantizing Systems," *IEEE Trans. Comm.*, vol. COM-17, October 1970, pp. 682–685.
5. Davisson, L. "Theory of Adaptive Data Compression," in *Advances in Communication Systems*, A. Balakrishman, ed., Academic, New York, 1966.
6. Huffman, D. "A Method for Constructing Minimum Redundancy Codes," *Proc. IRE*, vol. 40, May 1952, pp. 1098–1101.
7. Stiffler, J. *Theory of Synchronous Communication*, Part 3, Prentice-Hall, Englewood Cliffs, N.J., 1971.
8. Andrews, H. and Caspari, K. "A Generalized Technique for Spectral Analysis," *IEEE Trans. Comput.*, vol. C-19, January 1970, pp. 16–25.
9. Harmuth, H. *Transmission of Information by Orthogonal Functions*, Springer, New York, 1969.
10. Campanella, S. and Robinson, G. "A Comparison of Orthogonal Transformations for Digital Speech Processing," *IEEE Trans. Comm.*, vol. COM-19, December 1971, pp. 1645–1649.
11. Pratt, W. and Kane, J. "Hadamard Transform of Image Coding," *Proc. IEEE*, vol. 57, no. 1, January 1969, pp. 58–65.
12. Oppenheim, A. and Schaefer, R. *Digital Signal Processing*, Prentice-Hall, Englewood Cliffs, N.J., 1975.
13. Crowther, W. and Rader, C. "Efficient Coding of Vocoder Signals Using Linear Transformations," *Proc. IEEE*, vol. 54, November 1966, pp. 1594–1601.
14. Habibi, A. and Wintz, P. "Image Coding by Linear Transformations," *IEEE Trans. Comm.*, vol. COM-18, February 1971, pp. 56–62.
15. Pratt, W. "Transform Coding of Color Images," *IEEE Trans. Comm.*, vol. COM-18, February 1971, pp. 980–988.
16. Williams, R. *Communication System Analysis and Design*, Prentice-Hall, Englewood Cliffs, N.J., 1987.

PROBLEMS

6.1 (6.2) A baseband subsystem has a source producing a waveform with power spectrum $S_d(\omega)$ that is bandlimited white noise spectrum, with bandwidth B_d Hz and spectral level S_0 W/Hz. The carrier system effectively filters the baseband with channel filter $H_{bc}(\omega) = 1/(1 + j\tau\omega)$ and adds in a white noise interference of level N_{b0} W/Hz. (a) Determine the optimal baseband filter at the receiver for maximizing

SDR, assuming the source waveform is used directly as the baseband.
(b) Determine the optimal filters at the transmitter and receiver for
maximizing SNR for this system, assuming the transmitter power is
limited to P_m.

6.2   (6.2)   Given the baseband system in Figure P6.2, the additive spur is
a sine wave interference at frequency $(1\,\text{MHz} + 100\,\text{Hz})$ with a uniform distributed phase $(0, 2\pi)$. The source has a flat spectrum from
zero to $1\,\text{MHz}$. Sketch the receiver filter transfer function that
maximizes the SDR of the output source waveform.

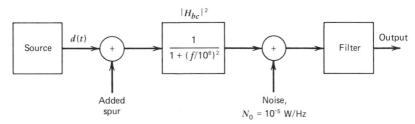

Figure P6.2.

6.3   (6.2)   A baseband filtering system (Figure 6.2) is designed with a
transmitter and receiver filter for maximum SNR. Assume that there
is no channel filter ($H_{bc} = 0$) and the system is designed that for a
source spectrum $S_{d_1}(\omega)$ and receiver noise $S_{n_1}(\omega)$. Suppose the true
source and noise spectra are instead $S_{d_2}(\omega)$ and $S_{n_2}(\omega)$. Show, in
integral form, how the resulting output SNR is altered.

6.4   (6.2)   Show that the maximal value of SNR in (6.2.8), under the
constraint of (6.2.7), is given by

$$\text{SNR}_{\max} = P_d P_m \left(\frac{1}{2\pi} \int_{-\infty}^{\infty} \frac{[S_d(\omega)S_{nb}(\omega)]^{1/2}\,d\omega}{|H_{bc}(\omega)|} \right)^{-2}$$

6.5   (6.2)   Formally carry out the minimization of (6.2.9), and derive the
results in (6.2.10).

6.6   (6.2)   Apply the results of Section 6.2 and determine the optimal
baseband filter that should be used to maximize SDR in (6.2.1) when
the source signal is a bandpass waveform over $(\omega_s - 2\pi B_d, \omega_s + 2\pi B_d)$.

6.7   (6.2)   Assume that $H_t(\omega)$ and $H_b(\omega)$ are given in Figure 6.2. Determine the optimal channel filter $H_{bc}(\omega)$ that maximizes SDR in
(6.2.1).

6.8 (6.3) There is a voice source with $P_d = 1$ and a bandwidth of 4 kHz. An RF carrier with an allowable RF bandwidth of 1 MHz is available. The RF carrier power just satisfies a threshold value of 15 dB over the noise in its RF bandwidth. Determine the demodulated source SNR_d that will occur for the following formats: (a) standard FM, (b) standard AM, (c) AM|FM, with a subcarrier of 100 kHz, (d) FM|FM with the same subcarrier and index $\beta = 3$.

6.9 (6.3) A bandlimited source of spectral level S_0 and bandwidth B_d Hz is to be transmitted over a baseband subsystem using an AM|FM format. The channel noise is white with level $N_0/2$. Channel filtering effects are negligible. (a) Using the results of Section 6.2, design the optimal receiver filter that should be placed after the FM demodulator so as to maximize the subcarrier to distortion rate prior to the AM demodulation. (b) Repeat for a transmitter–receiver pair, and comment on the required source preemphasis.

6.10 (6.3) A system uses AM/PM transmission over an ideal (no filtering), power limited baseband channel. The RF channel has additive white noise. Assume that the subcarrier frequency is 1 MHz and the modulated subcarrier spectrum is flat over B_{sc} Hz. Sketch how the modulated subcarrier should be filtered before and after RF modulation–demodulation in order to maximize the receiver subcarrier CNR.

6.11 (6.3) An AM|FM multiplexed system uses subcarrier frequencies at ω_1 and ω_2 ($\omega_2 > \omega_1$) with amplitudes k_1 and k_2, respectively. Assume satisfactory threshold operation, white noise interference, subcarrier bandwidths much less than subcarrier frequencies, and ideal (flat filter) functions. Assuming that the transmitter uses a preemphasis rule $C_i = C\omega_i^3$, determine: (a) the ratio of subcarrier SNR for the two channels ($\mathrm{SNR}_{\mathrm{sc}_1}/\mathrm{SNR}_{\mathrm{sc}_2}$), (b) the RF Carson rule bandwidth used (assume the rms carrier deviation). Then (c) generalize the result of (a) to an arbitrary preemphasis rule $C_i = C\omega_i^n$.

6.12 (6.4) Apply the results of Section 6.2 to the FDM|AM|FM system and determine the optimal preemphasis that should be used. (Do not assume the FM noise is flat over the AM spectra.)

6.13 (6.4) An FDM|AM|FM system uses subcarrier frequencies at 100, 300, and 600 kHz. An RF bandwidth of 4 MHz is available. The RF carrier power is just sufficient to provide a 15 dB SNR in the subcarrier bandwidths (all assumed identical). What is the maximum demodulated source SNR attainable?

6.14 (6.4) A group of source signals $d_1(t), d_2(t), \ldots, d_N(t)$, each having bandwidth B_d Hz, is multiplexed in an FDM|AM|PM system. Calculate the output SNR_d following subcarrier demodulation for each channel. Assume that all demodulators operate above threshold, ideal flat filters are used, and the receiver carrier noise is white. Identify all parameters used in your equations.

6.15 (6.4) Using the following parameters, design a set of four preselected FM subcarrier channels (center frequency and bandwidths) for source multiplexing:

$$\text{Lowest usable frequency, 370 Hz}$$

$$\Delta_{si} = 0.075\omega_{si}$$

$$B_{sc} = 2\Delta_{si}$$

Determine what source bandwidth B_d can be frequency modulated onto each channel, if a β_i of 5 is desired in all channels.

6.16 (6.4) An FM carrier uses an RF channel that is 36 MHz wide. The baseband is generated by modulating 4 kHz voice waveforms onto subcarriers by SSB-AM. The subcarrier bandwidths are placed side by side starting from zero frequency. The entire baseband is modulated onto the carrier with a modulation index of 4. How many voice waveforms can be sent simultaneously over the channel?

6.17 (6.5) Prove the converse of the sampling theorem: If $d(t)$ is time limited to T sec, then it can always be represented by its frequency samples taken every $1/T$ sec apart.

6.18 (6.5) A system reconstructs a waveform from its time samples $d(nT)$ by using the general summation

$$\hat{d}(t) = \sum_{n=-\infty}^{\infty} d(nT)h(t-nT)$$

(a) Show that the resulting waveform corresponds to the waveform $d(t)$ filtered by the transform of $h(t)$. (b) Use the result of part a to determine the frequency spectrum when $h(t)$ corresponds to a sample-and-hold circuit, that is, $h(t) = 1$, $0 \le t \le T$. (c) Repeat when a linear extrapolator is used. (A straight line is drawn between sample values.) *Hint:* First write the reconstructed waveform as in $\hat{d}(t)$ above, using the fact that

$$\hat{d}(t) = d(nT) + \{d[(n+1)T] - d(nT)\}(t-nT)$$

(d) Find the distortion effect that occurs with taking nonperfect waveform samples; that is

$$\hat{d}(nT) = \int_{-\infty}^{\infty} d(t)p(t - nT)\, dt$$

where $p(t)$ is a function over $(-T, T)$, centered at $t = 0$.

6.19 (6.5)   A source has power of 30 dBW and the spectrum

$$S_d(\omega) = \frac{C}{1 + [\omega/2\pi(5 \times 10^3)]^4}$$

The source is to be digitized with an SDR of 20 dB. The digitizing is to split the mean squared errors due to aliasing and quantization equally. Assuming that a uniform quantizer is optimal, estimate the bit rate at which the A-D source will operate.

6.20 (6.5)   An A-D converter is used to digitize a source having a uniform amplitude distribution over $(-A, A)$ volts, and a flat frequency spectrum out to 10 kHz. A 4 bit quantizer is used exactly covering $(-A, A)$ volts. Sketch a plot showing how the mean squared digitizing error, as a fraction of the source waveform power, varies with the output bit rate.

6.21 (6.5)   Show that the required rate for ideal bandpass sampling of a bandpass waveform with upper end frequency B_2 and bandwidth B is given by

$$f_2 = 2B\left(1 + \frac{k}{M}\right)$$

where M is the largest integer not exceeding B_2/B, and $k = (B_2/B) - M$.

6.22 (6.5)   A source of bandwidth 250 Hz is to be AM onto a subcarrier and then sampled at the rate of 1000 samples per second. Indicate the permitted subcarrier frequencies that may be used for sampling at the minimal rate.

6.23 (6.5)   Show that transforming a voltage sample that has probability density $p_d(x)$ by the transformation $y = \int_{-\infty}^{x} p_d(u)\, du$ will generate a uniformly distributed sample over $(0, 1)$.

6.24 (6.5)   Given $p_d(x) = ce^{-cx}$, $x \geq 0$. It is quantized by a two-level $(L = 2)$ uniform quantizer over $(0, 2c)$. Determine the MSQE.

6.25 (6.5)   A source signal voltage is uniform over ± 100 V. Its power spectrum is $S(\omega) = 1.3(\text{volt})^2/\text{Hz}$, $|\omega| \leq 2\pi(1\ \text{kHz})$. The signal is to be digitized and transmitted. It is sampled at rate 1990 samples per second. The quantizer has eight levels. Assuming perfect reconstruction, determine the resulting digital SDR at the receiver.

6.26 (6.5) A noisy source has an SNR of 7 dB. The source output is digitized and transmitted over a link that produces an overall SDR of 10 dB. (a) What will be the final reconstructed SNR of the source signal? (b) What must the system SDR be to ensure that the final SNR will be at least 5 dB?

6.27 (6.6) A signal $d(t)$ is time limited to T sec (but is not band-limited). A communicator wishes to digitize the waveform [send samples of some type to represent $d(t)$]. Using the sampling theorem *in the frequency domain*, explain (a) what type samples should be sent, (b) the reconstruction process at the receiver.

6.28 (6.7) Show that the sources in the example in Section 6.6 can be multiplexed with fewer total register stages by first multiplexing two sources in one PSC, and then multiplexing the output with the third source in another PSC.

6.29 (6.7) Design a parallel-series converter for sources with rates $R_1 = 100$ kbps, $R_2 = 45$ kbps, $R_3 = 6$ kbps and $R_4 = 3$ kbps. (a) What will be the bit rate at the PSC output? (b) What will the bit rate be if a synchronization bit is inserted every 200 bits?

6.30 (6.7) A false frame synchronization word detection will occur when random bits fill the correlator and happen to match the frame synchronization word. Show that if there are N bits in the synchronization word, and M of the N bits are needed for detection, then the probability of false synchronization is

$$\text{Prob false synchronization} = \frac{1}{2^N} \sum_{i=M}^{N} \binom{N}{i}$$

7

BINARY DIGITAL SYSTEMS

In Chapter 1 it was pointed out that a digital communication system operates by transmitting symbols (bits) over the communication link. This transmission is accomplished by first converting (encoding) the bits into binary waveforms, then transmitting them to the receiver by any of the RF modulation formats discussed previously. At the receiver the recovered baseband waveforms are then converted (decoded) back to the bits to complete the link. In a binary digital system the bits are transmitted one at a time over the link. In a block waveform encoded system, blocks of bits are encoded at one time for link transmission. In this and the next chapter the binary system is studied, while the block waveform system is considered in Chapter 9. Our primary concern is with the design of the encoding subsystem and the resulting channel decoding performance, the latter measured in terms of an average probability of recovering a bit in error. We are also interested in the way in which baseband power levels, bandwidth, and waveform structure influence this probability. Design techniques for reducing this bit error probability by proper decoder design and signal selection are also presented.

7.1 THE BINARY DIGITAL MODEL

The block diagram of a binary digital system is shown in Figure 7.1. At the transmitter, each source data bit is first encoded into a known signal for channel transmission. Thus there must be two available waveforms, one for a binary one and the other for a binary zero, to allow subsequent bit transmissions, each limited to a finite time interval. The bits are therefore converted, in sequence to the appropriate signal. This binary encoding operation can be represented symbolically as

$$1 \to s_1(t), \qquad 0 \le t \le T_b$$
$$0 \to s_0(t), \qquad 0 \le t \le T_b$$

(7.1.1)

Communication

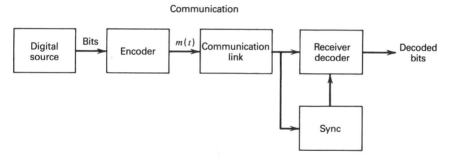

FIGURE 7.1. The binary digital communication block diagram.

That is, a binary one is converted to the waveform $s_1(t)$ whenever it occurs in the binary sequence, and a binary zero is converted to $s_0(t)$. Each signal is T_b sec long, where T_b is now the time to transmit a data bit and the resulting bit rate is $R_b = 1/T_b$. A bit sequence is therefore encoded into a sequence of bit waveforms, as discussed in Section 1.9, corresponding to a binary digital waveform. The spectral density of this waveform was computed in Section 2.1 and depended on the bit probability and the Fourier transforms of the bit waveforms in (7.1.1).

The waveform $m(t)$ is then transmitted over the communication link to the receiver. This is typically accomplished by either a direct transmission of $m(t)$, or by modulating $m(t)$ onto an RF carrier and transmitting to the receiver. At the receiver $m(t)$ is recovered and processed in the receiver decoder. The latter attempts to reconstruct the bit sequence by determining whether a zero or a one is transmitted during each T_b sec bit period. This is equivalent to deciding whether the known waveforms, $s_1(t)$ or $s_0(t)$, are being received during each T_b sec interval. A decoder decisioning error will therefore produce a bit error in the decoded sequence. Note that the decoder requirement is not to recover the bit waveforms, but simply to determine which one is being transmitted. For this reason channel decoding theory can be encouched within a framework of decision theory and hypothesis testing. We shall pursue this in Section 7.2.

The decoder is generally augmented by a synchronization subsystem that operates in conjunction with the decoder processing. This subsystem generates the necessary timing and synchronization information needed by the decoder, as we shall see. The design and performance of this subsystem is critical to the overall performance of the digital system, and in many cases it alone may limit the capabilities of the decoding system. Synchronization subsystem design is considered in Chapter 10.

For analytical purposes, the binary digital system can be modeled by the block diagram in Figure 7.2. During each T_b sec bit interval, $m(t)$ is either $s_1(t)$ or $s_0(t)$. Any RF modulation, transmission, and reception operations recover the waveform $m(t)$ while adding in the receiver noise $n(t)$, as

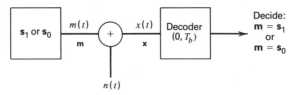

FIGURE 7.2. The binary decoding model.

discusssed in Chapter 5. The resulting waveform during a bit interval $(0, T_b)$, which can be any bit interval, is then

$$x(t) + m(t) + n(t), \qquad 0 \le t \le T_b \qquad (7.1.2)$$

This $x(t)$ is then processed in the decoder during $(0, T_b)$ in order to decode the bit for that interval. The decoder processing is then repeated during the next bit interval, and so forth.

Initially, the following assumptions are made:

1. Each successive bit is generated randomly, and therefore is equally likely a one or zero bit, independent of all other bits.
2. The signals selected to represent each bit are of equal energy:

$$\int_0^{T_b} s_i^2(t) \, dt = E_b, \qquad i = 0, 1 \qquad (7.1.3)$$

   Here E_b is the waveform bit energy, or the energy per bit.
3. The channel only adds in noise to the recovered encoded waveform. Later in Section 8.3 the effect of channel filtering and distortion is considered.
4. The additive decoder noise is a Gaussian noise process, with zero mean, and spectrally white with spectral level $N_0/2$ W/Hz. This is in accordance with our demodulation analysis in Chapter 5. Nonwhite noise is considered in Section 7.7.
5. The decoder has perfect timing and can delineate precisely the beginning and ending of each bit as it arrives at the receiver. Since this timing is provided by the synchronization subsystem in Figure 7.1, this assumes ideal time synchronization. Effects of imperfect timing are considered in Section 7.6.

A vector version of the system shown in Figure 7.2 can also be derived. Following our discussion of signal and noise expansions in Appendix C, we can consider some orthonormal set $\{\Phi_j(t)\}$, $j = 1, 2, \ldots, \nu$, defined over $(0, T_b)$, to be selected for waveform expansion. We can then consider both $s_i(t)$ and $n(t)$ to be generated from this expansion. Hence

$$s_i(t) = \sum_{j=1}^{\nu} s_{ij} \Phi_j(t), \qquad 0 \le t \le T_b \qquad (7.1.4a)$$

$$n(t) = \sum_{j=1}^{\nu} n_j \Phi_j(t), \qquad 0 \le t \le T_b \qquad (7.1.4b)$$

where s_{ij} and n_j are the resulting Fourier waveform coordinates. These coordinates then yield a vector version of each waveform:

$$\mathbf{s}_i = (s_{i1}, s_{i2}, \ldots, s_{i\nu}) \qquad (7.1.5a)$$

$$\mathbf{n} = (n_1, n_2, \ldots, n_\nu) \qquad (7.1.5b)$$

At the present time we shall not be concerned with the explicit orthonormal set to be used for the expansion, other than to assume that one set has been properly selected. Later we will give some specific examples that have a meaningful engineering interpretation.

The vectors in (7.1.5) now combine to produce the observable vector

$$\mathbf{x} = \mathbf{m} + \mathbf{n} \qquad (7.1.6)$$

in Figure 7.2, where $\mathbf{m}$ is either $\mathbf{s}_1$ or $\mathbf{s}_0$. Because of our ability to convert back and forth between the waveform domain and the vector domain, when expansions such as (7.1.4) are used, the decoder processing can be equivalently derived entirely in the vector domain.

On the basis of this receiver model in Figure 7.2, we can now attempt to determine the decoder processing on $x(t)$, or $\mathbf{x}$, in order to achieve efficient bit decoding. We first approach the decoder design from a decisioning point of view in order to mathematically establish the capabilities and limitations of general decoder algorithms. These will suggest realistic decoder mechanizations and structures that can be implemented in practical hardware.

7.2 OPTIMAL DECODING

The most important requirement to be satisfied by the decoder is to ensure that the individual bits are decoded reliably. Since there is additive noise at the decoder input, some of the bit decisions made by the decoder may be in error. An indication of this decoder performance is therefore given by the probability of the decoder making a decision error during a given bit, that is, the bit error probability. Our objective then is to determine the required decoder processing that will minimize this bit error probability. This requires us to first write a general expression for this probability.

Let PE_1 be the probability of making a bit error when a one-bit is sent, and let PE_0 be corresponding probability for a zero-bit. Since each bit is equally likely, the average bit error probability is then

$$PE = \tfrac{1}{2}PE_1 + \tfrac{1}{2}PE_0 \qquad (7.2.1)$$

The decoder should operate in such a way as to minimize PE. We shall show that, indeed, there is a mathematical algorithm that produces this minimum. Let

$P(i|\mathbf{x}) =$ probability that the decoder will select bit i ($i = 0, 1$) after observing $\mathbf{x}$ in Figure 7.2 during a particular bit interval. This probability depends on the decoder processing and will be determined by a way in which the decoder decides bits from $\mathbf{x}$.

$p(\mathbf{x}|i) =$ probability density of the observed vector $\mathbf{x}$ at the decoder when bit i is transmitted. Note that this is a probability density over the space of all vectors $\mathbf{x}$ produced at the decoder, and depends on the statistics of the channel noise.

It then follows that

$$PE_1 = \int_{\mathbf{X}} P(0|\mathbf{x})p(\mathbf{x}|1)\, d\mathbf{x} \qquad (7.2.2)$$

where $\mathbf{X}$ is the vector space of all ν-dimensional vectors $\mathbf{x}$. The integrand in (7.2.2) is simply the probability that the zero-bit will be decoded when $\mathbf{x}$ is observed, given that the one-bit is sent; that is, an error is made. The integral is the resulting average over all $\mathbf{X}$. Note that (7.2.2) is actually a ν-dimensional integration over a vector space. Likewise

$$PE_0 = \int_{\mathbf{X}} P(1|\mathbf{x})p(\mathbf{x}|0)\, d\mathbf{x} \qquad (7.2.3)$$

Since a decoding decision must be made each bit time, it necessarily follows that

$$P(1|\mathbf{x}) + P(0|\mathbf{x}) = 1 \qquad (7.2.4)$$

That is, with a probability of one either a one-bit a or zero-bit must be decided for any observed $\mathbf{x}$. With (7.2.2) and (7.2.3), PE can be written in integral form as

$$PE = \frac{1}{2}\int_{\mathbf{X}} P(0|\mathbf{x})(\mathbf{x}|1)\, d\mathbf{x}$$

$$+ \frac{1}{2}\int_{\mathbf{X}} P(1|\mathbf{x})p(\mathbf{x}|0)\, d\mathbf{x} \qquad (7.2.5)$$

Substitution with (7.2.4) allows this to be written more compactly as

$$PE = \frac{1}{2} \int_{\mathbf{x}} [p(\mathbf{x}|1) + P(1|\mathbf{x})(p(\mathbf{x}|0) - p(\mathbf{x}|1)] \, d\mathbf{x} \qquad (7.2.6)$$

Recall that our objective is to find the decoding procedure [which is embedded in the decoding probabilities $P(i|\mathbf{x})$] that minimizes (7.2.6). When written this way, the solution can be easily determined without direct integration by noting some properties of the integrand. Since the latter is always nonnegative, and since $P(i|\mathbf{x})$ is a probability that is bounded between $(0, 1)$, we can see that the integral in (7.2.6) is necessarily minimized if the integrand takes on its smallest value at every $\mathbf{x}$. The latter will occur if and only if

$$P(1|\mathbf{x}) = 1 \quad \text{at all } \mathbf{x} \text{ where } [p(\mathbf{x}|1) \geq p(\mathbf{x}|0)]$$
$$= 0 \quad \text{at all } \mathbf{x} \text{ where } [p(\mathbf{x}|1) < p(\mathbf{x}|0)] \qquad (7.2.7)$$

No other $P(1|\mathbf{x})$ can produce a smaller PE. Note that if $p(\mathbf{x}|0) = p(\mathbf{x}|1)$ for some $\mathbf{x}$, any value of $P(1|\mathbf{x})$ can be selected for that $\mathbf{x}$, so we arbitrarily set $P(1|\mathbf{x}) = 1$ for this case. Since $P(1|\mathbf{x}) = 1$ implies that we select the one-bit for that $\mathbf{x}$ with a probability of one, while $P(1|\mathbf{x}) = 0$ implies that we select the zero-bit with a probability of one, we can restate the decoding rule in (7.2.7) as follows:

Decide the one-bit if $\mathbf{x}$ is such that

$$p(\mathbf{x}|1) \geq p(\mathbf{x}|0)$$

Decide the zero-bit if $\mathbf{x}$ is such that $\qquad\qquad (7.2.8)$

$$p(\mathbf{x}|1) < p(\mathbf{x}|0)$$

This is an equivalent description of the optimal decoding rule. After observing $\mathbf{x}$ in a bit interval, it decodes the bit according to the preceding rule, repeating the rule for every bit interval. Hence, optimal bit decoding is based on a straightforward comparison test between the values of $p(\mathbf{x}|1)$ and $p(\mathbf{x}|0)$ at the observed $\mathbf{x}$.

This test can also be stated in alternate forms, such as

Decide the one-bit if

$$\Lambda(\mathbf{x}) = \frac{p(\mathbf{x}|1)}{p(\mathbf{x}|0)} \geq 1, \qquad \text{otherwise decide the zero-bit} \qquad (7.2.9)$$

The function $\Lambda(\mathbf{x})$ is called the *likelihood ratio* of the observed $\mathbf{x}$, and the test in (7.2.9) is commonly referred to as a *likelihood ratio test*, usually denoted simply

$$\Lambda(\mathbf{x}) \gtrless 1 \qquad (7.2.10)$$

with the direction of the inequality determining the bit. This means that optimal decoding can also be accomplished by evaluation of $\Lambda(\mathbf{x})$ in (7.2.9), followed by the test in (7.2.10). Since the right side of (7.2.10) plays the role of a threshold to which $\Lambda(x)$ is to be compared, the likelihood ratio test in (7.2.10) is sometimes called a *threshold test*. The likelihood ratio test in fact has been found to be an optimal hypothesis test for a more general class of criteria, in addition to minimization of error probability [1–4].

Of course, the bit decision depends only on whether Λ is above or below threshold. If f is any monotonic increasing function, as shown in Figure 7.3, it is clear that $\Lambda \lessgtr 1$ if $f(\Lambda) \lessgtr f(1)$. That is, the threshold test in (7.2.10) is identical to the threshold test

$$f[\Lambda(\mathbf{x})] \gtrless f[1] \qquad (7.2.11)$$

For example, $\log(\Lambda)$ is monotonic in $\Lambda, \Lambda > 0$, and the decoding test in (7.2.10) can be performed using instead

$$\log \Lambda(\mathbf{x}) \gtrless \log(1) = 0 \qquad (7.2.12)$$

In many applications the decoder processing is mathematically simplified by use of the alternate test in (7.2.12) instead of (7.2.10) or (7.2.8). The key point, of course, is that all these bit decisioning tests are identical, and all correspond to the optimal decoder rule. The test in (7.2.12) has the added advantage that the comparison is made to a threshold of zero, which is equivalent to determining whether $\log \Lambda(\mathbf{x})$ is positive or negative; that is, determining its sign. We remark that the discussion in this section is completely general, and holds for any probability density $p(\mathbf{x}|i)$. Our next object is to find practical hardware implementations that carry out these decoding tests.

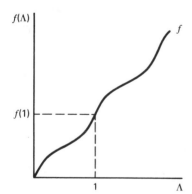

FIGURE 7.3. Monotonic function of Λ.

7.3 DECODING WITH GAUSSIAN NOISE

The optimal decoder processing for minimizing bit error probability in a binary digital system was given in (7.2.8) or (7.2.9). When the additive noise is a white Gaussian process, the required decoder implementation can be determined directly. The observable vector $\mathbf{x}$, corresponding to the waveform $x(t)$ over $(0, T_b)$, was shown in (7.1.6) to be

$$\mathbf{x} = \mathbf{m} + \mathbf{n}$$

where $\mathbf{m}$ is either signal vector $\mathbf{s}_1$ or $\mathbf{s}_0$ and $\mathbf{n}$ is a zero mean, Gaussian vector. Because $\mathbf{n}$ is the vector expansion of a white noise process, we know that its components are statistically independent and each has variance $N_0/2$ (see Appendix C.3). Thus the conditional densities of $\mathbf{x}$ become

$$p(\mathbf{x}|1) = \prod_{j=1}^{\nu} \frac{1}{(2\pi N_0/2)^{\nu/2}} \exp \frac{-\frac{1}{2}(x_j - s_{1j})^2}{N_0/2}$$

$$= \frac{1}{(2\pi N_0/2)^{\nu/2}} \exp\left[-\frac{1}{N_0} \sum_{j=1}^{\nu} (x_j - s_{1j})^2\right] \qquad (7.3.1a)$$

and

$$p(\mathbf{x}|0) = \frac{1}{(2\pi N_0/2)^{\nu/2}} \exp\left[-\frac{1}{N_0} \sum_{j=1}^{\nu} (x_j - s_{0j})^2\right] \qquad (7.3.1b)$$

The likelihood ratio in (7.2.9) then is

$$\Lambda(\mathbf{x}) = \frac{\exp\left[-(1/N_0)\sum_{j=1}^{\nu}(x_j - s_{1j})^2\right]}{\exp\left[-(1/N_0)\sum_{j=1}^{\nu}(x_j - s_{0j})^2\right]} \qquad (7.3.2)$$

Taking logs, and using the equivalent log test in (7.2.12), we restate the required decoder decision test as

$$\log \Lambda(\mathbf{x}) = \frac{1}{N_0} \sum_{j=1}^{\nu} [-(x_j - s_{1j})^2 + (x_j - s_{0j})^2] \gtrless 0 \qquad (7.3.3)$$

Expanding out the sums rewrites this as

$$\frac{2}{N_0} \sum_{j=1}^{\nu} x_j(s_{1j} - s_{0j}) + \frac{1}{N_0} \sum_{j=1}^{\nu} s_{0j}^2 - \frac{1}{N_0} \sum_{j=1}^{\nu} s_{1j}^2 \gtrless 0 \qquad (7.3.4)$$

With the conditions of equal signal energy in (7.1.3), and the zero threshold on the right, this simplifies further to

$$\sum_{j=1}^{\nu} x_j (s_{1j} - s_{0j}) \gtrless 0 \tag{7.3.5}$$

Thus the optimal decoder for additive white Gaussian noise requires, on observing the input vector $\mathbf{x} = \{x_j\}$, the computation of the test in (7.3.5). This decodes the bit depending simply on whether the left-hand side is positive or negative. Using the conversion from vector inner products to time correlation integrals [see (C.1.14)], the decoding procedure can equivalently be stated in terms of the receiver waveforms

$$\int_0^{T_b} x(t)[s_1(t) - s_0(t)]\, dt \gtrless 0 \tag{7.3.6}$$

The optimal decoder therefore should correlate (integrate) the observed waveform $x(t)$ with the difference waveform $s_1(t) - s_0(t)$ over each bit interval (recall perfect decoder timing is assumed). It is, in fact, somewhat remarkable that the optimal decoder processing, although formulated in the general context of error probabilities and likelihood ratios, reduces to the relatively simple correlator processor of (7.3.6). In particular, we note that the decoder involves only a linear correlation, and no other processor, linear or nonlinear, can produce a smaller PE with additive Gaussian noise. Note further that the decoding does not depend on the gain levels (coefficient in front of the integral) since only a sign test is required. By retracing our steps, we can further note that if the signals were not equal energy in (7.3.4), or if the bits were not equally likely in (7.1.1), the only effect is to bias the threshold in (7.3.6) away from zero (see Problem 7.6).

The optimal Gaussian decoder has an interesting geometric interpretation. Let us note the ν-dimensional signal vector space in Figure 7.4 in which $\mathbf{s}_1$ and $\mathbf{s}_0$ are plotted as unique vector points, according to their signal

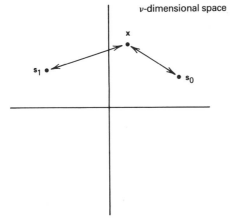

FIGURE 7.4. Signal vectors in ν-dimensional space.

coordinates. Now rewrite (7.3.3) as

$$\sum_{j=1}^{v} (x_j - s_{0j})^2 - \sum_{j=1}^{v} (x_j - s_{1j})^2 \gtrless 0 \qquad (7.3.7)$$

The summations on the left, when interpreted in this signal space, correspond to the squares of the distance from the observed vector point $\mathbf{x}$ to the vector point $\mathbf{s}_1$ and $\mathbf{s}_0$. The optimal decoder, performing the test in (7.3.5), is therefore equivalently decoding the bit based on whether the observed vector $\mathbf{x}$ is closer to $\mathbf{s}_1$ or $\mathbf{s}_0$ in signal space.

This gives an interesting mathematical flavor to a processing algorithm designed purely for hardware implementation. As we shall see later, this geometric interpretation aids us in understanding further improvements in system design.

7.4 DECODER IMPLEMENTATION AND PERFORMANCE

The optimal binary decoder must compute the correlation in (7.3.6) during each bit interval. However, implementation of this correlation can be constructed in several ways. The obvious way is a direct correlation of $x(t)$ with the difference waveform $s_1(t) - s_0(t)$, as in Figure 7.5a. Alternatively, we can define

$$y_i = \int_0^{T_b} x(t) s_i(t) \, dt , \qquad i = 0, 1 \qquad (7.4.1)$$

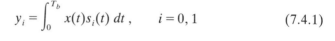

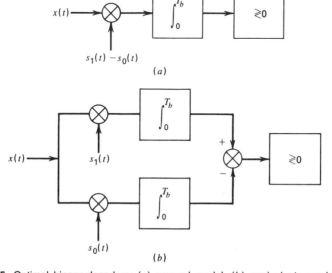

FIGURE 7.5. Optimal binary decoders: (a) general model, (b) equivalent correlator model.

and state the correlation test as

$$y = y_1 - y_0 \gtrless 0 \qquad (7.4.2)$$

This decoder can be implemented by the parallel system shown in Figure 7.5b, in which y_1 is computed in one channel and y_0 in the other. It requires the actual bit waveforms $s_1(t)$ and $s_0(t)$ to be actively generated, once every bit interval, and correlated with the decoder input. The output of the individual correlators is then subtracted for sign comparison. Because the receiver bit waveforms must be generated in exact time coherence with the received version of the same signals, the decoder mechanization illustrated in Figure 7.5 is often called a *synchronous*, or *coherent*, correlator. Since the comparison in (7.4.2) determines whether y_1 exceeds y_0, the parallel decoder is effectively deciding the bit based on whether the input waveform $x(t)$ correlates higher with $s_1(t)$ or with $s_0(t)$. This ties in with our geometric discussion in Figure 7.4.

As another approach, we can rewrite

$$y_1 = \int_0^{T_b} x(t)s_1(t) \, dt$$

$$= \int_0^{T_b} x(\rho)h_1(T_b - \rho) \, d\rho \qquad (7.4.3)$$

where $h_1(T_b - \rho) = s_1(\rho)$. This means that y_1 is equal to a sample at time $t = T_b$ of a filter whose input is $x(t)$ and whose impulse response is

$$h_1(t) = s_1(T_b - t) \qquad (7.4.4)$$

A filter with this specific response is called a *matched filter*. Note, however, that it is matched in a mirror imaged-shifted sense to the signal, as shown in Figure 7.6a; that is, the required impulse response is obtained by imaging $s_1(t)$ to $s_1(-t)$, then shifting by T_b sec. Similarly, we define

$$h_0(t) = s_0(T_b - t) \qquad (7.4.5)$$

as a matched filter for $s_0(t)$. The decoder shown in Figure 7.5b can therefore be replaced by the decoder in Figure 7.6b, corresponding to parallel channels of passive filters, each having the matched responses. The output of each filter is sampled at the end of each bit interval and the samples subtracted for sign comparison. The decoder can be redrawn with a single sampler as shown in Figure 7.6c. When redrawn this way, the decoder has the appearance of a filter–sampler construction. The input is filtered, by the parallel matched filters, subtracted, and the output sampled at the end of each bit time for decisioning. It should be emphasized that the exact filter

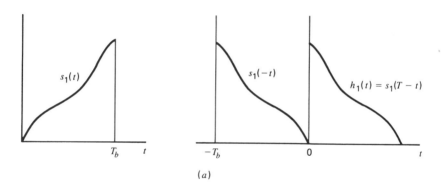

(a)

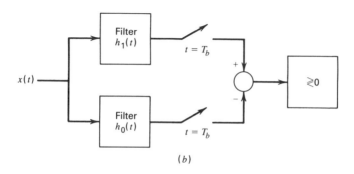

(b)

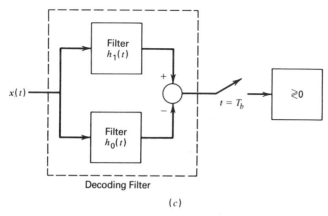

Decoding Filter

(c)

FIGURE 7.6. Matched filters: (a) impulse response, (b) decoder model, (c) filter–sampler model.

299

response $h_1(t)$ and $h_0(t)$ must be presented to each bit, which may require capacitor and inductor discharging from one-bit time to the next.

The choice of which decoder to implement is purely arbitrary, and is usually based on whether the matched filters are easier to construct, or whether the signal waveforms are easier to generate. Some common examples of each are presented in Section 7.5. The key point is that all implementations are theoretically identical, and preference is generally based on hardware convenience.

The performance of the decoders can be obtained by determining the resulting bit error probability PE. Since they perform identically, any one can be analyzed. Although PE can be determined directly from (7.2.6), it is easier to compute from its basic definition. Let us refer to the system in Figure 7.5a and determine the probability that it will be in error. With $y = y_1 - y_0$, let us substitute for $x(t)$, and write it explicitly as

$$y = \int_0^{T_b} x(t)[s_1(t) - s_0(t)] \, dt$$

$$= \int_0^{T_b} m(t)[s_1(t) - s_0(t)] \, dt + \int_0^{T_b} n(t)[s_1(t) - s_0(t)] \, dt \qquad (7.4.6)$$

This shows that the test variable y is composed of two components—one due to a correlation with the transmitted signal [which can be either $s_1(t)$ or $s_0(t)$], and one due to the additive Gaussian noise. The first term is simply a scalar, dependent on the signal waveforms. The second term is a Gaussian random variable (integral of a Gaussian process), zero mean [since $n(t)$ is zero mean] and has variance

$$\sigma_y^2 = \mathcal{E} \int_0^{T_b} n(t)[s_1(t) - s_0(t)] \, dt \int_0^{T_b} n(\rho)[s_1(\rho) - s_0(\rho)] \, d\rho$$

$$= \int_0^{T_b} \int_0^{T_b} \mathcal{E}[n(t)n(\rho)][s_1(t) - s_0(t)][s_1(\rho) - s_0(\rho)] \, dt \, d\rho$$

$$= \int_0^{T_b} \int_0^{T_b} R_n(t - \rho)[s_1(t) - s_0(t)][s_1(\rho) - s_0(\rho)] \, dt \, d\rho \qquad (7.4.7)$$

where $R_n(t - \rho) = \mathcal{E}[n(t)n(\rho)]$ is the correlation function of the decoder noise. Since this is assumed to be a white noise process, $R_n(t - \rho) = (N_0/2)\delta(t - \rho)$, we can easily integrate the delta function over ρ to produce

$$\sigma_y^2 = \left(\frac{N_0}{2}\right) \int_0^{T_b} [s_1(t) - s_0(t)]^2 \, dt \qquad (7.4.8)$$

Expanding the integrand, and inserting the energy constraint on the signals simplifies this to

$$\sigma_y^2 = \left(\frac{N_0}{2}\right)\left[E_b + E_b - 2\int_0^{T_b} s_1(t)s_0(t)\,dt\right]$$

$$= N_0\left[E_b - \int_0^{T_b} s_1(t)s_0(t)\,dt\right] \tag{7.4.9}$$

where again E_b is the energy in the waveforms. Thus y in (7.4.6) is a Gaussian random variable with variance σ_y^2 and having a mean value obtained from the first term in (7.4.6). This mean value depends on whether $s_1(t)$ or $s_0(t)$ is sent. Denote m_1 as the mean of y when $s_1(t)$ has been transmitted. Then

$$m_1 = \int_0^{T_b} s_1(t)[s_1(t) - s_0(t)]\,dt$$

$$= E_b - \int_0^{T_b} s_1(t)s_0(t)\,dt \tag{7.4.10}$$

Likewise, let m_0 be the mean value when $s_0(t)$ had been sent. Then

$$m_0 = \int_0^{T_b} s_0(t)[s_1(t) - s_0(t)]\,dt$$

$$= \int_0^{T_b} s_1(t)s_0(t)\,dt - E_b$$

$$= -m_1 \tag{7.4.11}$$

The decoder output variable y is Gaussian with a mean value of either m_1 or $-m_1$, depending on the transmitted bit during that interval. These two densities for y are shown in Figure 7.7. A bit error is made in decoding when a one-bit is sent if the decoder variable y is negative [the decoder will decide a zero bit according to (7.4.2)]. We see from Figure 7.7 that the

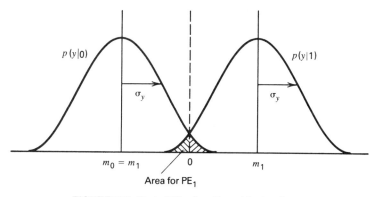

FIGURE 7.7. Probability densities of test voltage.

probability of this occurring is simply the probability associated with the Gaussian probability density $p(y|1)$ below zero. Hence

$$PE_1 = \int_{-\infty}^{0} \frac{1}{\sqrt{2\pi}\sigma_y} e^{-(y-m_1)^2/2\sigma_y^2} \, dy \qquad (7.4.12)$$

Similarly, the probability of deciding the incorrect bit when a zero-bit is sent is the area of $p(y|0)$ above zero. Since the means are equal and opposite, and the variances are the same, $PE_0 = PE_1$. The average bit error probability for any bit interval, when the optimum decoder is implemented, is then

$$PE = \tfrac{1}{2}PE_1 + \tfrac{1}{2}PE_0$$
$$= PE_1 \qquad (7.4.13)$$

Letting $u = (y - m_1/\sigma_y)$ in (7.4.12), we can write this as

$$PE = \frac{1}{\sqrt{2\pi}} \int_{-\infty}^{-m_1/\sigma_y} e^{-u^2/2} \, du$$
$$= \frac{1}{\sqrt{2\pi}} \int_{m_1/\sigma_y}^{\infty} e^{-u^2/2} \, du \qquad (7.4.14)$$

The function

$$Q(k) = \frac{1}{\sqrt{2\pi}} \int_{k}^{\infty} e^{-u^2/2} \, du \qquad (7.4.15)$$

is called the *Q-function*, or the *Gaussian tail function*, plotted and tabulated in Appendix B.1. We can therefore restate

$$PE = Q\left(\frac{m_1}{\sigma_y}\right) \qquad (7.4.16)$$

and the decoder error probability is obtained by evaluating the Q-function at the point m_1/σ_y. Substituting from (7.4.9) and (7.4.10), we obtain

$$PE = Q\left[\frac{E_b - \int_0^{T_b} s_1(t)s_0(t) \, dt}{N_0}\right]^{1/2} \qquad (7.4.17)$$

Decoder performance therefore depends on the bit energy E_b, the noise level N_0, and the signal waveshapes through the integral in (7.4.17). We often denote

$$\gamma = \frac{1}{E_b} \int_0^{T_b} s_1(t)s_0(t) \, dt \qquad (7.4.18)$$

as the normalized correlation coefficient of the signal pair $s_1(t)$ and $s_0(t)$. PE can be written in terms of this as

$$\mathrm{PE} = Q\left[\frac{E_b}{N_0}(1-\gamma)\right]^{1/2} \qquad (7.4.19)$$

We see that the minimal bit error probability that can be achieved by an optimal decoder in a binary digital Gaussian channel depends only on the bit energy to noise level ratio at the decoder inputs, E_b/N_0, and on the correlation of the signal pair used to encode the data bits. The only effect on performance of the choice of the signals used for the binary encoding is through the correlation parameter γ in (7.4.18).

7.5 SIGNAL SELECTION FOR BINARY DIGITAL SYSTEMS

We have assumed a pair of arbitrary signals, $s_1(t)$ and $s_0(t)$, to represent the binary bits. The only condition we placed on the signal waveform is that they have equal energy E_b. We then determined the optimal linear decoder for this signal pair. We found that when the optimal decoder is used, the bit error probability was given by (7.4.19), where γ was defined in (7.4.18). Thus the actual waveshape of the signals enters only through the parameter γ. We may first inquire if there is a best pair of signals to use, assuming that we can construct their proper matched filter. Clearly, from (7.4.19), with E_b and N_0 fixed, we want the most negative value of γ. From the Schwarz inequality, we note

$$|\gamma| \le \frac{[\int_0^{T_b} s_1^2(t)\,dt \int_0^{T_b} s_0^2(t)\,dt]^{1/2}}{E_b} = 1 \qquad (7.5.1)$$

We see that the most negative possible value is $\gamma = -1$. Furthermore, this occurs only when

$$s_0(t) = -s_1(t) \qquad (7.5.2)$$

That is, the signal pair are negatives of each other. Thus the antipodal binary digital waveforms discussed in Sections 1.10 and 2.1 are indeed optimal encoding waveforms. For these waveforms, (7.4.19) becomes

$$\mathrm{PE} = Q\left[\left(\frac{2E_b}{N_0}\right)^{1/2}\right] \qquad (7.5.3)$$

The curve is plotted in Figure 7.8 as a function of E_b/N_0. Note that PE now depends only on the energy ratio of the signals. Most importantly, we see that any pair of antipodal signals with the same energy is as good as any

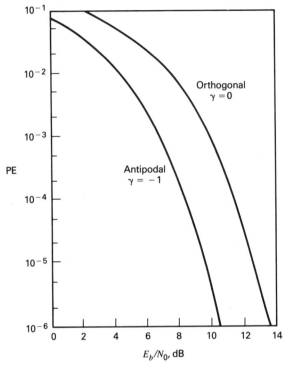

FIGURE 7.8. Bit error probability versus E_b/N_0 (γ = signal pair correlation).

other, in terms of the minimal possible PE. (However, there may be other reasons for selecting one signal pair over another, as we shall see.) Also included in Figure 7.8 is the curve for the signal pair such that

$$\int_0^{T_b} s_1(t)s_0(t)\, dt = 0 \qquad (7.5.4)$$

Such signals are said to be *orthogonal*. For orthogonal signal pairs, $\gamma = 0$ and the resulting PE curve requires twice the value of E_b to obtain the same error probability as in the antipodal case.

The optimality of the antipodal signals can also be interpreted in signal space. The argument of the Q-function in (7.4.19) contained the term

$$E_b(1 - \gamma) = E_b - \int_0^{T_b} s_1(t)s_0(t)\, dt$$

$$= \tfrac{1}{2} \int_0^{T_b} [s_1(t) - s_0(t)]^2\, dt \qquad (7.5.5)$$

which is half the distance squared between s_1 and s_0. Minimizing PE requires signals with maximum separation in vector space. From Figure 7.9 it is evident that for signals of equal energy (on a hypersphere), a pair of antipodal signals will have the maximal separation over all possible equal energy signal pairs. Their location is immaterial as long as they are antipodal.

Figure 7.8 shows that the decoder bit error probability can be determined by merely knowing the operating E_b/N_0. A given bit error probability can be achieved by guaranteeing that the E_b/N_0 is satisfactorily high. At PE = 10^{-3} (i.e., a one in a thousand chance of making a bit error, or about one out of every thousand bits will be in error), an $E_b/N_0 = 5.3$ or about 7 dB will be required; and at PE = 10^{-5}, $E_b/N_0 = 9.6$ or 9.7 dB. By recalling that $E_b = P_m T_b$, where P_m is the signal power, and $1/T_b$ is the bit rate in bits per second, we can also write this key energy ratio as

$$\frac{E_b}{N_0} = \frac{P_m T_b}{N_0} = \frac{P_m}{N_0 R_b} \tag{7.5.6}$$

Equation (7.5.6) has the form of a signal to noise power ratio, and for this reason E_b/N_0 is often referred to as the *decoding* SNR. It is, however, a specific SNR, with signal power referring to that occurring at the decoder input, and noise power referred to that collected in a bit rate bandwidth; that is, a bandwidth in Hz equal to the bit rate in bits/sec. It is precisely this fact—that a fairly reliable communication link can be maintained with a decoder SNR of about 7–10 dB (as opposed to the 12–16 dB demodulator SNR needed in an analog system—that has spurred continued interest in digital communication system design.

With antipodal signals, we see from (7.4.1) that y_0 will always be the exact negative of y_1, and the threshold test in (7.4.2) is always identical to the test $y = y_1 - y_0 = 2y_1 \lessgtr 0$. This means simply determining the sign of y_1.

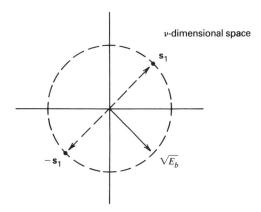

FIGURE 7.9. Antipodal signals in ν-dimensional vector space.

Hence the correlator (or matched filter) for $s_0(t)$ in Figures 7.5 and 7.6 can be eliminated. The optimal decoder simplifies to the single channel system shown in Figure 7.10. Antipodal signals therefore have the added advantage of a simpler decoder.

The bandwidth required to transmit the bits can be obtained from our spectral analysis of binary digital waveforms in Section 2.1. For antipodal signals with equally probable source bits, (2.1.15) showed that $m(t)$ has the spectrum

$$S_m(\omega) = \frac{1}{T_b} |F_s(\omega)|^2 \qquad (7.5.7)$$

The bandwidth required is that necessary to pass $F_s(\omega)$ without appreciable distortion. Note in particular that no harmonic lines appear in the spectrum, since (2.1.27) is satisfied.

Equations (7.4.16) and (7.4.19) assume that the optimal correlator or the matched filter response is used. Let us consider the effect of using in (7.4.16) an incorrect correlator $\hat{s}_1(t)$, due perhaps to the inability to construct properly the desired $s_1(t)$. We then see that

$$\frac{m_1}{\sigma_y} = \frac{\int_0^{T_b} s_1(t)\hat{s}_1(t)\, dt}{[(N_0/2) \int_0^{T_b} \hat{s}_1^2(t)\, dt]^{1/2}} \qquad (7.5.8)$$

If we multiply numerator and denominator by $\sqrt{E_b} = [\int_0^{T_b} s_1^2(t)\, dt]^{1/2}$, and rewrite, we see that this becomes

$$\frac{m_1}{\sigma_y} = \sqrt{\frac{2E_b}{N_0}} \left[\int_0^{T_b} \left(\frac{s_1(t)}{E_b} \right) \left(\frac{\hat{s}_1(t)}{E_{\hat{s}_1}} \right) dt \right] \qquad (7.5.9)$$

where $E_{\hat{s}_1}$ is the energy of the mismatched signal $\hat{s}(t)$. Hence the effect of using a mismatched correlator $\hat{s}_1(t)$, instead of $s_1(t)$, is to multiply E_b/N_0 by the bracketed terms in (7.5.9). The latter corresponds to a correlation of the normalized versions of $s_1(t)$ with $\hat{s}_1(t)$. We can therefore always account for improperly designed decoders by evaluating this integral. Since the waveforms $s_1(t)$ and $\hat{s}_1(t)$ are normalized, the integral is less than one, so a mismatch always reduces the decoding E_b/N_0 [not surprising, since only the proper matched filter maximizes the argument in (7.4.16)].

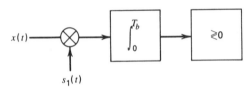

FIGURE 7.10. Antipodal signal decoder.

The integral factor in (7.5.9) is often called a decoder *implementation loss* and represents a degradation to the design value E_b/N_0. Alternatively, the operating link E_b/N_0 must be increased by this factor to maintain performance at the design E_b/N_0. Note that implementation losses enter the system link equations just as any RF transmission loss and must therefore be compensated by increases in power levels or antenna gains.

Some common types of antipodal signals used in modern binary digital systems are summarized in the following paragraphs.

Pulsed Waveforms (NRZ Signals)

$$s_1(t) = \left(\frac{E_b}{T_b}\right)^{1/2}, \qquad 0 \le t \le T_b$$
$$(7.5.10)$$
$$s_0(t) = -\left(\frac{E_b}{T_b}\right)^{1/2}, \qquad 0 \le t \le T_b$$

These bit waveforms correspond to antipodal pulses and are often called a *pulse code modulation* (PCM) signal pair (Figure 7.11a). Bit encoding

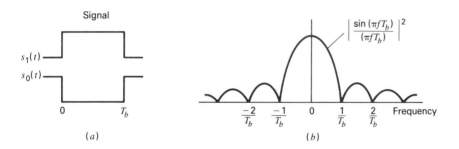

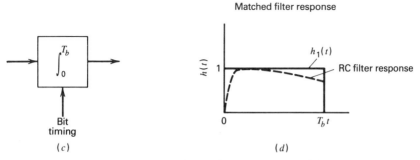

FIGURE 7.11. NRZ signals: (a) bit waveform, (b) spectral density, (c) correlator decoder, (d) matched filter.

therefore produces the pulse digital waveform depicted earlier in Figure 1.19b. As stated, these waveforms can be generated from bistable flip-flop devices that switch states according to the data bits. We see, therefore, that as an antipodal signal pair, pulse digital signaling is an example of an optimal encoding format for the white Gaussian noise channel. Since a string of like bits will leave the encoder in a fixed state with no output voltage change (appearing as a single, long constant signal), these signals are also called an NRZ or *nonreturn to zero* signal set. This may be a disadvantage in systems in which timing information is to be extracted from the bit waveform itself.

The pulse digital waveform has the frequency spectrum derived in (2.1.17), and is shown again in Figure 7.11b. The spectrum is that of a single T_b sec pulse, and is seen to have most of its energy concentrated about zero frequency out to a band of approximately $1/T_b$ Hz, with decreasing out-of-band spectral tails extending over the entire frequency axis.

Decoding is achieved by the synchronous correlation in Figure 7.10, which simplifies to that in Figure 7.11c. The multiplier can be removed, and we need only integrate over each bit interval. Such a device is often called an *integrate-and-dump* decoder. A passive matched filter for this signal will have impulse responses that are themselves pulses, as is evident from Figure 7.11d. Such impulse response can be well approximated by RC circuits with appropriate time constants.

Manchester Signals (Coded PCM Pulses)

$$
s_1(t) = \begin{cases} \left(\dfrac{E_b}{T_b}\right)^{1/2}, & 0 \le t \le \dfrac{T_b}{2} \\[2ex] -\left(\dfrac{E_b}{T_b}\right)^{1/2}, & \dfrac{T_b}{2} \le t \le T_b \end{cases}
$$

$$
s_0(t) = \begin{cases} -\left(\dfrac{E_b}{T_b}\right)^{1/2}, & 0 \le t \le \dfrac{T_b}{2} \\[2ex] \left(\dfrac{E_b}{T_b}\right)^{1/2}, & \dfrac{T_b}{2} \le t \le T_b \end{cases}
$$

$$(7.5.11)$$

These signals are a form of coded PCM signals (Figure 7.12a) and are referred to as *Manchester* signals. They are also called RZ (*return to zero*) signals, since they guarantee a zero crossing during each bit, which can be used to aid bit synchronization. Manchester signals can also be generated from flip-flop circuits that latch back from their initial state. They also can be generated by multiplying an NRZ bit waveform with a periodic square wave of frequency $1/T_b$ (Figure 7.12b).

The spectrum of Manchester signals can be obtained from (7.5.7), using

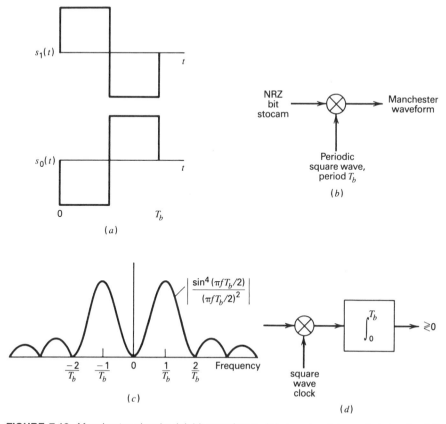

FIGURE 7.12. Manchester signals: (a) bit waveforms, (b) encoder, (c) spectral density, (d) correlator decoder.

the Fourier transform of the shifted coded pulses (Problem 7.17). The spectrum is shown in Figure 7.12c. The Manchester spectrum is concentrated away from zero frequency at $f = 1/T_b$, which is advantageous for synchronization and avoiding low frequency (60 cycle) interference. The synchronization potential is apparent from the waveshape, which shows that every Manchester bit contains a sign change exactly at bit center. This can be used to establish a synchronized receiver clock at precisely the bit time period. Subsystems for accomplishing this will be considered in Chapter 10.

Passive matched filters are slightly more difficult to construct, but decoder correlators can be constructed by multipling with the synchronized square wave and following with the integrate-and-dump integrators (Figure 7.12d). This can also be redrawn as a pair of clocked integrators operating over each half of the bit interval. It should be noted that Manchester signals are still a form of pulse digital waveforms, since an encoded sequence retains the property of having only the values $\pm\sqrt{(E_b/T_b)}$ at every t.

BPSK (Binary Phase Shift Keyed) Signals

$$s_1(t) = \left(\frac{2E_b}{T_b}\right)^{1/2} \sin(\omega_s t + \psi)$$

$$s_0(t) = -\left(\frac{2E_b}{T_b}\right)^{1/2} \sin(\omega_s t + \psi) \qquad (7.5.12)$$

$$= \left(\frac{2E_b}{T_b}\right)^{1/2} \sin(\omega_s t + \psi \pm \pi)$$

The signal corresponds to a burst of a carrier frequency to represent each data bit (Figure 7.13a). The signal $s_0(t)$ is the antipodal version of $s_1(t)$, but it is also the phase shifted version. Hence the waveform pair is actually a phase shift keyed encoding signal discussed in Section 2.5. Note that the phase shift must be exactly $\pm \pi$ to represent an antipodal set, but the

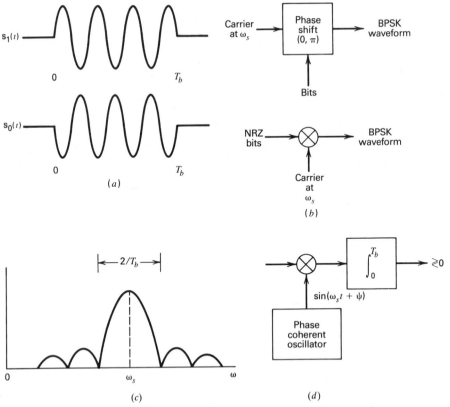

FIGURE 7.13. BPSK signals: (a) bit waveforms, (b) two types of encoder, (c) spectral density, (d) phase coherent decoder.

frequency ω_s and phase angle ψ are arbitary. The energy in the signals in (7.5.12) is

$$\frac{2E_b}{T_b} \int_0^{T_b} \sin^2 (\omega_s t + \psi) \, dt = \frac{2E_b}{T_b} \left[\frac{T_b}{2} - \frac{1}{2} \int_0^{T_b} \cos (2\omega_s t + 2\psi) \, dt \right]$$

$$(7.5.13)$$

and is exactly E_b if the frequency ω_s is harmonically related to $2\pi/T_b$, causing the integral to be zero, and is approximately E_b if $T_b \gg 2\pi/\omega_s$ (i.e., there are many cycles of ω_s within the bit interval T_b), causing the integral to be negligibly small.

BPSK waveforms are extremely popular because they are relatively easy to encode. From (7.5.12), they can be interpreted as the phase shifting of a free running encoder oscillator by the source bit stream, or equivalently as the amplitude modulation of the same oscillator with a PCM bit stream. Hence BPSK encoders can be implemented with binary $(0, \pi)$ phase shifters, or by multipliers, as shown in Figure 7.13b.

The spectrum of a BPSK signal corresponds to that of a carrier at frequency ω_s amplitude modulated by a baseband binary digital waveform at period T_b. Recall from (2.2.11) that this effectively shifts the spectrum of the PCM signal to the frequency ω_s. Thus the BPSK spectrum is approximately $2/T_b$ Hz wide, centered around the carrier frequency $\omega_s/2\pi$ (Figure 7.13c). The fact that we can arbitrarily locate the BPSK spectrum by control of ω_s is another advantage of BPSK systems. This lends itself nicely to an FDM multiplexing format for more than one digital source. By simply assigning different oscillator frequencies to each source at the encoder, we can space the bit modulated waveforms along the frequency axis.

The frequency ω_s can in fact be the RF carrier, corresponding to direct phase shift keyed modulation on to the carrier itself. This produces the PSK carrier considered in Section 2.5. The signal energy E_b in (7.5.6) now corresponds to the bit energy of the received RF carrier. The BPSK decoder therefore operates directly on the RF carrier, and the decoder input noise level N_0 is that occurring at the receiver input. The decoding E_b/N_0 can be determined directly from the link budget as

$$\frac{E_b}{N_0} = \frac{P_c T_b}{N_0} = \frac{C}{N_0 R_b}$$

$$(7.5.14)$$

where C/N_0 is computed as in (4.5.4).

Although ω_s and ψ in (7.5.12) are arbitrary, the receiver decoder must know this frequency and phase exactly in constructing matched filters or correlators. The operation of determining and maintaining this phase angle at the decoder is again a requirement for carrier phase referencing. A synchronous correlator would involve simply multiplying by a phase referenced local oscillator at the frequency ω_s, followed by a bit interval integration (Figure 7.13d). Thus the PSK decoder merely mixed the baseband and local oscillator signal continuously in time and integrates repeatedly over T_b sec intervals. The sequence of integrator output values is then used for bit decisioning. This combination of a mixer and integrate-and-dump filter appears as a cascade of an ideal phase referenced phase demodulator followed by an integrator for the resulting PCM signal. We emphasize that this combination is not an ad hoc decoder but is dictated by our matched filter theory.

The basic problem with PSK implementation is the necessity for phase referencing (i.e., knowing exactly the arbitrary phase ψ of the BPSK carrier with respect to the bit period). Consider, for example, a coherent correlator for PSK decoding that uses an inaccurate phase $\hat{\psi}$ in Figure 7.13d. We see that we have a mismatched correlator, and (7.5.9) indicates

$$
\begin{aligned}
\frac{m_1}{\sigma_y} &= \left(\frac{2E_b}{N_0}\right)^{1/2}\left[\int_0^{T_b} \frac{\sin(\omega_s t + \psi)}{(T_b/2)^{1/2}} \cdot \frac{\sin(\omega_s t + \hat{\psi})}{(T_b/2)^{1/2}}\, dt\right] \\
&= \left(\frac{2E_b}{N_0}\right)^{1/2}\left[\frac{2}{T_b}\int_0^{T_b} \frac{\cos(\psi - \hat{\psi})}{2}\, dt\right]
\end{aligned}
\tag{7.5.15}
$$

If the phase referencing error $\psi_e \overset{\Delta}{=} \psi - \hat{\psi}$ remains constant during the bit period, the resulting bit error is

$$
\mathrm{PE}(\psi_e) = Q\left[\left(\frac{2E_b}{N_0}\right)^{1/2}\cos\psi_e\right]
\tag{7.5.16}
$$

Hence the bit error probability is increased from its minimal value according to the cosine of the local oscillator phase referencing error. PSK therefore requires extremely accurate phase referencing to operate satisfactorily. If ψ_e varies with time during the bit interval, the integral in (7.5.15) must be evaluated. Unfortunately, the phase reference $\hat{\psi}$ is generated in practical systems by phase tracking loops with inherent noise, so that ψ_e evolves as a random process in time and (7.5.15) involves a stochastic integral. Even if it is assumed that ψ_e does not change with time, it should, in fact, be considered a random variable in (7.5.16). This means that PE is really a conditional error probability, conditioned on the random variable ψ_e, and we must determine the resulting average error probability by averaging over the randomness in the phase error ψ_e. Hence

$$
\mathrm{PE} = \int_{-\infty}^{\infty} \mathrm{PE}(\psi_e)p(\psi_e)\, d\psi_e
\tag{7.5.17}
$$

To compute the average, we use for the density of ψ_e the steady state error density,

$$
p(\psi_e) = \frac{1}{\sqrt{2\pi}\sigma_e}\, e^{-\psi_e^2/2\sigma_e^2}
\tag{7.5.18}
$$

where σ_e is the rms phase referencing error. Equation (7.5.17) is shown plotted in Figure 7.14 for several values of σ_e. As $\sigma_e \to 0$, the phase error density $p(\psi_e)$ approaches a delta function at $\psi_e = 0$, and (7.5.17) approaches the idealized antipodal error probability illustrated in Figure 7.8. However, as σ_e is increased, PE is increased at each value of E_b/N_0 because of the imperfect phase referencing used in the decoding. Note that the decoding performance may actually be significantly worse than that predicted by the ideal performance if the σ_e used for referencing is not adequate. Thus practical BPSK decoder design requires a careful interface of the detection and referencing subsystem performance.

Note that each curve in Figure 7.14 tends to flatten out for large values of E_b/N_0. This implies that imperfect phase referencing produces an irreducible detection error that cannot be overcome by simply increasing bit energy E_b. This can be seen by noting that if $|\psi_e| > \pi/2$, the cosine term in (7.5.16) changes sign and bit error will most likely be made. This implies

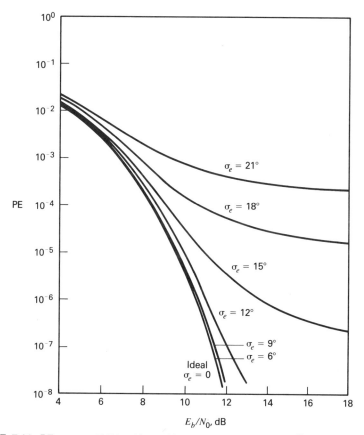

FIGURE 7.14. PE versus E_b/N_0 with nonideal phase referencing (σ_e^2 = variance of phase error).

$$PE(\psi_e) \to 1 \qquad \text{if} \qquad |\psi_e| > \frac{\pi}{2} \qquad (7.5.19)$$

Therefore, even if $E_b/N_0 \to \infty$, so that $PE(\psi_e) \to 0$ for $|\psi_e| \le \pi/2$, we still have

$$PE = \int_{|\theta_e| > \pi/2} (1)p(\psi_e)\, d\psi_e$$

$$= \text{Prob}\left[|\psi_e| > \frac{\pi}{2}\right] \qquad (7.5.20)$$

This value represents the asymptotic irreducible error probability. Thus BPSK decoding performance is ultimately limited by the capabilities of the referencing system alone.

We emphasize that the signal sets considered are all antipodal, and therefore all achieve the minimal PE when used with the same energy level and with the optimal decoder. Each signal set has characteristics, advantages, and disadvantages in terms of ease of generation, simplicity in decoding, baseband spectrum, and so on, that may favor one set over another in a particular application. In fact, in practice we may be willing to sacrifice the antipodal signal advantage for simplicity in encoder and decoder design. A common example of this is the following signal set.

Coherent Frequency Shift Keying (FSK)

$$s_1(t) = \left(\frac{2E_b}{T_b}\right)^{1/2} \sin(\omega_1 t + \psi)$$

$$s_0(t) = \left(\frac{2E_b}{T_b}\right)^{1/2} \sin(\omega_0 t + \psi) \qquad (7.5.21)$$

This signal pair corresponds to the frequency shift to ω_1 or ω_0, depending on the bit. Each bit is therefore sent as a T_b sec burst of a sine wave at one of two possible frequencies. The encoder can therefore be easily implemented by a single free running oscillator that is frequency switched. This also corresponds to frequency modulating the oscillator with a PCM waveform and a specified frequency deviation, and for this reason it is often referred to as PCM-FM. The signals are not antipodal, however. This can be seen by computing the signal correlation γ under the assumption $(\omega_1 + \omega_0) \gg 2\pi/T_b$. This produces

$$\gamma = \frac{1}{E_b} \int_0^{T_b} s_1(t)s_0(t)\, dt$$

$$= \frac{\sin[(\omega_1 - \omega_0)T_b]}{(\omega_1 - \omega_0)T_b} \qquad (7.5.22)$$

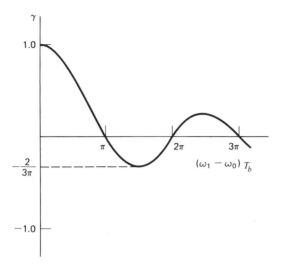

FIGURE 7.15. FSK signal correlation versus frequency separation.

The coefficient γ is shown in Figure 7.15 as a function of the normalized frequency difference $(\omega_1 - \omega_0)T_b$. The value γ never reaches minus one (required for the antipodal case) and approaches orthogonal signals as the frequency separation becomes large. However, we see that the most negative correlation value occurs when $(\omega_1 - \omega_0)T_b \cong 3\pi/2$, or when

$$\frac{\omega_1 - \omega_0}{2\pi} \cong 0.71\left(\frac{1}{T_b}\right) \qquad (7.5.23)$$

When this separation is used, $\gamma = -2/3\pi$, which effectively mutiplies E_b/N_0 by the factor $1 - \gamma = 1.21$. Thus the signal pair in (7.5.21), when used with the frequency separation in (7.5.23), is about 0.8 dB better than orthogonal signaling. That is, placing the signal frequencies close together, as dictated by (7.5.23), gives a slight advantage over wide frequency separation. This result occurs for any angle ψ, but a pair of coherent correlators (two must be used, one matched to each frequency, since the signals are not antipodal) must be phase synchronized to the angle ψ, just as in PSK. For this reason this signaling scheme is called *coherent FSK*.

When $(\omega_1 - \omega_0)T_b = \pi$, the FSK system operates at the first zero crossing in Figure 7.15, which is the minimal deviation that can be used for orthogonal binary signaling. This format is often called *minimum shift keying* (MSK). Note that for MSK, the frequency deviation between the one and zero bit is $1/2T_b$ Hz, which is one-half the bit rate frequency.

The frequency spectrum of an MSK waveform can be obtained by recognizing that it is a form of frequency shifted carrier, and we can use our spectral discussion in Section 2.4. The spectrum of the MSK waveform is sketched in Figure 7.16, along with the BPSK spectrum for the same bit rate. The MSK has a slightly wider main hump but reduced spectral tails.

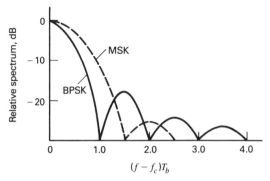

FIGURE 7.16. MSK signal spectrum.

7.6 EFFECT OF IMPERFECT BIT TIMING

We have assumed exclusively the existence of perfect bit timing in all our decoder analyses. This means that the location in time of each bit interval is known precisely at the decoder. In a practical system there are often errors in the decoder's location of the bit intervals, because of imperfections in the timing subsystem. These errors must be accounted for in system analysis.

Consider the situation depicted in Figure 7.17. Part *a* shows the actual time location of the arriving bit intervals. Part *b* depicts the decoder's interpretation of these interval locations, showing an offset by an amount δ sec. The decoder proceeds to correlate over each bit interval according to its timing schedule, unaware of the δ offset. This produces a correlation that overlaps the adjacent bit interval, and the resulting decoder bit decision is influenced by neighboring source bits. If we assume antipodal signals $\pm s(t)$, with the decoder in Figure 7.10, the situation in Figure 7.17 produces the signal sample value

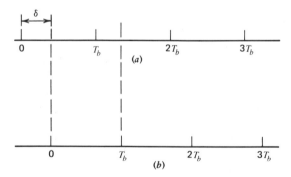

FIGURE 7.17. Bit timing error diagram.

$$z_1(T_b, \delta) = \int_0^{T_b} m(t)s(t)\, dt$$

$$= \pm \int_0^{T_b-\delta} s(t+\delta)s(t)\, dt \pm \int_{T_b-\delta}^{T_b} s(t+\delta - T_b)s(t)\, dt \quad (7.6.1)$$

where the first $\pm$ depends on the bit in $(0, T_b)$ and the second $\pm$ depends on the next bit. It is convenient to denote

$$\gamma(\tau) = \frac{1}{E_b} \int_0^{T_b} s(t+\tau)s(t)\, dt \qquad (7.6.2)$$

as the *signal correlation function* for the periodic signal $s(t)$. This allows us to write (7.6.1) as

$$z_1(T_b, \delta) = \pm \int_0^{T_b} s(t+\delta)s(t)\, dt = \pm E_b[\gamma(\delta)] \qquad (7.6.3)$$

if the bits are the same, and

$$z_1(T_b, \delta) = \pm E_b\left[\gamma(\delta) - \frac{2}{E_b} \int_{T_b-\delta}^{T_b} s(t+\delta - T_b)s(t)\, dt \right] \qquad (7.6.4)$$

if different. Thus bit timing errors causes the signal energy E_b to be effectively degraded by the bracketed term, depending on the sign of the adjacent bit. The resulting degradation depends on the value of δ and the correlation function of the antipodal signals. If the shift δ in Figure 7.17 had been in the opposite direction, the same result would occur, only the previous bit, instead of the next bit, would be involved. The bit error probability of an imperfectly timed decoder is obtained by averaging the PE over the possibility of the adjacent bits being the same or opposite. Hence the antipodal bit error probability with a timing offset δ is

$$PE(\delta) = \tfrac{1}{2} Q\left\{ \left(\frac{2E_b}{N_0}\right)^{1/2} [\gamma(\delta)] \right\}$$

$$+ \tfrac{1}{2} Q\left\{ \left(\frac{2E_b}{N_0}\right)^{1/2} \left[\gamma(\delta) - \frac{2}{E_b} \int_{T_b-\delta}^{T_b} s(t+\delta - T_b)s(t)\, dt \right] \right\}$$

$$(7.6.5)$$

Evaluation of (7.6.5) requires specification of the particular antipodal signals involved.

Example 7.1 (NRZ Signals). Let the antipodal signals be given by the NRZ pulsed set in (7.5.8). We immediately see that, for $s(t) = s_1(t)$,

$$\gamma(\tau) = 1$$

$$\frac{1}{E_b} \int_{T_b-\delta}^{T_b} s(t + \delta - T_b) s(t) \, dt = \frac{\delta}{T_b}$$

(7.6.6)

Thus

$$Z_1(T_b, \delta) = \begin{cases} E_b, & \text{same adjacent bit} \\ E_b\left(1 - \dfrac{2|\delta|}{T_b}\right), & \text{opposite adjacent bit} \end{cases}$$

(7.6.7)

The signal energy is not degraded for any δ if the adjacent bit is the same, and is reduced linearly with $|\delta|$ if the adjacent bit is opposite. The resulting bit error probability is then

$$\text{PE}(\delta) = \frac{1}{2} Q\left[\left(\frac{2E_b}{N_0}\right)^{1/2}\right] + \frac{1}{2} Q\left[\left(\frac{2E_b}{N_0}\right)^{1/2}\left(1 - \frac{2|\delta|}{T_b}\right)\right]$$

(7.6.8)

It is evident that timing offsets δ should be restricted to a small fraction of the bit interval T_b in order to prevent significant degradation.

Example 7.2 (Manchester Signals). For Manchester signals, it follows that

$$Z_1(T_b, \delta) = \begin{cases} E_b\left(1 - \dfrac{2|\delta|}{T_b}\right), & \text{same bit} \\ E_b\left(1 - \dfrac{4|\delta|}{T_b}\right), & \text{opposite bit} \end{cases}$$

(7.6.9)

and the bit error probability is then

$$\text{PE}(\delta) = \frac{1}{2} Q\left[\left(\frac{2E_b}{N_0}\right)^{1/2}\left(1 - \frac{2|\delta|}{T_b}\right)\right] + \frac{1}{2} Q\left[\left(\frac{2E_b}{N_0}\right)^{1/2}\left(1 - \frac{4|\delta|}{T_b}\right)\right]$$

(7.6.10)

Timing offsets are therefore more critical because of the coded pulse format of the Manchester signals.

Example 7.3 (BPSK Signals). When a BPSK signal set is used, decoding requires the phase coherent reference in Figure 7.13d. If the phase referencing subsystem has a phase error ψ_e, the BPSK carrier will be demodulated to an NRZ bit stream, with a $\cos \psi_e$ scaling factor, as in (7.5.15). The bit timing is then identical to Example 7.1. Thus

$$Z_1(T_b, \delta) = \begin{cases} E_b \cos \psi_e, & \text{same bit} \\ E_b \cos \psi_e \left(1 - \dfrac{2|\delta|}{T_b}\right), & \text{opposite bit} \end{cases} \qquad (7.6.11)$$

The resulting PE degradation in BPSK decoding due to both a phase reference error ψ_e and a bit timing error δ is then

$$\text{PE}(\psi_e, \delta) = \frac{1}{2} Q\left[\left(\frac{2E_b}{N_0}\right)^{1/2} \cos \psi_e\right] + \frac{1}{2} Q\left[\left(\frac{2E_b}{N_0}\right)^{1/2} \cos \psi_e \left(1 - \frac{2|\delta|}{T_b}\right)\right]$$

$$(7.6.12)$$

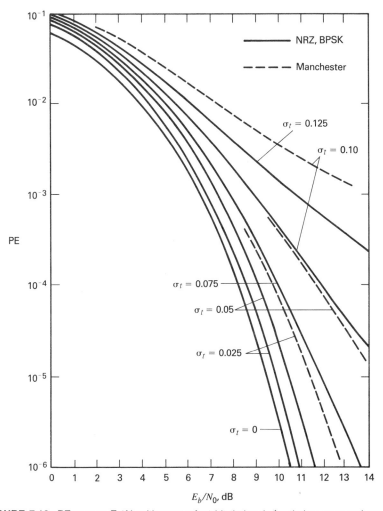

FIGURE 7.18. PE versus E_b/N_0 with nonperfect bit timing (σ_τ^c = timing error variance).

Thus both phase and timing errors degrade BPSK decoding, but in slightly different ways.

Bit timing, like phase referencing, is generally derived from noisy synchronization subsystems, and the timing error δ also evolves as a random variable. This means PE must be computed by an average over δ, similar to the phase error case in (7.5.16),

$$PE = \int_{-\infty}^{\infty} PE(\delta)p(\delta)\,d\delta \qquad (7.6.13)$$

If we again assume that δ is a zero mean Gaussian timing error with error variance σ_t^2, then

$$p(\delta) = \frac{1}{\sqrt{2\pi}\sigma_t} e^{-\delta^2/2\sigma_t^2} \qquad (7.6.14)$$

The result of integrating with (7.6.14) for NRZ case in (7.6.8) and the Manchester case in (7.6.10) is shown in Figure 7.18 for several values of σ_t^2. The NRZ result holds for BPSK if perfect phase referencing ($\psi_e = 0$) is assumed. It is again evident that bit timing errors cause PE degradation and that Manchester encoding is more susceptible to these errors. Note that only a few percent of rms timing error can cause significant PE loss. Further discussions of bit timing degradations can be found in References 5 and 6.

7.7   DECODING WITH NONWHITE NOISE

The results derived so far have been based on the condition that the decoder noise is a white noise process (i.e., has a flat spectrum over all receiver frequencies). As noted earlier, this is not always the case, as, for example, when digital subcarriers are decoded after FM demodulation. We may therefore inquire as to the modification that must be made to our earlier results when the decoder noise cannot be modeled as a white noise process.

Let the Gaussian decoder noise have the general power spectrum $S_n(\omega)$. We again assume additional signaling with $\pm s(t)$ and a filter–sampler decoder of the type in Figure 7.6c. Since the noise is nonwhite, the advantage of the matched filter decoder is no longer obvious. We therefore assume an arbitrary filter function $H(\omega)$, followed by a sampler and sign detector for bit decisioning, as shown in Figure 7.19. The effect of having nonwhite noise will alter the value of σ_y^2 used in our earlier development, and the optimal form for $H(\omega)$ is to be determined. The bit error probability for this decoder is still given by (7.4.16)

$$PE = Q\left[\left(\frac{m_1}{\sigma_y}\right)\right] \qquad (7.7.1)$$

where m_1 is the decoder signal sample value when $s(t)$ is sent and σ_y is the sample variance. From Figure 7.19 we obtain

$$m_1 = \frac{1}{2\pi} \int_{-\infty}^{\infty} H(\omega) F_s(\omega) e^{j\omega T_b} \, d\omega \qquad (7.7.2a)$$

and

$$\sigma_y^2 = \frac{1}{2\pi} \int_{-\infty}^{\infty} S_n(\omega) |H(\omega)|^2 \, d\omega \qquad (7.7.2b)$$

Our objective is to determine the decoder filter function $H(\omega)$ that achieves minimal PE in (7.7.1). This requires maximization of (m_1/σ_y), using (7.7.2). Thus we have

$$\frac{m_1}{\sigma_y} = \frac{\left|(1/2\pi) \int_{-\infty}^{\infty} H(\omega) F_s(\omega) e^{j\omega T_b} \, d\omega\right|}{\left[(1/2\pi) \int_{-\infty}^{\infty} |H(\omega)|^2 S_n(\omega) \, d\omega\right]^{1/2}} \qquad (7.7.3)$$

To establish the maximization it is convenient to denote

$$u_1(\omega) \overset{\Delta}{=} H(\omega)[S_n(\omega)]^{1/2}$$

$$u_2(\omega) \overset{\Delta}{=} \frac{F_s(\omega)}{\sqrt{S_n(\omega)}} e^{j\omega T_b} \qquad (7.7.4)$$

where $\sqrt{S_n(\omega)}$ is the function obtained by taking the square root of $S_n(\omega)$ at every ω. We then see that the numerator in (7.7.3) is simply

$$\left| \frac{1}{2\pi} \int_{-\infty}^{\infty} u_1(\omega) u_2(\omega) \, d\omega \right| \qquad (7.7.5)$$

Applying the Schwarz inequality to this numerator establishes that

$$\frac{m_1}{\sigma_y} = \frac{\left|(1/2\pi) \int_{-\infty}^{\infty} u_1(\omega) u_2(\omega) \, d\omega\right|}{\sigma_y}$$

$$\leq \frac{(1/2\pi) \left[\int_{-\infty}^{\infty} |u_1(\omega)|^2 \, d\omega \int_{-\infty}^{\infty} |u_2(\omega)|^2 \, d\omega\right]^{1/2}}{\left[(1/2\pi) \int_{-\infty}^{\infty} |u_1(\omega)|^2 \, d\omega\right]^{1/2}}$$

$$= \left[\frac{1}{2\pi} \int_{-\infty}^{\infty} \frac{|F_s(\omega)|^2}{S_n(\omega)} \, d\omega \right]^{1/2} \qquad (7.7.6)$$

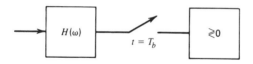

FIGURE 7.19. General filter–sampler bit decoder.

Furthermore, this upper bound occurs only if $u_1(\omega) = u_2^*(\omega)$, or if

$$H(\omega)\sqrt{S_n(\omega)} = \frac{F_s^*(\omega)}{\sqrt{S_n(\omega)}} e^{-j\omega T_b} \qquad (7.7.7)$$

where * denotes complex conjugate. Solving for the decoder filter function then yields

$$H(\omega) = \frac{F_s^*(\omega)}{S_n(\omega)} e^{-j\omega T_b} \qquad (7.7.8)$$

This is the required filter transfer function that minimizes PE for antipodal signaling in nonwhite noise. Note that the desired filter function now depends on the noise spectrum, as well as that of the signal $s(t)$. This means that, in the case of nonwhite noise, the desired filter is related to both the signal and the interfering noise. To obtain a physical interpretation of (7.7.8), it is convenient to rewrite as

$$H(\omega) = \left[\frac{1}{\sqrt{S_n(\omega)}} \right]\left[\frac{F_s^*(\omega)}{\sqrt{S_n(\omega)}} e^{-j\omega T_b} \right] \qquad (7.7.9)$$

The optimal decoder filter now appears as the cascade of two separate filters, each having transfer function given by the preceding bracketed terms. Without regard to physical realizability, we can interpret the first filter as one that filters the total baseband power spectrum with the function

$$\left| \frac{1}{\sqrt{S_n(\omega)}} \right|^2 = \frac{1}{S_n(\omega)} \qquad (7.7.10)$$

When applied to the noise portion of the baseband waveform, we see that this first filter generates an output spectrum that is of unit level for all ω. Thus the first filter effectively "whitens" the baseband noise and is referred to as a *whitening filter*. In the process of whitening the noise, however, the baseband signal $\pm s(t)$ is modified to a new signal $\pm \hat{s}(t)$, where $\hat{s}(t)$ has the transform

$$F_{\hat{s}}(\omega) = \left[\frac{1}{\sqrt{S_n(\omega)}} \right]F_s(\omega) = \frac{F_s(\omega)}{\sqrt{S_n(\omega)}} \qquad (7.7.11)$$

We now see that the second filter function [second bracket in (7.7.9)] is related to this spectrum. Recall that if a real function $F(t)$ has transform $F(\omega)$, then the function $f(-t)$ has transform $F(-\omega) = F^*(\omega)$, and $f(T_b - t)$ has transform $F^*(\omega)e^{-j\omega T_b}$. From this we see that the second filter function has an impulse response matched to the signal $s(t)$ having the transform in (7.7.11). That is, the second filter is a matched filter for the modified signal at the whitening filter output. The optimal decoder filter for nonwhite noise can therefore be considered as the cascade of a filter that first whitens the baseband noise followed by a filter that is matched to the modified signal produced by the whitening (Figure 7.20). Physically, the whitening filter may not be realizable as a separate filter [e.g., if $S_n(\omega) = 0$ beyond some upper cutoff frequency, the whitening filter must have infinite gain beyond this frequency]. However, when the required filter in (7.7.8) is constructed as a single filter function, the resulting filter is usually realizable or can at least by adequately approximated [7].

If the required filter function is implemented, then the resulting value in (7.7.6) is

$$\left(\frac{m_1}{\sigma_y}\right)_{\max} = \left[\frac{1}{2\pi} \int_{-\infty}^{\infty} \frac{|F_s(\omega)|^2}{S_n(\omega)} \, d\omega\right]^{1/2} \tag{7.7.12}$$

Note that the maximum value no longer depends only on the signal energy E_b, as for the case of white noise, but instead is integrably related to the spectrum of both signal and noise. This means the actual antipodal signal spectrum will determine the resulting value of the decoder performance in (7.7.1). (Compare this with the result for white noise, where we concluded that any antipodal signal set was as good as any other with the same energy.) We now see that one antipodal signal pair is preferable over another if it produces a larger value in (7.7.12). This immediately suggests that we attempt to determine the signal $s(t)$ such that (7.7.12) is maximized, under the condition that $s(t)$ have energy E_b. That is, we want to maximize over $F_s(\omega)$ the integral

$$\frac{1}{2\pi} \int_{-\infty}^{\infty} \frac{|F_s(\omega)|^2}{S_n(\omega)} \, d\omega \tag{7.7.13a}$$

subject to the energy constraint

$$\frac{1}{2\pi} \int_{-\infty}^{\infty} |F_s(\omega)|^2 \, d\omega = E_b \tag{7.7.13b}$$

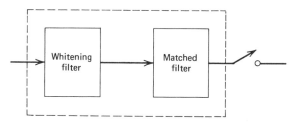

FIGURE 7.20. Matched filter decoder for nonwhite noise and antipodal signals.

We can proceed by rewriting (7.7.13a) by referring to Figure 7.21a and noting that (7.7.13a) corresponds to the signal energy at the output of a whitening filter $[H(\omega) = 1/\sqrt{S_n(\omega)}]$ when $s(t)$ is the input. In the vector domain this same output energy is that of the vector transformation in Figure 7.21b, corresponding to the transformation of s to ŝ through the matrix A. The matrix A must correspond to a whitening operation if Figure 7.21b is to be identical to Figure 7.21a. It is straightforward to show that if A is a whitening matrix, taking an input noise vector **n** with correlation matrix R_n to an output noise vector with correlation matrix equal to the identity matrix (unit level, white noise matrix), then A must have the form (Problem 7.23)

$$A = \begin{bmatrix} \lambda_1^{-1/2} & & & 0 \\ & \lambda_2^{-1/2} & & \\ & & \ddots & \\ 0 & & & \lambda_\nu^{-1/2} \end{bmatrix} \tag{7.7.14}$$

where $\{\lambda_i\}$ are the mean squared values of the input noise coordinates (see Appendix C). Since $\hat{\mathbf{s}} = A\mathbf{s}$, it follows that the energy of $\mathbf{s}_0$ is $\Sigma\,(s_i^2/\lambda_i)$, where s_i are the coordinates of **s**. We have therefore established the identity

$$\frac{1}{2\pi} \int_{-\infty}^{\infty} \frac{|F_s(\omega)|^2}{S_n(\omega)}\, d\omega = \sum_{i=1}^{\nu} \frac{s_i^2}{\lambda_i}. \tag{7.7.15}$$

The signal selection problem in (7.7.13) is now restated as

$$\max_{\mathbf{s}} \sum_{i=1}^{\nu} \frac{s_i^2}{\lambda_i} \quad \text{subject to} \quad \sum_{i=1}^{\nu} s_i^2 = E_b \tag{7.7.16}$$

It is easy to show that $\Sigma\,(s_i^2/\lambda_i) \le E_b/\lambda_{min}$, where λ_{min} is the smallest of all the λ_i, and the equality holds if s has all its energy placed in only those coordinates where λ_{min} occurs. That is, the transmitted antipodal signal vector pair should have all its energy concentrated in the coordinates where

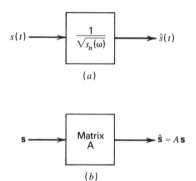

(a)

(b)

FIGURE 7.21. Whitening filters: (a) waveforms, (b) vector equivalent.

the noise is weakest. This means that the signal expansion in $(7.1.4a)$ will involve only those orthonormal functions corresponding to the minimum noise variance. Any combination of these specific functions can be used to produce the smallest PE. In particular, if only one coordinate has minimal noise variance, $s(t)$ should be identical to that orthonormal function (normalized to energy E_b). We are therefore signaling with a time waveform identical to an orthonormal function of the noise expansion.

Notice in (7.7.16) that if $\lambda_{\min} = 0$ (the noise has a zero coordinate somewhere), then the upper bound becomes infinite. This, in turn, means that the bit error probability in (7.7.1) is zero. Theoretically, perfect bit transmission can be achieved under this condition by proper design. Obviously, if we utilize a coordinate of the noise expansion where no noise is present, the decoder will always correctly decide if $s(t)$ or $-s(t)$ is being received by observing that one coordinate. Operation under such a condition is called *singular detection*. In other words, if any coordinate system can be determined for signal representation such that at least one coordinate contains no noise, singular detection can always be achieved by binary signaling over that particular coordinate.

REFERENCES

1. Weiss, L. *Statistical Decision Theory*, McGraw-Hill, New York, 1961.

2. Cramer, H. *Mathematical Methods of Statistics*, Princeton University Press, Princeton, N.J., 1957.

3. Helstrom, C. *Statistical Theory of Signal Detection*, Pergamon, New York, 1968.

4. Middleton, D. *Statistical Communication Theory*, McGraw-Hill, New York, 1960.

5. Holmes, J. *Coherent Spread Spectrum Systems*, Wiley, New York, 1982, Chapter 12.

6. Stiffler, J. *Theory of Synchronous Communications*, Prentice-Hall, Englewood Cliffs, N.J., 1971, Chapter 7.

7. Van Trees, H. *Detection, Estimation, and Modulation—Part 1*, Wiley, New York, 1968, Chapter 4.

PROBLEMS

7.1   (7.1)   Given two signal waveforms

$$s_1(t) = \text{Real} \{e_1(t)e^{j(2\pi/T_b)t}\}$$

$$s_0(t) = \text{Real} \{e_0(t)e^{j(2\pi/T_b)t}\}$$

where $e_1(t) = A$, $e_0(t) = Ae^{j\theta}$, $0 \le t \le T_b$. Show that the phase angle θ between the envelopes can be interpreted as the angle between $s_1(t)$ and $s_0(t)$ when plotted in vector space. *Hint:* $\sin(2\pi t/T_b)$ and $\cos(2\pi t/T_b)$ are orthogonal over $(0, T_b)$.

7.2 (7.2) A decisioning system must decide between the equally likely hypothesis H_0 and H_1 where H_0 is the hypothesis that an observed random variable x has density

$$p_0(x) = \frac{1}{\sqrt{2\pi}} e^{-(x^2/2)}$$

and H_1 is the hypothesis that x has a density

$$p_1(x) = \begin{cases} \frac{1}{3}, & 0 \le x \le 3 \\ 0, & \text{otherwise} \end{cases}$$

Determine the decoding rule for minimal decisioning error. (*Hint:* Treat the hypothesis as "signals" to be detected.)

7.3 (7.2) When a laser is pulsed on over $(0, T_b)$, it produces an average of K_s photoelectrons. Background noise adds in an additional average of K_b photoelectrons. The actual number of photons produced k is a random variable whose probability is related to the average count K by

$$P(k) = \frac{K^k e^{-K}}{k!}, \qquad k = \text{nonnegative integer}$$

(a) Determine the decoding test for deciding whether a laser is present by observing the photon count. (b) Write an expression for the probability of erring in making this decision.

7.4 (7.2) A dc signal of 6 V may or may not be present in additive noise. At any instant of time the noise has a uniform distribution from -4 to $+4$ V. (a) Determine the decoding test for detecting, with minimal error probability, when the dc is present using a single time sample. (b) Determine the error probability of the test.

7.5 (7.2) Two zero mean Gaussian noise sources are available, one with correlation function $R_1(\tau)$ and the other with $R_2(\tau)$. Either one may be turned on, equally likely, producing k time samples of the source. Determine the decoding test for deciding which source is on, using the output samples.

7.6 (7.3) Determine how the decoding test in (7.3.3) must be modified when the transmitted bits are not equally probable and the waveforms are not equal energy. Show this is equivalent to an adjustment of the threshold in (7.3.5).

7.7 (7.3) A binary communication system with additive white Gaussian noise having spectral level N_0 [denoted AWGN(N_0)] transmits a binary one by sending either waveform $f_1(t)$ or $f_2(t)$ (energy E_b, time T_b), selected by tossing a coin, and sends nothing for a binary zero. Sketch the decoder that minimizes bit error probability.

7.8 (7.3) A secret communication system sends a binary zero as either signal $f_1(t)$ or $f_2(t)$, selected by tossing a coin and sends a binary one as either signals $g_1(t)$ or $g_2(t)$, selected the same way. The signals are sent over AWGN(N_0) channel. Derive the decoder. Assume all signals have energy E_b and time T_b.

7.9 (7.3) A digital communication system [AWGN(N_0)] uses antipodal waveforms $f(t)$ of energy E_b and time length T_b. Bits are transmitted by two consecutive $f(t)$ waveforms. A student claims "The decoder should match filter each $f(t)$ waveform separately, deciding its sign. A bit is decoded only if its sign occurs in both waveforms." Is this correct? Explain.

7.10 (7.3) A binary communication system uses a single voltage sample to detect each bit. The signal sample values are $\pm s$. The communication channel typically adds in a Gaussian noise sample (zero mean, variance σ^2). However, there is a chance (with probability ρ) that the additive noise sample will instead be uniformly distributed over $\pm \sigma$. What is the required sample processing for minimizing PE?

7.11 (7.4) Let $s(t)$ be a signal, $0 \le t \le T_b$ with transform $F_s(\omega)$. Show that the matched filter for $s(t)$ has the frequency function $H(\omega) = F_s^*(\omega)e^{-j\omega T_b}$, where * denotes conjugate.

7.12 (7.4) Let $s(t)$ be a general signal bandlimited to B Hz and time limited to T_b sec. Show that a matched filter for any $s(t)$ of this class can always be designed as a tapped delay line with adjustable gains in conjunction with an ideal low pass filter. Show the final system. [*Hint:* Use the sampling theorem and the associated signal expansion for $s(t)$.]

7.13 (7.4) In using a pair of matched correlators in (Figure 7.5b) for binary decoding, show that if the binary signals $s_1(t)$ and $s_0(t)$ have identical waveshape over a subinterval τ in $(0, T_b)$, this τ sec of integration need not be included in the correlation operation.

7.14 (7.4) Show that the time waveform of the output of a filter matched to a signal $s(t)$ traces out the correlation function of $s(t)$ when the latter is applied at the input.

7.15 (7.5) A digital system samples at the rate of 1000 samples per second and quantizes to 16 possible levels. Assuming that the quantization words are sent by BPSK encoding, estimate the carrier bandwidth needed to send the encoded waveform.

7.16 (7.5) An integrating circuit is constructed as an RC low pass network having the impulse response $h(t) = (1/RC)e^{-t/RC}$, $0 \le t \le T_b$. Determine the decoding implementation loss when this is used as an integrate-and-dump decoder. Evaluate the loss in dB for $T_b/RC = 1$, 2, and 5.

7.17 (7.5) Compute the spectral density in (Figure 7.12c) for the Manchester coded antipodal PCM waveform.

7.18 (7.5) An NRZ code with T_b sec bits and amplitude A is FM onto a carrier and ideally frequency detected along with additive white Gaussian noise. The FM detector operates above threshold and is followed by an ideal low pass filter over bandwidth B Hz and a sampler that samples at $t = T_b$. (a) Assume $B \gg 1/T_b$ and write the expression for PE. (b) Will this be the minimal possible PE? Explain.

7.19 (7.5) An NRZ bit waveform is sent over a bandlimited AWGN(N_0, $\pm B$). Instead of a Bayes decoder, the bit is detected based on a sign determination of a voltage sample during each bit. Determine the PE of the system.

7.20 (7.5) A communication system sends the following fixed (uncoded) waveform:

$$c(t) = \begin{cases} 1, & 0 \le t \le \dfrac{T_b}{2} \\[2mm] 0, & \dfrac{T_b}{2} \le t \le T_b \end{cases}$$

along with an NRZ antipodal bit T_b sec long. Either $c(t)$ is sent first (over $0, T_b$) and the bit follows over $(T_b, 2T_b)$, or the bit is sent first $(0, T_b)$ and $c(t)$ follows $(T_b, 2T_b)$. The channel is WGN(N_0), and the data bit is equally likely. Explain the processing for deciding whether $c(t)$ precedes or follows the bit by observing over $(0, 2T_b)$.

7.21 (7.5) An antipodal signal set is selected as $+f(t)$ and $-f(t)$. A new signal set is formed by

$$s_1(t) = +f(t) + C$$
$$s_0(t) = -f(t) + C$$

where C is a positive constant. The channel is WGN(N_0). Student 1 claims, "The new signal set is not antipodal. Hence its PE will be higher (worse) than that of the first set." Student 2 claims, "But if C is made high enough, the second set has more energy than the first set and it will therefore have lower PE." Which student is correct? Explain your answer with equations.

7.22 (7.5) A binary communication link uses the waveforms in Figure P7.22 for sending bits over a AWGN(N_0) channel. Student 1 claims "The waveforms are simply Manchester antipodal waveforms with a constant level of $A/2$ added to them. If we first subtract out this

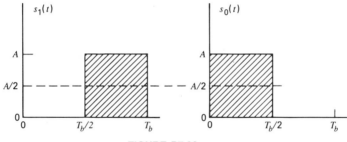

FIGURE P7.22.

constant (dc remove) at the receiver, the Manchester decoder can be used to minimize PE." Student 2 claims "The PE will be worse than if Manchester signals (signals shown with $A/2$ subtracted) were sent." Are the students correct? Prove your answer.

7.23 (7.7)   Refer to the analyses of Appendix C. (a) Write the correlation matrix of the KL coefficients of a noise process having ν components and eigenvalues $\{\lambda_i\}$. (b) Show that if the noise in part (a) is filtered by a system whose transformation matrix is given in (7.7.14), the output noise will correspond to a white noise process with a unit level spectral density.

7.24 (7.7)   Let a time signal $s(t)$ be represented by its time samples $[s(t_1), s(t_2), \ldots, s(t_N)]$. Assume that noise is added to $s(t)$ at the decoder input in such a way that one particular time point, say, t_j, has no noise at this instant. Explain why this will correspond to a case of singular detection. How should the encoder and decoder operate for this case?

8

NONCOHERENT AND BANDLIMITED BINARY SYSTEMS

In Chapter 7 the general binary digital system was developed, and the optimality of a synchronized bit decoder was presented. Performance was evaluated for the infinite bandwidth channel, where the effect of link filtering on the encoded waveforms can be neglected. In this chapter some alternative binary decoding procedures are considered, and the infinite bandwidth assumption is relaxed, requiring re-evaluation of performance and design for this bandlimited operation. Methods to improve error probability performance via error correction coding techniques will be considered later in the chapter.

8.1 NONCOHERENT BINARY SIGNALING

The fact that the BPSK signaling requires only the phase modulation of a free running oscillator makes it an attractive scheme for binary digital systems. However, the requirement to maintain phase synchronization over a long stream of bits is often difficult to accomplish. An alternative procedure that retains the encoding and spectral advantage is to use modulated carriers but to decode without using phase information. In this case the phase referencing is no longer required, but the optimality of true coherent detection is sacrificed. In this secton we examine this type of noncoherent binary operation.

Consider a binary digital system using the following encoder signals

$$s_1(t, \psi) = e_1(t) \cos(\omega_1 t + \psi), \qquad 0 \le t \le T_b$$

$$s_0(t, \psi) = e_0(t) \cos(\omega_0 t + \psi), \qquad 0 \le t \le T_b$$

$$(8.1.1)$$

where $e_i(t) \geq 0$ for all t, and represents the envelope of the signal $s_i(t)$. The frequencies ω_1 and ω_0 are arbitrary frequencies, and the phase ψ is the corresponding carrier phase angle during a bit period. If $\omega_1, \omega_0 \gg 2\pi/T_b$, the energy in the signals is again given approximately as

$$E_b = \tfrac{1}{2} \int_0^{T_b} e_i^2(t) \, dt \, , \qquad i = 0, 1 \tag{8.1.2}$$

A coherent decoder for this pair of signals would require exact knowledge of the phase ψ of the received signals. Consider instead a decoder that wishes to decode without ψ, instead considering ψ a completely random phase, uniformly distributed over $(0, 2\pi)$. The decoder must still decide between $s_1(t)$ and $s_0(t)$ in (8.1.1), where ψ is now random.

The optimal decoding processing is still given by (7.2.9), except the signal vectors depend on a random variable ψ. Denoting this vector dependence as $\mathbf{s}_i(\psi)$, the densities of Section 7.2 can be written by averaging over ψ. Thus

$$p(\mathbf{x}|i) = \int_0^{2\pi} p[\mathbf{x}|\mathbf{s}_i(\psi)]\left(\frac{1}{2\pi}\right) d\psi \tag{8.1.3}$$

The required decoding test in (7.2.9) is now modified to

$$\Lambda(\mathbf{x}) = \frac{(2\pi)^{-1} \int_0^{2\pi} p[\mathbf{x}|\mathbf{s}_1(\psi)] \, d\psi}{(2\pi)^{-1} \int_0^{2\pi} p[\mathbf{x}|\mathbf{s}_0(\psi)] \, d\psi} \geq 1 \tag{8.1.4}$$

The function $\Lambda(\mathbf{x})$ is referred to as the generalized likelihood ratio since it involves averaged probability densities.

For the case of additive white Gaussian noise the integration can be performed. From (7.3.1),

$$
\begin{aligned}
p[\mathbf{x}|\mathbf{s}_i(\psi)] &= C_1 \exp\left\{ -\frac{1}{N_0} \sum_{j=1}^{v} [x_j - s_{ij}(\psi)]^2 \right\} \\
&= C_1 \exp\left\{ -\frac{1}{N_0} \int_0^{T_b} [x(t) - s_i(t, \psi)]^2 \, dt \right\}
\end{aligned}
\tag{8.1.5}
$$

where $C_1 = 1/(2\pi N_0/2)^{v/2}$. Expanding the exponent, and combining the terms that do not involve the i index, and separating out the terms involving ψ, produces

$$p[\mathbf{x}|\mathbf{s}_i(\psi)] = C_2 \exp\left[\frac{2}{N_0} \int_0^{T_b} x(t) s_i(t, \psi) \, dt \right] \tag{8.1.6}$$

with $C_2 = C_1\{\exp -[\int_0^T x^2(t) \, dt + E_b]\}$. The averaged density in (8.1.3) is then

$$p(\mathbf{x}|i) = C_2 \int_0^{2\pi} \exp\left[\frac{2}{N_0} \int_0^{T_b} x(t)s_i(t, \psi)\, dt\right]\left(\frac{1}{2\pi}\right) d\psi \qquad (8.1.7)$$

Substituting from (8.1.1) and expanding the exponent,

$$\int_0^{T_b} x(t)e_i(t) \cos(\omega_i t + \psi)\, dt = (\cos \psi) Z_i - (\sin \psi) Y_i \qquad (8.1.8)$$

where

$$Z_i = \int_0^{T_b} x(t)e_i(t) \cos(\omega_i t)\, dt \qquad (8.1.9a)$$

$$Y_i = \int_0^{T_b} x(t)e_i(t) \sin(\omega_i t)\, dt \qquad (8.1.9b)$$

The expansion allows a separation of the ψ variable and the observed input $x(t)$. Note that Z_i and Y_i can be computed directly from $x(t)$ without knowing ψ. By defining

$$E_i^2 = Z_i^2 + Y_i^2$$
$$\phi_i = \tan^{-1} \frac{Y_i}{Z_i} \qquad (8.1.10)$$

we can simplify further to

$$(\cos \psi) Z_i - (\sin \psi) Y_i = E_i \cos(\psi + \phi_i) \qquad (8.1.11)$$

The integral in (8.1.7) is then

$$p(\mathbf{x}|i) = C_2\left(\frac{1}{2\pi}\right) \int_0^{2\pi} e^{(2E_i/N_0) \cos(\psi + \phi_i)}\, d\psi$$
$$= C_2 I_0(E_i) \qquad (8.1.12)$$

where we have inserted the definition of the Bessel function (see Appendix A.25),

$$I_0(\beta) = \frac{1}{2\pi} \int_0^{2\pi} e^{\beta \cos \alpha}\, d\alpha \qquad (8.1.13)$$

Thus the required decode test in (8.1.4) reduces to $I_0(2E_1/N_0) \gtrless I_0(2E_0/N_0)$. Furthermore, since $I_0(\beta)$ is monotonic for $\beta > 0$, this can be simplified to

$$E_1 \gtrless E_0 \qquad (8.1.14)$$

From the observed $x(t)$ during a bit interval, the optimal noncoherent decoder should compute E_1 and E_0, using (8.1.9) and (8.1.10), and decide the bit based on the larger of the two. A block diagram of this decoder is shown in Figure 8.1. The parameter E_1 is computed from the input $x(t)$ using a pair of simultaneous quadrature correlations with $e_1(t) \cos (\omega_1 t)$ and $e_1(t) \sin (\omega_1 t)$, followed by a squaring of the resultant integrations. A similar pair of channels is needed to generate E_0. Note that the true phase ψ is not used for decoding, but the decoder is significantly more complicated than the coherent correlator that would be used if ψ was known (or estimated). The optimal decoder is in fact nonlinear, as evident by the squaring operation in each arm.

The quadrature correlators shown in Figure 8.1 can be somewhat simplified. Consider the channel in Figure 8.2, composed of a matched bandpass filter [matched to the signal $e_1(t) \cos (\omega_1 t)$] followed by an ideal envelope detector and a sample taken at the end of each bit interval. The matched filter output is

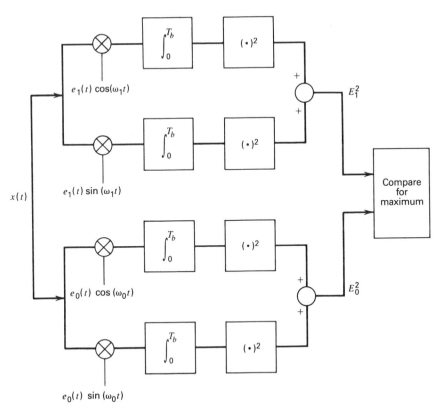

FIGURE 8.1. Noncoherent quadrature correlator for bit decoding.

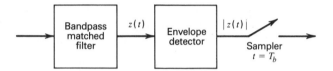

FIGURE 8.2. Envelope detector channel.

$$z(t) = \int_0^t x(\rho)h(t-\rho)\, d\rho$$

$$= \int_0^t x(\rho)[e_1(T_b - t + \rho)\cos[\omega_1(T_b - t + \rho)]\, d\rho$$

$$= \cos[\omega_1(T_b - t)]Z(t) - \sin[\omega_1(T_b - t)]Y(t) \qquad (8.1.15)$$

with

$$Z(t) = \int_0^t x(\rho)e_1(T_b - t + \rho)\cos(\omega_1\rho)\, d\rho$$

$$Y(t) = \int_0^t x(\rho)e_1(T_b - t + \rho)\sin(\omega_1\rho)\, d\rho \qquad (8.1.16)$$

The waveform $z(t)$ can be written as

$$z(t) = \text{Real}\{A(t)e^{j\omega_1(T_b - t)}\} \qquad (8.1.17)$$

where $A(t) = Z(t) + jY(t)$ is the complex envelope. The output of the ideal envelope detector is

$$|z(t)| = |A(t)| = [Z^2(t) + Y^2(t)]^{1/2} \qquad (8.1.18)$$

If we sample $|z(t)|$ at $t = T_b$ we have

$$|z(t)|_{t=T_b} = [Z^2(T_b) + Y^2(T_b)]^{1/2}$$
$$= [Z_1^2 + Y_1^2]^{1/2}$$
$$= E_1 \qquad (8.1.19)$$

Thus the sampled output of the system illustrated in Figure 8.2 produces the required test variable E_1. A similar system will produce E_0. The resulting noncoherent decoder will have the form of Figure 8.3. Decoding is therefore achieved by first filtering for the two signals, and then determining which has the largest envelope sample. Thus the system in Figure 8.3 performs a comparison test but uses only the signal envelopes for matching, rather than

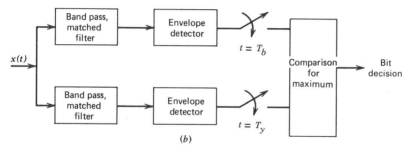

FIGURE 8.3. Equivalent noncoherent decoder.

the complete signal. Note that the carrier phase is lost in the envelope detection processes.

The noncoherent bit error probability is more complicated now since the decoder is nonlinear, because of the envelope detection operation. In particular, sampler statistics are no longer Gaussian even when the input noise is Gaussian, as it was for the linear decoder. The bit error probability is then

$$\text{PE} = \tfrac{1}{2}\text{Prob}\,[E_1 > E_0 \,|\, \text{one-bit sent}]$$
$$\tfrac{1}{2}\text{Prob}\,[E_0 > E_1 \,|\, \text{zero-bit sent}] \qquad (8.1.20)$$

Since $E_i = [Z_i^2 + Y_i^2]^{1/2}$, the computation of PE requires the joint statistics of the variables (Z_1, Y_1, Z_0, Y_0). From (8.1.9) it can be shown that in general these are correlated Gaussian variables with nonzero means that depend on ψ [Problem 8.1]. However PE simplifies for the case of orthogonal signals in (8.1.1); that is, when

$$\int_0^{T_b} s_1(t, \psi)s_0(t, \psi)\, dt = 0 \qquad (8.1.21)$$

In this case the Z and Y Gaussian variables become pairwise uncorrelated, and the incorrect Z_i and Y_i have zero mean values. The variables E_1 and E_0 are therefore statistically independent, and (8.1.20) reduces to

$$\text{PE} = \int_0^\infty p_1(\xi_1) \int_{\xi_1}^\infty p_2(\xi_2)\, d\xi_2\, d\xi_1 \qquad (8.1.22)$$

where $p_1(\xi_1)$ and $p_2(\xi_2)$ are the densities of the correct and incorrect E_i, respectively. These densities will have the form

$$p_1(\xi_1) = \frac{\xi_1}{\rho^2}\, e^{-(\xi_1^2 + \rho^4)/2\rho^2} I_0(\xi_1)\,, \qquad \xi_1 \geq 0 \qquad (8.1.23a)$$

$$p_2(\xi_2) = \frac{\xi_2}{\rho^2} e^{-\xi_2^2/2\rho^2}, \qquad \xi_2 \geq 0 \qquad (8.1.23b)$$

$$\rho^2 = \frac{2E_b}{N_0}$$

and $I_0(x)$ is again the Bessel function in (8.1.13). Equation (8.1.23a) is the so-called Rice density associated with the probability density of an envelope sample of narrowband Gaussian noise with an additive signal term of energy E_b. Equation (8.1.23b) is the Rayleigh density for an envelope sample of

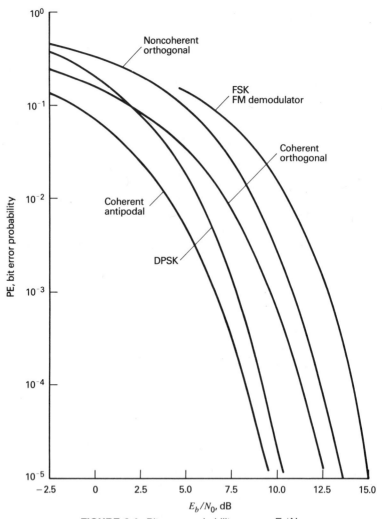

FIGURE 8.4. Bit error probability versus E_b/N_0.

narrowband Gaussian noise alone. When (8.1.23) is used in (8.1.22) the result integrates to (Problem 8.2)

$$PE = \tfrac{1}{2} e^{-(E_b/2N_0)} \tag{8.1.24}$$

where E_b is the signal energy in (8.1.2). Equation (8.1.24) is plotted in Figure 8.4 as noncoherent orthogonal. The coherent results of Figure 7.8 are also included. Note that noncoherent orthogonal binary systems are about 4 dB worse than coherent antipodal at $PE = 10^{-3}$, and roughly 3 dB worse at $PE = 10^{-5}$. This represents the price being paid for not making use of the phase angle in (8.1.1); that is, using coherent decoding. It is interesting that at low error probabilities noncoherent orthogonal performs about the same as coherent orthogonal. We can examine the orthogonality assumptions in (8.1.21) by substituting from (8.1.1)

$$\int_0^{T_b} s_1(t, \psi) s_0(t, \psi)\, dt \cong \tfrac{1}{2} \int_0^{T_b} e_1(t) e_0(t) \cos(\omega_1 - \omega_0) t\, dt \tag{8.1.25}$$

Note that the integration does not depend on ψ, as long as $\omega_1, \omega_0 \gg 2\pi/T_b$. Any pair of signals such that (8.1.25) is zero will constitute a pair of orthogonal signals. Particular examples of such signals, useful for noncoherent operation, are presented in the following paragraphs.

Pulse Position Modulation (PPM) Pulses

$$
e_1(t) = \begin{cases} \left(\dfrac{4E_b}{T_b}\right)^{1/2}, & 0 \le t \le \dfrac{T_b}{2} \\[2ex] 0, & \dfrac{T_b}{2} \le t \le T_b \end{cases}
$$

$$
e_0(t) = \begin{cases} 0, & 0 \le t \le \dfrac{T_b}{2} \\[2ex] \left(\dfrac{4E_b}{T}\right)^{1/2}, & \dfrac{T_b}{2} \le t \le T_b \end{cases}
\tag{8.1.26}
$$

$$\omega_1 = \omega_0$$

The signals use carrier bursts at the same frequency, but each located in one of two possible time intervals, one representing a one, the other a zero, as shown in Figure 8.5a. The signals are orthogonal and have equal energy but require time gating of an oscillator burst. Each requires a transmission bandwidth wide enough to pass a $T_b/2$ sec burst of oscillator frequency, and therefore a bandwidth of approximately $4/T_b$ Hz about the carrier. The decoder matched filters correspond to tuned bandpass filters, gated during

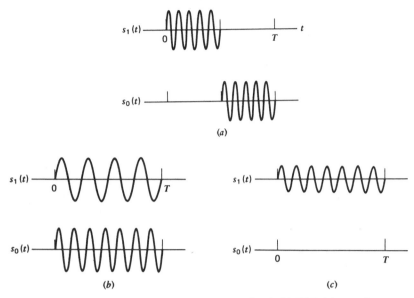

FIGURE 8.5. Noncoherent signaling formats: (a) PPM, (b) FSK, (c) on–off keying.

the time interval of the corresponding signal. Comparison of the envelope samples at $t = T_b/2$ and $t = T_b$ yields the bit decision, with the bit error probability given in (8.1.24). Note that the decoder bases its decision on which of two time intervals has the largest envelope sample.

Frequency Shift Keying (FSK)

$$e_1(t) = e_0(t) = \left(\frac{2E_b}{T_b}\right)^{1/2}, \qquad 0 \le t \le T_b$$

$$\omega_1 - \omega_0 = 2\pi\left(\frac{k}{T_b}\right), \qquad k = \text{integer}$$

$$(8.1.27)$$

The preceding signal pair (Figure 8.5b) is identical to the FSK signals in (7.5.21), except that we do not assume knowledge of the phase angle ψ. By separating the frequencies by an exact multiple of the bit period, we can form signals that are orthogonal, have equal energy, and have the same advantage of ease of encoder generation. Whereas in coherent FSK the phase angle ψ had to be exactly phase synchronized at the receiver, the signals here produce orthogonal signals without any knowledge of the value of ψ. The bandwidth is the same as that for coherent FSK in Section 7.5. For this signal pair, the individual channel matched filters in Figure 8.3 correspond to tuned bandpass filters. Hence the noncoherent decoder for FSK binary transmission corresponds to a receiver that attempts to determine which of two frequency channels produces the larger envelope sample. The frequencies are separated enough so that the transmitted frequency will not be received in the incorrect channel. (Contrast this result with the case of coherent FSK, where we purposely place the signaling frequencies close together to take advantage of slight negative correlation.) The bandpass filters must be matched to the envelopes of the FSK signals, and therefore have bandwidths of approximately $2/T_b$, centered at ω_1 and ω_0. The ease of encoding again makes FSK extremely popular for noncoherent binary signaling.

A problem with this noncoherent FSK decoder is the necessity of knowing the frequencies ω_1 and ω_0 exactly. Since the bandpass filters are only $2/T_b$ Hz wide, Doppler effects and associated oscillator instabilities may shift the expected signaling frequencies outside the decoder filter bandwidths. In practice this is compensated for by increasing the bandpass filter bandwidth so as to be sure to encompass the frequency uncertainty. This means that the filters are no longer truly matched, and the input noise to the decoder channels is increased. Analytically, the envelope densities in (8.1.23) are no longer valid, and the envelope sample comparison becomes a generalized energy comparison. That is, each channel of the decoder actually measures the received energy of its channel, and bit decisioning is now based on the channel with the most energy. The densities in (8.1.23) are now replaced by families of chi-squared densities [1], and bit error probability requires the integration over these densities. Although the resulting PE no longer integrates to a result as simple as (8.1.24), some results are known [1], and fairly accurate approximations have been developed. Equation (8.1.24) can also be used to approximate the resulting PE, provided that a modification to the effective E_b/N_0 is made. Figure 8.6 shows how the true parameter E_b/N_0 must be degraded, as the bandpass filter bandwidth is increased, before using in (8.1.24) to obtain an approximation to the true PE. For $B = 2/T_b$, the system of course reduces to the noncoherent envelope decoder considered earlier.

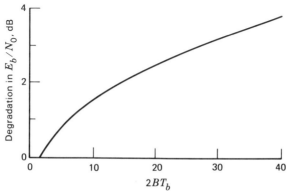

FIGURE 8.6. Equivalent increase in E_b to account for bandpass filter bandwidth in noncoherent FSK decoding (B = filter bandpass bandwidth).

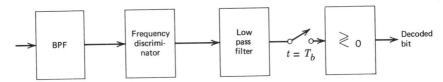

FIGURE 8.7. FSK discriminator decoder.

The two channel system in Figure 8.3 must be used to decode noncoherent FSK with the minimal PE. However, another technique often used is actually to frequency detect the digital FSK modulation using a standard frequency discriminator. The decoder treats the FSK signal as a carrier frequency between ω_1 and ω_0. This FSK carrier is then FM demodulated to recover the PCM modulation and bit decisions are based on the sign of the recovered bits. Such a system is shown in Figure 8.7 and is called an *FSK frequency discriminator decoder*. The system is again a form of noncoherent decoder since the subcarrier phase angle ψ does not affect the FM detection. The low pass filter can be a bit integrator to provide matched filtering of the demodulated PCM signal, or it can be a simple low pass filter for rejecting demodulator noise. The prediscrimination bandpass filter must be wide enough so as not to bandlimit the FSK signal, while reducing noise for the discriminator threshold.

As the FM modulation index is increased (wider frequency separation) there is an increase in the demodulated signal energy (i.e., the FM improvement). However, the predetection bandpass filter bandwidth must increase, letting in more noise. The optimal value of frequency separation and bandwidth to use depends on the operating E_b/N_0, but an exact analysis is complicated by the fact that the output of the frequency discriminator is not truly a Gaussian process. The general procedure is to first set the BPF noise bandwidth B_n so as to obtain an input CNR at the given E_b/N_0 [$\text{CNR} = (E_b/N_0)/B_n T_b$] that satisfies the FM threshold. The frequency separation is then increased to yield the lowest PE. Although the result depends on the filter shape, a rough rule of thumb is that the required frequency separation is about $0.8B_n$.

The resulting PE for discriminator decoding, using the preceding separation, is included in Figure 8.4. At each E_b/N_0, operation was adjusted for a discriminator threshold of 12 dB. We see that discriminator decoding does slightly poorer than noncoherent decoding, although the actual difference will depend on the choice of threshold value.

On–Off Keying

$$e_1(t) = \left(\frac{2E_b}{T_b}\right)^{1/2}, \qquad 0 \le t \le T_b$$

$$e_0(t) = 0$$

(8.1.28)

Here a binary zero is transmitted as no signal, whereas a one is transmitted as a T_b sec carrier burst of frequency ω_1 (Figure 8.5c). The subcarrier frequency and phase can be arbitrary, and the signal pair is always orthogonal. The required bandwidth is only half that of PPM signaling. The signals, however, do not have equal energy, and require continual on–off keying of the encoder carrier oscillator (i.e., not continuous operation). Since one signal is zero, only a single channel decoder in Figure 8.3 is needed. Bit decisioning is achieved by comparing the envelope sample to a threshold voltage. Since the signals do not have equal energy, the bit error probability in (8.1.24) must be modified to determine decoder performance. In particular, the probability of an error for a zero-bit is different from that of a one-bit. When a zero is sent, only noise is received and the error probability is given by the probability that an envelope sample of bandpass filtered Gaussian noise exceeds a threshold value of Y. This becomes

$$\mathrm{PE}|\text{zero-bit} = \int_{Y'}^{\infty} p(\xi_2)\, d\xi_2 \qquad (8.1.29)$$

where Y' is the normalized threshold $Y/(2E_b/N_0)^{1/2}$ and $p(\xi_2)$ is the envelope sample density in (8.1.23b). Similarly, the probability of an error when a one is sent is given by the probability that an envelope sample of a sine wave in Gaussian noise is less than the threshold. Thus

$$\mathrm{PE}|\text{one-bit} = \int_{0}^{Y'} p(\xi_1)\, d\xi_1 \qquad (8.1.30)$$

where $p(\xi_1)$ is given in (8.1.23a). Both of these integrals can be written in terms of the Marcum Q-function, defined as

$$Q(a, b) \overset{\Delta}{=} \int_{b}^{\infty} \exp\left[-\frac{a^2 + x^2}{2} \right] I_0(ax) x\, dx \qquad (8.1.31)$$

The average bit error probability for noncoherent on–off keying can therefore be written as

$$\mathrm{PE} = \frac{1}{2} [\mathrm{PE}|\text{zero-bit}] + \frac{1}{2} [\mathrm{PE}|\text{one-bit}]$$

$$= \frac{1}{2} Q[0, Y'] + \frac{1}{2} \left\{ 1 - Q\left[\left(\frac{2E_b}{N_0} \right)^{1/2}, Y' \right] \right\} \qquad (8.1.32)$$

Equation (8.1.32) is plotted in Figure 8.8 for Y' = 0.5. The noncoherent orthogonal result (FSK or PPM) with the same signal energy E_b is superimposed. At the higher E_b/N_0 values, the on–off system suffers from the lack of usable signal energy during the zero bit transmission. The Q function has been tabulated [2] and recursive computational methods have been developed for its evaluation [3].

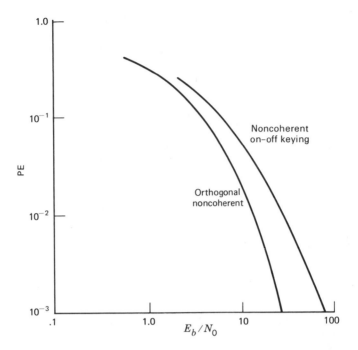

FIGURE 8.8. PE for noncoherent on–off signaling. Threshold $\gamma = 0.5 \, (2E_b/N_0)^{1/2}$.

8.2 DIFFERENTIAL PSK BINARY SIGNALING

The noncoherent encoding scheme avoided the necessity of achieving and maintaining the phase coherence required in the coherent BPSK system. Another way to bypass this problem while still performing binary encoding is by the use of *differential* PSK (DPSK). In this scheme, antipodal PSK signaling is again used, but the bits are encoded by relating them to the previous bit. If a binary one is to be sent during a bit interval, it is sent as a PSK signal with the same phase as the previous bit. If a binary zero is sent, it is transmitted with the opposite phase (i.e., the negative version) of the previous bit. Thus ones and zeros are transmitted as a phase change or no phase change from the previous bit. The source sequence is therefore being sent as a sequence of PSK signals, but the phase of each is obtained by the preceding rule. As an example, if we denote the PSK signals as

$$f_1(t) = \left(\frac{2E_b}{T_b}\right)^{1/2} \cos\left(\omega_s t + \psi\right), \qquad 0 \le t \le T_b$$

$$f_0(t) = \left(\frac{2E_b}{T_b}\right)^{1/2} \cos\left(\omega_s t + \psi \pm \pi\right), \qquad 0 \le t \le T_b$$

$$(8.2.1)$$

and if we assume that the last bit sent was $f_1(t)$, then the bit sequence 1 0 0 1 will be sent as the waveform sequence $\{f_1, f_0, f_1, f_1\}$. Note that we continue to send one bit every T_b sec (except for the first bit, which requires an initial reference waveform) and is therefore a form of binary encoding.

A rigorous approach to DPSK analysis is to recognize that we are really transmitting each bit with the binary signal pair

$$s_1(t) = \begin{cases} (f_1, f_1), \\ \text{or} \qquad\qquad -T_b \le t \le T_b \\ (f_0, f_0), \end{cases}$$

$$\qquad\qquad\qquad\qquad\qquad\qquad\qquad\qquad\qquad (8.2.2)$$

$$s_0(t) = \begin{cases} (f_1, f_0), \\ \text{or} \qquad\qquad -T_b \le t \le T_b \\ (f_0, f_1), \end{cases}$$

where f_i is the signal $f_i(t)$ in (8.2.1) and (f_i, f_j) denotes $f_i(t)$ followed by $f_j(t)$. The first T_b sec of each waveform is actually the last T_b sec of the previous waveform. Note that $s_1(t)$ and $s_0(t)$ each can have either of two possible forms. We see these signals all have energy $2E_b$, and the signal pair has correlation

$$\int_{-T_b}^{T_b} s_1(t)s_0(t) \, dt = \int_{-T_b}^{0} [\pm f_1(t)]^2 \, dt - \int_{0}^{T_b} [\pm f_1(t)]^2 \, dt = 0 \quad (8.2.3)$$

for any combination. The DPSK signal set in (8.2.2) therefore represents a pair of orthogonal signals, $2T_b$ sec long, with arbitrary phase angle ψ. Decoding of these signals should correspond to noncoherent envelope detection with four channels matched to each of the possible envelope waveforms, two representing a one and two representing a zero. However, the two waveforms representing each bit are negatives of each other, and the envelope sample of each will be the same. Hence we need only construct a single noncoherent channel for $s_1(t)$ and a single channel for $s_0(t)$, each matched to either of the two noncoherent waveforms in (8.2.2), as in Figure 8.9a. The noncoherent decoders for $s_1(t)$ and $s_0(t)$ in (8.2.2) are shown in Figures 8.9b and 8.9c, respectively, obtained by modifying Figure 8.1. This decoder can be redrawn and simplified, however, by noting that if we define

$$y_c(t) = \int_{t-T_b}^{t} x(t) \cos(\omega_s t) \, dt$$

$$\qquad\qquad\qquad\qquad\qquad\qquad\qquad (8.2.4)$$

$$y_s(t) = \int_{t-T_b}^{t} x(t) \sin(\omega_s t) \, dt$$

then

$$E_1^2 = [y_c(0) + y_c(T_b)]^2 + [y_s(0) + y_s(T_b)]^2 \qquad\qquad (8.2.5)$$

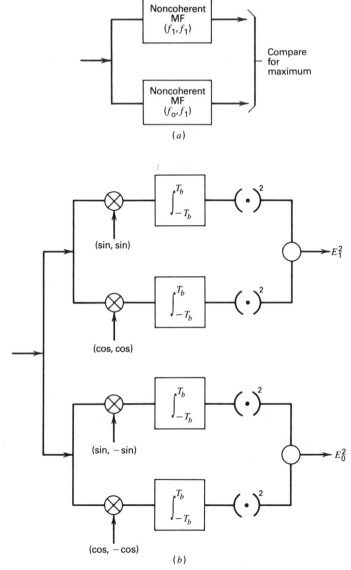

FIGURE 8.9. DPSK optimal decoders: (a) general two-channel decoder, (b) quadrature correlator decoder for BPSK.

Similarly, the noncoherent decoder for $s_0(t)$ in (8.2.2) would compute

$$E_0^2 = [y_c(0) - y_c(T_b)]^2 + [y_c(0) - y_c(T_b)]^2 \qquad (8.2.6)$$

A comparison of $E_1^2 \gtrless E_0^2$, as required by the optimal decoder, would therefore be identical to the test,

$$y_c(0)y_c(T_b) + y_s(0)y_s(T_b) \gtrless 0 \tag{8.2.7}$$

Hence the optimal DPSK decoder can be reduced to that in Figure 8.10, where samples of $y_c(t)$ and $y_s(t)$ are processed sequentially for bit decisioning.

The test in (8.2.7) is in fact equivalent to deciding if the carrier phase has changed between the time intervals $(-T_b, 0)$ and $(0, T_b)$. Since the phase angle of the vector $[y_c(T_b) + jy_s(T_b)]$ is a measurement of the carrier phase angle during $(0, T_b)$, it follows that

$$\text{phase angle of } [(y_c(0) + jy_s(0))(y_c(T_b) - jy_s(T_b)]$$

$$= \text{difference of phase angle between } (-T_b, 0) \text{ and } (0, T_b) \tag{8.2.8}$$

The bit decision should depend on whether this phase angle difference is closer to 0 or to π radians. This is equivalent to determining if the real part of the complex product in the bracket in (8.2.8) is positive or negative, which is precisely the test performed in (8.2.7).

To obtain the bit error probability for the DPSK decoding test in (8.2.7), or equivalently the test $E_1 \gtrless E_0$ in (8.2.6), we first note that

$$y_c(0) + y_c(T_b) = \int_{-T_b}^{T_b} x(t) \cos(\omega_s t) \, dt \tag{8.2.9}$$

When a one bit is sent [f_1 followed by f_1 in (8.2.2)] the integral is a Gaussian random variable with mean $\sqrt{2E_b T_b} \cos \psi$ and variance $(N_0/2)(2T_b/2) = N_0 T_b/2$. Likewise, $y_s(0) + y_s(T_b)$ is Gaussian with mean $\sqrt{2E_b T_b} \sin \psi$ and the same variance. Combining as in (8.2.5), and following the development in (8.1.19) and (8.1.23a), we see that E_1 is a Rician random variable with noncentrality parameter $\sqrt{2E_b T_b}$. The variables $[y_c(0) - y_c(T_b)]$ and $[y_s(0) - y_s(T_b)]$ have zero mean, and each has the variance $N_0 T_b/2$, so that

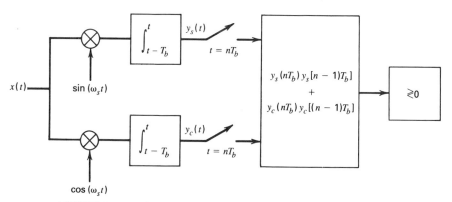

FIGURE 8.10. DPSK decoder equivalent to that shown in Figure 8.9b.

E_0 in (8.2.6) has a Rayleigh density. The probability that $E_1 < E_0$ when a one bit is sent during $(0, T_b)$ will integrate as in (8.1.24) to produce (see Problem 8.2) PE $= \frac{1}{2} \exp(-\rho^2/4)$, where now $\rho^2 = (\sqrt{2E_bT_b})^2/(N_0T_b/2) = 4E_b/N_0$. The resulting DPSK bit error probability is then

$$\text{PE} = \tfrac{1}{2}e^{-E_b/N_0} \tag{8.2.10}$$

where E_b is the bit energy in T_b sec. This result is superimposed in Figure 8.4. We see that DPSK achieves the same PE as noncoherent orthogonal at half the bit energy E_b. Thus DPSK is 3 dB better than noncoherent operation, while still avoiding the phase referencing operation. At low $(\text{PE} \le 10^{-5})$ error probabilities, DPSK is only about 1 dB poorer than coherent PSK without the necessity of phase referencing subsystems. However, the result must be accepted with care, since PE may not be a good measure of performance with DPSK, since errors tend to accumulate in pairs. That is, a burst of noise during a single T_b sec interval will likely cause two bits to be decoded in error.

An apparent simplification of DPSK can be observed by noting that decoding can be achieved by correlating each T_b sec waveform with the previous bit, as shown in Figure 8.11. Without noise the signal correlation produces the value $+\sqrt{E_b}$ is a one is sent, and correlates to $-\sqrt{E_b}$ if a binary zero is sent, for any value of the phase angle ψ. Since these correlation values are identical to the signal correlation values produced in coherent PSK, it at first appears that DPSK is providing exact coherent operation without phase referencing. However, when noise is added, it appears with both bits, and performance is worse than that of coherent PSK. This can be attributed to the fact that two noisy waveforms are actually being correlated in Figure 8.11, and the idealized values of $\pm E_b$ are not really being achieved.

If we argue that the integration variable y in Figure 8.11 is approximately Gaussian (since the integration bandwidth $1/T_b$ is generally much less than the bandpass noise bandwidth), we can estimate PE performance by com-

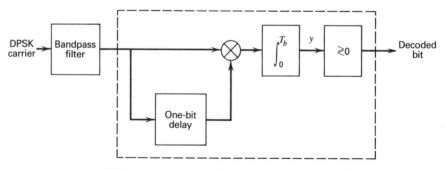

FIGURE 8.11. Suboptimal bit delay correlator for DPSK.

puting the mean and variance of y. If this is done (Problem 8.5), the PE for the system in Figure 8.11 is approximately

$$\text{PE} = Q\left[\frac{E_b^2}{E_b/N_0 + N_0^2/4}\right]^{1/2}$$

$$= Q\left\{\left(\frac{E_b}{N_0}\right)^{1/2}\left[\frac{1}{1 + (N_0/4E_b)}\right]^{1/2}\right\} \qquad (8.2.11)$$

The inner bracket term therefore accounts for the degradation due to the noisy correlation. When $E_b/N_0 \gg 1$,

$$\text{PE} \approx Q\left[\left(\frac{E_b}{N_0}\right)^{1/2}\right]$$

$$\approx \frac{1}{\sqrt{2\pi}(E_b/N_0)^{1/2}} \, e^{-(E_b/2N_0)} \qquad (8.2.12)$$

Hence the delay-and-correlate DPSK decoder shown in Figure 8.11 is about 2 dB worse than the optimal DPSK decoder in Figure 8.10, at high $(E_s/N_0 \approx 10)$ energy levels.

8.3 BANDLIMITING EFFECTS

We have been assuming that the bit signals at the decoder input are identical to the signals formed at the transmitter encoder. We know from practice that this again is an idealized condition and that, in fact, decoder signals will often be distorted during transmission. This most predominant cause is the filtering applied to the signal during modulation, transmission, and reception. The most severe filtering generally comes from baseband filtering either just prior to modulation at the transmitter, or following demodulation at the receiver. For PCM pulsed signals, this usually corresponds to a low pass filter designed to pass only a bandwidth approximately equal to the bit rate in hertz. For BPSK signals, which basically represent an oscillator amplitude modulated by a PCM pulsed sequence, the filtering is caused by bandpass filters at the carrier or subcarrier level. By our AM–filter conversion in Section 2.6 we can again consider this to be an effective low pass filtering, with a bandwidth equal to half the bandpass bandwidth, on the PCM sequence itself.

Bandlimiting and filtering causes two primary effects—waveform distortion and bit spreading. The first causes removal or reduction of portions of the bit waveform spectrum, thereby reducing its energy content for decoding. The second produces waveform spreading in time (recall the basic inverse relation between signal bandwidth and time extent), causing the signal in one bit interval to interfere with the bits in the subsequent intervals. This latter effect is referred to as *intersymbol interference*.

At the baseband level, intersymbol interference is often analyzed in terms of an "eye diagram," obtained by superimposing multiple bit intervals on top of each other. This will produce the diagram one would observe on an oscilloscope, having a sweep rate equal to the bit rate. Eye diagrams are sketched in Figure 8.12. If the PCM bit waveform was ideal, the resulting eye diagram would show the superposition of rectangular pulse waveforms with arbitrary bits, as in Figure 8.12a. The superposition of all positive bits produces the upper rectangle, and all negative bits, the lower rectangle. When bandlimiting occurs, the pulses are distorted and spread in time, and the superposition of the tails over subsequent bit intervals produces the diagram in Figure 8.12b. If the pulse stretching is severe, the ideal square diagram in Figure 8.12a begins to fill in with the pulse tails, effectively "closing the eye," as in Figure 8.12c. Hence an increase in the thickness of the top of the eye pulse exhibits a measure of the degree of intersymbol interference.

Likewise, if the bit time interval is not identical from one bit to the next, the resulting superposition spreads the zero crossing lines at the bit edges, as shown in Figure 8.12c. The width of the zero crossing line, relative to the

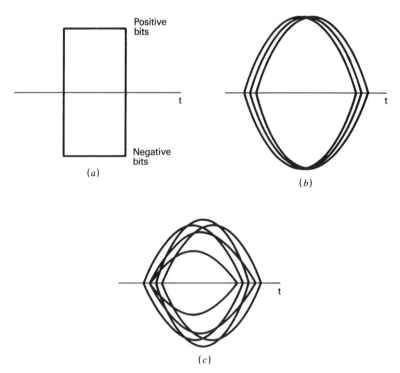

FIGURE 8.12. Eye diagrams for NRZ bit stream: (a) ideal bits, (b) with moderate intersymbol interference, (c) with heavy interference and timing errors.

total bit interval, is therefore an indication of the timing accuracy of the baseband waveform.

To examine both these effects consider again the link model in Figure 8.13a, in which $H_{bc}(\omega)$ represents the equivalent filtering applied to the transmitted bit waveforms up to the decoder input and $n(t)$ represents white Gaussian noise at the decoder input. We assume the encoder signals consist of antipodal NRZ waveforms, which we write as $\pm\sqrt{(E_b/T_b)}w(t)$, where

$$w(t) = \begin{cases} 1, & 0 \le t \le T_b \\ 0, & \text{elsewhere} \end{cases} \tag{8.3.1}$$

The decoder is taken as the coherent integrate-and-dump matched filter associated with the encoder signals. We can write the filtered decoder input during any specific bit interval $(0, T_b)$ as

$$x(t) = \left(\frac{E_b}{T_b}\right)^{1/2} \sum_{j=0}^{\infty} d_j s(t + jT) + n(t) \tag{8.3.2}$$

where $d_j = \pm 1$, d_0 is the bit during $(0, T_b)$, and d_j is the bit j intervals in the past. The waveform $s(t)$ represents the filtered version of the pulse $w(t)$ in (8.3.1). Since $s(t)$ is no longer of infinite bandwidth, it is no longer limited to T_b sec and is instead spread in time (Figure 8.13b). Note that this will have the transform

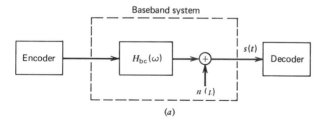

(a)

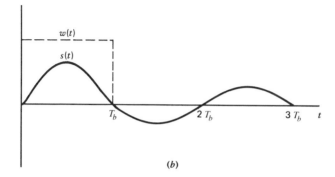

(b)

FIGURE 8.13. Filtered binary digital system: (a) block diagram, (b) filtered bit waveforms.

$$F_s(\omega) = W(\omega)H_{hc}(\omega) \qquad (8.3.3)$$

where $W(\omega)$ is the transform of $w(t)$. The integrate-and-dump decoder produces the sample value

$$y(T_b) = \int_0^{T_b} x(t)\, dt$$

$$= \sqrt{E_b T_b}\, m_0 + \sqrt{E_b T_b}\, m_I + \int_0^{T_b} n(t)\, dt \qquad (8.3.4)$$

where

$$m_0 = \frac{1}{T_b} \int_0^{T_b} d_0 s(t)\, dt \qquad (8.3.5a)$$

$$m_I = \frac{1}{T_b} \sum_{j=1}^{\infty} d_j \int_0^{T_b} s(t+jT)\, dt \qquad (8.3.5b)$$

The first term in (8.3.4) is the effect of the present bit, whereas m_I accounts for the combined intersymbol interference from past bits. The last term is the integrated decoder noise and represents a zero mean Gaussian variable with variance $N_0 T_b/2$. Note that m_I is itself a random variable dependent on the random data bits. In fact, m_I evolves as a discrete density, generated as the sum of weighted binary variables, with the weighting dependent on the filtered bit waveform. Thus $y(T_b)$ is a random variable composed of the signal mean (first term) plus the sum of a Gaussian noise variable and a discrete interference variable m_I. Clearly, $y(T_b)$ is no longer Gaussian. To directly determine the resulting bit error probability of the decoding test, it is necessary to recompute PE based on the new (non-Gaussian) statistics of $y(T_b)$. Although this procedure has been pursued [4, 5], it is hindered by the complexity of the exact statistics (requiring a convolution of the Gaussian noise density with the discrete density of the m_I).

A more fruitful approach is to compute PE in steps; that is, first conditioning on the bit sequence, then averaging over all possible sequences. If we assume a particular bit history $\mathbf{d} = (d_0, d_1, d_2, \ldots)$, then $y(T_b)$ is Gaussian with a mean value $\sqrt{E_b T_b}(m_0 + m_I)$ and a variance $N_0 T_b/2$. For this bit sequence $\mathbf{d}$ the decoder error probability PE($\mathbf{d}$) follows as

$$\mathrm{PE}(\mathbf{d}) = Q\left[\left(\frac{2E_b}{N_0} \right)^{1/2} (m_0 + m_I) \right] \qquad (8.3.6)$$

Here m_0 accounts for the waveform distortion, while m_I is the intersymbol effect of previous bits, adding a constant term to the present mean m_0. This constant depends on the channel filter pulse response $s(t)$ and on the particular sequence of previous bits. This term may have the same sign as the m_0 term, thereby aiding detection, or it may be opposite in sign, causing further decoding degradation. To determine the exact filter effect on the bit error probability of a given bit, it is necessary to average (8.3.6) over all prior bit sequences $\{\mathbf{d}_i\}$. If we assume that all bits are equal likely, and the intersymbol interference extends for K bits, then $p(\mathbf{d}) = 1/2^k$ for all $\mathbf{d}$, and the average PE becomes

$$\text{PE} = \frac{1}{2^k} \sum_{i=1}^{2^K} Q\left\{\left(\frac{2E_b m_0^2}{N_0}\right)^{1/2} [1 + I(\mathbf{d}_i)]\right\} \tag{8.3.7}$$

where

$$I(\mathbf{d}_i) = \frac{m_I}{m_0} \tag{8.3.8}$$

is the normalized interference variable produced by the bit sequence $\mathbf{d}_i = (d_{i_1}, d_{i_2}, d_{i_k})$ in m_I in (8.3.5b). This can also be written as

$$I(\mathbf{d}_i) = \sum_{j=1}^{K} d_{ij} c_j \tag{8.3.9}$$

with

$$c_j = \int_0^{T_b} s(t + jT_b)\, dt \Big/ \int_0^{T_b} s(t)\, dt$$

The resulting bit error probability can therefore be computed by a direct summation in (8.3.7). Clearly, K, the filter memory, cannot be too large for this computation to be feasible, since the number of terms grows exponentially with K.

Figure 8.14a shows the result of filtering a baseband PCM bit waveform with an nth order Butterworth filter [see Equation (1.7.6)] for several values of n, with a 3 dB bandwidth B_m of $1/T_b$. The pulse distortion and pulse spreading into adjacent bit intervals is evident. Figure 8.14b shows the corresponding PE degradation for $n = 5$ and several bandwidths. This shows how E_b must be increased to maintain performance at a desired PE with nonideal filtering.

An alternative method, when K is large, is by the use of moment theory [6–8]. In this approach (8.3.7) is approximated by a sum involving fewer terms that will asymptotically approach the true result. The idea is to recognize that (8.3.7) is actually an average over a random variable I. We seek to replace the true density of I by an approximate density that is simpler to integrate. We therefore consider I to have the density

$$p(I) = \sum_{j=1}^{N} W_j \delta(I - I_j) \tag{8.3.10}$$

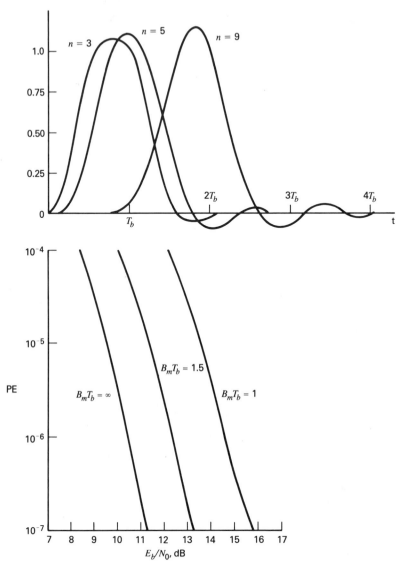

FIGURE 8.14. Bandpass filtered BPSK bits; Chebychev filters: (a) filtered bit waveforms (n = order filter), (b) PE degradation, with filter bandwidth.

corresponding to a discrete density at the points $\{I_j\}$ with weights $\{W_j\}$. The latter are to be selected so that the random variable I in (8.3.10) has the same statistical moments as the true I in (8.3.8). It is easy to establish that with independent data bits the latter has a kth moment $\mu_d(k) \Sigma_j c_j^k$, where $\mu_d(k)$ is the kth moment of each binary data variable d_i. The density in (8.3.10) has the kth moment $\Sigma_j W_j I_j^k$. Equating moments up to $k = 2N$ requires

$$\sum_{j=1}^{N} W_j I_j^k = \mu_d(k) \sum_{j=1}^{N} c_j^k, \qquad k = 1, 2, \ldots, 2N \qquad (8.3.11)$$

This corresponds to $2N$ equations with $2N$ unknowns (I_j, W_j) that can be solved. The resulting PE, using the approximating density in (8.3.10), with the solutions (I_j, W_j) from (8.3.11) is then

$$\text{PE} \approx \sum_{j=1}^{N} W_j Q\left[\left(\frac{2E_b m_0^2}{N_0}\right)^{1/2}(1 + I_j)\right] \qquad (8.3.12)$$

The advantage is that only N terms (instead of 2^k) need be computed. However, N must be large enough to give the desired accuracy in computing PE. Some convergence bounds and error estimates for N can be found in Reference 6. A larger N requires more equations to solve (8.3.11), although recursive algorithms can be developed for aiding this computation.

A simpler indication of the effect is to examine (8.3.7) for the worst case bit sequence only; that is, the past bit sequence for which m_I causes the largest degradation to m_0. The worst case sequence can be easily determined in (8.3.5b) (Problem 8.7), and the resulting value of m_0 and m_I can be

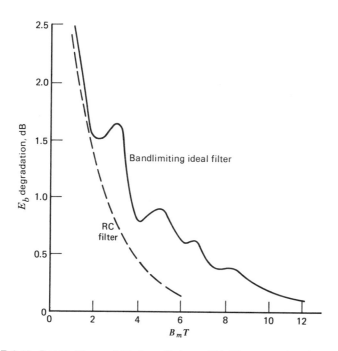

FIGURE 8.15. Bandlimiting and filtering effects on PE. Worst case interference. ($B_m = 3$ dB bandwidth of low pass filter.)

evaluated by inverse transforming (8.3.3) and integrating in (8.3.5). For a bandlimiting channel filter,

$$H_{bc}(\omega) = 1 , \qquad |\omega| \le 2\pi B_m$$

$$= 0 , \qquad \text{elsewhere} \qquad\qquad (8.3.13)$$

this result has been computed [9] and is shown in Figure 8.15 as a function of B_m. Also included is the worst case result for a single pole RC filter with 3 dB bandwidth B_m, making use of the results of Problem 1.24. Both results clearly exhibit the worst case degradation to E_b as B_m is narrowed. The corresponding error probability in (8.3.6) yields a worst case result, and although somewhat pessimistic, it does indicate an upper bound to the true performance.

8.4   DESIGNING FOR THE BANDLIMITED CHANNEL

The degradation due to the channel filter is caused by the mismatch between the decoder and bandlimited signals. We may therefore inquire whether, given a known bandlimiting channel filter $H_{bc}(\omega)$ (either as part of the transmission channel or imposed by system constraints), there is a formal design procedure that best compensates for the filtering. To examine this, consider again the bandlimiting filter of (8.3.13). Let us allow an arbitrary encoder signal $\pm z(t)$ to be transmitted for antipodal signaling and again assume that the decoder has the general form in Figure 7.19 composed of a filter $H(\omega)$ followed by a sampler. We would like to determine an optimal decoder filter and encoder signal $z(t)$ for combating the channel bandlimiting and yielding minimal error probability. Let $Z(\omega)$ be the transform of $z(t)$ and assume it to be confined to energy E_b. The output of the decoding filter can again be written as in (8.3.2)

$$y(t) = \sum_{j=0}^{\infty} d_j s(t + jT_b) + n_y(t) \qquad\qquad (8.4.1)$$

where now $s(t)$ is the filtered version of $z(t)$ and therefore has the transform

$$F_s(\omega) = Z(\omega) H_{bc}(\omega) H(\omega)$$

$$= Z(\omega) H(\omega) , \qquad |\omega| \le 2\pi B_m \qquad\qquad (8.4.2)$$

Note that $F_s(\omega)$ is also bandlimited to B_m Hz because of the channel filtering. The noise $n_y(t)$ has the spectrum

$$S_{ny}(\omega) = S_n(\omega) |H(\omega)|^2 \qquad\qquad (8.4.3)$$

At the sampling instant, $t = T_b$, $y(T_b)$ is

$$y(T_b) = d_0 s(T_b) + \sum_{j=1}^{\infty} d_j s[(j+1)T_b] + n_y(T_b) \qquad (8.4.4)$$

Again we have the effect of the present bit (first term), the intersymbol effect of past bits (summation), and the baseband noise. However, we immediately see that the entire intersymbol effect depends only on $s(t)$ at $t = jT_b$, $j \geq 2$. This means that any bandlimited waveform $s(t)$ such that

$$s(jT) \begin{cases} = 0, & j \geq 2 \\ \neq 0, & j = 1 \end{cases} \qquad (8.4.5)$$

automatically removes intersymbol interferences. If the waveform is not bandlimited, then, of course, any waveform confined to $(0, T_b)$ would suffice, such as those considered throughout this chapter. Since the sinc function

$$\text{sinc}\left(\frac{\pi(t - T_b)}{T_b}\right) \triangleq \frac{\sin\left[(\pi(t - T_b)/T_b\right]}{\pi(t - T_b)/T_b} \qquad (8.4.6)$$

satisfies (8.4.5), a general solution occurs if $s(t)$ has the form

$$s(t) = s_a(t) \, \text{sinc}\left[\frac{\pi(t - T_b)}{T_b}\right] \qquad (8.4.7)$$

where $s_a(t)$ is an arbitrary time function. The transform of (8.4.7) is the convolution of the transform of the arbitrary function $s_a(t)$ and the transform of (8.4.6). That is,

$$F_s(\omega) = F_{s_a}(\omega) \oplus F_T(\omega) \qquad (8.4.8)$$

where $F_T(\omega) = 1$, $|\omega| \leq 2\pi/2T_b$, and zero elsewhere. We see that $s(t)$ has a bandwidth of B_m Hz, as required by (8.4.2), only if $B_m \geq 1/2T_b$, and only if $s_a(t)$ has a bandwidth of $B_m - (1/2T_b)$ Hz. In other words, the convolution of a bandlimited spectrum of width $1/2T_b$ with one bandlimited to $B_m - (1/2T_b)$ Hz would produce an overall spectrum bandlimited to B_m Hz. Thus, if we are transmitting T_b sec bits over a channel bandlimited to B_m Hz, with $B_m \geq 1/2T_b$, then intersymbol interference can be completely removed by *any* decoder filter output waveform of the form in (8.4.7), provided $s_a(t)$ is a time function bandlimited to $B_m - (1/2T_b)$ Hz. This means that for any $B_m \geq 1/2T_b$, there are an infinite number of solutions that can be used for intersymbol elimination, one for each possible waveshape $s_a(t)$ whose bandwidth is $B_m - (1/2T_b)$ Hz.

Of particular importance is the fact that a channel bandwidth B_m no larger than $1/2T_b$ is needed. If $B_m = 1/2T_b$, then $s_a(t)$ has zero bandwidth (delta function at $\omega = 0$) and corresponds to a constant. The decoder filter output corresponds to signals of the form of (8.4.6), in which each bit has a single peak voltage value when all other bit waveforms pass through zero

(Figure 8.16). A bit transmission rate of $1/T_b$ bits/sec is therefore maintained with no intersymbol effect. Conversely, we can state that a channel bandlimited to B_m can theoretically transmit at a rate of $2B_m$ bits/sec without intersymbol effects. This is referred to as the *Nyquist rate* of data transmission. Of course, the intersymbol interference is removed only if we sample exactly at each $t = iT_b$. A slight offset in the decoder timing generates intersymbol values that can accumulate rather quickly. Thus transmission at the Nyquist rate places severe requirements on decoder timing. For this reason it may be advantageous to use a larger B_m for a given time T_b (or, conversely, a lower rate for a given bandwidth B_m) in order to utilize $s_a(t)$ in (8.4.7) to cause a faster decay in the interference values caused by timing offsets. [Roughly speaking, $s_a(t)$ adds a decaying factor to $s(t)$ that decreases as the reciprocal of its bandwidth.] Thus, by proper selection of $s_a(t)$, timing requirements can be eased at the expense of information rate. The amount by which the bandwidth B_m exceeds the minimal (Nyquist) bandwidth of $1/2T_b$ is called the *excess* bandwidth.

It should also be pointed out that the frequency function corresponding to the signal in (8.4.6), used with communicating at the Nyquist rate, is that of an ideal filter. This means that the combined effect of all the channel filtering on the transmitted pulse waveform must produce a waveform with an ideal frequency function at the decoder input, a result that can be approached only asymptotically. Thus binary transmission with no excess bandwidth presupposes ideal frequency functions. By allowing some excess, the required function can depart from the idealized result and can therefore be more realistically constructed. A convenient frequency function of this type that produces a waveform as in (8.4.7) is

$$F_s(\omega) = \begin{cases} 1, & 0 \le \omega \le \dfrac{\pi}{T_b}(1-\alpha) \\ \dfrac{1}{2}\left[1 - \sin\dfrac{T_b}{2\alpha}\left(\omega - \dfrac{\pi}{T_b}\right)\right], & \dfrac{\pi}{T_b}(1-\alpha) \le \omega \le \dfrac{\pi}{T_b}(1+\alpha) \end{cases}$$

$$(8.4.9)$$

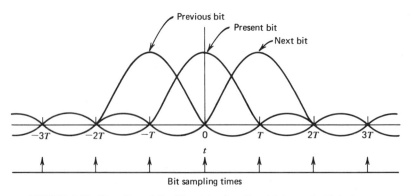

FIGURE 8.16. Signaling at the Nyquist rate without intersymbol interference.

with $0 \leq \alpha \leq 1$. This frequency function is sketched in Figure 8.17a, and its corresponding time function is shown in Figure 8.17b. We note that this pulse function passes through zero at all multiples of T_b for any α, as required for eliminating the intersymbol interference. The parameter α indicates the amount of bandwidth excess used, with $\alpha = 0$ corresponding to the ideal filter function and $\alpha = 1$ producing a slower rolloff and a doubling of the Nyquist bandwidth. However, as $\alpha \to 0$, the pulse function is more oscillatory and spread, rendering exact timing mandatory. As $\alpha \to 1$, the response decays quickly and therefore produces signals that are easier to handle. We emphasize that (8.4.9) is just one of many possible frequency functions whose transform eliminates intersymbol interference.

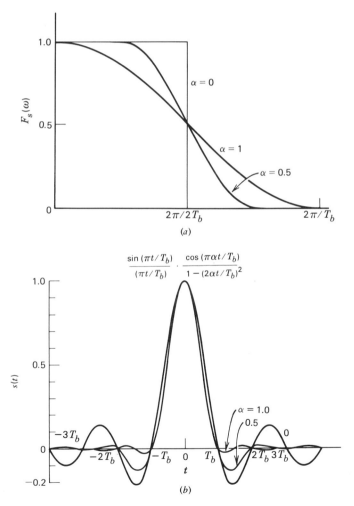

FIGURE 8.17. Desired decoding pulse for zero intersymbol interference: (a) spectrum, (b) bit waveform.

Assume now that the decoder signal frequency function has been selected for a bandwidth B_m and bit time T_b. We now must find combinations of $Z(\omega)$ and $H(\omega)$ to produce this $F_s(\omega)$. With intersymbol interference removed, decoding performance will depend on only the ratio

$$\frac{m_1}{\sigma_y} = \frac{s(T_b)}{\sigma_{ny}} \tag{8.4.10}$$

where, from (8.4.3),

$$\sigma_{ny}^2 = \frac{1}{2\pi} \int_{-\infty}^{\infty} S_n(\omega) |H(\omega)|^2 \, d\omega \tag{8.4.11}$$

This immediately suggests a design procedure whereby m_1/σ_y is maximized with respect to $|H(\omega)|^2$, subject to a constraint on the energy in $z(t)$, that is, on the condition

$$\int_{-\infty}^{\infty} |Z(\omega)|^2 \, d\omega = \int_{-\infty}^{\infty} \left| \frac{F_s(\omega)}{H(\omega)} \right|^2 \, d\omega = \text{constant} \tag{8.4.12}$$

The problem is now identical to the minimization considered in Section 6.2. The solution for the decoder filter follows as

$$|H(\omega)|^4 = \frac{|F_s(\omega)|^2}{S_n(\omega)}, \qquad |\omega| \le 2\pi B_m \tag{8.4.13}$$

The required $Z(\omega)$ needed to produce the desired $F_s(\omega)$ is then

$$Z(\omega) = \frac{KF_s(\omega)}{H(\omega)}, \qquad |\omega| \le 2\pi B_m \tag{8.4.14}$$

where K is adjusted to satisfy the energy constraint in (8.4.12). We therefore achieve a maximization in (8.4.10) (minimum error probability) without intersymbol interference. Note that $H(\omega)$ in (8.4.13) is basically achieving noise reduction, whereas the transmitted encoder signal $z(t)$, whose transform is given in (8.4.14), is selected to generate the desired $s(t)$ in (8.4.5). Our results, of course, reduce to all our earlier conclusions if we have no bandwidth restrictions. In this case $B_m \to \infty$ and $s(t)$ can be selected to be any waveform time limited to T_b sec. We therefore always satisfy (8.4.5), and the optimal decoding filter reduces to our nonwhite noise matched filter derived earlier in (7.6.9). Further discussions of bandlimiting on digital data transmissions can be found in References 10 and 11.

8.5   PARTIAL RESPONSE SIGNALING

So far we have treated intersymbol interference as an effect that must be eliminated. This requires the design of ideal filter functions and perfectly timed systems if transmission is to be with Nyquist bandwidths or requires excess bandwidth if more realistic filters are used. An alternate approach is to use realizable rolloff filters with the Nyquist bandwidth and, rather than eliminate intersymbol effects, produce a controlled amount of interference. The decoder can then be designed around this effect. Suppose that we consider a decoder signal waveform $s(t)$ that, at the bit sampling instants, has the values

$$s(jT_b) = \begin{cases} s_0, & j = 0, 1 \\ 0, & j \geq 2 \end{cases} \qquad (8.5.1)$$

Note that since two samples of $s(t)$ are nonzero we are now allowing some intersymbol interference to be generated from the previous bit (but no other bit). This means that in sampling a given bit for decoding, the sample value of the previous bit is added (Figure 8.18). Thus, in the absence of noise, the present sample value will be $\pm 2s_0$ if the previous bit matched the present bit, and will be zero if not. Having already decided the previous bit, a decision can be made on the present bit by deciding whether the signal sample value is zero or $\pm 2s_0$ (the sign determined from the previous bit decision). Thus the decoder must decide whether the sample is above or below $\pm s_0$, rather than the usual binary decision (whether the sample is positive or negative). The former usually incurs a penalty in bit error probability, but has the advantage of using signals with the nonideal Nyquist bandwidth. This encoding–decoding procedure is referred to as *partial response* signaling (since the ideal Nyquist filter response is not being entirely utilized), or sometimes called *duobinary* signaling [12].

A waveform that satisfies the condition in (8.5.1) while requiring a nonideal Nyquist bandwidth is the waveform whose transform is

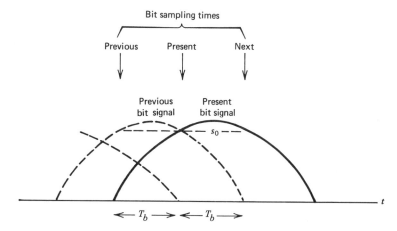

FIGURE 8.18. Partial response signaling. One-bit interference.

$$F_s(\omega) = \begin{cases} 2T_b \cos\left(\dfrac{\omega T_b}{2}\right), & |\omega| \le \dfrac{\pi}{T_b} \\ 0, & \text{elsewhere} \end{cases} \qquad (8.5.2)$$

and is shown in Figure 8.19a. Note the frequency function occupies a Nyquist bandwidth of $1/2T_b$ Hz with no excess, while the time transform (Figure 8.19b) has unit sample values at $\pm T_b/2$, with zero values at all other samples T_b seconds apart.

For the case of white noise $[S_n(\omega) = N_0/2]$, $H(\omega)$ in (8.4.13) and $Z(\omega)$ in (8.4.14) become

$$H(\omega) = [\cos(\omega T_b/2)]^{1/2}, \qquad\qquad |\omega| \le \dfrac{2\pi}{2T_b} \qquad (8.5.3)$$

and

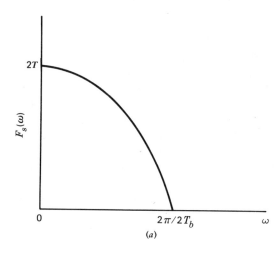

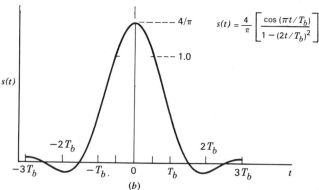

FIGURE 8.19. Partial response signal pulse: (a) frequency spectrum, (b) time function.

$$Z(\omega) = \left(\frac{E_b \pi T_b}{2}\right)^{1/2} [\cos{(\omega T_b/2)}]^{1/2}, \quad |\omega| \le \frac{2\pi}{2T_b} \qquad (8.5.4)$$

Here $Z(\omega)$ is normalized to energy E_b. The signal transform at the decoder sampler is then $H(\omega)Z(\omega)$ and, from Figure 8.19, will have the time sample value at $\pm T_b/2$ of

$$s_0 = \sqrt{E_b \pi/8T_b} \qquad (8.5.5)$$

The noise sample at each sample time has the variance in (8.4.11),

$$\sigma_{ny}^2 = \left(\frac{N_0}{2}\right) \frac{1}{2\pi} \int_{-2\pi/2T_b}^{2\pi/2T_b} \cos{\frac{\omega T_b}{2}} \, d\omega$$

$$= \frac{N_0}{\pi T_b} \qquad (8.5.6)$$

The probability that the sample value will be incorrectly decided as a zero or not will then be

$$\text{PE} = Q\left[\left(\frac{s_0^2}{\sigma_{ny}^2}\right)^{1/2}\right]$$

$$= Q\left[\left(\frac{\pi^2 E_b}{8N_0}\right)^{1/2}\right] \qquad (8.5.7)$$

This then becomes the probability of decoding the present bit with partial response signals when the previous bit had been decoded correctly. If there were no intersymbol interference (wide bandwidths) with the same signal energy per bit the coherent decoding PE would be $Q(2E_b/N_0)^{1/2}$. Hence partial response signaling at the Nyquist rate causes a $2/(\pi^2/8) = 1.62 = 2.1$ dB penalty in required power.

Partial response signals can be combined with amplitude modulation for combating bandlimiting bandpass channels as well. The direct amplitude modulation of the partial response signal having transform $Z(\omega)$ onto a carrier produces a resulting carrier spectrum that occupies a bandpass bandwidth of twice the Nyquist band, or $2(1/2T_b) = 1/T_b$ Hz wide. Coherent carrier mixing to baseband followed by the $H(\omega)$ filtering and bit sampling will then produce the same PE performance in (8.5.7).

8.6 CODING AND ERROR CORRECTION

We have been considering binary digital transmission in which data bits are transmitted individually over the link. Optimal and suboptimal decoders were studied and shown to produce the PE performance in Figure 8.4. In particular, we showed that the coherent, antipodal type of signaling produced the best performance among all binary signaling schemes.

Suppose, however, that the performance of even this optimal method is not satisfactory for a specific application. We may require a smaller PE at a given E_b/N_0, or we may want a specified PE at a smaller E_b/N_0. This means that we seek operation within the region of Figure 8.4, bounded by the optimal binary performance we previously derived. Since no binary scheme can improve on coherent, antipodal signaling, operation within the optimal performance appears to be unattainable by any direct binary digital system. However, improved PE performance can occur by resorting to some form of coding. In coded systems, bits are transmitted and decoded in blocks rather than one at a time. Block coding can be obtained in one of two basic ways: (1) channel coding (also referred to as *error correction*), which is utilized with a binary transmission link; and (2) block waveform encoding, which requires a complete modification of the transmit encoder and receiver decoder. Channel coding is considered here, while block waveform encoding is presented in Chapter 9.

Channel coding operates by inserting a level of coding prior to the actual binary waveform encoding considered in this chapter, as shown in Figure 8.20. The channel coder converts the data bits to new binary symbols, which we call *chips*. The binary chips are then transmitted over the link via standard binary encoding and decoding. At the receiver the decoded chips are then further processed via digital logic circuitry to reconstruct the data bits. This additional level of channel coding helps to improve the PE of the overall link.

To visualize this, consider a simple version of channel coding in which each data bit is repeated, say, three times prior to transmission. This means, for example, that the bit sequence 101 would be coded into the binary sequence 1 1 1 0 0 0 1 1 1 during channel encoding. Since the latter symbols do not truly represent data bits (a group of 3 represents only one data bit), we call them chips instead of bits. The chips are now sent over a binary channel by any of the waveform encoding methods considered, sending one chip at a time in sequence. Thus the actual digital communication link is not altered. Suppose the link power levels are such that the resulting chip error probability is PE_c, obtained from curves as in Figure 8.4. After the chip

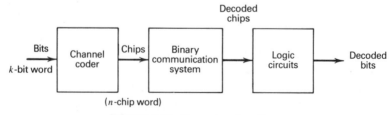

FIGURE 8.20. Channel coding diagram.

decoding, the receiver logic reconstructs the data bits by using the fact that 3 in a row represent the true bit (it is assumed that synchronization extends over the chip sequence so that the chip groupings for each bit are known). It is natural for the receiver logic to use "majority rule," deciding the bit corresponding to which decoded chip appears most often in each group. A data bit error will occur only if at least 2 chips are in error. Considering all the ways in which exactly 2 or 3 chip errors can occur in a group of 3 chips, the bit error probability is then

$$PE = 3PE_c^2(1 - PE_c) + (PE_c)^3 \tag{8.6.1}$$

Since PE_c is generally small ($\leq 10^{-2}$), $PE \approx 3PE_c^2$. Hence the bit error probability is significantly less than the PE_c that would occur if the bits were sent directly ($PE = PE_c$). We have therefore improved the overall bit performance with the 3 for 1 channel coding. In fact, if exactly one chip error occurred in a chip word, the resulting bit word would still be decoded correctly. In effect, the channel coding has "corrected" for the forward link chip error.

The disadvantage, of course, is that we have sent only one bit in 3 chip times, so the data rate is only one-third of the link chip rate. Conversely, we would have to increase the chip rate (increase link bandwidth) by 3 to maintain the data rate with the new PE. Hence, PE improvement is obtained at the expense of data rate or bandwidth.

Simply repeating bits is, in fact, not the best way to improve performance. Instead, we can consider channel coding blocks of data bits into blocks of chips. Consider the example summarized in Table 8.1. A block of 2 bits is coded into the block of 5 chips as shown. Decoding is accomplished by comparing the five decoded chips (which may have errors) with each of the four chip words, and selecting the word that differs in the fewest chips. The chip words have the property that a single chip error in decoding will still be recognized as the correct chip word (since the decoded word will still lie closest to the correct word). The correct block of 2 bits will therefore be decided, and we have again corrected for the transmission chip error. However, we have achieved this with a coding rate of only 5 chips/2 bits = 2.5 (instead of the 3 in the previous case).

TABLE 8.1 Bit Word Encoding Example for Two-Bit Words

Bit Word	Five Chip Coded Word
00	00000
01	01110
10	10111
11	11001

We can view the bit repeating example as an elementary coding scheme in which we are appending two extra chips to the original bit to form the repeated chip word. These extra chips are often referred to as *parity* or *check* symbols, and they allow for the eventual chip word correction. The example in Table 8.1 shows three check symbols being added to the original two bits to form the chip words. To generalize , a block of k bits can be coded into n chips by adding $n - k$ check symbols to the original k bits. If the check symbols are properly selected, the chip word will have inherent word correction capability.

This correction capability can now be unified into a basic error correction coding theory [13–17]. To review this briefly, we define a code set as the set of all chip blocks used by the channel coder for its output. The chip words in Table 8.1 constitute such a code set. We define the *Hamming distance* between any two chip words as the number of positions where corresponding bits differ. The code set distance is the smallest distance among all pairs of codes in the set. (The code set in Table 8.1 has distance 3.) A code set with distance d can always correct $(d - 1)/2$ chip errors and can make a bit word error only with $d - 1$ or more chip errors.

The code set in Table 8.1 is referred to as a *(5, 2) code set*, since it uses words of five chip symbols to encode the original two bits. The notion of code set distance allows us to generalize error correction coding to arbitrary (n, k) codes. We see that we will be basically interested in the relation between n, k and the code set distance d. For a given n and k, we would like to find the particular code set with the largest distance. On the other hand, for a specific distance (error correcting capability), we would like to determine the smallest (n, k) combination (simplest coding). The development of the theory of this mapping is beyond our scope here, but for properly selected code (mapping) rules, the coding distance d will be related to the number of check symbols, that is, to the difference $n - k$. The larger this difference, the more chip errors can be made before data bit errors occur. Unfortunately, the larger the values of n and k, the more complex is the implementation of the required coding and decoding hardware. Some of the more common block codes typically used in modern digital systems are listed in Table 8.2. This shows the relationship between the chip block size n, the bit block size k, and the code distance d for each type of block code.

The channel coding operation is physically implemented by interconnecting read-in/read-out shift registers, as shown in Figure 8.21. The data bits are first shifted sequentially into the k-stage input register. When the register is filled, the stored bits are logically combined to form the n chips that fill the n-stage output register. The combining logic is based on the type of code and converts the k bits to the n chips by determining the $n - k$ check symbols that are to be appended to the original data bits. The chip word is then clocked out of the output register chip by chip for link transmission. As this register is emptying, the next block of k bits is shifted into the input register. Clearly the read-out chip rate must be n/k times faster than the bit

TABLE 8.2 Common Block Code Parameters

Type of Code	Chip Block Size, n	Bit Block Size, k	Code Distance, d
Hamming	$2^l - 1$, $l = 2, 3, \ldots$	$2^l - 1 - l$	3
Extended Hamming	2^l	$2^{l-1} - l$	4
Maximal length	$2^l - 1$, $l = $ integer	l	2^{l-1}
Reed–Muller	2^m, $m = $ integer	$\sum_{j=0}^{r} \binom{m}{j}$, $r = 1, 2, 3, \ldots, r < m$	2^{m-r}
Golay	23	12	7
	24	12	8
BCH	$2^m - 1$, $m = $ integer	$2^m - 1 - mt$, $t = 1, 2, 3, \ldots$	$2t + 1$
Reed–Solomon	$m(2^m - 1)$, $m = $ integer	$m(2^m - 1) - 2mt$, $t = $ integer	$2t + 1$

read-in rate to maintain continuous coding. Hence channel coding rates are directly related to register clocking rates, register size, and the speed at which logical functions can be performed.

Coding performance can be determined by considering again the generalized coder of Figure 8.20. A block of k data bits is coded into a block of n chips, $n > k$, with an (n, k) code set that has distance d. There are n/k chips per bit, and the code rate is defined as

$$ r = \frac{k}{n} \tag{8.6.2} $$

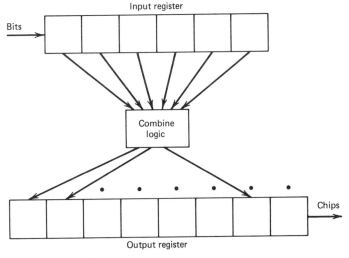

FIGURE 8.21. BLock coder implementation.

Each consecutive block of k data bits is converted into a block of n chips, according to the prescribed coding rule. The chips are then sent over the binary channel one at a time. If the chips are sent at a rate R_c chips/sec, the transmitted data rate is

$$R_b = rR_c \quad \text{bits/sec} \tag{8.6.3}$$

Similarly, if each chip is sent with energy E_c, the effective bit energy is

$$E_b = E_c/r \tag{8.6.4}$$

The probability that a transmitted chip will be decoded in error depends on the chip binary transmission format, as was indicated in Figure 8.4. For example, with a BPSK format, the chip error probability is given by

$$\text{PE}_c = Q\left[\left(\frac{2E_c}{N_0}\right)^{1/2}\right]$$

$$= Q\left[\left(\frac{2E_b r}{N_0}\right)^{1/2}\right] \tag{8.6.5}$$

Since the code set has distance d, a chip word will be corrected if $t \stackrel{\Delta}{=} (d-1)/2$ or less chip errors occur during transmission. Hence the probability of a chip word error PWE is then

$$\text{PWE} = \sum_{j=t+1}^{n} \binom{n}{j}(\text{PE}_c)^j(1 - \text{PE}_c)^{n-j} \tag{8.6.6}$$

Since PE_c is generally small ($<10^{-3}$), the summation is dominated by the term corresponding to the smallest value of j, and we can accurately approximate as

$$\text{PWE} \approx \binom{n}{t+1}(\text{PE}_c)^{t+1} \tag{8.6.7}$$

Clearly, the larger the code distance d, the lower the probability of a chip word error. If a chip word error is made, it will most likely have d out of its n chips in error, and the individual chip error probability is approximately $(d/n)\text{PWE}$. This means that, on the average, there will be $(k/n)(d\text{PWE})$ bits in error for each k bit block. This identifies the average bit error probability as

$$\text{PE} = \left(\frac{d}{n}\right)\text{PWE} \tag{8.6.8}$$

Thus the data bit error probability of the block coded system can be computed directly from the chip word error probability and the code parameters. It can be shown that this PE can be accurately bounded by [8]

$$PE \le 2^{-n(r_0 - r)} \qquad (8.6.9)$$

where r_0 is the chip *cutoff rate*, defined by

$$r_0 = 1 - \log_2 \left[1 + \sqrt{4PE_c(1 - PE_c)}\right] \qquad (8.6.10)$$

The bit error probability can therefore be easily estimated directly from the code parameters. Note that as long as the code rate r is less than the cutoff rate r_0 [the latter dependent on the chip energy through (8.6.5) and (8.6.10), the bit error probability can be reduced by increasing the chip block length n (and therefore also increasing data block length k, since $k = rn$). This simply reiterates the fact that for a fixed code rate k/n, the larger the values of k and n, the larger is the difference $n - k$ and the more chip errors can be tolerated. The number of computations needed to convert the decoded chips into the decoded bits, however, increases as 2^k. Thus decoding complexity and required processing speed increases exponentially with error correction block lengths.

Figure 8.22 plots the bound in (8.6.9) as a function of block code length k for several values of r/r_0. Typically, $PE_c \le 10^{-3}$, for which r_0 in (8.6.10) is about 1, and the ratio r/r_0 is approximately equal to the code rate r in

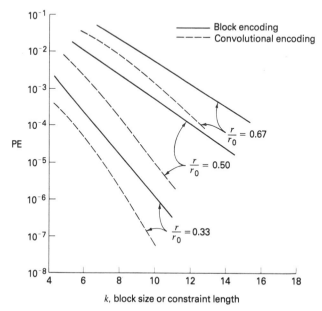

FIGURE 8.22. PE bound versus block or constraint length k (r = code rate, r_0 = channel cut-off rate).

(8.6.2). Note the continual improvement in performance as higher levels of coding are introduced at a fixed rate. However, as stated earlier, the number of computations needed also increases exponentially with k. Thus another price to be paid for PE improvement via channel coding is in the decoding complexity that must be implemented. As digital processing capabilities and computation speeeds increase through technological advances, higher levels of block lengths can be physically implemented, leading directly to digital performance improvement.

An alternative to the block coding is the use of *convolutional coding* [6, 8, 18]. Convolutional coding is another technique for coding data bits into transmission chips, such that subsequent chip decoding leads to improved bit decoding. Rather than coding each disjoint data block into disjoint chip blocks, each chip is generated instead through a continuous combination of past bits. As the input bits are shifted into the convolutional coder, the output chips are generated continuously in real time rather than in blocks.

A convolutional coder is also constructed from shift registers and logic combiners, as shown in Figure 8.23. The source bits to be coded are shifted into the register in a clocked sequence and dumped out at the end. The register stages are binary summed in V separate modulo 2 adders, each connected to certain register taps. At preselected clock times, a commutating switch reads out the output of the V adders in sequence, forming a V bit word at this clock time. As the source sequence shifts in, the clocking of the commutator is repeated and another V bit output occurs, with the succession of V bit words forming the encoded output. The commutator clocking is performed after, say, k bits have shifted into the register (and k bits have been dumped out the end). For each k source bits there are therefore V

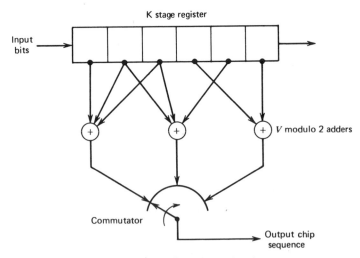

FIGURE 8.23. Convolutional coder.

output bits produced, and the coder operates at an output rate of V/k times the input rate. This is called a *rate k/V convolutional coder*. Typically, $k = 1$ (a commutation performed after each input source bit) and $V = 2$ or 3. When coding is accomplished in this manner, the output sequence can be written as a convolution of the input sequence and a coder weighting function, hence the name convolutional coding.

As a given source bit shifts through a K stage register we see that it will affect $(K/k)V$ successive output bits via the register summation logic. This parameter, KV/k, is called the *constraint length*, or *memory time*, of the coder. Recall that in block coding each input bit influenced only the chips in one specific output block. Hence constraint length in convolutional codes is somewhat similar to block size in block coding. The manner in which V adders are connected determines the structure of the output sequence. Various studies have been made to evaluate the distance properties of convolutional codes generated by specific register tap connections [18].

Convolutional decoding is achieved at the receiver by detecting each chip in sequence, and decoding the data bits based on the processing of the resulting chip sequence. Optimal decoding is obtained by means of a *Viterbi decoding algorithm* [19, 20] which unconvolves the bits from the decoded chips. Convolutional decoding algorithms are primarily *search* algorithms that attempt to keep track of the most probable bit sequences that could have produced the decoded chip sequence.

The convolving property of convolutional coding produces codes that can be decoded with slightly better bit error performance than previous block codes with the same rate and constraint length. Performance analysis is more difficult, however, due to the interlaced nature of the coded bits. It has been shown [8, 19] that for a code rate r and constraint length k there exists a convolutional code with bit error probability bounded by

$$PE \leq \frac{2^{-k(r_0/r)}}{[1 - 2^{-(r_0/r - 1)}]^2} \tag{8.6.11}$$

where r_0 is again given in (8.6.10). Figure 8.22 also shows (8.6.11) for convolutional codes, showing the potential advantage over fixed-length block codes. Convolutional decoding can also be improved by "soft-decisioning" decoding algorithms, which allow the decoder to achieve a finer granularity during the chip processing. This results in an improved chip cutoff rate of approximately

$$r_0 = 1 - \log_2 [1 + e^{-rE_b/N_0}] \tag{8.6.12}$$

as compared to (8.6.10). This soft-decision rate yields the same values of r_0 in (8.6.10) with about 2 dB less E_b. Hence, when entering the convolutional coding curves in Figure 8.22, and reading PE_b at a specific constraint length k and ratio r/r_0, we can achieve the resulting performance with 2 dB less bit

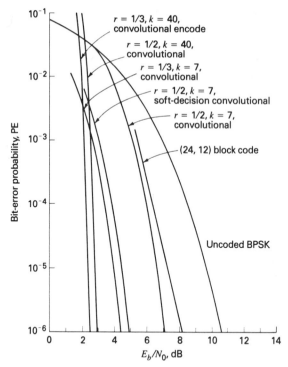

FIGURE 8.24. PE versus E_b/N_0 with channel coding.

energy if soft-decisioning decoding is used. Soft-decision processing, however, further increases the required decoder complexity.

Channel coding, with either block codes or convolutional codes, directly improves the bit-error probability. Figure 8.24 shows the expected improvement over standard BPSK obtained by coding with various code rates and block sizes or constraint lengths k as a function of channel E_b/N_0. This improvement can be interpreted as a lower PE at the same E_b/N_0, or the same PE at lower E_b/N_0, as higher levels of coding are used. Again, the reciever processor complexity required in the block decoding or Viterbi decoder algorithms also increases with the length k, placing practical limits on achievable performance. It is expected that this thrust toward faster decoding processing and reduced PE will continue throughout the future of digital communications.

8.7 CODING FOR BURST ERRORS

The error probability improvements obtained via the error correction codes of Section 8.6 were based on the assumption that random chip errors will occur throughout the transmission. Thus there is an extremely low probability that more than one or two chip errors will occur in a given chip word. If, however, the communication channel is a fading, or multipath, link, then a deep power fade over a relatively long time period can cause an error burst to occur. This means a consecutive string of chips, perhaps 100 to 1000 long, transmitted to the receiver during the fade will tend to be decoded erroneously. The insertion of coding, designed to have a correction capability of only a few errors per chip word, would be useless during the fade. To use a more powerful code having the correction capability to cover the expected burst length would require a prohibitively long block length and extensive decoder processing. Furthermore, for the majority of the time (when the fade doesn't occur) the excessive coding hardware is unnecessary and inefficient.

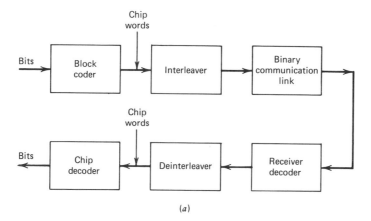

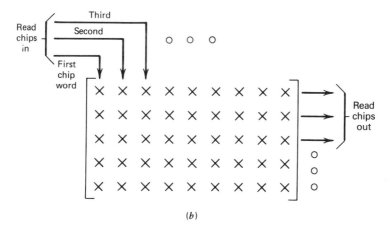

FIGURE 8.25. Channel coding with interleaving and deinterleaving: (*a*) block diagram, (*b*) interleaver matrix for read-in and read-out (x denotes chips).

A solution to this burst error problem is the use of *interleaving* and *deinterleaving*. The latter corresponds to an additional level of relatively simple digital processing inserted at both the transmitter and receiver to combat the error bursts. An interleaver and deinterleaver is simply a matrix of read-in/read-out shift registers in which chips can be sequentially read in to fill rows or columns, then read out sequentially by either rows or columns.

An interleaved digital system is shown in Figure 8.25*a* and operates as follows. The source data bits are first coded into *n*-chip words by standard block coding operations. The sequence of chip words then fill the *columns* of transmit interleaver (Figure 8.25*b*). When the interleaver matrix is filled, its *rows* are read-out in order, and the resulting chip sequence transmitted to the receiver over the binary communication link. At the receiver the decoded chips are used to fill the *rows* of the deinterleaver. When the latter is filled, its *columns* are read out in sequence to regenerate the original coded chip sequence generated at the transmitter. Except for the additional time to fill and read out the interleaver and deinterleaver, the overall link is theoretically transparent to their presence.

Now assume a fade occurs during chip transmission, lasting over N consecutive chips. This means that N consecutive entries of some row of the deinterleaver are incorrect (assuming no other chip error occur), as shown in Figure 8.26, for example. When the chips are read-out by columns, however, the resultant chip words will have at most one error. Hence a relatively simple error correcting code used to form the transmitter chip words can correct for the entire N chip fade. (If the N chips straddled two rows, then a slightly stronger code is needed, but this event will occur with

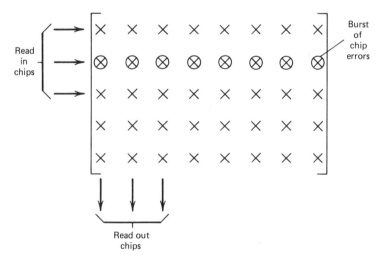

FIGURE 8.26. Deinterleaver matrix with error burst.

low probability if the row length greatly exceeded the fade length.) This interleaving permits simple codes to combat long fades.

It is clear from Figure 8.26 that the height (number of rows) of the interleaving matrix must accommodate the coded chip word size, while the width (number of columns) should exceed the expected fade duration. As a rough rule, an interleaver with width w and height h, storing code words with correction capability t, can correct for fades of length wt. However, the interleaving matrices cannot be too large since it must fill in a time period shorter than the time between fades (so only one fade will occur during each interleaving time). In addition, large interleavers increase the delay time required to read in and read out the registers. Hence, interleaver design requires careful balance of fade duration and fade rate, and generally reduces to a tradeoff of burst length correction versus overall link delay.

REFERENCES

1. Urkowitz, H. "Energy Detection of Unknown Deterministic Signals," Proc. IEEE, Vol. 17, April 1967, pp. 523–531.

2. Marcum, C. "Tables of the Q Function," *Rand Corp. Res. Memo*, **RN-399**, January 1950.

3. Brennan, L. and Reed, I. "A Recursive Method for Computing the Q Function," *IEEE Trans. Inf. Theory*, vol. IT-11, April 1965, p. 312.

4. Shimbo, O. and Celebiler, R. "Probability of Error due to Insymbol Interference in Digital Systems," *IEEE Trans. Comm. Tech.*, vol. COM-19, 1971, pp. 113–119.

5. Ho, E. and Yeh, Y. "A New Approach for Evaluating Error Probability with Intersymbol Interference," *Bell Syst. Tech. J.*, vol. 49, 1970, pp. 2249–2265.

6. Benedetto, S., Biglieri, E. and Castellani, V. *Digital Transmission Theory*, Prentice-Hall, Englewood Cliffs, N.J., 1987.

7. Yao, K. and Biglieri, E. "Moment Error Bounds for Digital Communication," *IEEE Trans. Inf. Theory*, vol. IT-26, 1980, pp. 454–464.

8. Viterbi, A.J. and Omura, J.K. *Principles of Digital Communication and Coding*, McGraw-Hill, New York, 1979.

9. Hartmann, H. "Degradation of SNR Due to Filtering," *IEEE Trans. Aerosp. Electron. Syst.*, vol. AES-4, January 1969, pp. 22–32.

10. Ziemer, R. and Tranter, W. *Principles of Communications*, Houghton Mifflin, Boston, Mass., 1976.

11. *Transmission Systems for Communications*, 4th ed., Bell Telephone Laboratories, Winston-Salem, N.C., 1970.

12. Lucky, R., Salz, J. and Weldon, E. *Principles of Data Transmission*, McGraw-Hill, New York, 1968.

13. Lin, S. *An Introduction to Error Correcting Codes*, Prentice-Hall, Englewood Cliffs, N.J., 1970.

14. Lin, S. and Costello, D. *Error Control Coding*, Prentice-Hall, Englewood Cliffs, N.J., 1983.

15. McEliece, R. *Theory of Information and Coding*, Addison-Wesley, Reading, Mass., 1977.

16. MacWilliams, F. and Sloane, N. *Theory of Error Correcting Codes*, North-Holland, Amsterdam.

17. Clark, G.C. and Cain, J.B. *Error-Correction Coding for Digital Communications*, Plenum, New York and London, 1981.

18. Bussgang, J. "Properties of Binary Convolutional Codes Generators," *IEEE Trans. Inf. Theory*, vol. IT-11, January 1965, pp. 90–100.

19. Viterbi, A.J. "Error Bounds for Convolutional Codes and an Asymptotically Optimum Decoding Algorithm," *IEEE Trans. Inf. Theory*, vol. IT-13, 1967, pp. 260–269.

20. Forney, D. "The Viterbi Algorithm," *Proc. IEEE*, vol. 61, March 1973.

PROBLEMS

8.1 (8.1) Let $x(t) = e_1(t) \cos(\omega_1 t + \psi) + n(t)$,

$$0 \leq t \leq T_b, \qquad \omega_1 T_b \gg 1$$

where ψ is a random variable and $n(t)$ is a white zero mean Gaussian process of one-sided level N_0. Define

$$\binom{Z_1}{Y_1} = \int_0^{T_b} x(t) e_1(t) \binom{\cos}{\sin} [\omega_1 t] \, dt$$

$$\binom{Z_0}{Y_0} = \int_0^{T_b} x(t) e_0(t) \binom{\cos}{\sin} [\omega_0 t] \, dt$$

(a) Find the mean values and variances of Z_1, Y_1, Z_0, and Y_0, conditioned on a given ψ. (b) Find the covariance of the pairs (Z_1, Y_1), (Z_1, Z_0), (Z_0, Y_0), (Y_1, Y_0), and (Z_1, Y_0), conditioned on a given ψ. (c) Write the conditional joint probability densities

$$p(Z_1, Y_1 | \psi) \quad \text{and} \quad p(Z_0, Y_0 | \psi)$$

(d) Define

$$E_i = (Z_i^2 + Y_i^2)^{1/2}$$

and write the densities $p(E_i | \psi)$, $i = 0, 1$.

8.2 (8.1) Verify (8.1.24) by formally integrating (8.1.22) using (8.1.23) and showing that the resulting integration is

$$\text{PE} = \tfrac{1}{2}e^{-\rho^2/4}$$

Hint: use the fact that the integral of (8.1.23a) is 1.

8.3 (8.1)   Consider the detection of a pulse. If the pulse is present, it is received as

$$s(t) = b\cos(\omega_0 t + \psi), \qquad 0 \le t \le T_b, \; \omega_0 T_b \gg 1$$

where ψ is a random phase and b is a random amplitude. Derive the integral for the test for detecting the pulse in AWGN, assuming that ψ is uniform and b is independent of ψ with density $p(b)$.

8.4 (8.1)   Consider the signal set in (8.1.1) with $e_i(t) = A$, and let ψ be a random phase with the density

$$p(\psi) = C_1 e^{-C_2 \cos \psi}, \qquad |\psi| \le \pi$$

From knowledge of the classical noncoherent solution (uniform phase) in Section 8.1, derive a realization of the required decoder.

8.5 (8.2)   Consider the DPSK decoder in Figure 8.11. Assume that an energy per bit E_b, zero mean white Gaussian noise at the input, and the noise at the BPF output in two consecutive bit intervals is uncorrelated. (a) Determine the mean and variance of y. (b) Assume that y is Gaussian and write the PE of the DPSK decoder above.

8.6 (8.3)   Let PE depend on a random variable D, and denote it as $\text{PE}(D)$. Show that the mean PE, $\overline{\text{PE}}$, can be estimated as

$$\overline{\text{PE}} = \text{PE}(\bar{D}) + \frac{\text{Var } D}{2}\,\text{PE}''(\bar{D})$$

where the primes denote derivatives, and $\bar{D}$ is the mean of D. Begin by expanding $\text{PE}(D)$ in a Taylor series about $\bar{D}$.

8.7 (8.3)   Show that the worst case degradation occurs in (8.3.4) when the sequence of prior bits $(d_1, d_2, \ldots)$ are defined by

$$d_i = \mp d_0 \quad \text{if} \quad \left[\int_0^{T_b} s(t + iT)\,dt\right]\left[\int_0^{T_b} s(t)\,dt\right] \ge 0$$

Show that for this condition,

$$T_b(m_0 + m_I) = \int_0^{T_b} s(t)\,dt - \sum_n \left| \int_{nT_b}^{(n+1)T_b} s(t)\,dt \right|$$

8.8 (8.6) Show that for any (n, k) code, the minimum distance must satisfy the inequality

$$2^k \le \frac{2^n}{\displaystyle\sum_{i=0}^{(d_{min}-1/2)} \binom{n}{i}}$$

8.9 (8.6) A linear (n, k) code set is one formed by starting with k distinct binary words of length n and adding all module two linear combinations to these k to form the entire 2^k set. Show that for any linear code set, the minimum Hamming distance of the set is equal to the minimum number of ones occurring in all words of the set (excluding the all zero word).

9

BLOCK WAVEFORM
DIGITAL SYSTEMS

In digital systems using block waveform encoding, source bits are transmitted in blocks of bits in sequence to the receiver. These blocks of bits are often referred to as *binary words*. Each word is encoded into a distinct baseband waveform, and receiver decoding of the entire word is accomplished by attempting to recognize which waveform is being decoded during each block interval. In this chapter we study the basic charcterics, design procedures, and performance tradeoffs in practical block waveform digital systems. One of the most basic questions facing a system designer is whether block waveform encoding or standard binary encoding should be used. This decision requires that the designer be aware of the prime advantages of these techniques and the cost or penalty, if any, required to obtain these advantages. We shall emphasize these points throughout our investigation.

9.1 THE BLOCK WAVEFORM SYSTEM

The general block waveform digital system is shown in Figure 9.1. The data bits are separated into consecutive blocks, or bit words, and the entire block is encoded into a waveform for transmission. Since a block of k bits can take $M = 2^k$ different forms, there must be M distinct signals available in the encoder. We assume that each such waveform is T_w sec long, where T_w now represents the word time, or the time to transmit k bits. For example, the block encoding operation for $k = 2$ would be represented by

$$\begin{aligned} 11 &\to s_1(t)\,, \\ 10 &\to s_2(t)\,, \\ 01 &\to s_3(t)\,, \\ 00 &\to s_4(t)\,, \end{aligned} \qquad 0 \le t \le T_w \qquad (9.1.1)$$

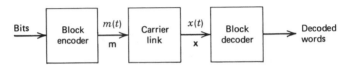

FIGURE 9.1. Block waveform encoding system.

Each consecutive block of two bits would be encoded into a waveform $s_i(t)$ according to (9.1.1). The block waveform encoder therefore produces a baseband signal $m(t)$ corresponding to a sequence of T_w sec waveforms generated from the sequence of k-bit blocks.

The encoded waveform is then sent over the carrier link and demodulated to recover the baseband. The decoder processes the received baseband during each T_w sec word, deciding the received waveform $s_i(t)$ and decoding the corresponding k-bit block. A correct decision now represents the correct decoding of k consecutive bits. A decision error selects an incorrect bit word. Note that with block waveform encoding, the encoder and decoder deal with a set of M waveforms $\{s_i(t)\}$, as contrasted with block channel coding described in Section 8.6, in which the transmission is still binary.

Decoding timing for the block intervals is provided by the auxiliary timing subsystem. We again assume (1) each bit combination during a block is equally likely; (2) the baseband noise is additive, Gaussian, and white with one-sided level N_0; and (3) the baseband signals satisfy the energy constraint

$$\int_0^{T_w} s_i^2(t)\, dt = E_w, \qquad i = 1, 2, \dots, M \qquad (9.1.2)$$

where E_w now represents the block, or word, energy.

The bandwidth required in the baseband subsystem to transmit the encoded signal can be computed just as in (2.1.25). If $s_i(t)$, $i = 1, 2, \dots, M$, are the M encoder baseband signals to be used and $F_{s_i}(\omega)$ is their individual transform and P_i their probability of being used, then the baseband spectrum is given by (Problem 9.1).

$$S_m(\omega) = \frac{1}{T_w} \sum_{i=1}^M P_i |F_{si}(\omega)|^2 - \frac{1}{T_w} \left[\sum_{i=1}^M P_i F_{si}(\omega) \right]\left[\sum_{i=1}^M P_i F_{si}^*(\omega) \right]$$

$$+ \frac{2\pi}{T_w^2} \sum_{j=-\infty}^{\infty} \left| \sum_{i=1}^M P_i F_{si}\left(\frac{2\pi j}{T_w}\right) \right|^2 \delta\left(\omega - \frac{2\pi j}{T_w}\right) \qquad (9.1.3)$$

The spectrum is therefore directly related to the individual signal transforms and contains the harmonic lines, just as with the binary transmission.

The optimal decoding problem can be formulated exactly as in the binary case in Chapter 7. If PWE_i is the probability a word decisioning error is made when word i is sent, the average probability of a word error by the decoder is

$$\text{PWE} = \frac{1}{M} \sum_{i=1}^{M} \text{PWE}_i \qquad (9.1.4)$$

For a vector observation $\mathbf{x}$ in $(0, T_w)$ at the decoder input, we can denote $P(i|\mathbf{x})$ as the probability of selecting block i when $\mathbf{x}$ is received, and let $p(\mathbf{x}|i)$ as the probability density of $\mathbf{x}$ when block i is sent. Then

$$\text{PWE}_i = \sum_{\substack{j=1 \\ j \neq i}}^{M} \int_{\mathbf{x}} P(j|\mathbf{x}) p(\mathbf{x}|i) \, d\mathbf{x} \qquad (9.1.5)$$

and

$$\sum_{j=1}^{M} P(j|\mathbf{x}) = 1 \qquad \text{for all } \mathbf{x} \qquad (9.1.6)$$

This means that we write

$$\text{PWE} = \frac{1}{M} \sum_{i=1}^{M} \sum_{\substack{j=1 \\ j \neq i}}^{M} \int_{\mathbf{x}} P(j|\mathbf{x}) p(\mathbf{x}|i) \, d\mathbf{x}$$

$$= \frac{1}{M} \sum_{i=1}^{M} \int_{\mathbf{x}} \sum_{j \neq i}^{M} P(j|\mathbf{x}) p(\mathbf{x}|i) \, d\mathbf{x}$$

$$= \frac{1}{M} \int_{\mathbf{x}} \sum_{i=1}^{M} [1 - P(i|\mathbf{x})] p(\mathbf{x}|i) \, d\mathbf{x} \qquad (9.1.7)$$

The set of $P(i|\mathbf{x})$ that will minimize the integral is that that minimizes the integrand at every $\mathbf{x}$. This requires that the set of probabilities $P(i|\mathbf{x})$, subject to the constraint of (9.1.6), maximize

$$\sum_{i=1}^{M} P(i|\mathbf{x}) p(\mathbf{x}|i) \qquad \text{at every } \mathbf{x} \qquad (9.1.8)$$

The maximization occurs for the solution

$$P(i|\mathbf{x}) = \begin{cases} 1 & \text{for the } i \text{ such } p(\mathbf{x}|i) \text{ is maximum} \\ 0 & \text{for all other } i \end{cases} \qquad (9.1.9)$$

Hence the optimal block decoder, on observing $\mathbf{x}$, should select the block corresponding to the largest value of $p(\mathbf{x}|i)$ over all i. This requires the decoder to compute all $p(\mathbf{x}|i)$ and compare for the maximum. (If there is more than one maximum, any one of the maxima may be used.) By our monotonic argument in Figure 7.3, we can equivalently select the maxima among $f[p(\mathbf{x}|i)]$ for any monotonic function f. In particular, we can compare $\log p(\mathbf{x}|i)$ functions.

9.2 BLOCK DECODING FOR THE GAUSSIAN CHANNEL

When the communication system in Figure 9.1 is an additive, Gaussian, white noise (spectral level, N_0) channel, the optimal decoder can be determined, as in Section 7.3. The observable $\mathbf{x}$ is a ν-dimensional Gaussian random vector, with its mean dependent on the transmitted waveform. From (7.3.1a), this has the density

$$
\begin{aligned}
p(\mathbf{x}|i) &= \frac{1}{(2\pi N_0/2)^{\nu/2}} \, e^{-\Sigma_{j=1}^{\nu} (x_j - s_{ij})^2/N_0} \\
&= C e^{(2/N_0) \Sigma_{j=1}^{\nu} x_j s_{ij}} \\
&= C e^{(2/N_0) \int_0^{T_w} x(t) s_i(t)\, dt}
\end{aligned}
\tag{9.2.1}
$$

where $\{s_{ij}\}$ are the vector coordinates of waveform $s_i(t)$ and C is a constant not dependent on i. The optimal decoder that must select the maximum of $p(\mathbf{x}|i)$ over i is equivalent to one that selects the maximum among the parameters

$$
y_i = \int_0^{T_w} x(t) s_i(t)\, dt \,, \qquad i = 1, 2, \ldots, M
\tag{9.2.2}
$$

We recognize (9.2.2) as the coherent correlation of the input $x(t)$ with the waveform $s_i(t)$ over the block interval $(0, T_w)$. The optimal decoder therefore has the form shown in Figure 9.2, in which each parallel channel computes one of the $\{y_i\}$, and a decoding decision is based on the largest of these. Physically, the decoder again operates by having each channel of the

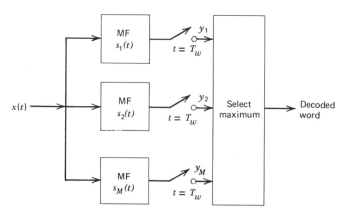

FIGURE 9.2. Block decoder.

decoder look for a particular signal of the encoder set, and the largest value is selected as the received signal. Note that the maximum y_i in (9.2.2) corresponds to the minimal value of the quantity

$$\int_0^{T_w} [x(t) - s_i(t)]^2 \, dt = \int_0^{T_w} x^2(t) \, dt + E_w - 2y_i \qquad (9.2.3)$$

over all i. The integral on the left is the integrated squared difference between the noisy input $x(t)$ and the signal $s_i(t)$. Hence optimal block decoding (selecting the largest y_i) is equivalent to determining to which $s_i(t)$ the input is "closest." This decoder processing is repeated every word interval to decode the sequence of received blocks.

Performance of the optimal decoder can be obtained by evaluating the word error probability PWE in (9.1.4). To do this, we examine the behavior of each decoder channel during the decoding operation. The decoder input is written specifically as

$$x(t) = m(t) + n(t) \qquad (9.2.4)$$

as in (7.1.2). The ith channel computes y_i in (9.2.2). When $m(t) = s_j(t)$ (i.e., the jth signal is sent), the sample output y_i is

$$y_i|s_j = E_w \gamma_{ij} + \int_0^{T_w} n(t)s_i(t) \, dt \qquad (9.2.5)$$

where

$$\gamma_{ij} \stackrel{\Delta}{=} \frac{1}{E_w} \int_0^{T_w} s_i(t)s_j(t) \, dt , \qquad i \neq j$$
$$\gamma_{ii} = 1 \qquad (9.2.6)$$

The parameter γ_{ij}, $i \neq j$, is the cross-correlation coefficient of the signals $s_i(t)$ and $s_j(t)$. We see that $y_i|s_j$ is a Gaussian random variable with mean

$$\mathscr{E}[y_i|s_j] = E_w \gamma_{ij} \qquad (9.2.7)$$

and a pair of them have covariance

$$\mathscr{E}\{[y_i|s_j - \mathscr{E}(y_i|s_j)][y_k|s_j - \mathscr{E}(y_k|s_j)]\}$$
$$= \int_0^{T_w} \int_0^{T_w} \mathscr{E}[n(t)n(\rho)]s_i(t)s_k(\rho) \, dt \, d\rho$$
$$= \int_0^{T_w} \int_0^{T_w} \frac{N_0}{2} \delta(t - \rho)s_i(t)s_k(\rho) \, dt \, d\rho$$
$$= \left(\frac{N_0}{2}\right) E_w \gamma_{ik} \qquad (9.2.8)$$

The random variables $\{y_i|s_j\}$ are therefore not statistically independent. The probability of an error in decoding is the probability that an incorrect channel sample value exceeds the correct one. Thus

$$\text{PWE}|j = 1 - \text{Prob}\,[\,y_j|s_j \geq y_i|s_j\,], \qquad i = 1, 2, \ldots, M, \qquad i \neq j \qquad (9.2.9)$$

The probability on the right is simply the probability that one Gaussian variable exceeds $M - 1$ other Gaussian variables but requires an integration over an M-dimensional Gaussian density; that is,

$$\text{PWE}|j = 1 - \int_{-\infty}^{\infty} \int_{-\infty}^{y_j} \cdots \int_{-\infty}^{y_j} p(y_1, y_2, \ldots, y_M|s_j)\, dy_1\, dy_2 \cdots dy_j$$

$$(9.2.10)$$

when $p(y_1, y_2, \ldots, y_M|s_j)$ is the conditional joint Gaussian density of $y_1, \ldots, y_M$, given signal $s_j(t)$ is received. (Note the order in the integration, where the variable y_j is integrated last.) For our case the joint Gaussian density of the $\{y_i\}$ becomes

$$p(y_1, y_2, \ldots, y_M|s_j) = \frac{1}{(\pi E_w/N_0)^{M/2} |\det \Gamma|^{1/2}} \exp\left[-\tfrac{1}{2} Y_j^{\text{tr}} \Gamma^{-1} Y_j\right][2/E_w N_0]$$

$$(9.2.11)$$

Here Γ is the M by M covariance matrix [matrix of covariance values in (9.2.8)],

$$\Gamma = [\gamma_{ij}] \qquad (9.2.12)$$

$\det \Gamma$ is its determinant, Γ^{-1} its inverse, tr denotes *transpose*, and Y_j is the column vector

$$Y_j = \begin{bmatrix} y_1 - \mathcal{E}(y_1|s_j) \\ y_2 - \mathcal{E}(y_2|s_j) \\ \vdots \end{bmatrix} \qquad (9.2.13)$$

Integration of (9.2.10) for each s_j and insertion in the sum in (9.1.4) yields the desired average word error probability. Unfortunately, a general integration of the form (9.2.10) is difficult because of the dependence among the (y_i) and the necessity to invert the matrix Γ. Nevertheless, we do see that the resulting value of PWE depends only on the parameters E_w and N_0, and the matrix of signal cross-correlation values γ_{ij}. Thus the selection of the signal set for channel encoding influences the resulting system performance only through its cross-correlation values. In the next section we present several types of common signal sets and the matrices they generate in (9.2.12).

In many cases, we find exact calculation for PWE difficult to determine and we must often settle for approximation, simulation, or bounding methods. A simple bound often applied in block encoding analysis is the "union bound." This involves use of the fact that if $x_1, \ldots, x_M$ are M random numbers, the probability that x_1 is less than the $M-1$ remaining numbers is bounded from above by the union (sum) of probabilities that x_1 is less than each x_j individually (Problem 9.3). If the numbers $\{x_i\}$ correspond to the (y_i) when a particular word is sent, this union rule implies

$$\text{PWE}|j \leq \sum_{\substack{q=1 \\ q \neq j}}^{M} \text{Prob}\,[(y_j \leq y_q)|j] \tag{9.2.14}$$

where the probability is conditioned on the jth word being sent. The average word error probability then satisfies

$$\text{PWE} \leq \frac{1}{M} \sum_{j=1}^{M} \sum_{\substack{q=1 \\ q \neq j}}^{M} \text{Prob}\,[(y_j \leq y_q)|j] \tag{9.2.15}$$

By properly pairing terms, we can then write (9.2.15) as

$$\text{PWE} \leq \frac{1}{M} \sum_{j=1}^{M} \sum_{\substack{q=1 \\ q \neq j}}^{M} \{\tfrac{1}{2}\,\text{Prob}\,[(y_j \leq y_q)|j] + \tfrac{1}{2}\,\text{Prob}\,[(y_q \leq y_j)|q]\} \tag{9.2.16}$$

The term in the braces is the probability of erring in attempting to decide between a pair of the y_i and is therefore the error probability associated with a binary test. Since the latter can usually be evaluated, (9.2.16) is a useful upper bound to PWE. Its value is in the fact that we are guaranteed that performance must be at least this good. Since (9.2.16) involves a binary error probability, many of our results in the preceding chapter are immediately applicable for establishing this bound. For example, our discussion in Figure 7.9 showed that the error probability in deciding between two signals, $s_i(t)$ and $s_j(t)$, can be written as

$$\text{PE} = Q\!\left[\left(\frac{E_w d_{ij}^2}{2N_0}\right)^{1/2}\right] \tag{9.2.17}$$

where d_{ij} is the normalized distance between the signals (Figure 9.3)

$$d_{ij}^2 = \frac{1}{E_w} \int_0^{T_w} [s_i(t) - s_j(t)]^2\, dt \tag{9.2.18}$$

Hence (9.2.16) simplifies to

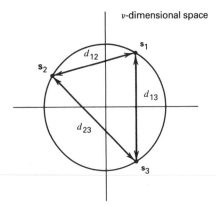

FIGURE 9.3. Signal vectors in ν-dimensional space.

$$\text{PWE} \le \frac{1}{M} \sum_{\substack{i \\ i \ne j}}^{M} \sum_{j}^{M} Q\left[\left(\frac{E_w d_{ij}^2}{2N_0} \right)^{1/2} \right] \qquad (9.2.19)$$

Furthermore, if we replace each d_{ij} by its minimum value over all signal pairs,

$$d_{\min} = \min_{\substack{i,j \\ i \ne j}} d_{ij} \qquad (9.2.20)$$

we retain the bound in (9.2.19) as

$$\text{PWE} \le \frac{1}{M} \sum_{\substack{i \\ i \ne j}}^{M} \sum_{j}^{M} Q\left[\left(\frac{E_w d_{\min}^2}{2N_0} \right)^{1/2} \right]$$

$$= \left(\frac{M^2 - M}{M} \right) Q\left[\left(\frac{E_w d_{\min}^2}{2N_0} \right)^{1/2} \right]$$

$$= (M - 1) Q\left[\left(\frac{E_w d_{\min}^2}{2N_0} \right)^{1/2} \right] \qquad (9.2.21)$$

This produces a simple upper bound to block waveform encoding performance that depends on only the minimum distance between the signals of the encoding set. Thus the signals of the set that are closest together in vector space determines the performance bound.

9.3   BLOCK ENCODING SIGNALING SETS

When using a coherent decoder with white noise interference, we have found that the resulting performance in terms of PWE will depend only on the signal correlation values γ_{ij}. We denoted the matrix of the correlation values by Γ in (9.2.12). Since only the correlation values determine performance, all signal set generating the same matrix Γ perform the same. We therefore label signal sets by their corresponding matrix Γ. In the following we list some examples of signal sets commonly used in block encoded systems.

Orthogonal Signals

The set of M signals $s_i(t)$, $i = 1, 2, \ldots, M$ for which

$$\begin{aligned} \gamma_{ij} &= 0, & i \neq j \\ \gamma_{ii} &= 1, & i = j \end{aligned} \qquad i, j = 1, 2, \ldots, M \qquad (9.3.1)$$

constitute an orthogonal signal set. The matrix Γ then becomes an Mth order identity matrix

$$\Gamma = \begin{bmatrix} 1 & & 0 \\ & \ddots & \\ 0 & & 1 \end{bmatrix} \triangleq I_M \qquad (9.3.2)$$

Orthogonal signal sets have the property that any pair of signals from the set are pairwise orthogonal signals, as defined in (7.5.4).

Biorthogonal Signals

The set of M signals containing $M/2$ pairs of antipodal signals of which each pair is orthogonal is called a *biorthogonal* signal set. Thus each signal of the set is the negative of one other signal and orthogonal to all others. By proper ordering of the signals the correlation matrix can always be partitioned into the form

$$\Gamma = \begin{bmatrix} I_{M/2} & -I_{M/2} \\ -I_{M/2} & I_{M/2} \end{bmatrix} \qquad (9.3.3)$$

A set of M biorthogonal signals can always be formed by starting with $M/2$ distinct orthogonal signals and adding the negative of each to form the total set.

Equally Correlated Signals

The set of M signals whose cross-correlation among all pairs is identical constitutes a set of equally correlated signals. The correlation matrix is then

$$\Gamma = \begin{bmatrix} 1 & & \gamma \\ & \ddots & \\ \gamma & & 1 \end{bmatrix} \qquad (9.3.4)$$

where

$$\gamma = \frac{1}{E_w} \int_0^{T_w} s_i(t)s_j(t) \, dt \,, \qquad i \neq j \tag{9.3.5}$$

is the common cross-correlation value. Orthogonal signals are a special case where γ is zero. From our binary study we would expect best performance to occur when γ has its most negative value. However, there is a limit to how negative we can make γ. This can be observed by noting that the signal formed by summing all the signals of the set must have nonnegative energy. This means that

$$\int_0^{T_w} \left(\sum_{i=1}^M s_i(t) \right)^2 dt = \int_0^{T_w} \left[\sum_{i=1}^M s_i^2(t) + \sum_{j=1}^M \sum_{i=1}^M s_i(t)s_j(t) \right] dt$$

$$= ME_w + (M^2 - M)\gamma E_w$$

$$= [M + M(M-1)\gamma]E_w$$

$$\geq 0 \tag{9.3.6}$$

Solving for γ shows that it is necessary that

$$\gamma \geq -\frac{1}{M-1} \tag{9.3.7}$$

Thus M equally correlated signals can have a cross-correlation value no more negative than that given in (9.3.7). Signal sets having this minimal cross-correlation value for γ are called *simplex signal sets*, or *transorthogonal signal sets*. Note that for $M = 2$ (binary encoding case), simplex signals reduce to antipodal signals. It can be shown [1, 2] that for equally correlated signals, and any M,

$$\frac{\partial \text{PWE}}{\partial \gamma} \geq 0 \tag{9.3.8}$$

This means that indeed γ should be as negative as possible for minimizing word error probability. Hence simplex signal sets achieve the minimal value of PWE over all equally correlated sets. For this reason simplex signals are always of interest and general methods for generating them are desired. This will be considered subsequently. We point out that for $M \gg 1$, $\gamma \approx 0$, and simplex signals are approximately orthogonal signals when the set is large.

There remains only to find appropriate signal waveforms corresponding to the desired correlation matrices Γ. There is a general procedure for deriving a waveform set from a given matrix Γ. This generalization is achieved by resorting to signal representations using orthonormal functions as in (7.1.4a). By this approach the problem of generating signals of any particular type can be unified into a simpler procedure of selecting orthonormal functions. Let $\Phi_q(t)$, $q = 1, 2, \dots, \nu$, again be an arbitrary set of orthonormal functions over $(0, T_w)$ as defined in (7.1.4) and Appendix C.1, and write each of the signals of a block encoding set as

$$s_i(t) = \sum_{q=1}^{\nu} s_{iq} \Phi_q(t), \qquad i = 1, 2, \dots, M \qquad (9.3.9)$$

where the s_{iq} are the signal coordinates associated with each component function and form the signal vectors

$$\mathbf{s}_i = (s_{i1}, s_{i2}, \dots, s_{i\nu}) \qquad (9.3.10)$$

as in (9.2.1). As we know from Appendix C, the correlation between the waveforms is

$$\gamma_{ij} = \frac{1}{E_w} \int_0^{T_w} s_i(t) s_j(t) \, dt = \frac{1}{E_w} \sum_{q=1}^{\nu} s_{iq} s_{jq} \qquad (9.3.11)$$

and is therefore equivalent to the inner product of the signal vectors. Furthermore, signals of fixed energy are confined to a hypersphere in vector space, and the inner product is therefore directly related to the angle between the vectors [see (C.1.15)]. Thus a given correlation matrix corresponds to properly spaced vectors on a hypersphere in ν-dimensional signal space. In fact, there may be many distinct vector sets with the same correlation matrix. For example, a rotaton of a vector set on a hypersphere, while preserving the same spacings, will not change the correlations. The conversion of a vector set to a waveform set then only requires using the vector coordinates as the coefficients of an orthonormal set in (9.3.9). Since the orthonormal set is arbitrary, many different waveform sets can be derived from a specific vector set, all with the same correlation properties.

A popular way of building the signal vectors is with binary vectors (vectors with binary coefficients ± 1). Let $\mathbf{h}_i$ be a ν-dimensional binary vector having the binary coefficients $h_{ij} = \pm 1$. Then the correlation between any two such vectors can always be expressed as

$$\gamma_{ij} = \sum_{q=1}^{\nu} h_{iq} h_{jq} = \sum_{\mathcal{N}} (1) - \sum_{\mathcal{D}} (1) \qquad (9.3.12)$$

$\mathcal{N}$ and $\mathcal{D}$ are the number of components where $\mathbf{s}_i$ and $\mathbf{s}_j$ have like and opposite symbols and $\mathcal{N} + \mathcal{D} = \nu$. Thus the correlation coefficients of binary vectors depend only on the number of positions where the vectors are the same and opposite, when placed side by side. This reduces the selection of a signal set to simply finding binary (± 1) sequences with the proper number of like and opposite bits.

A signal vector set with binary vectors can always be written as a $\nu \times M$ matrix of ± 1 values whose rows represent each signal vector. For an orthogonal signal set, $v = M$, and we require an $M \times M$ binary square matrix such that any pair of rows differ in exactly $M/2$ columns. This requires M to be even. A general method for deriving such orthogonal binary matrices is as follows. Denote the 2×2 matrix H_1 as

$$H_1 \overset{\Delta}{=} \begin{bmatrix} 1 & -1 \\ 1 & 1 \end{bmatrix} \qquad (9.3.13)$$

The preceding is a trivial example of two orthogonal vectors. Now define the nth order extension of H_1 as

$$H_n = \begin{bmatrix} H_{n-1} & \hat{H}_{n-1} \\ H_{n-1} & H_{n-1} \end{bmatrix} \qquad (9.3.14)$$

where $\hat{H}_{n-1}$ represents the complement of H_{n-1}, obtained by replacing ± 1 by ∓ 1. It is easy to see H_n will always have orthogonal rows and can therefore serve as the binary vector set of orthogonal signals. The matrix is called a *Hadamard* matrix.

With the orthogonal binary matrix H_n derived, the orthogonal signals can be generated as shown in Figure 9.4. Let the binary entries of H_n be $\{h_{ij}\}$, and let $s(t)$ be any signal defined over $(0, T_w/M)$ with energy E_w/M. The orthogonal signals can then be formed as

$$s_i(t) = \sum_{j=1}^{M} h_{ij} s[t - j(T_w/M)], \qquad i = 1, 2, \ldots, M \qquad (9.3.15)$$

corresponding to a sequence of M binary waveforms $\{\pm s(t)\}$ formed from each $\mathbf{h}_i$. It is straightforward to show that the set $\{s_i(t)\}$ is always orthogonal (Problem 9.8). The waveform $s(t)$ is completely arbitrary, however, and can

$$H_n = \begin{bmatrix} 1 & -1 & 1 & 1 & -1 & -1 \end{bmatrix} \longleftarrow \text{row } i$$

$$\mathbf{s}_i = (1, -1, 1, 1, -1, -1)$$

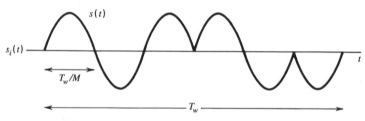

FIGURE 9.4. Generation of orthogonal waveforms.

be any convenient waveform that is easy to generate. For example, $s(t)$ can be simply a T_w/M sec burst of a carrier, and each $s_i(t)$ in (9.3.15) is then an M-ary sequence of BPSK waveforms. Thus an arbitrarily large class of orthogonal waveforms can be generated from a basic binary signal set.

From this discussion we see that the basic waveform encoding operation is now divided into separate and distinct tasks: (1) finding the appropriate vector set for the encoding and (2) finding an orthonormal function set that generates waveforms suitable for signal transmission. The former involves finding vector sets with an inherent correlation property, whereas the latter is influenced by engineering considerations (ease of generation, bandwidth utilization, processing convenience, etc.). This dichotomy allows us to separate block encoding into a cascade of two separate operations, as shown in Figure 9.5. In the first operation the block of bits is converted to the appropriate vector, and in the second the vector is converted to its corresponding waveform. Sometimes, in the literature, the first block is considered to perform the word encoding, and the second block is referred to as an *encoding modulator*, in which the encoding vector is modulated onto the given orthonormal set to form the time waveform.

An advantage of formulating the encoding operation in vector space is that many interesting and far-reaching notions in encoding theory have been developed. In particular, it allows precise derivation of the relation between ν and M of the signal set. For example, a set of M signals that are orthogonal must be described by orthogonal vector points. However, a vector space of at least M dimensions is needed to have M distinct orthogonal vector points on a hypersphere of fixed radius. Thus $\nu \geq M$, and we must therefore have at least M orthogonal components (expansion functions) in order to describe an M-ary orthogonal signal set. Just as important, however, is the fact that we do not need any more than M components to form this set.

For simplex signals, vectors are needed whose inner products are all given by $\gamma = -1/(M-1)$. This means that the signal vectors must have the property of vector points that are equally spaced on the hypersphere with a separation given by

$$[2E_w(1-\gamma)]^{1/2} = \left[2E_w\left(1+\frac{1}{M-1}\right)\right]^{1/2} = \left[\frac{2E_w M}{M-1}\right]^{1/2} \quad (9.3.16)$$

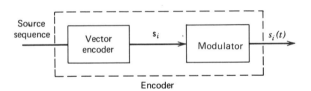

FIGURE 9.5. Generalized waveform generator.

The last term on the right also happens to be the distance between vertices of a geometric regular simplex[†] of M vertices inscribed on a hypersphere of radius $\sqrt{E_w}$. Hence design of simplex signal sets can be equated to determining regular simplex surfaces (from which the signal set gets its name) on hyperspheres and associating a signal vector with each of its vertices. From geometry [1] it is known that regular simplex surfaces of M vertices can only be constructed in spaces of dimension $\nu \geq M - 1$. This means the dimension of the signal need be no larger than $M - 1$.

Biorthogonal signals are any set of $M/2$ antipodal signals that are mutually orthogonal. Since it requires only one signal dimension to generate the positive and negative version of a vector, we can see that a biorthogonal vector set requires only $M/2$ coordinates, that is, $M/2$ orthogonal functions. This reduction in signal space size by a factor of 2 over orthogonal signal sets often gives biorthogonal signals significant hardware advantage.

Representing signal waveforms by orthonormal expansions also gives us more insight into decoder simplification. Formally, the coherent decoder computes the parallel correlation of the input $x(t)$ with a coherent version of each of the M signal waveforms, as defined in (9.2.3). However, use of the expansion in (9.3.9) allows us to write the ith channel correlation instead as

$$y_i = \int_0^{T_w} x(t)s_i(t)\, dt$$

$$= \sum_{q=1}^{\nu} s_{iq} \int_0^{T_w} x(t)\Phi_q(t)\, dt \qquad (9.3.17)$$

The integrals are common to all i. Therefore, the only difference between y_i and y_j is in the manner in which these integrals are weighted prior to summing. Thus the M channel coherent decoder can be reduced to an equivalent operation in which the ν integral values in (9.3.17) are computed simultaneously and the resulting outputs weighted and summed to form each y_i (Figure 9.6a). Only ν correlators are now needed, followed by M parallel store-and-sum circuits. The decoder is therefore greatly simplified if $\nu \ll M$, but the arithmetic operations of storing and summing generally require some type of decoder quantization. This reduces the accuracy of the $\{y_i\}$ computation. (Recall that the $\{y_i\}$ will be compared for the maximum.) In addition, the summing must be done after the T_w sec integrations and within the next word time. This introduces decoding delay and may limit the word transmission rate if the speed of computation is constrained.

An important and useful special case occurs if the orthonormal functions are pulsed sequences similar to Figure 9.4. The signaling waveforms therefore correspond to sequences of ν pulses, each pulse with amplitude s_{ij}, spanning the word interval T_w. The integrals in (9.3.17) define disjoint

---

[†]A simplex is a closed surface formed from the intersection of hyperplanes. A regular simplex has equal surface areas for all its outer faces.

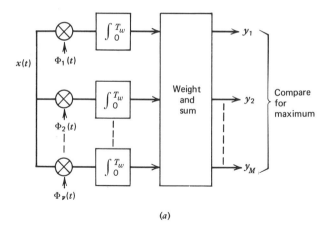

(a)

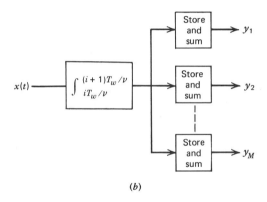

(b)

FIGURE 9.6. Alternative block decoder implementation: (a) general orthonormal functions, (b) pulsed functions.

integrations, each over successive T_w/ν intervals. The integrations can therefore be performed by a single integrator operated and read out successively (Figure 9.6b). Each output of the integrator is weighted in the M store-and-sum circuits and added to the previous sum. The $\{y_i\}$ are therefore computed in real time, and no bit delay occurs. Typically, decoder quantization is used to perform the summations, and the degree of quantization is important here. In two-interval quantization only the sign of the integrator output is stored. Since the integration is over exactly one pulse of the binary waveform, two-interval quantization is equivalent to making a binary decision concerning the polarity of that pulse, that is, detecting its binary symbol. Thus two level decoder quantization is equivalent to simply storing the bit decisions of the binary waveform prior to making a word decision. This type of operation is called *hard decisioning*. In high level quantization, the integrator output is quantized more accurately in computing the necessary y_i. This is referred to as *soft decisioning*. Decoder simplification of the form of Figure 9.6*b* makes the handling of word lengths on the order of $k = 10$ bits ($M = 1024$) quite practical, and such systems have been implemented.

9.4 PERFORMANCE OF BLOCK CODED SYSTEMS

Evaluation of word error probability for particular signal sets formally requires computation of the multidimensional Gaussian integral in (9.2.10). For general signal sets this is hampered by the necessity to invert Γ in (9.2.11). For the more common signal sets of Section 9.3, these probabilities have been extensively investigated in the literature, and various simplifying techniques have been inserted to aid evaluation. We review these results here.

Orthogonal Signals

When the signal set is orthogonal, the straightforward evaluation of PWE becomes somewhat tractable. This is due to the fact that for the orthogonal case of (9.3.1), the joint density in (9.2.11) factors into a product of individual Gaussian densities. Stated alternatively, for the orthogonal case the Gaussian random variables $y_i|s_j$ become uncorrelated (and therefore independent), leading to the density simplification. Furthermore, each such independent Gaussian variable has variance

$$\sigma^2 = \frac{N_0 E_w}{2} \tag{9.4.1}$$

and mean values in (9.2.7) of

$$\mathscr{E}\{y_j|s_j\} = E_w$$
$$\mathscr{E}\{y_i|s_j\} = 0, \qquad i \neq j \tag{9.4.2}$$

Thus (9.2.10) can be simplified to

$$\text{PWE}|j = 1 - \int_{-\infty}^{\infty} G\left(y_j, E_w, \frac{N_0 E_w}{2}\right) \left[\prod_{\substack{i=1 \\ i \neq j}}^{M} \int_{-\infty}^{y_j} G\left(y_i, 0, \frac{N_0 E_w}{2}\right) dy_i\right] dy_j \tag{9.4.3}$$

where

$$G(y_j, m, \sigma^2) \stackrel{\Delta}{=} \frac{1}{\sqrt{2\pi}\sigma} \exp\left[-\frac{(y_j - m)^2}{2\sigma^2} \right] \qquad (9.4.4)$$

The integral over each y_i can be written in terms of the Q-function, and (9.4.3) simplifies to

$$\mathrm{PWE}|j = 1 - \int_{-\infty}^{\infty} G(y, 0, 1)\left[1 - Q\left(y + \sqrt{\frac{2E_w}{N_0}} \right) \right]^{M-1} dy \qquad (9.4.5)$$

where we have substituted $y = (y_j - E_w)/(N_0 E_w/2)^{1/2}$ as the integration variable. Since the right side of (9.4.5) does not depend on j, (9.4.5) is also the average word error probability PWE, obtained from substituting into (9.1.5). Thus we have established

$$\mathrm{PWE} = 1 - \int_{-\infty}^{\infty} \left[1 - Q\left(y + \sqrt{\frac{2E_w}{N_0}} \right) \right]^{M-1} \frac{e^{-y^2/2}}{\sqrt{2\pi}} dy \qquad (9.4.6)$$

We emphasize that (9.4.6) assumes orthogonal, equiprobable, equal energy signals and white noise. Note that PWE depends only on the two parameters M and E_w/N_0. It is immediately evident that since the bracket in (9.4.6) is less than one, PWE increases as M gets larger for a fixed E_w/N_0. We see from the preceding discussion that system performance apparently degrades as we increase block size. This can be attributed to the fact that as we increase M for a given E_w/N_0, we are increasing the number of possible signals occurring in a fixed T_w sec block time. The decoder therefore has a more difficult decision and tends to make more errors. However, it may not be meaningful to compare block systems of different values M at the same value of E_w/N_0 because the higher M systems, while possibly making more word errors, are sending more bits in the same T_w sec interval. This suggests that the bit transmission rate be normalized before comparisons are made. That is, we compare PWE when all systems transmit source bits at the same rate, while operating at the same signal power P_m. The transmission bit rate of an M-ary system is

$$R_b = \frac{k}{T_w} = \frac{\log_2 M}{T_w} \quad \text{bits/sec} \qquad (9.4.7)$$

The right-hand side increases with M for fixed T_w. To have all M-ary systems operate at the same bit rate R_b, it is necessary to increase T_w as M increases. That is, a particular M-ary system should use a block time T_M, given by

$$T_M \stackrel{\Delta}{=} \frac{\log_2 M}{R_b} \qquad (9.4.8)$$

This means that different M systems must operate with different block time intervals if they are to have the same rate R_b. However, this also means that for a given M,

$$\frac{E_w}{N_0} = \frac{P_m T_M}{N_0} = \left[\frac{P_m}{N_0 R_b}\right] \log_2 M = \left(\frac{E_b}{N_0}\right)k \qquad (9.4.9)$$

Since the bracket involves all fixed parameters, E_w/N_0 increases as $\log_2 M$, and the word energy is k times the bit energy. Thus M-ary systems operating at fixed bit rate R_b and transmitting power P_m should be compared at values of the parameter E_w/N_0 as given above.

It is often desirable to convert word error probability in block coding to an equivalent bit error probability. This can be accomplished by determining the probability that a given bit of the word will be incorrect after incorrect decoding. With orthogonal signals the incorrectly decoded word is equally likely to be any of the remaining $M - 1$ words, and a given bit will be decoded as any of the bits in the same position of each other word. In M equally likely patterns a given bit will be a one or a zero $M/2$ times. The change of it being the incorrect bit in the $(M - 1)$ incorrect patterns is then $(M/2)/(M - 1)$. This means that the probability of a given bit being in error is then this probability times the probability that the word was in error. Hence the equivalent bit error probability PE is related to the word error probability PWE by

$$\text{PE} = \frac{1}{2}\left(\frac{M}{M - 1}\right)\text{PWE} \qquad (9.4.10)$$

when using orthogonal signaling.

When the adjustment in PWE is made in (9.4.9), the curves in Figure 9.7 are derived. These show PWE as a function of E_b/N_0, where $E_b = E_w/k = P_m/R_b$ is the effective bit energy at the decoder. The PWE can be converted to bit error probability PE by (9.4.10). The curve for $k = 1$ corresponds to binary orthogonal. The results now show an actual improvement in performance with increasing block size. Thus increasing block size yields lower bit error probabilities when transmitted at the same rate with the same bit energy. This block waveform encoding has therefore permitted us to improve performance over the standard binary transmission, just as we achieved with coding and error correction. The improvement however has now been obtained with a revised encoder and decoder in the communication link.

A curve has been included for the case $k = \infty$, obtained by examining the limiting behavior of (9.4.6), when (9.4.9) is substituted. By properly taking the limits as $M \to \infty$ (Problem 9.9), we obtain

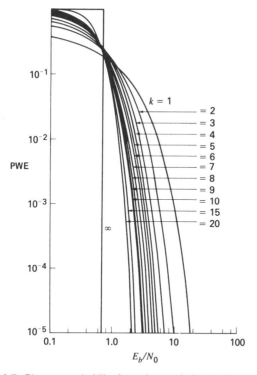

FIGURE 9.7. Bit error probability for orthogonal signals (k = block size).

$$\text{PWE} \xrightarrow[M \to \infty]{} \begin{cases} 1, & \text{if } \dfrac{P_m}{N_0 R_b} < \log_e 2 \\[3mm] 0, & \text{if } \dfrac{P_m}{N_0 R_b} \geq \log_e 2 \end{cases} \qquad (9.4.11)$$

Thus continual increase of the block size produces uniform PWE improvement, approaching the limit in (9.4.11). We see that as long as the energy per bit is large enough to satisfy (9.4.11), continual increase of M will eventually allow digital transmission with zero error probability. This means that we approach *perfect* transmission, in spite of the decoder noise, by using infinitely long block sizes, under the preceding conditions.

The result has another interesting interpretation. The condition required in (9.4.11) can also be stated

$$R_b \leq \frac{P_m}{N_0 \log_e 2} = \frac{P_m}{N_0(0.69)} \qquad (9.4.12)$$

This now appears as a condition on the rate R_b at which we transmit. The term on the right is the infinite bandwidth channel capacity (theoretical maximum allowable bit rate) for an additive white Gaussian noise channel having signal power P_m. Hence (9.4.11) and (9.4.12) state that we can transmit with zero error probability over a noisy channel as long as the transmitted bit rate is less than the channel capacity. We immediately see the physical drawbacks to this theorem (which are often obscured in theoretical derivations). To achieve the perfect transmission predicted by this theorem we must transmit with infinite block lengths. This means the decoder becomes infinitely complex and, since word interval $T_M \rightarrow \infty$, an infinite delay occurs in the decoding of any particular bit. We point out that, on the other hand, if the energy per bit is not large enough, as required by (9.4.11), we are certain to make word errors if we continually increase block size.

Although the preceding argument indicates the theoretical optimality of block encoding with infinite word lengths, the advantage of block encoding vis-à-vis binary encoding with finite word lengths is not obvious. To investigate this tradeoff, consider a specific binary word of k bits. This can be sent as k binary antipodal tansmissions, or by a single k bit block transmission. In the binary case the probability of not receiving every bit in the word correctly is

$$\text{PWE} = 1 - \left[1 - Q\left(\frac{2E_b}{N_0}\right)^{1/2} \right]^k \tag{9.4.13}$$

where E_b is the energy per bit. Since the Q-function can be read from Figure 7.6, the binary PWE can be easily evaluated for any k. An orthogonal block coded system, using words of length k and transmitting at the same bit rate, has its PWE given by (9.4.6) at the same value of $E_b = P_m/R_b$. When these results are compared, the block coded system produces the smaller PWE. This effect can best be exhibited by determining how much less bit energy is needed in block coding to achieve the same PWE as a binary system sending the same number of bits. The result is shown in Figure 9.8 for several values of PWE, as a function of word size k. Note that a power savings of about 5 dB is obtained using only $k = 5$ bit blocks. The improvement increases more slowly for higher k values, being only about 6 dB for $k = 10$. This comparison illustrates a basic fact of bit detection. A binary system sending a k bit interval can be envisioned as having a decoding scheme that makes k separate bit decisions to decode the word. A block coded system, on the other hand, integrates the total word energy and makes a single decision at the word end. We see clearly that the single decision, using all the available energy, is superior to a sequence of subword decisions. This is even more apparent as we let $k \rightarrow \infty$. The block coded system has the asymptotic behavior of (9.4.11). However, (9.4.13) indicates that PWE $\rightarrow 1$ for a binary system for any value of E_b. This means that we are making so many bit

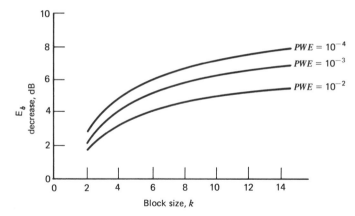

FIGURE 9.8. E_b reduction with block waveform encoding relative to binary encoding to achieve same PWE (k = block size).

decisions to decode a long word that we are in fact bound eventually to make an error.

The PWE for the orthogonal signal set can be approximated for large M by use of the union bound of (9.2.16). Since the binary test between any two decoding channels is equivalent to an orthogonal coherent test, we have

$$\text{PWE} \le \frac{M^2 - M}{M} \left[Q\left(\frac{E_w}{N_0}\right)^{1/2} \right] \qquad (9.4.14)$$

This serves as a simple upper bound, which becomes increasingly accurate as M and E_w/N_0 is increased. Note that by use of the bounds in (B.1.14), (9.4.9), and (9.4.10) we can establish that for large M, the block coded bit error probability is bounded by

$$\text{PE} \le \frac{M}{4} e^{-(E_w/N_0)}$$

$$= (\tfrac{1}{4}) \exp - \left[\left(\frac{P_m}{R_b N_0}\right) - \log_e 2 \right] \log_2 M \qquad (9.4.15)$$

We again see that as $M \to \infty$, the bit error probability PE, with orthogonal block encoding, must necessarily go to zero as long as (9.4.12) is true. This again verifies perfect transmission under these conditions.

Biorthogonal Signals

Evaluation of PWE using (9.2.10) for the biorthogonal signal set of (9.3.3) is hindered by the singularity of the matrix Γ, preventing matrix inversion.

This can be avoided by recalling that M biorthogonal signals are actually $M/2$ pairs of antipodal signals. The M channel decoder can be reduced to $M/2$ channels, since a channel matched to any particular signal is likewise negatively matched to its negative version. Word decisions can be made by first determining which channel sample y_i has the largest magnitude, then determining the polarity of this sample. For a decoder implemented in this way only $M/2$ outputs y_i are needed, and PWE can be obtained directly from this reduced system, thus avoiding the inversion problem. It can be shown [3] that the PWE for this system is given by

$$\text{PWE} = 1 - \int_{-(2E_w/N_0)^{1/2}}^{\infty} \frac{e^{-x^2/2}}{\sqrt{2\pi}} \left[1 - 2Q\left(x + \left(\frac{2E_w}{N_0}\right)^{1/2}\right)\right]^{(M/2)-1} dx$$

(9.4.16)

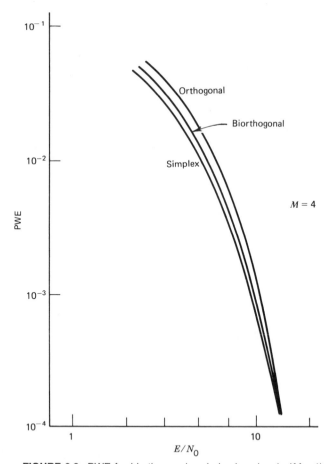

FIGURE 9.9. PWE for biorthogonal and simplex signals ($M = 4$).

Equation (9.4.16) is plotted in Figure 9.9 for the case $M = 4$. Performance is comparable to that of orthogonal signals, but we retain the advantage of a reduced decoder. An accurate approximation in biorthogonal signaling can be obtained from the union bound in (9.2.16). In this case we have $(M/2) - 1$ orthogonal binary tests, and one antipodal test, for determining the bound. Hence

$$\text{PWE} \cong (M - 2)Q\left[\left(\frac{E_w}{N_0}\right)^{1/2}\right] + Q\left[\left(\frac{2E_w}{N_0}\right)^{1/2}\right] \qquad (9.4.17)$$

The bound is quite accurate at $\text{PWE} \leq 10^{-2}$ and can be easily evaluated to estimate performance.

For biorthogonal signals the bit error probability is no longer related to the word error probability by (9.4.10) because when a word error is made it is no longer equally likely to be any of the remaining words, as in the orthogonal case. Instead the bit error distribution depends on the manner in which the word error occurs. If the decoder decides the correct maximum magnitude bit but errs in the polarity decision, every bit of the word is incorrect. If the decoder decides the wrong magnitude, bit errors are equally distributed. The probability of a given bit being in error must then be obtained by averaging over the probability of each of these types of error. This is given approximately by

$$\text{PE} \cong Q\left[\left(\frac{2E_w}{N_0}\right)^{1/2}\right] + \left(\frac{M - 2}{2}\right)Q\left[\left(\frac{E_w}{N_0}\right)^{1/2}\right] \qquad (9.4.18)$$

which becomes increasingly accurate with large M and E_w/N_0. Biorthogonal signals have the same limiting performance as $M \to \infty$ as orthogonal signals.

Equally Correlated Signals

The word error probability for signal sets that have equal cross-correlation can be obtained from the known results of the orthogonal case. This follows since equally correlated signals with energy E_w and cross-correlation γ generate the same PWE as an orthogonal set with energy $E_w(1 - \gamma)$ (Problem 9.10). The PWE for the latter is obtained from (9.4.6) and Figure 9.7. Hence we need only adjust the effective energy in determining the corresponding PWE for a given M. For simplex signals, $\gamma = -1/(M - 1)$, and the effective energy is increased to $E_w(1 - \gamma) = E_w M/M - 1$. Performance of simplex signals is therefore better than for orthogonal signals because of the slight increase in effective energy but this advantage becomes negligible at large M. A plot of the simplex signal set performance for $M = 4$ is superimposed in Figure 9.9.

Other types of M-ary signaling, such as M level differential encoding, have also been proposed for digital systems. We shall not pursue their study here. The interested reader is referred to the discussion of such systems in References 3 and 4.

9.5 POLYPHASE (MPSK) SIGNALING

An important class of block signals are those obtained from expansions into the $\nu = 2$ orthonormal quadrature set

$$\Phi_1(t) = \left(\frac{2}{T_w}\right)^{1/2} \cos \omega_s t \, ,$$

$$0 \le t \le T_w \qquad (9.5.1)$$

$$\Phi_2(t) = \left(\frac{2}{T_w}\right)^{1/2} \sin \omega_s t \, ,$$

where ω_s is an arbitrary carrier or subcarrier frequency. The signals of the set are all of the form

$$s_i(t) = s_{i1}\left(\frac{2}{T_w}\right)^{1/2} \cos \omega_s t + s_{i2}\left(\frac{2}{T_w}\right)^{1/2} \sin \omega_s t \qquad (9.5.2)$$

where $E_w = s_{i1}^2 + s_{i2}^2$. If we convert the coordinates (s_{i1}, s_{i2}) to the new coordinate θ_i by the transformation

$$\theta_i = \tan^{-1}\left(\frac{s_{i2}}{s_{i1}}\right), \qquad \begin{array}{l} s_{i1} = \sqrt{E_w} \cos \theta_i \\ s_{i2} = \sqrt{E_w} \sin \theta_i \end{array} \qquad (9.5.3)$$

then (9.5.2) can be rewritten as

$$s_i(t) = \left(\frac{2E_w}{T_w}\right)^{1/2} [\cos \theta_i \cos \omega_s t + \sin \theta_i \sin \omega_s t]$$

$$= \left(\frac{2E_w}{T_w}\right)^{1/2} \cos [\omega_s t - \theta_i] \qquad (9.5.4)$$

The signals are therefore phase shifted sine waves, and the signal vectors become phasors in polar coordinates (Figure 9.10). Each signal becomes a point on a circle of radius $\sqrt{E_w}$ at angle θ_i. A set of M signals appears as M such points, and is called a *polyphase*, or *multiple phase shift keyed* (MPSK), signal set. MPSK signals have cross-correlation

$$\gamma_{iq} = \cos (\theta_i - \theta_q) \qquad (9.5.5)$$

and therefore depend only on the angular difference between the signals. If the phase angles are equally spaced over $(0, 2\pi)$, then $\theta_i = 2\pi i/M$, and

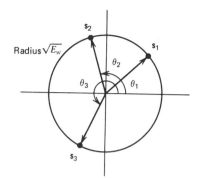

FIGURE 9.10. Signal vectors in two-dimensional space.

$$\gamma_{iq} = \cos\left[\frac{2\pi}{M}(i-q)\right] \qquad (9.5.6)$$

Note that polyphase signal sets are not, in general, equally correlated. For $M = 2$, equally spaced polyphase signals reduce to the antipodal PSK signals of Section 7.3. For $M = 3$, they would constitute a simplex signal set. For $M = 4$, they correspond to four phasors, spaced every 90° in $(0, 2\pi)$. Such a signal set is referred to as a *quadraphase* or *QPSK* set. QPSK signaling, which has become extremely important in modern digital systems, is considered in more detail in Section 9.6.

Since the phasor signals represent a T_w sec burst of sinusoidal signal, the bandwidth required is identical to that of a PSK signal. Thus quadrature signals have a frequency spectrum

$$S_m(\omega) = E_w\left[\frac{\sin(\omega - \omega_s)T_w/2}{(\omega - \omega_s)T_w/2}\right]^2 \qquad (9.5.7)$$

This occupies a bandwidth of approximately $2/T_w$ Hz, around the carrier frequency ω_s. Note that the bandwidth does not depend on M. This is a basic advantage of polyphase signaling, since the required bandwidth does not increase as the size of the signal set increases. Decoding is achieved by coherent correlation of the receiver baseband signal with each member of the set, selecting the largest correlator output as the signaling phase state. Each correlator computes

$$y_i = \int_0^{T_w} x(t)s_i(t)\,dt$$

$$= [X\cos\theta_i + Y\sin\theta_i] \qquad (9.5.8)$$

where

$$X \stackrel{\Delta}{=} \int_0^{T_w} x(t) \cos \omega_s t \, dt$$

$$Y \stackrel{\Delta}{=} \int_0^{T_w} x(t) \sin \omega_s t \, dt \qquad (9.5.9)$$

We see that (9.5.9) is identical to writing

$$y_i = C \cos (\theta_i - \theta) \qquad (9.5.10)$$

where $C^2 = (X^2 + Y^2)$ and

$$\theta \stackrel{\Delta}{=} \tan^{-1} \frac{Y}{X} \qquad (9.5.11)$$

The decoding decision as to which y_i is largest is therefore equivalent to determining which $(\theta_i - \theta)$ is smallest, over all i. That is, determining which θ_i the θ is closest to. The decoder can therefore be reduced to a pair of channels in which X and Y in (9.5.9) are separately determined, and the decoder decisioning is based only on a computation of θ in (9.5.11) and a comparison to $\{\theta_i\}$ (Figure 9.11). In essence, the decoder detects the phase of the input signal [as defined in (9.5.11)] for decisioning. Note that the decoder uses quadrature mixing followed by an integrate-and-dump detector. The quadrature signals, $\cos \omega_s t$ and $\sin \omega_s t$, could be generated from a common free-running oscillator in which the quadrature component is achieved by an additional 90° phase shift. However, this oscillator must be perfectly phase coherent with the received oscillator frequency to compute X and Y in (9.5.9). This is often called a *coherent quadrature correlator*.

The PWE can be evaluated by formally computing the signal correlation matrix from (9.5.6) and again evaluating (9.2.10). However, a simpler procedure is to determine PWE by computing the probability that the angle θ is not closer to the correct θ_i. For the case of equally spaced values for θ_i over $(0, 2\pi)$, this has been determined [3] to be

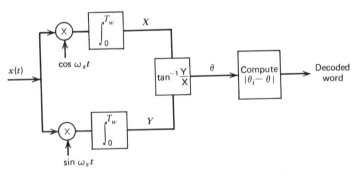

FIGURE 9.11. Decoder for polyphase signals.

$$\text{PWE} = \frac{M-1}{M} - \frac{1}{2}\,\text{Erf}\left[\left(\frac{E_w}{N_0}\right)^{1/2}\sin\left(\frac{\pi}{M}\right)\right]$$

$$-\frac{1}{\sqrt{\pi}}\int_0^{(E_w/N_0)^{1/2}\sin(\pi/M)} \exp\left[-y^2\right]\text{Erf}\left[y\cot\left(\frac{\pi}{M}\right)\right]dy \quad (9.5.12)$$

where $\text{Erf}(x) = 1 - 2Q(\sqrt{2}x)$. The word error probability PWE in (9.5.12) is plotted in Figure 9.12 as a function of the normalized bit energy $(E_b = E_w/\log_2 M)$ for several values of M. For $M = 2$, the signal set is antipodal and (9.5.12) reduces to our earlier result. When $M \gg 1$, PWE is adequately approximated by

$$\text{PWE} \cong 2Q\left[\left(\frac{2E_w}{N_0}\right)^{1/2}\sin\left(\frac{\pi}{M}\right)\right] \quad (9.5.13)$$

We immediately see a basic disadvantage to polyphase signaling with large M. The signal vector points (phasors on the circle of $\sqrt{E_w}$) become crowded together and accurate detection becomes difficult. This effect is most obvious by noting that for large M, $\sin(\pi/M) \approx \pi/M$ so that the required

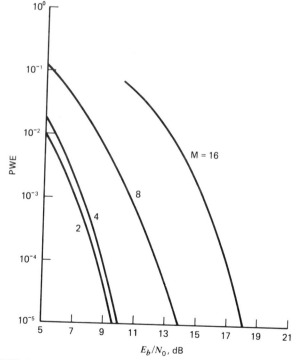

FIGURE 9.12. Word error probability versus E_b/N_0 for MPSK signal set.

signal energy E_w must be increased as M^2 to maintain a given PWE. For this reason, polyphase signals are generally restricted to low values of M ($M = 4$, 6, 8). Although these values of M do not correspond to long block sizes and therefore may not produce significant improvement over binary transmission, polyphase signals do allow increased bit rate without increase in the required transmission bandwidth.

The conversion of MPSK word error probabilities to bit error probabilities depends on the manner in which the bit words are mapped onto the $M = 2^k$ phases and on how the decoder word errors occur. If the encoder mapping is such that the bit words corresponding to adjacent phase states differ by only one bit (called *Gray encoding*) and if MPSK decoding errors always select the adjacent phase state as the incorrect word, then the bit error probability PE and MPSK word error probability are related by

$$PE = \frac{PWE}{\log_2 M} \qquad (9.5.14)$$

This merely assumes that when MPSK phase decoding errors are made, any single bit of the bit word can be in error.

9.6   QUADRATURE PHASE SHIFT KEYING (QPSK)

A particularly important class of MPSK signaling is the $M = 4$ case, referred to as quadrature phase shift keying (QPSK). When written as (9.5.4), this corresponds to signals of the form

$$s_i(t) = A \cos(\omega_s t - \theta_i + \psi), \qquad 0 \le t \le T_w \qquad (9.6.1)$$

where A and ψ are carrier amplitude and phase constants and θ_i takes on one of four phases separated by 90°. Assume that these phases are (45°, 135°, −135°, −45°) as shown in Figure 9.13a. Two bits of information are sent every word time T_w with one of the phases θ_i and with waveform energy $E_w = A^2 T_w/2$. Note that QPSK signals can also be viewed as two pairs of antipodal phasors placed orthogonally to each other, and therefore also represent a biorthogonal signal set.

The QPSK waveform can be written as, using (9.5.2) and (9.5.3),

$$s_i(t) = \frac{A}{\sqrt{2}} m_c \cos(\omega_s t + \psi) + \frac{A}{\sqrt{2}} m_s \sin(\omega_s t + \psi)$$

$$0 \le t \le T_w \qquad (9.6.2)$$

where (m_c, m_s) will be ±1 depending on the value of θ_i. Figure 9.13b shows the combinations corresponding to each θ_i. The time alignment of the bit

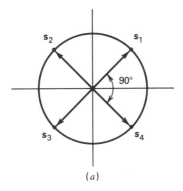

(a)

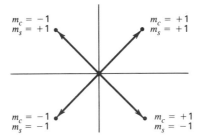

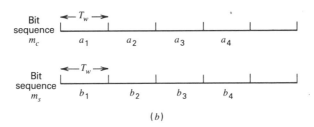

(b)

FIGURE 9.13. The QPSK signal set: (a) vector signals, (b) quadrature combinations.

pair is also shown. When written as in (9.6.2), each signal has the form of a binary modulated cosine and sine wave, each similar to BPSK. Thus a QPSK signal set can be interpreted as either a carrier wave in which a two-bit word is encoded on to one of four phases every T_w sec, or as a sum of two separate BPSK signals in which each of the bits in the two-bit word are used to separately binary phase modulate the quadrature components of the same carrier every T_w sec. Although the two interpretations are equivalent, they lead to two different encoder implementations. Figure 9.14a corresponds to (9.6.1) and shows an encoder formed by phase modulating the input carrier according to the two-bit word. Figure 9.14b shows the overall encoder as two separate BPSK encoders using the quadrature components of the same carrier, with the outputs summed to form (9.6.2). Both implementations have been used extensively in practice. The phase shifter suffers from the fact that a large (190°) phase shift may be required in shifting between words, while the BPSK combiner requires more hardware.

The QPSK carrier in (9.6.1) has word energy $E_w = A^2 T_w/2$. However, it sends two bits in T_w sec, and therefore has a bit time T_b of $T_w/2$, and an equivalent bit energy of

$$E_b = \frac{A^2 T_b}{2} = \frac{A^2 T_w}{4} = \frac{E_w}{2} \tag{9.6.3}$$

Each individual BPSK term in (9.6.2) uses a quadrature bit time of T_w sec and therefore has a bit energy of $E_b = (A/2)^2 T_w = A^2 T_w/4$, as in (9.6.3). In other words, each of the BPSK components has a bit energy equal to the bit energy of the QPSK carrier itself.

The spectrum of the QPSK carrier is given by the modified version of (9.5.7)

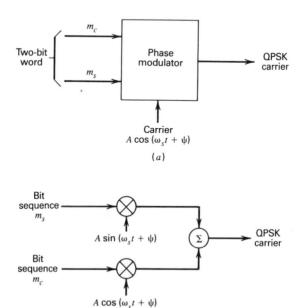

FIGURE 9.14. QPSK encoders: (a) phase shifter, (b) quadrature sum.

$$S_c(\omega) = A^2 T_w \left[\frac{\sin\left[(\omega - \omega_s)T_w/2\right]}{(\omega - \omega_s)T_w/2} \right]^2 \qquad (9.6.4)$$

which is equivalent to summing the individual spectra of each BPSK component and shifting to the carrier frequency. Since identical bit waveforms are used in each quadrature component, the QPSK carrier has the same spectrum as the BPSK of each component. This means the additional quadrature modulation is obtained "free of charge" as far as spectral extent is concerned. In the same spectral band as BPSK, a QPSK carrier can carry an additional data sequence. Since communication links are invariably bandwidth-constrained, this doubling of the bit rate is extremely advantageous. Note that QPSK has a main hump bandwidth of $2/T_w$ Hz, or equivalently, $2/2T_b = 1/T_b$ Hz, depending on whether it is expressed in symbol times or carrier bit times. A BPSK carrier with the same bit rate requires a bandwidth of $2/T_b$, or twice the QPSK bandwidth.

Optimal QPSK decoding can be achieved with the MPSK decoder shown in Figure 9.11, which essentially determines the received phase. The QPSK word error probability is then obtained from the PWE in (9.5.12) which, for $M = 4$, reduces to

$$\text{PWE} = 2Q\left[\left(\frac{E_w}{N_0}\right)^{1/2}\right] - Q^2\left[\left(\frac{E_w}{N_0}\right)^{1/2}\right] \qquad (9.6.5)$$

Since QPSK can be considered two BPSK carriers in parallel, a question arises as to whether it can be optimally decoded in this way also. Consider again the QPSK phase angles θ_i in Figure 9.13a (45°, 135°, −135°, −45°). Deciding which θ_i is closest to θ in (9.5.11) is identical to determining which quadrant θ lies in. However, this quadrant will be determined by simply the signs of X and Y. If $X, Y \geq 0$, then $0 \leq \theta \leq 90$; if $X < 0$, $Y > 0$, then $180 \geq \theta \geq 90$, and so on. This sign detection is simply a binary test performed separately on X and Y with a separate zero threshold test over $(0, T_w)$. Thus optimal QPSK decoding is in fact achieved with separate BPSK decoding on each quadrature bit in (9.6.2). The overall decoder will have the form shown in Figure 9.15, composed of a coherent quadrature cor-

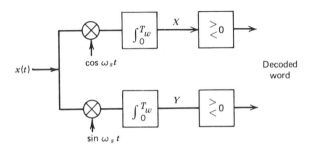

FIGURE 9.15. QPSK decoder.

relator, with synchronous detectors operating in each arm of the decoder. Because of the orthogonality of the quadrature carrier components, each decoding arm is not affected by the presence of the other.

As a separate BPSK decoder, each arm in Figure 9.15 will operate with a bit error probability dependent on the bit energy of that quadrature arm. From (9.6.2), this is $E_b = (A/2)^2 (T_w/2) = A^2 T_w/4 = E_w/2$. The BPSK bit error probability per arm is then

$$PE = Q\left[\left(\frac{2E_w/2}{N_0}\right)^{1/2}\right] \tag{9.6.6}$$

The word error probability on the two-bit block is then the probability that both quadrature bits are not correct. Thus

$$PWE = 1 - (1 - PE)^2$$
$$= 2PE - PE^2 \tag{9.6.7}$$

which is identical to (9.6.5). Note that the quadrature symbol energy used to determine PE in (9.6.6) is identical to the carrier bit energy obtained from (9.6.3). This means bit PE for QPSK can be obtained by looking up the BPSK error probability at the bit energy of the QPSK carrier.

A difficulty with QPSK is the necessity of making large (180°) phase changes almost instantaneously in changing between certain phase states. Physical requirements on phase modulators prevent these instantaneous changes, so that the bit rate is effectively slowed down, reducing an inherent advantage of QPSK. This problem can be somewhat combated by using so-called offset QPSK encoding. If we again envision QPSK as a bior-thogonal encoding scheme, we can consider the first bit of every two-bit block of source bits to be a binary sequence modulated on one quadrature signal in (9.6.1), and the second bit of each block as another sequence for the other quadrature component. We list these sequences as

$$\mathbf{m}_c = \{a_1, a_2, a_3, \ldots\}$$
$$\mathbf{m}_s = \{b_1, b_2, b_3, \ldots\} \tag{9.6.8}$$

where a_i, b_i, are ±1. Each pair of bits (a_j, b_j) therefore determine the phase angle θ_i in QPSK via

$$\theta_i = \tan^{-1}\left(\frac{b_j}{a_j}\right) \tag{9.6.9}$$

The sequences $\mathbf{m}_c$ and $\mathbf{m}_s$ must therefore be in bit alignment. To obtain offset QPSK, we continue to use the same quadrature encoding, except that we offset one sequence in (9.6.8), by one-half of a bit period from the other (Figure 9.16). The transmitted phase is still given by (9.6.9), but halfway

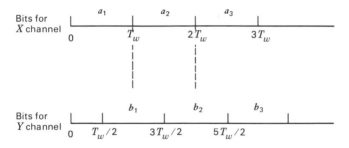

FIGURE 9.16. Offset QPSK bit alignment.

during a bit interval b_j changes to b_{j+1} while a_j remains the same (or vice versa). Either the next bit b_{j+1} has the same polarity as b_j, in which case θ_j does not change, or it has opposite polarity, in which case θ_j jumps to the adjacent quadrant. Hence a change of at most ±90° can occur every half bit period by staggering, or offsetting, the bit sequence in (9.6.8). This achieves control of the maximum phase change, limiting it to a 90° phase shift. The quadrature decoders in Figure 9.14 must also be offset by $T_w/2$ for individual bit decisioning. Since the channels are orthogonal, the offset quadrature signal will not affect individual bit detections of the in phase channel.

Just as in PSK, QPSK decoders must be phase referenced. To determine the effect of imperfect referencing, consider (9.5.9) when the decoder quadrature signals have an added phase shift ψ_e relative to the proper phase (assumed zero). Then

$$X(\psi_e) = \int_0^{T_w} x(t) \cos \left[\omega_s t + \psi_e\right] dt$$

$$= \int_0^{T_w} x(t)[\cos \psi_e \cos \omega_s t - \sin \psi_e \sin \psi_s t] \, dt$$

$$= X \cos \psi_e - Y \sin \psi_e \qquad (9.6.10)$$

and, similarly,

$$Y(\psi_e) = Y \cos \psi_e + X \sin \psi_e \qquad (9.6.11)$$

where X, Y are the values generated from perfect referencing. Thus the effect of a phase error ψ_e is to convert the desired values of X, Y to $X(\psi_e)$, $Y(\psi_e)$. In addition to the typical cosinusoidal degradation characterizing phase referencing errors in PSK, we now see that ψ_e also causes a cross-coupling into the orthogonal channel, proportional to $\sin \psi_e$. This cross-coupling can be either destructive or constructive, depending on the bit signs. To determine its effect on bit error probability PE, we average over both possibilities. Since X and Y have mean values of $\pm(E_w T_w/2)^{1/2}$, this becomes

$$PE(\psi_e) = \frac{1}{2} Q\left[\left(\frac{E_w}{N_0}\right)^{1/2}(\cos \psi_e + \sin \psi_e)\right]$$
$$+ \frac{1}{2} Q\left[\left(\frac{E_w}{N_0}\right)^{1/2}(\cos \psi_e - \sin \psi_e)\right] \qquad (9.6.12)$$

When ψ_e is a random phase error, $PE(\psi_e)$ must be averaged over ψ_e to obtain the resulting bit error probability PE, as in (7.5.16). When ψ_e is taken to be Gaussian, with zero mean and variance σ_e^2, the result for the QPSK case is shown in Figure 9.17. The plots indicate higher PE degradations than with standard BPSK for the same phase error variances. Hence QPSK systems are degraded more by phase reference errors than BPSK systems and generally must be designed with better referencing accuracy.

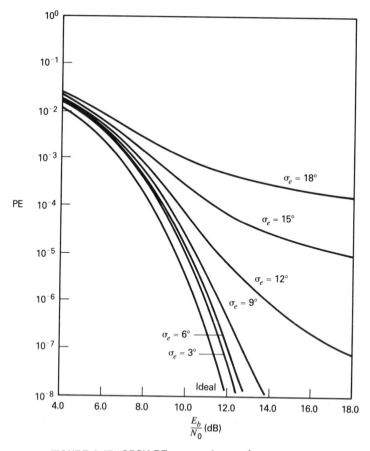

FIGURE 9.17. QPSK PE versus phase reference error.

In offset QPSK, bit transitions in the other channel may occur at the middle of the bit interval of each channel. If no transition occurs, performance during that bit is identical to QPSK. When a transition occurs, the cross-coupling term (involving sin ψ_e) changes polarity at midpoint, and the cross-coupling during the first half bit is cancelled by that during the second half of the interval. Hence performance is identical to that with no interference at all, and therefore is the same as for independent binary PSK channels with phase reference error ψ_e. Since a bit transition occurs with a probability of $\frac{1}{2}$, the offset QPSK bit error probability with phase error ψ_e is

$$\mathrm{PE}_{\mathrm{OQPSK}}(\psi_e) = \tfrac{1}{2}\mathrm{PE}_{\mathrm{QPSK}}(\psi_e) + \tfrac{1}{2}\mathrm{PE}_{\mathrm{BPSK}}(\psi_e) \qquad (9.6.13)$$

where $\mathrm{PE}_{\mathrm{QPSK}}(\psi_e)$ is given in (9.6.12) and $\mathrm{PE}_{\mathrm{BPSK}}(\psi_e)$ is given in (7.5.15). This result shows that offsetting the bits will produce slightly less PE degradation than standard QPSK with the same phase referencing error.

Although offset QPSK achieves an advantage in phase shift reduction, its frequency spectrum will still be fairly wide because of effective pulse modulation in each quadrature channel by the $\mathbf{m}_c$ and $\mathbf{m}_s$ sequences in (9.6.8). This means that any transmission bandlimiting effects filtering off the tails of the spectrum must be accounted for. (Remember the prime advantage of QPSK is that it allows doubling of the bit rate with the same spectrum as BPSK, so its prime application is in bandlimited situations.) This suggests that an advantage can be attained by using modulation formats in each quadrature channel that are not perfectly pulsed by that avoid significant spectral tails and energy loss caused by bandlimiting. Consider using rounded, half-sine wave pulses as the modulating signals in each quadrature channel, as shown in Figure 9.18a. The modified encoder is shown in Figure 9.18b. We let T_w be the symbol time per channel, and again offset the $\mathbf{m}_c$ and $\mathbf{m}_s$ sequences by $T_w/2$. The resultant offset QPSK waveform is then

$$c(t) = A \sum_k a_k \cos\left[\frac{\pi}{T_w}(t - kT_w)\right]\cos(\omega_s t + \psi)$$

$$+ A \sum_k b_k \sin\left[\frac{\pi}{T_w}(t - kT_w)\right]\sin(\omega_s t + \psi) \qquad (9.6.14)$$

However, by applying trigonometric identities, this can be written equivalently as

$$c(t) = \begin{cases} \cos\left(\omega_s t + a_i \dfrac{\pi}{2T_w} t + \psi_{i1}\right), & iT_w \leq t \leq \left(i + \dfrac{1}{2}\right)T_w \\[2mm] \cos\left(\omega_s t + b_i \dfrac{\pi}{2T_w} t + \psi_{i2}\right), & \left(i + \dfrac{1}{2}\right)T_w \leq t \leq (i+1)T_w \end{cases} \qquad (9.6.15)$$

where ψ_{i1} and ψ_{i2} are constant phase angles, depending on the data bits. We

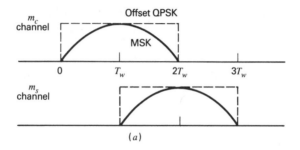

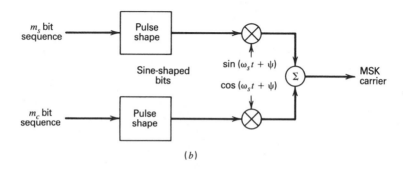

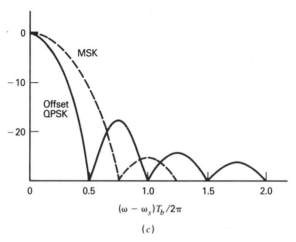

FIGURE 9.18. Sine wave OQPSK (MSK): (a) waveforms, (b) MSK encoder, (c) spectral comparison to QPSK.

see that (9.6.15) is equivalent to a carrier at frequency ω_s, frequency shifted between $\pm 1/4T_w$ Hz according to the bit stream $(a_1, b_1, a_2, b_2, \ldots)$. In Section 7.4 we defined this as a minimum shift keyed (MSK) signaling format. Thus offset QPSK signals with half-sine modulating bit pulses in each quadrature channel are identical to binary MSK signaling with the same bit stream and a controlled phase variable. The spectrum of MSK signals was shown in Figure 7.11, and in Figure 9.18c, it is compared to the corresponding spectrum of pulsed offset QPSK. Note that MSK signals have a slightly wider main hump but fewer out-of-band tails than for offset QPSK. Thus offset QPSK with sine pulses is advantageous in bandlimited channels, provided that the slightly larger spectral hump can be accommodated. The offset QPSK coherent decoders now correspond to matched correlators in each quadrature channel, each properly matched to the offset half-sine carrier burst representing each bit (Figure 9.19). Since phase coherent decoding is used in each quadrature channel, the MSK bits will be decoded with the same PE performance as any other antipodal system at the same E_b. Hence PE values for phase coherent MSK can be obtained from the BPSK (or QPSK) curves. It should be pointed out that since MSK is a form of frequency shift keying, it can also be decoded by noncoherent FSK detectors if phase synchronization happens to be lost.

Unbalanced QPSK is a modified version of QPSK in which the power levels and bit rates of each quadrature component are not identical. The unbalanced QPSK (UQPSK) carrier has the form

$$c(t) = Am_c(t) \cos(\omega_s t + \psi) + Bm_s(t) \sin(\omega_s t + \psi) \qquad (9.6.16)$$

where $A \neq B$, and $m_c(t)$ and $m_s(t)$ are NRZ binary waveforms with bit times T_A and T_B, respectively. We define the component powers $P_A = A^2/2$, $P_B = B^2/2$, and the unbalance power ratio,

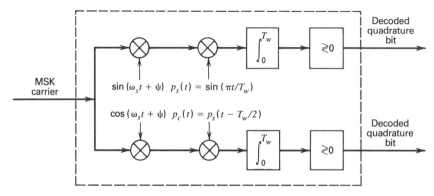

FIGURE 9.19. MSK decoder.

$$r = \frac{A^2}{B^2} = \frac{P_A}{P_B} \qquad (9.6.17)$$

The total carrier power is again

$$P_c = P_A + P_B \qquad (9.6.18)$$

and each component power can be written as a fraction of the total power

$$P_A = \left(\frac{r}{r+1}\right)P_c$$

$$P_B = \left(\frac{1}{r+1}\right)P_c \qquad (9.6.19)$$

Since the bit times are unequal, the component bit energies will be

$$E_A = P_A T_A, \qquad E_B = P_B T_B \qquad (9.6.20)$$

For equal bit energy in each channel, $E_A = E_B$, we require $P_A T_A = P_B T_B$, or

$$r = \frac{P_A}{P_B} = \frac{T_B}{T_A} = \frac{R_A}{R_B} \qquad (9.6.21)$$

Thus, for equal bit energy per channel, the power ratio should equal the data ratio. That is, the higher rate channel should have the most power. We therefore can transmit unequal bit rates with the same bit energies by using UQPSK with the proper power ratio.

Decoding performance with phase referencing errors is now slightly more complicated, because the cross-coupling involves different bit times and power levels. Consider Figure 9.20, showing $R_A > R_B$ so that there are

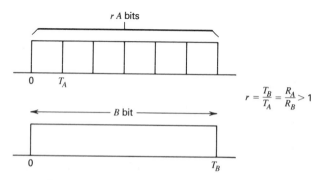

FIGURE 9.20. Unbalanced QPSK bit alignment.

$r = T_B/T_A$ bits in the A channel for every bit in the B channel. Assuming an integrated-and-dump decoder in each channel and a phase referencing error of ψ_e, each A bit will have either a $\pm B$ bit superimposed, with the power level of the weaker channel. Here the bit error probability of the A channel is then

$$\begin{aligned}
\mathrm{PE}_A &= \frac{1}{2} Q\left[\frac{AT_A}{\sqrt{N_0 T_A/2}} \cos \psi_e \pm \frac{BT_A}{\sqrt{N_0 T_A/2}} \sin \psi_e \right] \\
&= \frac{1}{2} Q\left[\left(\frac{2E_A}{N_0}\right)^{1/2} \cos \psi_e \pm \left(\frac{2E_B}{N_0}\right)^{1/2} \left(\frac{R_B}{R_A}\right)^{1/2} \sin \psi_e \right] \\
&= \frac{1}{2} Q\left[\left(\frac{2E_A}{N_0}\right)^{1/2} \left(\cos \psi_e \pm \frac{1}{\sqrt{r}} \sin \psi_e \right) \right]
\end{aligned}$$
(9.6.22)

where $\pm$ in the argument means that Q must be evaluated with each sign separately and summed. We see that the higher power channel in UQPSK has a reduced crosstalk due to the power imbalance. The phase error couples in a lower power bit onto a higher power bit, and its effect is therefore reduced.

In the B channel, the integration over a bit time integrates over r separate A bits coupled in by the phase error ψ_e (see Figure 9.19). The effect will depend on the number of A bits that match or mismatch the B bit. Since these occur equally likely,

$$\begin{aligned}
\mathrm{PE}_B &= \frac{1}{2^r} \sum_{i=1}^{r} \binom{r}{i} Q\left[\frac{BT_B}{\sqrt{N_0 T_B/2}} \cos \psi_e + (r-2i) \frac{AT_B}{\sqrt{N_0 T_B/2}} \sin \psi_e \right] \\
&= \frac{1}{2^r} \sum_{i=1}^{r} \binom{r}{i} Q\left[\left(\frac{2E_B}{N_0}\right)^{1/2} \left(\cos \psi_e + \left(\frac{r-2i}{\sqrt{r}}\right) \sin \psi_e \right) \right]
\end{aligned}$$
(9.6.23)

This is imply the average bit error probability given i mismatched bits, averaged over all i. The worst case occurs when $i = r$, which contributes a term

$$\mathrm{PE}_B = \frac{1}{2^r} Q\left[\left(\frac{2E_B}{N_0}\right)^{1/2} (\cos \psi_e - \sqrt{r} \sin \psi_e) \right]$$
(9.6.24)

Thus the reference error crosscouples the higher power A bits onto the lower power B channel but has the advantage that the crosscoupled bits are effectively averaged over all the possible bit patterns. This partially compensates for the power levels, as witnessed by the reduction in PE due to $1/2^r$ term in (9.6.24). That is, the worst case in (9.6.23) occurs less often. It is therefore not evident whether the A or B channel is more degraded by phase referencing errors in UQPSK, and both PE_A and PE_B must be evaluated for comparison.

9.7 CPFSK AND PHASE TRAJECTORY DECODING

In Section 2.4 it was pointed out that frequency shifting of a carrier with a binary digital waveform produces an angle modulated carrier whose phase function followed a specific time trajectory. This notion can be generalized to the transmission of block encoded waveforms by associating a different trajectory with each data block. Hence each k-bit word is encoded into a carrier whose phase traces out a specific trajectory, and decoding is achieved by deciding which trajectory is occurring. The encoding is accomplished by frequency shifting the carrier in the proper manner. This produces a carrier waveform that is purely angle modulated with a phase modulation that is always continuous in time (no phase jumps). This type of modulation is referred to as *continuous phase frequency shift keying* (CPFSK). The smoothness of the phase trajectory produces a minimal bandwidth (reduced spectral tails) in the frequency domain. Hence CPFSK encoding is important in strict bandlimited digital communications.

The encoding process is shown in Figure 9.21. Each k-bit word is converted to one of $M = 2^k$ binary sequences $\mathbf{a}_i = \{a_{ij}\}$ of length ν, where

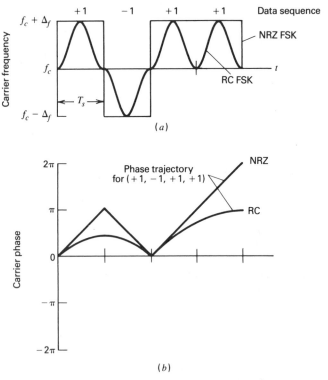

FIGURE 9.21. CPFSK: (*a*) frequency modulation, (*b*) resulting phase trajectory (RC = raised cosine).

$a_{ij} = \pm 1$. The sequence a_i then frequency shifts a carrier producing the waveform

$$c_i(t) = A \cos \left[\omega_c t + (\pi h/T_s) \sum_{j=1}^{\nu} a_{ij} q(t - jT_s) + \psi \right] \qquad (9.7.1)$$

where $h = 2\Delta_f T_s$,

$$q(t) = \int_0^t s(\rho) \, d\rho \, , \qquad 0 \le t \le T_s \qquad (9.7.2)$$

and $s(t)$ is the frequency modulation waveform of length T_s sec. If the modulation $s(t)$ is a pulse, $q(t)$ is given in (2.4.21) and the waveform is a standard FSK carrier with separation $2\pi h/T_s$. When $h = 0.5$ ($\Delta_w = 2\pi/4T_s$), the carrier is a MSK carrier, defined earlier in (9.6.15). Figure 9.21a shows an MSK carrier with particular $\mathbf{a}_i$ sequence, and the corresponding phase trajectory of $c_i(t)$ is shown in Figure 9.21b. A different vector $\mathbf{a}_i$ will produce a different trajectory. The spectrum of $c_i(t)$ is that of the MSK carrier, shown in Figure 9.18.

By modifying (rounding) the frequency pulse $s(t)$, one can produce an even smoother phase variation $q(t)$ in (9.7.2). A popular technique is the use of raised cosine (RC) pulses, defined in Figure 2.3. Figure 9.21 shows the FSK carrier and the corresponding modified trajectory it produces. The smoother phase variation leads to an even further reduction of the spectrum over the MSK waveform, as shown in Figure 9.22.

Phase tracking decoding of CPFSK is obtained by attempting to determine which phase trajectory is being received during word time $(0, T_w)$, $T_w = \nu T_s$. This is achieved by correlating the received noisy carrier with a phase coherent carrier having each possible phase trajectory and deciding which has the largest correlation y_i. The probability of erring in deciding between the correct y_i and an incorrect y_j will depend on the distance between these carrier waveforms, given by

$$d_{ij}^2 = \frac{1}{E_w} \int_0^{T_w} (c_i(t) - c_j(t))^2 \, dt$$

$$= 2 \left[1 - \int_0^{T_w} \cos \left[(\pi h/T_s) \sum_r (a_{ir} - a_{jr}) q(t - rT_s) \right] dt \right] \qquad (9.7.3)$$

The distance depends on the $\mathbf{a}_i$ sequences, the waveform $q(t)$, and the deviation coefficient h, which are effectively observed through the cosine function. Performance of the CPFSK decoding can be bounded as in (9.2.21) by

$$\text{PWE} \le (M - 1) Q \left[\left(\frac{d_{\min}^2 E_w}{2N_0} \right)^{1/2} \right] \qquad (9.7.4)$$

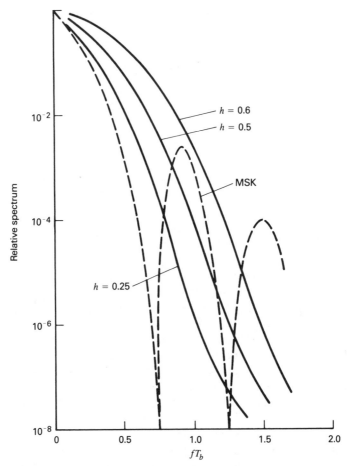

FIGURE 9.22. Spectral comparison (h = modulation index, RC = raised cosine).

where $d_{\min}$ is the normalized minimum distance d_{ij} over all i and j, $i \neq j$. Thus PWE can be determined by estimating $d_{\min}$ over all sequences $\mathbf{a}_i$. For a given phase function $q(t)$ and separation h, this usually requires numerical computations. Figure 9.23 shows some error probability bounds computed in this way for the pulsed and raised cosine modulations. The RC system has slightly poorer performance than the MSK case, but achieves the spectral advantages of Figure 9.22. By increasing the separation h, the RC performance is improved, but at the expense of increased bandwidth. Further studies of other modulation pulses, and improved sequence selection can be found in References 4–7.

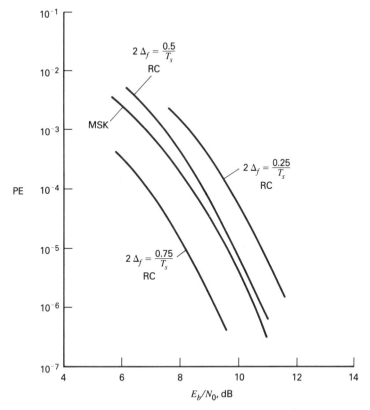

FIGURE 9.23. PE versus E_b/N_0 for CPFSK encoding.

9.8 NONCOHERENT SIGNALING (MFSK)

Our discussion of block coded systems has considered only coherent decoding. In each case it was tacitly assumed that each member of the block signaling set can be generated coherently at the receiver. This means that all block encoded carrier signals must be phase synchronized at the decoder, which may complicate the overall receiver design. For this reason there is interest in using MFSK signals with noncoherent decoders, that is, decoders that perform by ignoring the phase information. Such a noncoherent M-ary decoder can be constructed as an extension of the binary, noncoherent envelope detectors of Section 8.1. Such a system is shown in Figure 9.24.

The signal set is given by

$$s_i(t) = \left(\frac{2E_w}{T_w}\right)^{1/2} \sin\left[\omega_s t + \frac{2\pi i}{T_w} t + \psi\right], \qquad i = 1, 2, \dots, M$$

$$0 \le t \le T_w \qquad (9.8.1)$$

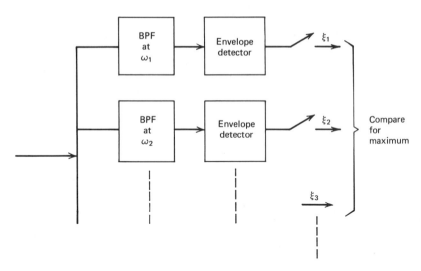

FIGURE 9.24. Noncoherent MFSK block decoder.

corresponding to a set of M frequencies $1/T_w$ Hz apart, so as to be orthogonal over $(0, T_w)$ for any arbitrary phase angle ψ. The selection of ω_s allows us to locate the frequency set anywhere on the frequency axis. Note that the FSK signal set occupies a total bandwidth of M/T_w Hz, which must be provided in the transmitter and receiver subsystems. The decoder utilizes the noncoherent decoding bank in Figure 9.24. The bandpass filters are tuned matched filters for each frequency and the envelope samples $\{\xi_i\}$ are used for comparison decisioning as in the binary case. In essence, the decoder is simply a frequency detector composed of a filtering bank, and word decisioning is based on which frequency is being received after each T_w sec. By using envelope samples, the phase angle ψ in (9.8.1) does not affect decisioning.

Performance of the noncoherent MFSK decoder is obtained by evaluating the probability that the correct envelope sample is not the largest. Since the frequencies are separated so as to produce orthogonal signals, the desired probability is the probability that the envelope sample of a filtered sine wave plus Gaussian noise does not exceed $M - 1$ envelope samples of Gaussian noise alone. The probability density of these samples was given earlier in (8.1.23), and it follows that

$$\text{PWE} = 1 - \int_0^\infty ue^{-(u^2+\rho^2)/2}I_0(u\rho)\left[\int_0^u \eta e^{-\eta^2/2}\, d\eta\right]^{M-1} du \quad (9.8.2)$$

where $\rho^2 \overset{\Delta}{=} 2E_w/N_0$ and E_w is defined in (9.8.1). Equation (9.8.2) is the noncoherent MFSK equivalent of the coherent orthogonal result in (9.4.6). The integration can be carried out by noting that

$$\left[\int_0^u \eta e^{-\eta^2/2}\, d\eta\right]^{M-1} = [1 - e^{-u^2/2}]^{M-1}$$

$$= \sum_{q=0}^{M-1} (-1)^q \binom{M-1}{q} e^{-qu^2/2} \qquad (9.8.3)$$

Substitution then yields

$$\text{PWE} = 1 - e^{-\rho^2/2} \sum_{q=0}^{M-1} (-1)^q \binom{M-1}{q} \int_0^\infty u e^{-u^2(q+1)/2} I_0(\rho\mu)\, du$$

$$= 1 - e^{-\rho^2/2} \sum_{q=0}^{M-1} (-1)^q \binom{M-1}{q} \frac{\exp\left[(\rho^2/2)/(1+q)\right]}{1+q} \qquad (9.8.4)$$

When the parameter ρ^2 is normalized to the same bit rate R_b using (9.4.9), and PWE is converted to bit error probability using (9.4.10), the curves in Figure 9.25 are generated. The results exhibit the same uniform improvement with M as in the coherent case, except that the noncoherent performance is several decibels poorer at each M. Note that we again obtain the limiting behavior of (9.4.11) as $M \to \infty$.

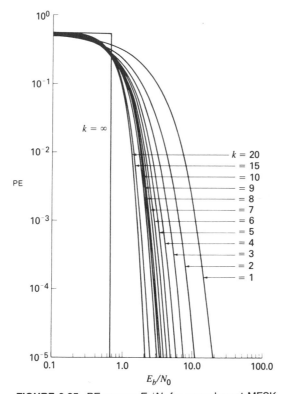

FIGURE 9.25. PE versus E_b/N_0 for noncoherent MFSK.

9.9 AMPLITUDE SHIFT KEYING (ASK)

Another popular block encoding signaling scheme uses pulse transmission with *amplitude shift keying* (ASK). In this format, a pulse is sent with one of M distinct amplitudes to represent each of the M block words of the set. Block encoding is achieved by encoding the words into specific pulse amplitudes. ASK is therefore a form of baseband PCM transmission where more than two amplitudes are used. For this reason ASK signaling is also referred to as *pulse amplitude* modulation (PAM) signaling. The prime advantage of ASK is that the bandwidth needed to send any level signal need be no larger than that needed to send a single pulse. Hence any size signal set M can be accommodated within a single pulse bandwidth, and ASK has bandwidth advantages similar to polyphase signaling.

ASK differs from the previous signaling schemes of Section 9.3 since the signals of the set no longer have equal energy. The ASK signal set can be denoted as

$$s_i(t) = s_i, \qquad 0 \le t \le T_w, \qquad i = 1, 2, \ldots, M \qquad (9.9.1)$$

where $\{s_i\}$ is a set of distinct pulse amplitudes, assumed to be equally spaced with width Δ, and symmetrical about zero, as shown in Figure 9.26a. The ith signal has energy $s_i^2 T_w$, and the entire signal set has average energy

$$E_w = \frac{1}{M} \sum_{i=1}^{M} s_i^2 T_w = \frac{2 T_w}{M} \sum_{i=1}^{M/2} \left[\frac{(2i-1)\Delta}{2} \right]^2$$

$$= \frac{\Delta^2 T_w (M^2 - 1)}{12} \qquad (9.9.2)$$

ASK word decoding is accomplished by integrating over the received baseband pulse, and determining which value of $s_i T_w$ is closest to the integrator value (Problem 9.17). A word decoding error will occur if the decoder noise causes the integrator value to lie closer to an amplitude value different from that transmitted. For the symmetrical ASK signal set, a decoding error will occur if the integrated decoder noise has a value exceeding $\pm \Delta T_w/2$ for the inner amplitudes, exceeds $\Delta T_w/2$ for the most negative amplitude, and is less than $-\Delta T_w/2$ for the highest amplitude. For additive Gaussian white noise with level N_0, the integrated noise has variance $N_0 T_w/2$. The probability of a word decision, assuming all levels equally likely, is then

$$\text{PWE} = \frac{2}{M} \left[(M-2)Q\left(\frac{\Delta T_w/\sqrt{2}}{\sqrt{N_0 T_w}} \right) + Q\left(\frac{\Delta T_w/\sqrt{2}}{\sqrt{N_0 T_w}} \right) \right]$$

$$= 2\left(\frac{M-1}{M} \right) Q\left(\frac{\Delta T_w/2}{\sqrt{N_0 T_w}} \right) \qquad (9.9.3)$$

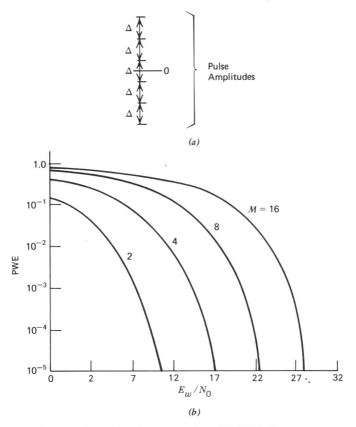

(a)

(b)

FIGURE 9.26. ASK signaling: (a) pulse amplitudes, (b) PWE (E_w = average energy per pulse).

We can rewrite this in terms of the average pulse energy by substituting from (9.9.2)

$$\text{PWE} = 2\left(\frac{M-1}{M}\right)Q\left[\left(\frac{6E_w}{(M^2-1)N_0}\right)^{1/2}\right] \qquad (9.9.4)$$

Equation (9.9.4) is plotted in Figure 9.26b as a function of the average pulse energy to noise level E_w/N_0 for several values of M. We note the obvious degradation as the number of pulse levels M is increased while maintaining fixed bit energy. Conversely, if PWE is to remain constant, it is necessary to increase average energy as the number of pulse amplitudes is increased. We emphasize that this effect is uniform in M, and we always degrade performance by increasing M with ASK signaling.

ASK signals occupy a frequency spectrum corresponding to a single pulse, and therefore extend over approximately $1/T_w$ Hz. Amplitude shifting can also be used with carrier waveforms, in which T_w sec bursts of a carrier are used with multiple amplitudes for the block signals. This produces ASK carriers that occupy a carrier bandwidth that is $2/T_w$ Hz wide around the carrier frequency. The resulting PWE performance is again given by (9.9.4) with E_w representing the average carrier word energy. Since ASK carriers shift their amplitude from word to word, they represent a form of

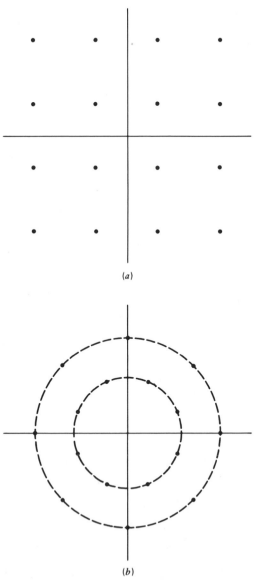

(a)

(b)

FIGURE 9.27. Signal vector sets: (a) QASK, (b) combined ASK/PSK.

amplitude modulation that must be preserved by the link. For example, no limiting can be inserted.

A further modification is to use quadrature signaling, as in (9.6.2), in which separate ASK is used on each quadrature component of a carrier. Each component is separately adjusted in amplitude each word time, and summed to form the carrier signal. This permits twice the number of bits to be sent simultaneously over the same bandwidth as a single ASK carrier link. Phase coherent decoding of the quadrature signals will then avoid the interference from the orthogonal channel. Such a system is called *quadrature ASK* (QASK) and is the amplitude modulated version of QPSK.

If a four level ASK signaling format is used on each quadrature component, the 16-ary carrier signal vector set in Figure 9.27*a* is generated. Each vector point corresponds to a particular amplitude level placed on each carier component during each word time. Thus a four-bit word can be encoded on to the QASK signal set shown. The procedure can be extended to larger signal sets as well.

Decoding performance of QASK signals is generally estimated via the union bound analysis in (9.2.16), or by bounding with the signal distance procedures in (9.2.21). Hence minimum distance between adjacent vector points becomes a key parameter in these *M*-ary vector constellations. The objective then is to attempt to spread and separate the vector points as much as possible, while maintaining average energy constraints.

Amplitude shift keying can also be combined with phase shift keying to generate ASK|PSK carriers [8–12]. During each word time the carrier has both its phase and amplitude adjusted according to the data word. Since MPSK signal vectors lie on circles of fixed radius, the combined ASK|PSK carrier signals will have vector points spaced on sets of circles of different radii. This generates modified vector constellations, as shown, for example,

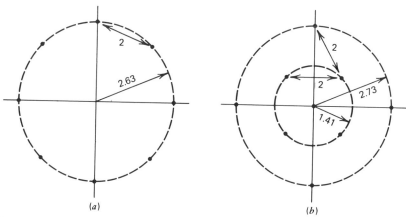

FIGURE 9.28. Eight-point signal sets with equal minimum distance: (*a*) 8 PSK, (*b*) 8-ary ASK/PSK.

in Figure 9.27*b*. Here the signal set corresponding to two-level amplitude adjustment of eight-phase PSK is shown, and the repositioning of the 16-point vector set generated by the combined modulation is evident. By controlling vector points via the modulation format, minimal vector spacing can sometimes be increased, or conversely, a specified minimal separation can be obtained at lower energy levels. For example, Figure 9.28*a* shows an 8PSK signal set with energy $E_w = (2.63)^2 = 6.92$, while Figure 9.28*b* shows an 8-point combined ASK|PSK signal set having the same minimal separation but with an average energy of only $(1.41)^2/2 + (2.73)^2/2 = 4.72$, which is 1.6 dB lower than the 8PSK energy.

ASK, QASK and combined ASK|PSK carriers have the disadvantage that their operation requires variable peak power levels. In addition, they require phase coherency, accurate timing, and suffer from power degradation and intersymbol interference if the transmitted signals are distorted and bandlimited. Further modifications in the form of partial response shaping of the QASK pulses can be employed to combat these latter effects [13].

9.10 BANDWIDTH EFFICIENCY AND DIGITAL THROUGHPUT

The previous section compared signaling formats at fixed information rates and their ability to trade off performance (PE) and bit energy at the expense of system bandwidth. By expending bandwidth, a given level of performance was achievable at the minimum possible energy level while always communicating at a fixed bit rate. Such tradeoffs are important when power is at a premium while system bandwidth is readily available. There is also, however, a significant interest in links in which channel bandwidth is constrained, and power may be expendable, as in satellite links. In this case, the objective is to design systems that transmit the most information over the bandlimited channel, at the expense of transmitter power, while maintaining a fixed level of performance. A comparison measure for such bandlimited operation is the *bandwidth efficiency*, or the *digital throughput*, defined as the ratio of the bit rate transmitted R_b to the bandwidth B used:

$$\text{Throughput} = \frac{R_b \text{ bits/sec}}{B \text{ Hz}} \qquad (9.10.1)$$

This throughput factor indicates the number of bits that can be transmitted per cycle of channel bandwidth. The larger the throughput, the more bits can be transmitted for a given link bandwidth. However, recall the operating bit rate also determines the decoding performance PE through the energy per bit parameter

$$\frac{E_b}{N_0} = \frac{P_c T_b}{N_0} = \frac{P_c}{N_0 R_b} \qquad (9.10.2)$$

This means that throughput and PE are inherently related. If we increase the bit rate by simply sending bits at a faster rate through a fixed bandwidth channel, we will eventually introduce bit distortion, and a higher E_b/N_0 in (9.10.2) will be needed to produce the same PE. This means that the power P_c must be increased to maintain the PE at the higher rate. We therefore expect that throughput can always be increased, but at the expense of increased carrier power. In communication systems operating under a bandwidth constraint, this tradeoff may be desirable in order to get higher data rates. The objective then is to find signaling formats that provide the best tradeoff of carrier power for throughput.

A standard BPSK binary system sends bits at a rate $R_b = 1/T_b$ bps and requires a carrier bandwidth of approximately $B_c \cong 2/T_b$ Hz. The throughput is then

$$\frac{R_b}{B_c} = \frac{1/T_b}{2/T_b} = \frac{1}{2} \quad \frac{\text{bps}}{\text{Hz}} \tag{9.10.3}$$

The required carrier power is $P_c = Y(N_0 R_b)$ where Y is the value of E_b/N_0 needed to achieve the desired PE. Figure 9.29 plots throughput versus normalized required carrier power for $PE = 10^{-4}$ for several classes of signaling sets. Also included is the upper bound on achievable throughout

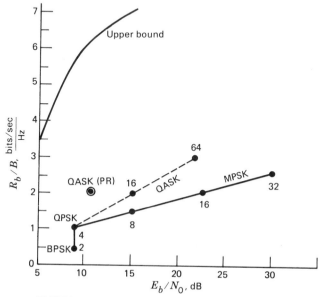

FIGURE 9.29. Throughput versus E_b/N_0. $PE = 10^{-4}$.

performance based on theoretical derivations.[†] Figure 9.29 shows the throughput point for BPSK. If we used QPSK, the bit rate is doubled for the same bandwidth, and the throughput increases to 1 bps/Hz. The required carrier power remains the same, based on our discussion in Section 9.6. This means, for example, if 10 Mbps is to be transmitted, BPSK requires 20 MHz of bandwidth, but QPSK needs only 10 MHz.

To further increase the throughput, it is necessary to increase the number of bits encoded onto the carrier without increasing the carrier bandwidth. One way that this can be achieved is by increasing the number of phase states in MPSK. With MPSK the throughput is

$$\frac{(\log_2 M)/T_w}{B_c} = \frac{\log_2 M}{2} \quad \frac{\text{bps}}{\text{Hz}} \tag{9.10.4}$$

However, the carrier power must be increased approximately as M^2 as noted in (9.5.13). Figure 9.29 shows how the throughput increases with higher level MPSK encoding.

Digital amplitude modulation (ASK) are also efficient throughput systems since additional amplitude levels can be encoded without increasing pulse bandwidth. By using QASK the throughput can then be doubled, achieving an advantage over MPSK, as included in Figure 9.29.

By employing partial response signaling, bandwidth can be reduced to twice the Nyquist bandwidth, further increasing the efficiency for QASK signals, as shown for the case $M = 4$. This is achieved at a slight increase in power level, and further improvements are possible by resorting to higher level partial response signals [12, 13]. This demonstrates that any attempt to increase throughput will invariably require an increase in decoder complexity. Bandwidth efficiency curves, as in Figure 9.29, are also convenient for signal selection. For example, if we are designing a system that is to send 90 Mbits through a 45 MHz subcarrier channel, a throughput factor of 2 is required. We see that 16-level MPSK, 4-level (per quadrature component) QASK, and 2-level QASK with partial response are candidates. The tradeoff in required power and complexity must then be made.

MFSK systems transmit $\log_2 M$ bits with each T_w sec frequency burst, but requires a channel bandwidth that increase as $M(2/T_w)$ Hz. Hence throughput for MFSK is $\log_2 M/2M$ and actually decreases with increasing M. Likewise, insertion of channel encoding, while improving PE, reduces the throughput by the coding rate.

The inherent signal design problem for improved bandwidth efficiencies

---

[†]A well-known result from information theory [14–16] states that the maximum data rate R_b achievable over a communication channel having bandwidth B, signal power P, and white noise of level N_0 is $R_b = B \log [1 + (P/N_0 B)]$. This implies that the maximum throughput R_b/B satisfies $R_b/B = \log [1 + (E_b/N_0)(R_b/B)]$. The value of R_b/B satisfying this equation at each value of E_b/N_0 is the upper bound curve plotted in Figure 9.29.

can also be described from our earlier signal space concepts, where the signals again correspond to vector points. Any digital system that transmits $\log_2 M$ bits in T_w sec, using a bandwidth B always has throughput of $\log_2 M / BT_w$. For fixed M, we see that signals with the smaller BT_w products are most bandwidth efficient. In general, signals with dimension ν require bandwidths approximately $B \approx \nu / T_w$, and the throughput is

$$\text{Throughput} = \frac{\log_2 M}{\nu} \tag{9.10.5}$$

Signals with high dimension are therefore not as bandwidth efficient as low dimension signals. The minimal number of dimensions we wish to consider is two, and the objective is to use as large a number M of signal vector points as possible in this space. This demonstrates the importance of the class of QASK and combined ASK|PSK signal sets in Figure 9.27 for producing high throughput performance. We emphasize the direct contrast here with the waveform encoding objective of Section 9.3. There the objective was to use as high a dimension (bandwidth) as possible to allow large signal separations with a given power level. In bandwidth efficiency design, the objective is to pack as many signals in as few dimensions as possible, while achieving sufficient signal spacing through increased power levels and modulation selection.

9.11 COMBINED CODING AND WAVEFORM SIGNALING

In Section 8.6 it was pointed out that the use of channel coding (converting bits to chips) can improve the error probability of a binary link. The primary disadvantage was that the chip rate, or bandwidth, had to increase to maintain the desired data bit rate. In this chapter it was shown that waveform encoding of blocks of bits can also improve a communication link. While some waveform signaling formats improved bit error probability, others permitted increasingly larger block sizes to be transmitted over bandlimited channels. The most important of this latter type were the class of MPSK and MASK waveforms, which directly improved throughput, but required significant power increases to maintain PE performance. A question then arises as to whether some form of channel coding can possibly be combined with bandwidth-efficient waveform encoding to produce a communication link that is also efficient in required power.

In order to achieve this, the channel coding must provide for the coding of sequences of data words into sequences of waveform encoded signals. Decoding is then based on the entire sequence rather than deciding one word at a time. Sequence coding allows the insertion of coding memory, so that signal selection for a particular data word depends on the past words (that is, on the previous "states" of the coder). This memory helps to increase the separability of the sequences. The design objective then is to find systematic ways of deriving the necessary coding and the choice of signals so that this sequence decoding improves error probability without expanding bandwidth.

A way to accomplish this was proposed by Ungerboeck [17, 18], using the concept of signal set partitioning in conjunction with convolutional trellis coding. The details will not be covered here, but the method can be summarized as follows. The idea is to begin with a bandwidth-efficient signal

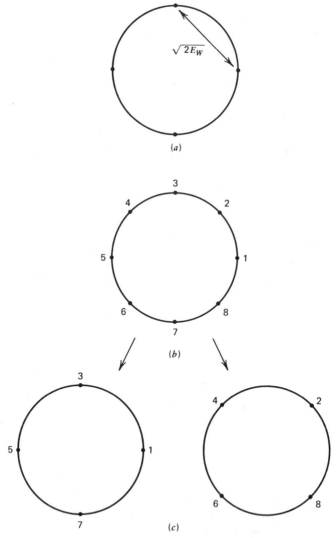

FIGURE 9.30. Example of set partitioning: (*a*) standard QPSK, (*b*) 8 PSK, (*c*) partitioned version of (*b*).

set larger than that needed to encode each data word alone. This set is then partitioned into subsets, and only signal combinations within each subset that have largest minimum distance (vector separation) are used. The coding operation then selects sequences of the candidate signals from the family of subsets. Decoding is achieved in two steps: (1) by making preliminary decisions among the candidate signals within each subset, then (2) deciding among the sequence of subsets that could have occurred via the coding. This latter operation is carried out similar to the Viterbi search algorithms, and eventually decide the most likely sequence based on the preliminary decisions. The procedure results in sequence decisions that are made from signals having a greater minimum distance then if independent block waveforms were transmitted.

The design of the required convolutional code is not easily described. Most of the usable codes that have been reported have been derived by heuristic code selection, using some basic rule for optimally selecting signal paths while transiting state diagrams. The interested reader may wish to pursue this topic in References 11, 18–21.

A simple example of the set of partitioning is shown in Figure 9.30. Two-bit words are to be transmitted in sequence. If a direct word-by-word waveform encoding is used, then each word will be coded into a QPSK signal set as in Figure 9.30a (four signals in two-dimensional space). A decision would then be made among the QPSK signal vectors each word time, with a minimum signal distance of $\sqrt{2E_w}$. Now consider beginning with the 8PSK signal set of Figure 9.30b and partitioning into the two subsets of Figure 9.30c. The four signals to be used during any word time will be an antipodal signal from each subset. For example, the signal vector points $(1, 5, 2, 6)$ represent one set, while $(3, 7, 4, 8)$ would represent another. Other combinations of the same signal pairs can be denoted. Within each subset a maximal distance (separation of $2\sqrt{E_w}$) decision will be made. The coding operation at the transmitter determines which signal from which subset should be selected during each word time, based on the data bit sequence. All of the 8PSK signals will eventually be used, when considering all the possible sequences that can occur.

The transmitter encoder has the overall block diagram shown in Figure 9.31a. The sequence of two-bit words are convolutionally coded into a corresponding sequence of three-chip words via the convolutional coder with rate 2/3. Each chip word is then waveform encoded into one of the eight phase states (Figure 9.31b), and transmitted as an 8PSK carrier waveform each word time. The receiver decoder makes the antipodal phase comparison from the known candidate pairs in Figure 9.30c, and stores the resulting signal "score" concerning the possible order in which the phase subsets occurred. A sequence decision concerning the most likely phase states is then made. The final decision is made from among sequence of signals having average distance greater than the $\sqrt{2E_w}$ if independent word coding was used. This translates directly to improved decoding performance.

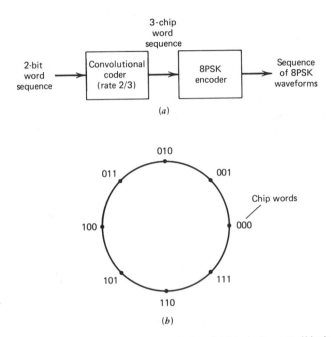

FIGURE 9.31. Coder for two-bit, 8 PSK transmission: (*a*) block diagram, (*b*) phase mapping.

The system in Figure 9.31 can be extended in complexity by using a higher number of phases (more signal vector points) and by coding sequences of longer words (for example, four-bit words instead of two-bit words). The basic concept remains the same, with the convolutional coder mapping the bit words onto the partitioned phase states. Classes of QASK or ASK|PSK signal constellations can be used as alternatives to the MPSK signal sets.

Figure 9.32 compares the resulting bit error probability for sequence coding, predicted from simulation and theoretical bounds, as reported in the literature [4, 18]. The uncoded QPSK curve refers to the PE obtained if independent two-bit word-by-word encoding is used. The result of the sequence coding is shown for a rate 2/3 convolutional coder using a two-bit, 8PSK system (Figure 9.31) and a more complex four-bit, 16PSK system. Note that at $PE = 10^{-4}$, approximately 3 dB improvement is gained over uncoded QPSK with 8PSK, and this can be extended to about 5 dB with 16PSK.

It is to be emphasized that the power improvement is achieved without increasing bandwidth, since the transmission of 8PSK and 16PSK signals require the same bandwidth as uncoded QPSK. Instead, the improvement is obtained by using a signal set larger than that needed. That is, the improvement is obtained by inserting redundancy in the signal set, and only

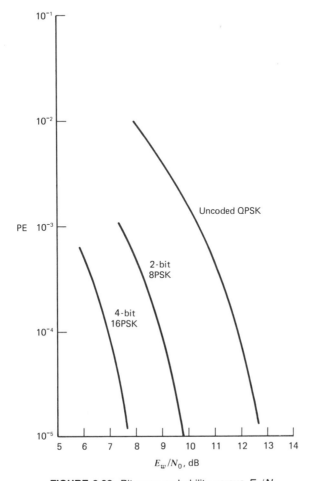

FIGURE 9.32. Bit error probability versus E_w/N_0.

using a fraction of the available signals each word time. For this reason, this combination of sequence coding and set partitioning is referred to as *signal space coding*. Contrast this with the error correction described in Section 8.6, where the redundancy appeared in the extra chips that were sent. While the latter required excess bandwidth, the receiver still required only a binary decoder. Signal space coding requires no increase in bandwidth, but complicates the decoder design, since an increasingly larger number of signal waveforms must be handled by the receiver decoder. This again illustrates the primary technological direction of modern digital communications, in which extensive decoder processing is accepted as a viable tradeoff for bandwidth and power efficiency of the communication channel.

REFERENCES

1. Weber, C. *Elements of Detection and Signal Design*, Springer-Verlag Inc., New York, 1987, Chap. 14.

2. Viterbi, A. *Principles of Coherent Communications*, McGraw-Hill, New York, 1966, Sect. 8.8.

3. Lindsey, W. and Simon, M. *Telecommunications System Design*, Prentice-Hall, Englewood Cliffs, N.J., 1973.

4. Benedetto, S., Biglieri, E. and Castellani, V. *Digital Transmission Theory*, Prentice-Hall, Englewood Cliffs, N.J., 1987.

5. Aulin, T. and Sundberg, C. "Continuous Phase Modulation—Part I," *IEEE Trans. Comm.*, vol. COM-29, 1981, pp. 196–209.

6. Ungerboeck, G. "Channel Encoding with Multi-phase Signals," *IEEE Trans. Inf. Theory*, vol. IT-28, 1982, pp. 55–67.

7. Anderson, J.B. and Taylor, D.O. "A Bandwidth-Efficient Class of Signal–Space Codes," *IEEE Trans. Inf. Theory*, vol. IT-24, November 1978, pp. 703–712.

8. Thomas, C.M., Weldner, M. and Durrani, S. "Digital Amplitude Phase Keying with M-ary Alphabets," *IEEE Trans. Comm.*, vol. COM-22, February 1974, pp. 168–180.

9. Salz, J., Sheehan, J. and Paris, D. "Data Transmission by Combined AM-PM," *Bell Syst. Tech. J.*, September 1971, pp. 174–193.

10. Miyauchi, K., Seki, S. and Ishio, H. "New Techniques for Generating and Detecting Multilevel Signals," *IEEE Trans. Comm.*, vol. COM-24, February 1976, pp. 263–267.

11. Forney, D. "Efficient Modulation for Bandlimited Channels," *IEEE Trans. Selected Areas Comm.*, vol. SAC-3, Sept. 1984, pp. 632–647.

12. Lender, A. "Correlative Techniques and Applications to Digital Radio," in *Digital Communications—Microwave Applications*, K. Feher, Prentice-Hall, Englewood Cliffs, N.J., 1981, Chap. 7.

13. Lucky, R., Salz, J. and Weldon, E. *Principles of Data Transmission*, McGraw-Hill, New York, 1968.

14. Gallagher, R. *Information Theory and Reliable Communications*, McGraw-Hill, New York, 1969.

15. Wozencraft, W. and Jacobs, I. *Principles of Communication Engineering*, Wiley, New York, 1965.

16. Abramson, N. *Information Theory and Coding*, McGraw-Hill, New York, 1963.

17. Ungerboeck, G. "Channel coding with multilevel/phase signals," *IEEE Trans. Information Theory*, vol. IT-28, Jan. 1982, pp. 55–67.

18. Ungerboeck, G. "Trellis coded modulation with redundant signal sets—Part 1 and Part 11," *IEEE Communication Magazine*, Vol. 25, No. 2, February 1987, pp. 5–22.

19. Wei, L.F. "Rotationally invariant convolutional channel coding with expanded signal space—Part I: 180 degrees," *IEEE Trans. Selected Areas in Comm.*, vol. SAC-2, Sept. 1984, pp. 659–672.

20. Wei, L.F. "Rotationally invariant convolution channel coding with expanded

signal space—Part II: nonlinear codes," *IEEE Trans. Selected Areas in Comm.*, vol. SAC-2, Sept. 1984, pp. 672–686.

21. Calderbank, A.R. and Mazo, J.E. "A new description of trellis codes," *IEEE Trans. Information Theory*, vol. IT-30, Nov. 1984, pp. 784–791.

PROBLEMS

9.1   (9.1) Compute the encoded waveform spectrum in (9.1.3) by formulating the problem as in (2.1.25), determining the equivalent of the correlation $R_m(\tau)$ in (2.1.12), and then Fourier transforming.

9.2   (9.2) Show how the PWE in (9.1.4) must be altered if the y_i are quantized before comparison for block decoding. Consider $M = 2$ only.

9.3   (9.2) Prove formally the union bound inequality; that is, prove that if $x_1, x_2, \ldots, x_m$ are random numbers, then

$$\text{Prob}\,[x_1 \leq x_2, x_3, \ldots, x_m] \leq \sum_{i=2}^{m} \text{Prob}\,[x_1 \leq x_i]$$

9.4   (9.2) Derive a lower bound on the PWE in (9.1.4) by proving the following sequence of inequalities:

$$\text{PWE} \geq \frac{1}{M}\,\text{PWE}_{i_0}$$

$$\geq \frac{1}{M} \begin{bmatrix} \text{binary probability of error} \\ \text{between signal } i_0 \text{ and the} \\ \text{closest signal } s_j, \; j \neq i_0 \end{bmatrix}$$

$$\geq \frac{1}{M} \begin{bmatrix} \text{binary probability of the} \\ \text{two closest signals} \end{bmatrix}$$

where i_0 is any particular value of i. (b) Use part (a) and (9.2.17) to show that the minimal distance of the signal set provides both an upper bound and a lower bound (with different coefficients) on PWE.

9.5   (9.2) Let $D_{ij}^2 \overset{\Delta}{=} |\mathbf{s}_i - \mathbf{s}_j|^2$ be the distance between two signals in vector space. Define d_m as the minimum D_{ij}, $i \neq j$, and $\bar{D}$ as the average D_{ij}, $\bar{D}^2 = (\sum_i^M \sum_j^M D_{ij}^2)/M^2$. (a) Show that for M signals on a sphere of radius $\sqrt{E_w}$

$$\bar{D}^2 = 2E_w - 2|\mathbf{s}|^2$$

where $\mathbf{s}$ is the centroid of the signal set. (The vector $\mathbf{s}$ has for its qth component $\sum_{i=1}^{M} s_{iq}/M$, where s_{ij} is the jth component of vector $\mathbf{s}_i$.) (b) Under what condition on the set is $\bar{D}^2$ maximized? (c) Show that $\bar{D}^2 \geq d_m^2[(M-1)/M]$. (d) Use part c to prove $d_m^2 \leq 2E_w M/M - 1$ for any signal set.

9.6   (9.2) A signaling interval $(0, T)$ is divided into L Δ sec intervals. A message is sent by placing voltage pulses in a set of Q intervals. M different messages are sent by selecting M different sets of Q intervals. The interval voltage samples are observed in white Gaussian noise. Determine the optimal test for decoding messages using an observation of the interval voltage samples. Explain simply in words how the decoder operates.

9.7   (9.3) Prove that in two-dimensional space, an equal energy simplex signal set corresponds to the three vertices of a triangle inscribed within a circle.

9.8   (9.3) Show that the signals in (9.3.15) are orthogonal if the vectors h are orthogonal.

9.9   (9.4) Substitute (9.4.9) into (9.4.6) and prove (9.4.11) by taking the limit as $M \to \infty$. First prove that

$$\underset{M \to \infty}{\text{Lim}} \{(M-1) \log_e [1 - \tfrac{1}{2} \text{Erfc}(x + \beta \log_2 M)]\}$$
$$= \begin{cases} -\infty, & \text{if } \beta < \log_e 2 \\ 0, & \text{if } \beta > \log_e 2 \end{cases}$$

9.10 (9.4) If a constant voltage c is added to every decoder output sample y_i in block decoding, the same PWE will occur since the c will not affect which is maximum. Consider an equally correlated signal set $\{s_i(t)\}$ with correlation γ and energy E. Let c have the value $c = \int_T^{T+\delta} An(t)\, dt + A^2\delta$ for some δ. (a) Show that $(y_i + c)$ would be the decoder output for a signal set

$$s_i'(t) = \begin{cases} s_i(t), & 0 \le t \le T \\ A, & T \le t \le T + \delta \end{cases}$$

(b) Show this new signal set, defined over $(0, T+\delta)$ has correlation $\gamma_{ii} = E + A^2\delta$ and crosscorrelation $\gamma_{ij} = E\gamma + A^2\delta$. (c) Show that for the choice $A^2\delta = -E\gamma$ the new set is an orthogonal set with energy $E(1-\gamma)$. (d) What can we conclude about the PWE of the two signal sets $\{s_i(t)\}$ and $\{s_i'(t)\}$?

9.11 (9.4) A signal takes the form of a unit step function, where the step can occur at one of the time instants $t_1 < t_2 < t_3 \cdots t_k$, all equally likely. The step function is observed in white Gaussian noise (spectral level N_0 W/rps) at the sampling instants $q_1, q_2, \ldots, q_k$ where $t_i < q_i < t_{i+1}$. (a) Determine the opimal test for determining where the step has occurred. (b) Which parameters will determine the ability to resolve two different step times, assuming a "signal distance" criterion?

9.12 (9.4) A coherent M-ary system uses binary waveforms (N consecutive pulses of $+1$ and -1 amplitude) as its signal set. A student claims "to decode with minimal PE, the receiver should detect each of the N pulses one at a time, decide ± 1 for each, then decode which binary waveform sequence is being received from the N decisions." Is this correct? Explain simply in words.

9.13 (9.4) A communication system is to transmit k bits in T_w sec. (a) What bandwidth is required if each bit is sent as an antipodal BPSK waveform? (Use main-hump bandwidth.) (b) What bandwidth is required if the k bits are first coded into orthogonal $(0, 1)$ binary vectors, then each vector symbol $(0, 1)$ is sent as an antipodal BPSK waveform, with the overall data rate remaining the same?

9.14 (9.6) Show that a QPSK system achieves twice the bit rate per carrier bandwidth that a binary PSK system does but requires twice the carrier power for the same PE.

9.15 (9.6) A communication link operates at 10 kbps with a received signal power of 9 μW and a receiver noise level $N_0 = 10^{-10}$ W-sec. Determine the following error probabilities for the stated system: (a) PE for binary, NRZ PSK signal set; (b) PWE for QPSK; (c) PE for QPSK; (d) PWE for M-ary PSK ($M = 32$); (e) PWE for $M = 4$ simplex set.

9.16 (9.8) A noncoherent decoder for MFSK can be constructed as follows. During each word time $(0, T_w)$ we time sample the decoder input waveform $x(t)$ generating the N voltage samples $x(i/2B)$. We then compute the discrete Fourier transform function

$$\sum_{i=1}^{N} x\left(\frac{i}{2B}\right) e^{-j(2\pi i q / 2BT_w)}$$

for different q. Explain how and why this can noncoherently decode the MFSK signal set.

9.17 (9.9) A pulse is received with one of M amplitudes $\{s_i\}$ and width T_w, along with additive white Gaussian noise. The noisy signal is integrated for T_w sec producing a voltage y. (a) Show that [Prob pulse amplitude $= s_i | y$] $\geqq$ Prob [pulse   amplitude $= s_j | y$] if $(s_i y - s_i^2 T_w) \geqq$ $(s_j y - s_j^2 T_w)$. (b) Show that this is equivalent to determining if y is closer to $s_i T_w$ or to $s_j T_w$. (c) Use these facts to define a maximum likelihood decoding procedure for ASK signaling.

9.18 (9.10) An M-ary pulse position modulated (MPPM) system communicates by placing a τ sec carrier pulse in one of $M \tau$ sec slots to represent a data word of $\log_2 M$ bits. Use the main hump bandwidth of the carrier to determine the MPPM throughput.

10

FREQUENCY ACQUISITION AND SYNCHRONIZATION

In our earlier discussions of RF carrier processing we tacitly assumed that the received RF frequency was known fairly accurately at the receiver. This allowed the RF front end filter to be properly tuned to the carrier and allowed RF mixing with a local oscillator to position the received carrier directly in the proper IF frequency band. The advantage of this is that the IF filter bandwidth can be made as narrow as possible, reducing the total noise involved in the demodulation processing. In many forms of practical systems, however, frequency instabilities and Doppler effects during transmission may prevent knowledge of the RF frequency location when communication first begins, and system requirements may prevent manual local frequency tuning. For this reason, most types of systems must perform an automatic *frequency acquisition* and *tracking* operation prior to any communication transmission. By frequency acquisition we refer to the operation of internally generating a local frequency identical to, or a fixed distance away from, the actual frequency of the received carrier. By frequency tracking we refer to the operation of continually maintaining this local frequency in synchronism with the received frequency, even though the latter may be undergoing variation during communication. Once acquisition has been successful and tracking initiated, communication can begin, and our demodulation and decoding analysis in Chapters 5–9 is then applicable.

In addition to the mandatory frequency acquisition, a digital system has an additional requirement for bit and word timing, and possibly for phase referencing in phase coherent systems. Bit and word timing is generally provided by receiver bit clock synchronized to the data. By phase referencing we refer to the general operation of providing a local carrier frequency in exact, or almost exact, phase synchronism with a received carrier (or subcarrier) at the same frequency. We have shown in earlier analysis several applications where phase synchronism allows specific performance advantages. Phase referencing may be desired at the RF, IF, or subcarrier levels, depending on system implementation. Phase referencing may be obtained simultaneously for all three stages from a single phase referencing subsystem or may be obtained individually at each stage by separate subsystems.

These digital synchronizing operations are provided by a synchronization subsystem that operates in conjunction with the receiver demodulation and decoding, as was shown back in Figure 1.2. The design of this synchronization system is becoming increasingly more important in modern communication systems, and in many cases its accuracy actually becomes the ultimate limit to the decoding performance. In addition, the synchronization system may be used for purposes other than simply aiding digital decoders. For example, the reference carrier may be used for Doppler tracking or carrier retransmission, while the bit clock may be used for establishing coherent timing between multiple receivers of a satellite constellation. Often these auxiliary operations may place stricter requirements and specifications on synchronization performance than the decoding alone. In this chapter we consider some of the typical implementations and hardware used for this synchronization operation.

10.1 FREQUENCY ACQUISITION

In communication systems using RF–IF conversion it is necessary that the received and local carrier frequencies be within the IF bandwidth of each other. If the frequencies are too widely separated for the IF and no provision is made to reduce the separation, no IF processing will take place. Frequency acquisition can be achieved by manually adjusting the local frequency or can be achieved by an automatic frequency acquisition loop (FAL).

A typical FAL block diagram is shown in Figure 10.1. The loop uses a

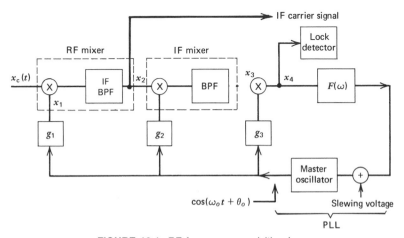

FIGURE 10.1. RF frequency acquisition loop.

single oscillator, called a *master* oscillator, and multiplies or divides its frequency to the RF and IF. The RF is used to mix the received carrier from the antenna to the IF stage, which is then mixed with the IF to a tracking frequency. The master oscillator is then translated to the tracking frequency, where it is phase detected with the received carrier in a phase lock loop (PLL). When in lock, the PLL forces the master oscillator to operate at the required frequency for maintaining the RF, IF, and tracking frequencies. This is referred to as a *double conversion*. FAL, or simply called a "long" loop. When the FAL alone is used for acquiring the RF frequency, the operation is called *quiescent acquisition*. However, in some instances the master oscillator and received frequency may be so far apart that no quiescent acquisition can take place. In this case a *slewing* signal must be applied to the loop oscillator to force the frequencies into near alignment, from which the automatic tracking can take place. This slewing operation is called *frequency searching*. With frequency searching often an auxiliary circuit called a *lock detector* is provided for recognizing when the frequencies are close to being aligned. The lock detector disengages the slewing operation and allows the PLL tracking to be initiated. In typical systems the lock detector monitors the PLL error voltage and uses a threshold test for lock-in recognition.

To design an acquisition loop properly, it is necessary to understand its basic quiescent operation. To determine these, we write the FAL loop equations throughout the system, where signals refer to points in Figure 10.1. Let the received RF signal be written as

$$x_c(t) = A \sin \left[\omega_c t + \theta(t) + \psi(t) \right] + n(t) \qquad (10.1.1)$$

as in Chapter 3. Here A is the RF carrier amplitude, $\theta(t)$ represents phase or frequency modulation, and $n(t)$ is the RF noise. We assume only an angle modulated carrier, but the result extends to AM if we interpret (10.1.1) as the carrier component only. The function $\psi(t)$ again accounts for the extraneous phase or frequency variation and is now important to our discussion. If the received carrier has a frequency offset from the desired RF frequency ω_c, it must be represented by $\psi(t)$. Thus if the frequency offset is Ω rps, then in (10.1.1)

$$\psi(t) = \Omega t \qquad (10.1.2)$$

If the frequency is varying or if oscillator phase noise is present, the analytical expression for $\psi(t)$ must be suitably modified. We denote the master oscillator signal as

$$x_0(t) = \cos \left[\omega_o t + \theta_o(t) \right] \qquad (10.1.3)$$

where the amplitude level is absorbed into the mixer gain. The master oscillator output is frequency translated (multiplied or divided) by the frequency factor g_1 to the RF, producing the RF mixer signal

$$x_1(t) = \cos\left[g_1\omega_o t + g_1\theta_o(t)\right] \qquad (10.1.4)$$

according to our discussion in Section 3.2. The output of the RF mixer is then

$$x_2(t) = K_{m_1} A \sin\left[(\omega_c - g_1\omega_o)t + \theta + \psi - g_1\theta_o\right] + n_{IF}(t) \qquad (10.1.5)$$

where K_{m_1} is the mixer gain, $n_{IF}(t)$ is the mixed IF noise, and the t has been dropped from the phase functions. Recall that if the RF bandwidth is much wider than the bandwidth of θ_o in (10.1.3), the noise $n_{IF}(t)$ is simply a frequency shifted version of the RF bandpass noise. Simultaneously, the master oscillator is frequency translated by g_2 to provide the mixing signal for converting the IF carrier to the PLL tracking frequency. The loop input is then

$$x_3(t) = K_{m_1} K_{m_2} A \sin\left\{\left[\omega_c - (g_1 + g_2)\omega_o\right]t + \theta + \psi - (g_1 + g_2)\theta_o\right\} + n_2(t)$$

$$(10.1.6)$$

If we now desire the PLL to force the master oscillator to track the carrier in (10.1.6), it is necessary that $g_3\omega_o = \left[\omega_c - (g_1 + g_2)\omega_o\right]$, or

$$g_1 + g_2 + g_3 = \frac{\omega_c}{\omega_o} \qquad (10.1.7)$$

In this case the PLL has a mixer output given by

$$x_4(t) = K_m A \sin\left[\theta + \psi - (g_1 + g_2 + g_3)\theta_o\right] + n_m(t) \qquad (10.1.8)$$

where $K_m = K_{m_1} K_{m_2} K_{m_3}$ is the mixer gain product and $n_m(t)$ is the low frequency mixer noise process described in Section 4.8. Inserting (10.1.7) yields

$$x_4(t) = K_m A \sin\left[\theta + \psi - \left(\frac{\omega_c}{\omega_o}\right)\theta_o\right] + n_m(t) \qquad (10.1.9)$$

Comparison of (10.1.9) with the basic PLL system equation [cf. (5.6.3)] shows that this tracking phase error function is identical to that generated in the equivalent system shown in Figure 10.2. The latter is a modified PLL equivalent system containing the actual PLL low pass filter, the total FAL mixer gains, and an effective phase multiplication by the ratio ω_c/ω_o. The loop is nonlinear in that it involves the sine of the phase tracking error. Thus the double conversion RF acquisition loop in Figure 10.1 is equivalent in its

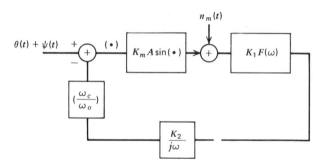

FIGURE 10.2. The equivalent FAL loop.

tracking capability to a modified, nonlinear PLL. Note that the loop does not depend on the actual values of the frequency translation parameters g_1, g_2, and g_3. This means that the FAL can be generalized to other implementations, corresponding to different combinations of these parameters. For example, letting $g_3 = 1$ corresponds to phase tracking directly at the master oscillator frequency ω_o. Letting $g_2 = 0$ corresponds to eliminating the intermediate IF frequency mixing and converting the system to a single conversion FAL. Letting $g_1 = 0$ produces an FAL that frequency locks to the IF instead of the RF. In all cases the performance is governed by the loop model in Figure 10.2. The selection of the master oscillator frequency ω_o is generally based on hardware considerations. The lower its value, the less it must be deviated for RF frequency acquisition, but the larger the frequency multiplication to RF that must be provided.

To evaluate acquisition behavior, we need only study the equivalent system. Note that the effective loop input is the RF carrier phase variation $\theta(t) + \psi(t)$. The RF phase variation tracking error generated in the FAL is then defined as

$$\psi_e(t) \overset{\Delta}{=} \psi(t) - \left(\frac{\omega_c}{\omega_o}\right)\theta_o(t) \qquad (10.1.10)$$

The corresponding RF frequency error in radians per second is then

$$\omega_e(t) \overset{\Delta}{=} \frac{d\psi_e}{dt} = \frac{d}{dt}\left[\psi(t) - \left(\frac{\omega_c}{\omega_o}\right)\theta_o(t)\right] \qquad (10.1.11)$$

In our earlier analysis of tracking loops we assumed the phase error $\psi_e(t)$ was small, which linearized the loop. During frequency acquisition, however, we cannot initially impose this assumption. In fact, we are primarily interested in the quiescent locking procedure specifically, when the frequency separation is large. In this case we must treat the FAL as a true nonlinear system, and evaluate its dynamical behavior from the basic nonlinear differential equation describing its loop. The equation can be easily generated from (10.1.11), using the fact that

$$\frac{d\theta_o}{dt} = K_1 K_2 F\{AK_m \sin(\theta + \psi_e) + n_m(t)\}$$

$$= AK_m K_1 K_2 F\{\sin(\theta + \psi_e)\} + K_1 K_2 F\{n_m(t)\} \qquad (10.1.12)$$

Here $F\{x(t)\}$ represents the output time function of the loop filter $F(\omega)$ when its input is $x(t)$, K_2 is the sensitive gain of the master oscillator, and K_1 is the gain of the loop filter. When (10.1.12) is inserted into (10.1.11), we obtain the basic system differential equation

$$\frac{d\psi_e}{dt} + AKF\{\sin(\theta + \psi_e)\} = \frac{d\psi}{dt} - K_1 K_2\left(\frac{\omega_c}{\omega_o}\right) F\{n_m(t)\} \qquad (10.1.13)$$

where K is now the total component gain around the entire loop,

$$K = K_M K_1 K_2\left(\frac{\omega_c}{\omega_o}\right)$$

$$= K_{m_1} K_{m_2} K_{m_3} K_1 K_2\left(\frac{\omega_c}{\omega_o}\right) \qquad (10.1.14)$$

where it is to be understood that a single conversion FAL would have the appropriate mixer gain (K_{m_1} or K_{m_2}) deleted. Frequency acquisition therefore reduces to simply an investigation of the nonlinear equation in (10.1.13). The forcing function of the system is the extraneous phase variation $\psi(t)$, relative to the desired RF carrier frequency, and the filtered mixer noise. Note the manner in which the carrier modulation influences behavior, effectively appearing as a variable coefficient in the system equation. To handle the modulation term analytically it is convenient to expand the sine term as

$$\sin(\theta + \psi_e) = \sin[\psi_e(t)] \cos[\theta(t)] - \cos[\psi_e(t)] \sin[\theta(t)] \qquad (10.1.15)$$

If the $\theta(t)$ corresponds to a modulating subcarrier with amplitude Δ and frequency ω_s outside the bandwidth of the loop filter, it is best to expand (10.1.15) using Bessel functions, as in (2.4.10). The low pass filtering operation of F in (10.1.13) then passes only the zero order term, yielding

$$AKF\{\sin(\psi_e + \theta)\} \cong AKJ_0(\Delta)F\{\sin\psi_e\} \qquad (10.1.16)$$

In this case the modulation merely suppresses the carrier amplitude A, effectively changing its value to

$$A_c = AJ_0(\Delta) \qquad (10.1.17)$$

This is simply the amplitude of the carrier component of (10.1.1), and we see that only this amplitude enters the system differential equation. This reflects the fact that only the carrier component of the RF signal is being tracked by the loop. The system equation in (10.1.13) becomes

$$\frac{d\psi_e}{dt} + A_c KF\{\sin(\psi_e)\} = \frac{d\psi}{dt} - \left(\frac{K}{K_m}\right) F\{n_m(t)\} \qquad (10.1.18)$$

If the modulation $\theta(t)$ is composed of a sum of N modulating subcarriers, with deviations Δ_i, respectively, then (10.1.17) is instead

$$A_c = A \prod_{i=1}^{N} J_0(\Delta_i) \qquad (10.1.19)$$

If the RF carrier modulation was a unit level binary waveform (e.g., a binary code or digital PCM waveform), modulated directly onto the carrier with phase deviation $\pm\Delta_c$, then (10.1.15) becomes, from (2.5.7),

$$-r(t) \sin \Delta_c \cos \psi_e(t) + (\cos \Delta_c) \sin [\psi_e(t)] \qquad (10.1.20)$$

Since the first term will typically have time variations outside the loop filter bandwidth, only the second term affects the loop equation. This means A is effectively converted to the value

$$A_c = A \cos \Delta_c \qquad (10.1.21)$$

This again corresponds to the suppressed carrier amplitude due to the binary waveform. Lastly, if $\theta(t)$ is a random modulating waveform, then $\cos \theta(t)$ and $\sin \theta(t)$ in (10.1.15) must be treated as a random process, requiring an analysis similar to Section 2.5. Taking only the constant portion of the $\cos \theta(t)$ term, and assuming that $\psi_e(t)$ is small, we have

$$\sin [\theta(t) + \psi_e(t)] \approx |\Psi_\theta(1)| \sin \psi_e(t) - \sin \theta(t) \qquad (10.1.22)$$

where $\Psi_\theta(\omega)$ is the first order characteristic function of $\theta(t)$. This means that we can write

$$A_c = A|\Psi_\theta(1)| \qquad (10.1.23)$$

and the second term in (10.1.22) can be incorporated with the right side of (10.1.18), forming a new effective noise term

$$n_m(t) + (A_c K) \sin \theta(t) \qquad (10.1.24)$$

In this case the random carrier modulation has suppressed the carrier

TABLE 10.1   Tracking Carrier Amplitudes[a]

Modulation Format	A_c
No modulation	A
Subcarrier, deviation Δ	$AJ_0(\Delta)$
K subcarriers, deviation Δ_i	$A\prod_{i=1}^{K} J_0(\Delta_i)$
Binary waveform, deviation Δ_c	$A(\cos \Delta_c)$
Binary waveform (Δ_c) plus K subcarriers, (Δ_i)	$A(\cos \Delta_c)\prod_{i=1}^{K} J_0(\Delta_i)$
Random, characteristic function $\Psi_\theta(\omega)$	$A|\Psi_\theta(1)|$

[a] A = RF carrier amplitude; $A^2/2$ = RF carrier power.

component, as we know from (2.5.17), while adding an effective noise interference to the desired loop tracking operation.

In summary, we see that any modulation superimposed on the carrier during frequency acquisition acts to reduce the RF carrier amplitude A to an effective value of A_c in the loop dynamical equation. The various forms of A_c are summarized for convenience in Table 10.1. These values of A_c are merely the amplitudes of the RF carrier component, and the discussion here has shown that only this component is being used for frequency acquisition.

10.2   DYNAMICAL ANALYSIS OF FREQUENCY ACQUISITION

We wish to examine the frequency acquisition behavior described by (10.1.13). Initially we concentrate only on the properties of the actual frequency synchronization itself and neglect the effect of the added noise $n_m(t)$. We initially assume no modulation term $\theta(t)$, contending that such modulation will be placed outside the filter bandwidth or that no modulation will be transmitted during the frequency acquisition operation. We assume that the received and local RF frequencies differ by Ω rps, as in (10.1.2), so that

$$\frac{d\psi}{dt} = \Omega \qquad (10.2.1)$$

Under these conditions, quiescent frequency acquisition is described by the differential equation

$$\frac{d\psi_e(t)}{dt} + A_c KF\{\sin \psi_e(t)\} = \Omega \qquad (10.2.2)$$

Let us consider first the case of a simple first order loop, where the filter is effectively removed, that is, $F\{\sin \psi_e\} = \sin \psi_e$. In this case (10.2.2) becomes

$$\frac{d\psi_e(t)}{dt} + A_cK \sin \psi_e(t) = \Omega \tag{10.2.3}$$

The frequency error of the FAL, as given in (10.1.11), is then given specifically by

$$\omega_e(t) = \Omega - A_cK \sin [\psi_e(t)] \tag{10.2.4}$$

To determine frequency behavior, we must solve (10.2.3) for $\psi_e(t)$ and insert the solution in (10.2.4). We immediately note a basic property of the solution. If the frequency error $\omega_e(t)$ is to be driven to zero (i.e., the frequency acquisition is to occur), then it is necessary that $\psi_e(t)$ approach a final value ψ_{e_0} such that $\Omega - A_cK \sin \psi_{e_0} = 0$, or

$$\psi_{e_0} = \sin^{-1}\left(\frac{\Omega}{A_cK}\right) + i2\pi, \qquad i = 0, 1, 2, \ldots \tag{10.2.5}$$

This, however, has a solution only if $\Omega/A_cK \le 1$. Since the loop noise bandwidth of a first order loop is given by $B_L \overset{\Delta}{=} A_cK/4$ (see Table 5.1), we can restate this condition as

$$\Omega \le 4B_L \tag{10.2.6}$$

Thus a first order frequency tracking loop eventually acquires an RF frequency offset Ω rps only if (10.2.6) is satisfied. This requires that Ω be within the loop noise bandwidth. If the frequency offset is too large, no steady state acquisition is achieved. If (10.2.6) is satisfied, the system achieves a steady state of zero frequency error and an RF phase error given (10.2.5). This means the RF frequencies have been brought into synchronism and the local RF carrier tracks the received RF carrier with the phase error ψ_{e_0}.

Actual loop behavior is best illustrated by examining the relation between the frequency variable ω_e and the phase error ψ_e, as given by (10.2.4). A plot of the equation $\omega_e = \Omega - A_cK \sin \psi_e$ is shown in Figure 10.3, indicating the behavior of the instantaneous value of ω_e as a function ψ_e. In nonlinear analysis this is called a *phase plane* diagram, and equivalently illustrates how ψ_e and its derivative are related. The arrows on the curve indicate directions of phase variation with time, since they correspond to positive and negative derivatives of $\psi_e(t)$. At any instant of time, $\psi_e(t)$ and the corresponding value of $\omega_e(t)$ can be located on the diagram. As $\psi_e(t)$ traces out its solution with time [obtained by solving (10.2.3)] along the abscissa, following the arrow directions, $\omega_e(t)$ traces out its behavior along the ordinate. If (10.2.6) is satisfied, the phase plane plot crosses the $\omega_e = 0$ axis at the steady state values of (10.2.5). In this case the solution trajectory begins at a point on the curve corresponding to the initial condition of $\psi_e(t)$ and moves according

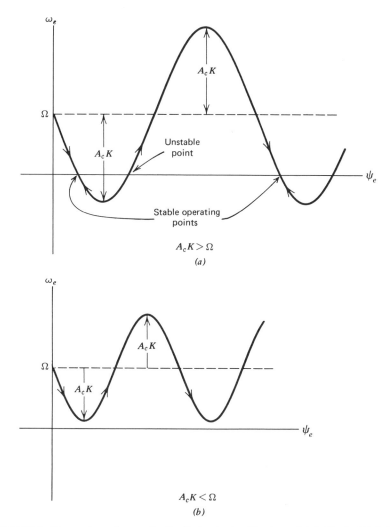

FIGURE 10.3. Phase plane diagram for the first order tracking loop: (a) $A_c K \geq 0$, (b) $A_c K < 0$.

to the arrows to the nearest solution point of (10.2.5). Note that not all solution values of (10.2.5) become steady state values for ψ_{e_0}, since the arrows force operation away from certain solution points. (The latter points are called *unstable* solution points.) When (10.2.6) is not satisfied, as shown in Figure 10.3b, no $\omega_e = 0$ solution points exist, and the solution for $\psi_e(t)$ slides continually to the right, increasing with time. This means the phase error is undergoing continuous 2π phase changes, and the loop is said to be *slipping cycles*. The error frequency $\omega_e(t)$ therefore hovers about Ω, implying that the FAL is not approaching frequency acquisition.

An important parameter in frequency acquisition is the time required to achieve synchronization, the latter called the *acquisition time* of the system. To determine acquisition time for nonlinear systems, it is necessary to know how fast we move along the phase plane trajectories. For our first order system this requires a specific solution for $\psi_e(t)$ in (10.2.3). The equation is a classical first order nonlinear differential equation, often called the *pendulum equation*, since it describes the angle motion of a hanging pendulum. When (10.2.6) is satisfied, solutions exist in the form of log functions (Problem 10.3). Although the solution approaches the steady state only asymptotically, the phase error traces through a cycle in Figure 10.3a in a time period of approximately several multiples of $A_c K$ [1]. Thus first order loop acquisition time is about $5/B_L$ sec.

In general, acquisition loops have some type of loop filtering. Let us extend our discussion to the case where

$$F(\omega) = \frac{j\omega\tau_2 + 1}{j\omega\tau_1 + 1} \qquad (10.2.7a)$$

and

$$\tau_1 = \frac{A_c K}{\omega_n^2}, \qquad \tau_2 = \frac{2\zeta}{\omega_n} - \frac{1}{A_c K} \qquad (10.2.7b)$$

corresponding to a second order loop model and a second order differential equation in (10.2.2). As $\tau_1 \to \infty$, (10.2.7) approaches an integrator, and the loop behaves as the double integrating loop discussed in Table 5.1. To determine the system equation, we convert (10.2.7) to a differential operator in which the filter output is related to its input by

$$\left(1 + \tau_1 \frac{d}{dt}\right) \text{output} = \left(1 + \tau_2 \frac{d}{dt}\right) \text{input} \qquad (10.2.8)$$

If we operate on both sides of (10.2.2) with $(1 + \tau_1 d/dt)$ and insert (10.2.8), we have

$$\left(1 + \tau_1 \frac{d}{dt}\right)\frac{d\psi_e}{dt} + A_c K\left(1 + \tau_2 \frac{d}{dt}\right)\sin\psi_e = \left(1 + \tau_1 \frac{d}{dt}\right)\Omega \qquad (10.2.9)$$

Expanding and taking derivatives yields the second order system equation

$$\tau_1 \frac{d^2\psi_e}{dt^2} + (1 + A_c K\tau_2 \cos\psi_e)\frac{d\psi_e}{dt} + A_c K\sin\psi_e = \Omega \qquad (10.2.10)$$

The preceding equation must be solved for $\psi_e(t)$, and the frequency error is then obtained by differentiating $\psi_e(t)$. We immediately see from (10.2.10) that if $\omega_e(t) = 0$ is to be a solution, it is necessary that $\theta_e(t)$ have a steady state value ψ_{e_0} such that $A_c K\sin\psi_e = \Omega$, or

$$\psi_{e_0} = \sin^{-1}\left(\frac{\Omega}{A_c K}\right) + i2\pi, \qquad i = 0, 1, 2, \ldots \qquad (10.2.11)$$

No such solution can exist unless[†]

$$\Omega \leq A_c K \qquad (10.2.12)$$

We again see a limitation to the maximum frequency offset Ω that can possibly be acquired by quiescent frequency acquisition. This maximum frequency offset depends on the loop gain, and increasing K extends this acquisition range. Since acquisition loops are typically designed with fixed natural frequencies ω_n, we can rewrite (10.2.12) from (10.2.7b),

$$\Omega \leq \tau_1 \omega_n^2 \qquad (10.2.13)$$

We now see that for $\tau_1 \to \infty$, the frequency range becomes infinite, and the FAL has the capability of acquiring any RF frequency. To further examine second order loop acquisition, let us consider specifically the case where $\tau_1 \gg 1$. By dividing (10.2.10) by τ_1 and letting $\tau_1 \to \infty$, the system equation becomes

$$\frac{d^2\psi_e}{dt^2} + (2\zeta\omega_n \cos \psi_e)\frac{d\psi_e}{dt} + \omega_n^2 \sin \psi_e = 0 \qquad (10.2.14)$$

To determine the frequency-phase behavior (phase plane trajectories), we note from (10.2.14) that since $\omega_e = d\psi_e/dt$,

$$\frac{d\omega_e}{d\psi_e} = \frac{d\omega/dt}{d\psi_e/dt} = -(2\zeta\omega_n)\cos\psi_e - \frac{\omega_n^2 \sin \psi_e}{\omega_e} \qquad (10.2.15)$$

The functional behavior of ω_e with ψ_e is obtained from exact solution of (10.2.15). Note that for large values of ω_e, (10.2.15) has the approximate behavior

$$\omega_e \approx -2\zeta\omega_n \sin \psi_e + \text{constant} \qquad (10.2.16)$$

and therefore varies sinusoidally with ψ_e (Figure 10.4a), following the upper trajectory to the right and the lower to the left. To determine the amount of frequency change that will occur, we multiply both sides of (10.2.15) by $\omega_e\, d\psi_e$ and integrate over an integer number of cycles of ψ_e, yielding

---

[†]Although a solution cannot exist if (10.2.12) is not satisfied, acquisition may not necessarily occur if it is. It has been shown [1] that acquisition actually requires $\Omega \leq A_c K [4\zeta\omega_n/K]^{1/2}$. The latter is referred to as the *pull in range* of the acquisition loop.

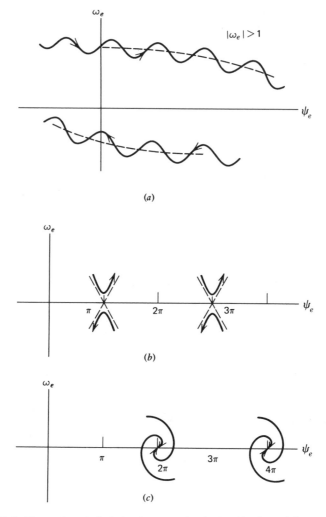

FIGURE 10.4. Phase plane trajectories for second order tracking loop: (a) $\omega_e \gg 1$, (b) $\omega_e = 0$, unstable saddle point, (c) $\omega_e = 0$, spiral node.

$$\omega_e^2(\psi_{e_2}) - \omega_e^2(\psi_{e_1}) = -2\zeta\omega_n \int_{\psi_{e_1}}^{\psi_{e_2}} \omega_e(\theta) \cos\theta \, d\theta \qquad (10.2.17)$$

since the last term in (10.2.15) will integrate to zero. Here ψ_{e_1}, ψ_{e_2} are the initial and final phase angles. From integrating by parts, we can establish

$$\int_{\psi_{e_1}}^{\psi_{e_2}} \omega_e(\theta) \cos\theta \, d\theta = \int_{\psi_{e_1}}^{\psi_{e_2}} \sin\theta \, d\omega_e(\theta) \qquad (10.2.18)$$

and from (10.2.15)

$$\int_{\psi_{e_1}}^{\psi_{e_2}} \sin\theta \, d\omega_e = -2\zeta\omega_n \int_{\psi_{e_1}}^{\psi_{e_2}} \cos\theta \sin\theta \, d\theta - \omega_n^2 \int_{\psi_{e_1}}^{\psi_{e_2}} \frac{\sin^2\theta}{\omega_e(\theta)} \, d\theta \quad (10.2.19)$$

Since the first term on the right is zero, we use (10.2.17), (10.2.18), and (10.2.19) to conclude

$$\omega_e^2(\psi_{e_2}) - \omega_e^2(\psi_{e_1}) = -2\zeta\omega_n^3 \int_{\psi_{e_1}}^{\psi_{e_2}} \frac{\sin^2\theta}{\omega_e(\theta)} \, d\theta \quad (10.2.20)$$

Although (10.2.20) does not easily produce a solution, we do note the following. For any $\omega_e > 0$, the right-hand side is negative, which means that ω_e must decrease as ψ_e increases. For $\omega_e < 0$, ω_e becomes less negative as ψ_e decreases. Hence as ψ_e follows the trajectories in Figure 10.4a, it must decrease in magnitude, and the phase plane trajectories must continually converge toward the abscissa, as shown.

When ω_e approaches zero, its solution behavior is governed by the *singular points* [2, 3] of (10.2.15). (Singular points are the phase plane values where $d\omega_e/d\psi_e = 0/0$.) Singular points of (10.2.15) occur for $\omega_e = 0$ and

$$\psi_{e_0} = i\pi, \quad i = 0, 1, 2, \ldots \quad (10.2.21)$$

To determine solutions in the vicinity of such points, (10.2.15) can be linearized to form the equation $d\omega_e/d\psi_e \approx \omega_n^2(\psi_e - \psi_{e_0})/\omega_e$. From this it can be established [3] that for i odd, an unstable saddle point (Figure 10.4b) exists, whereas for i even, a stable spiral node (Figure 10.4c) exists. Thus as the frequency error is reduced and the system approaches the singular points of (10.2.21), the points for i odd represent unstable points from which solution trajectories are reflected away, whereas for i even, solution trajectories spiral in. The steady state solutions for the system therefore correspond to $\omega_e = 0$ and $\psi_{e_0} = 0, 2\pi, 4\pi$, and so on. Thus frequency acquisition is achieved, starting from an initial large frequency offset Ω rps, by following the oscillating trajectories in Figure 10.4a in which the frequencies are slowly pulling together. During this time the phase is continually changing by 2π rad and the loop is slipping cycles. As the trajectories approach the abscissa, the cycle slipping terminates and the loop spirals into its stable position.

The time to acquire with this system is determined by the amount of cycle slipping that occurs and the time needed to achieve the final phase state after the frequency difference has been reduced to zero. The latter phase pull in time is on the order of several loop time constants and is therefore approximately $2/\omega_n$ sec, where ω_n is the loop natural frequency. The time needed to reduce the frequency error can be obtained from (10.2.20). Setting $\psi_{e_1} = 0$, $\omega_e(0) = \Omega$, and $\omega_e(\psi_{e_2}) = 0$ shows that

$$\Omega^2 = 2\zeta\omega_n^2 \int_0^{\psi_{e_2}} \frac{\sin^2\theta}{\omega_e(\theta)} \, d\theta \quad (10.2.22)$$

where ψ_{e_2} is the accumulated phase shift during the frequency acquisition. To approximate the integral, we replace $\sin^2 \theta$ by its maximum value and recall $\omega_e = d\psi_e / dt$ so that

$$\Omega^2 \cong 2\zeta\omega_n^3 \int_0^{\psi_{e_2}} \frac{d\theta}{d\theta / dt} = 2\zeta\omega_n^3 \int_0^{T_{ac}} dt$$

$$= 2\zeta\omega_n^3 T_{ac} \qquad (10.2.23)$$

where T_{ac} is the frequency acquisition time (time to accumulate the ψ_{e_2} phase shift). Hence we have

$$T_{ac} \cong \frac{\Omega^2}{2\zeta\omega_n^3} \qquad (10.2.24)$$

We see that although the loop can theoretically acquire any frequency offset Ω, the quiescent frequency acquisition time increases as Ω^2. Equation (10.2.24) is plotted in Figure 10.5 for several values of damping factor ζ. Note that if the initial offset is significantly outside the loop bandwidth, an

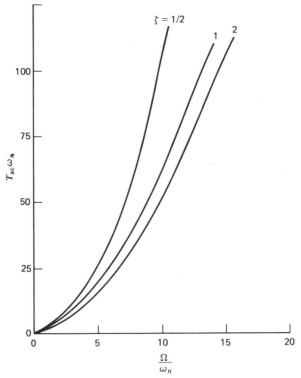

FIGURE 10.5. Acquisition time versus initial frequency offset Ω (ω_n = loop natural frequency, ζ = loop damping factor).

appreciably long time may occur before acquisition is achieved. We also see that the acquisition time varies inversely as the cube of the loop natural frequency, suggesting that loop bandwidths should be made extremely large. Although this indeed is the preferable design, the actual bandwidth size must be traded against the noise effects within the loop, which we have temporarily neglected in our current study.

10.3 FREQUENCY SEARCHING WITH OSCILLATOR SLEWING

When the initial frequency uncertainty Ω is large, we have found that quiescent frequency acquisition may require a long time or may not be accomplished at all. In these cases, a slewing voltage must be inserted at the oscillator input (Figure 10.1) to aid the acquisition operation. This slewing voltage is in the form of a linearly increasing voltage such that if the FAL loop were opened, its RF frequency would increase as βt Hz. The parameter β is called the *RF slewing rate* and has units of rps/sec. In addition, β may be positive or negative, depending on the direction in which we slew. We are interested in determining the effect of a slewing voltage on the frequency acquisition operation.

We first note that inserting a slewing voltage with rate β in the loop is identical to inserting a phase signal $\beta t^2/2$ in the RF output phase $(\omega_c/\omega_o)\theta_o(t)$. This in turn adds a signal $-\beta t^2/2$ to the phase tracking error, which is identical to adding the same term to the forcing function $\psi(t)$ in Figure 10.2. This alters the system differential equation in (10.2.2) by adding the same term, $-\beta t^2/2$, to the right-hand side. It then becomes

$$\frac{d\psi_e}{dt} + A_c KF\{\sin \psi_e\} = \Omega - \beta t \qquad (10.3.1)$$

The slewing term βt adds to or subtracts from the original value of Ω, depending on the direction of the slewing signal. For a second order loop with $\tau_1 \gg 1$, the system equation now becomes

$$\frac{d^2\psi_e}{dt^2} + (2\zeta\omega_n \cos \psi_e)\frac{d\psi_e}{dt} + \omega_n^2 \sin \psi_e = -\beta t \qquad (10.3.2)$$

If we recompute (10.2.20) with the slewing inserted, and integrate over an integer number n of phase error cycles, we find

$$\omega_e(n2\pi) - \omega_e(0) = 2\zeta\omega_n \int_0^{2n\pi} \frac{-\beta - \omega_n^2 \sin^2 \theta}{\omega_e(\theta)} d\theta \qquad (10.3.3)$$

We see that if the frequency error is positive, it will not decrease during n phase cycles if β is negative [in the opposite direction of $\omega_e(0)$] and if

$|\beta| > \omega_n^2$. That is, if the slewing rate is too fast and in the wrong direction, the frequency error is continually increased and the FAL cannot acquire the RF carrier frequency in spite of two loop integrations. If $|\beta| < \omega_n^2$, there exists the possibility of frequency acquisition even through we are slewing in the wrong direction. In this case the loop dynamics may be sufficient to acquire the RF carrier by internally compensating for the slewing signal.

When the slewing is in the proper direction, the frequency error will be reduced even without the loop dynamics. In this case the original frequency uncertainty Ω, no matter how large, will eventually be driven toward zero by the slewing operation. The important dynamical question concerns the behavior of the loop as the instantaneous frequency error passes through zero and the slewing continues to be applied. To determine this, we resort to our linear theory and examine linear loop behavior as the phase input forcing function has the form $\beta t^2/2$. The linear system equivalent to (10.3.2), obtained by setting $\cos \psi_e = 1$ and $\sin \psi_e = \psi_e$, will have the solution

$$\psi_e(t) = \left(\frac{\beta}{K}\right)t + \frac{\beta}{\omega_n^2} \tag{10.3.4}$$

This implies that an offset RF frequency error of β/K rad/sec will exist in the FAL. To reduce this to zero, it is necessary to remove the slewing signal (i.e., set $\beta = 0$) as the loop passes through the zero error position. This removal is provided by the local detector subsystem in Figure 10.1. Although we do not pursue lock detector design here (see Problem 10.4), we point out that the response time of the lock detector should be fast enough to recognize FAL lockup as it occurs. Hence the lock detector should have a response time approximately equal to the acquisition loop bandwidth.

The preceding statements concerning the frequency search behavior, as the FAL passes through frequency lock, were based on the assumption that the loop had time to operate linearly. Since the loop will have a nonzero response time in its linear mode, it is necessary that the slewing rate β not be too fast for the system. That is, the slewing signal must cause the frequency error to be within the loop bandwidth long enough for the loop dynamics to acquire the RF carrier. If RF slewing is at the rate of β rps/sec, the FAL frequency error spends approximately ω_n/β sec in its loop bandwidth. If we assume a response time of $2/\omega_n$ sec is required, then the slewing rate must be such that $\omega_n/\beta \geq 2/\omega_n$, or

$$\beta \leq \frac{\omega_n^2}{2} \quad \text{rad/sec}^2 \tag{10.3.5}$$

Hence RF slewing rates are limited by the loop natural frequency.[†] Faster

---

[†]Slewing may also be constrained by hardware considerations that prevent "pulling" an oscillator frequency too quickly. However, remember that β refers to the RF slewing rate, which benefits from the frequency multiplication of the acquisition system.

slewing rates may cause the FAL to pass through the RF carrier frequency without acquisition. When slewing at the maximum rate in the correct direction, the time required to reduce an initial offset of Ω to zero is

$$T_{ac} = \frac{\Omega}{\beta} = \frac{\Omega}{(\omega_n^2/2)} \tag{10.3.6}$$

Compare this result with (10.2.24) for quiescent acquisition. For $\zeta = \frac{1}{2}$, we find quiescent acquisition is actually preferable to frequency searching if $\Omega \leq 2\omega_n$. That is, frequency searching is not advantageous if the expected frequency uncertainty is well within the FAL bandwidth. Again we see the obvious tradeoff of improved dynamical performance of large loop bandwidths (no slewing necessary) versus the increased loop noise.

Our discussion of frequency searching has been based purely on the response capabilities of a noiseless FAL. When noise is added to our model in Figure 10.2, it circulates within the loop, causing random phase effects. This random phase interferes with the normal tracking operations of the loop and increases the chance of the loop not acquiring during the slewing operation. Hence the addition of noise reduces the maximum allowable slewing rate for attaining a given acquisition probability. A rigorous analysis relating allowable rates and the amount of loop noise is somewhat difficult, although simulation, laboratory testing, and approximate analysis have been performed. A result often quoted [4] for the maximum slewing rate β_{max} is the empirical formula,

$$\beta_{max} = \frac{\omega_n^2}{2} \left[1 - \frac{1}{\sqrt{CNR_L}} \right], \qquad CNR_L \geq 1 \tag{10.3.7}$$

where $CNR_L = A_c^2/2N_0 B_L$ represents the RF carrier to noise ratio in the loop noise bandwidth. As the loop CNR decreases, the maximum slewing rate must be decreased to maintain a desirable acquisition probability. This of course leads directly to an increase in the acquisition time of (10.3.6).

10.4 CARRIER PHASE REFERENCING

After the RF or IF frequencies are brought into frequency alignment, the carrier can then be phase aligned by the FAL. This is referred to as *receiver phase synchronization*, and we say that the receiver is phase coherent with the transmitter.

An important problem is to continue to maintain phase synchronization once it has been achieved from the acquisition operation. This requires that the loop oscillator continually maintain its phase position relative to the carrier being tracked, in spite of variations that may occur in its phase or frequency. The most important indicator of the receiver's phase referencing

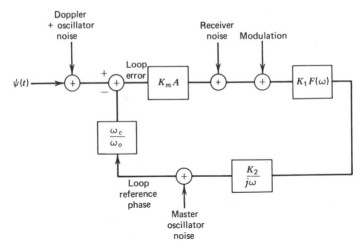

FIGURE 10.6. RF phase lock loop with tracking disturbances, linear model.

performance is the loop phase error, which determines how well the loop is maintaining phase coherence. This error is usually measured by the mean squared loop phase error, as developed back in Chapter 5.

The primary factors that tend to destroy the receiver phase synchronization are (1) dynamical tracking errors, as the received carrier phase tends to vary and the FAL oscillator attempts to follow; (2) front end noise that enters the loop; (3) phase noise on both the transmitter and receiver oscillators that cause inherent phase fluctuations; and (4) modulation inserted on the carrier during communications operations. For the most part, the effect of these disturbances can best be estimated from linear analysis. We assume a linear tracking loop, obtained by linearizing the equivalent loop model in Figure 10.2 and inserting the various disturbances as effective loop phase inputs at the appropriate place in the loop. Such a model is shown in Figure 10.6. Since the loop is linear, the effect of each such input can be determined separately, and then combined to determine the overall effect on the phase referencing error. Note the loop in Figure 10.6 has the loop gain function

$$H(\omega) = \frac{A[K_m K_1 K_2 (\omega_c/\omega_o) F(\omega)/j\omega]}{1 + [A K_m K_1 K_2 (\omega_c/\omega_o) F(\omega)/j\omega]} \qquad (10.4.1)$$

and therefore depends on the loop gains, the loop filter, and the frequency multiplication factor.

Dynamic Tracking Errors

As the input carrier phase $\psi(t)$ dynamically changes in time, the tracking loop in Figure 10.1 must continually track these variations so as to maintain phase coherence (small loop phase errors). The loop error transform due to input phase variations $\psi(t)$ is given by

$$\Phi_e(\omega) = [1 - H(\omega)]\Phi_\psi(\omega) \qquad (10.4.2)$$

where $\Phi_\psi(\omega)$ is the transform of $\psi(t)$ and $H(\omega)$ is the loop gain function in (9.4.1). In practice, the extraneous carrier phase variations of most concern are those due to offset carrier phase shifts, Doppler, and Doppler rates. These tend to be purely deterministic effects and usually quite predictable (or at least boundable). Such deterministic phase effects contribute a deterministic function to the total loop error, given by the inverse Fourier transform of (10.4.2). Formal inversion of $\Phi_e(\omega)$ produces the exact error time response due to $\psi(t)$. Such responses involve a transient term, followed by a steady state solution. Since the steady state is reached in a relatively short time, an adequate indication of the loop response to $\psi(t)$ can be achieved by examining only the steady state value. The steady state value of this deterministic error time response (which occurs after loop transients have diminished) can be obtained by appealing to the final value theorem of transform theory (Section A.2)

$$\begin{aligned} \text{Steady state values of the}& \\ \text{time response of the loop error} &= \lim_{j\omega \to 0} j\omega[1 - H(\omega)]\Phi_\psi(\omega) \\ \\ &= \lim_{j\omega \to 0}\left[\frac{(j\omega)^2\Phi_\psi(\omega)}{j\omega + A_c KF(\omega)}\right] \qquad (10.4.3) \end{aligned}$$

The response caused by deterministic extraneous phase shifting therefore depends on the form of $\psi(t)$ and the type of loop filtering $F(\omega)$. For the important types of extraneous phase shift mentioned, the resulting steady state error in (10.4.3) is listed in Table 10.2, for the various loop filter functions shown. The loop therefore tracks the stated extraneous effects with zero, constant, or infinite steady state loop errors. The zero error means that the loop VCO will eventually lock onto the carrier phase shift. A

TABLE 10.2 Steady State Tracking Errors

			Steady State Error	
$\psi(t)$	Cause	$\Phi_\psi(\omega)$	$F(\omega) = 1$	$F(\omega) = \dfrac{j\omega + b}{j\omega}$
ψ_0	Constant phase offset	$\dfrac{\psi_0}{j\omega}$	0	0
$\omega_d t$	Constant Doppler shift	$\dfrac{\omega_d}{(j\omega)^2}$	$\dfrac{\omega_d}{A_c K}$	0
$\dfrac{\alpha t^2}{2}$	Increasing Doppler shift	$\dfrac{\alpha}{(j\omega)^3}$	∞	$\dfrac{\alpha}{A_c K}$

constant error means that the loop VCO will track with the correct frequency but have a constant offset phase error. An infinite steady state error means that the loop cannot track the frequency shift, and the input carrier will eventually "pull away" from the loop VCO phase with time, approaching an infinite phase error in the steady state (i.e., as $t \to \infty$). Note that the loop with the more complicated filter yields better phase tracking capabilities. The ability to track higher order polynomials of the extraneous phase can be related to the number of integrations within the loop. (Recall that an inherent phase integration is already provided by the loop oscillator.) We see, for example, that we need a loop with a total of two integrations to track an increasingly Doppler shift with constant phase error. Since the VCO inherently provides one integration this added integration must be provided by the loop filter.

Consider specifically the case of a Doppler offset ω_d. This means that the received carrier RF or IF frequency is offset from the loop VCO center frequency by ω_d rps. The results in Table 10.2 indicate that if a first order loop $[F(\omega) = 1]$ is used, the VCO frequency is driven to that of the carrier, but an offset phase error of $\omega_d / A_c K$ occurs. This phase error can be written in terms of the loop noise bandwidth in Table 5.1, since $B_L = A_c K / 4$. Thus the Doppler tracking phase error is small if $\omega_d / 4 B_L \ll 1$, or $B_L \gg \omega_d / 4$ (i.e., if the loop noise bandwidth is large enough to encompass the Doppler offset). If an integrating filter is used, the carrier tracking VCO is driven to the exact Doppler frequency and phase of the carrier.

It is interesting to compare our result in Table 10.2 with our earlier result in (10.2.11), where we concluded that a first order loop always tracks a frequency offset with a phase error $\sin^{-1}(\Omega / A_c K)$. Table 10.2 indicates a phase error of $\Omega / A_c K$. We now see that the results are similar only if $\Omega / A_c K \ll 1$, the latter being the condition for the loop to be linear, which is the assumption under which Table 10.2 was derived.

Note that a linear frequency drift cannot be tracked by an imperfect second order loop (nonintegrating loop filter) and will produce a steady state phase error that increases in time. If the linear frequency variation is due only to oscillator drift, its rate α will typically be quite small, and the error contribution may be negligible over most communication periods. If the variation is due to transmitter or receiver acceleration, however, it may be necessary to compensate to prevent the FAL from breaking lock. Compensation can occur by injecting oscillator slewing to cancel the Doppler rate or by using transmitter frequency corrections. Note that the constant phase error caused by the Doppler rate can always be reduced by increasing the loop gain K.

Before reaching the steady state values in Table 10.2, the loop undergoes a transient. If this transient becomes too large, the linear assumption of the loop may be violated, and the loop may, in fact, lose lock. By far the most serious transient effect for a second order loop is a step in frequency. Figure 10.7 shows the loop error transient response to a step change Δf for

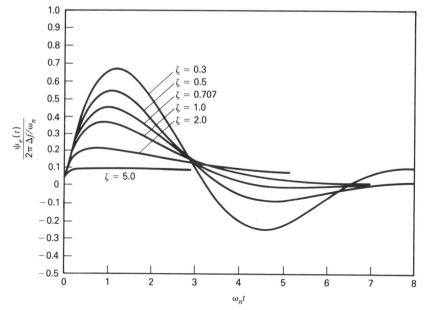

FIGURE 10.7. Second order loop response to step change in frequency Δf ($\zeta =$ loop damping).

different loop damping factors. Although the steady state response is zero (the loop will eventually pull in), the peak transient may be significantly large to cause loss of lock. The peak transient is directly related to the frequency stop size Δf, and keeping the peak transient less than $30°$ requires $\Delta f\, \omega_n \leq 0.55$. The frequency step at which this is violated, called the *loop drop-out frequency*, is approximately

$$\Delta f_{do} \cong 0.55\omega_n = 2.61 B_L \qquad (10.4.4)$$

Thus, if the tracking loop is in lock, and a frequency step change occurs in the carrier at the loop input exceeding the value in (10.4.4), the peak loop transient can force the loop out of its linear range.

Receiver Noise

Receiver noise, resulting from front end and multiplier noise, enters the tracking loop and causes random errors within the loop. This random error has the variance computed earlier in (5.7.10). For white noise and a linear loop model, this becomes

$$\sigma_n^2 = \frac{N_0 B_L}{P_c} \qquad (10.4.5)$$

where $P_c = A_c^2/2$ is the power of the carrier component being used for referencing and N_0 is the combined spectral level of all the equivalent receiver noise. The loop bandwidth B_L is the noise bandwidth of the overall RF tracking loop in Figure 10.6. This means that B_L must be computed from the RF loop gain function in (10.4.1), using our earlier result in (5.8.11). Since the function includes the frequency multiplication factor, B_L represents the loop noise bandwidth presented to the RF carrier. We can again denote

$$\text{CNR}_L = \frac{P_c}{N_0 B_L} \tag{10.4.6}$$

as the carrier to noise ratio in the loop bandwidth, as we did in (5.8.9), and introduce the carrier bit energy $E_b = P_c T_b$. This means that (10.4.5) can be written as

$$\sigma_n^2 = \frac{1}{\text{CNR}_L} = \frac{B_L T_b}{E_b/N_0} \tag{10.4.7}$$

To determine precisely the effect of receiver noise in an operating tracking loop, we require knowledge of the actual probability density of the random phase error produced by the noise. Since the input noise is Gaussian, linear theory predicts that the phase error due to the noise will also be Gaussian with the variance given in (10.4.7). More accurately, however, the tracking loop is truly nonlinear, described by the differential equation in (10.1.13), and a more rigorous analysis of noise effects requires further study of the basic nonlinear equation. Results of such nonlinear studies are included in References [1, 5]. Although a rigorous investigation is beyond our scope here, we point out several results associated with nonlinear loops and white Gaussian input noise. The steady state probability density of the loop phase error for a first order nonlinear loop has been found to have the form

$$p(\psi_e) = \frac{e^{\rho \cos \psi_e}}{2\pi I_0(\rho)}, \qquad |\psi_e| \le \pi \tag{10.4.8}$$

where $I_0(\rho)$ is the zero order Bessel function of argument ρ, and $\rho = \text{CNR}_L$ in (10.4.6). This solution is plotted in Figure 10.8 for several values of ρ. Note that the density indeed approaches a Gaussian shape with variance $1/\rho$ for large ρ, as predicted by linear theory, but tends to spread as ρ is decreased (see Problem 10.11). For higher order loops it has been shown that (10.4.8) serves a good approximation for ρ values larger than about 5, as long as the B_L used to define ρ is taken to be the actual loop noise bandwidth. The mean squared value of the phase error in (10.4.8) is plotted in Figure 10.9 and compared to the mean square phase error $\sigma_n^2 = 1/\rho$ in

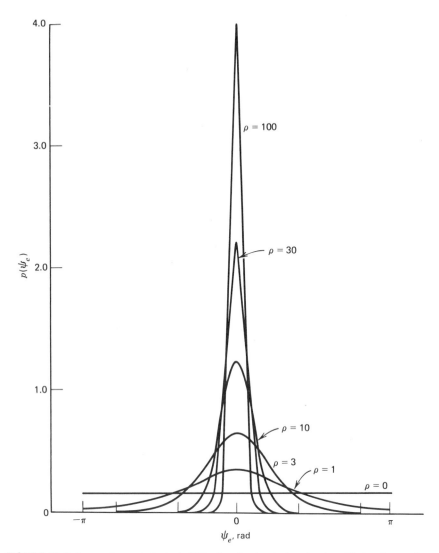

FIGURE 10.8. Loop phase error probability density due to input noise. First order nonlinear loop.

(10.4.7) predicted by linear theory. We see that linear theory accurately predicts mean squared phase error performance for $\rho \geq 5$. For smaller values, the system departs from its linear performance, and the more accurate density in (10.4.8) must be used.

In practice, we also find that tracking loop noise actually causes other effects than simply generating the mean squared phase error in (10.4.5). If we recall our phase plane trajectories in Section 10.2, we see that stable

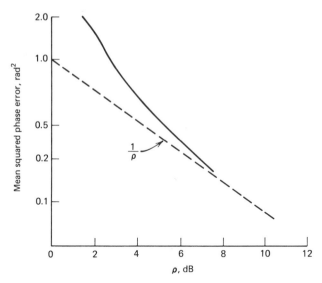

FIGURE 10.9. Loop mean squared phase error versus loop CNR$_L$.

phase lock points are actually dispersed along the phase error axis. The frequency tracking dynamics in the absence of noise caused us to slide into one of these stable points and remain there. In an actual system, however, external disturbances such as noise may intermittently force us out of a stable state and into a new phase stable position. (This would correspond, for example, to sliding over the sinusoidal hump in Figure 10.3a and pulling into the adjacent phase lock position.) During this transient the phase error changes by multiples of 2π, corresponding to *cycle slipping*, and we say that the loop has "lost lock" temporarily. As the phase error slides into the new stable position, the loop regains phase lock. Although the total time out of lock may be quite short, phase referencing operations using the loop oscillator are temporarily interrupted. In addition, loss of lock is exhibited throughtout the receiver processing as phase impulses, or phase "clicks," superimposed on oscillator waveforms. If these clicks occur too often, interference in subsequent receiver operations may become significant.

An important parameter associated with cycle slipping is the mean time over which loss of lock occurs, that is, the expected time between cycle slips. Simulation and experimental studies [6, 7] have attempted to predict mean cycle slipping time in tracking loops. Figure 10.10 shows a plot of estimated mean slip times for a second order loop, as a function of loop CNR ρ in (10.4.6) for several values of loop damping factor ζ. Theoretical studies of nonlinear loops using diffusion theory [8], and more recently, renewal theory [9], have verified these estimates. Note that the mean time between cycle slips decreases rapidly as ρ is decreased, and such performance may be the ultimate limitation on operating CNR. Note also the increase in slip time

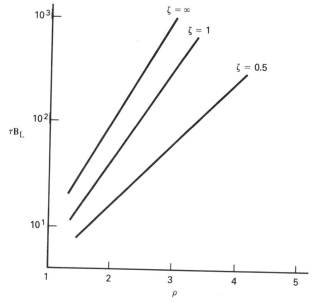

FIGURE 10.10. Average time between cycle slips versus loop CNR_L.

with damping factor, with an order of magnitude increase occurring as ζ increases to infinity. In the latter case, the second order loop is performing practically as a first order loop and the lock points are of a more stable nature. For $\zeta = \frac{1}{2}$, the behavior in Figure 10.10 is adequately approximated by the equation

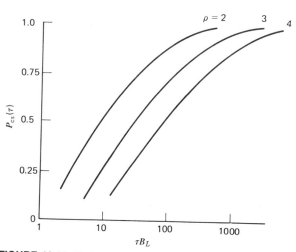

FIGURE 10.11. Probability of a cycle slip in τ sec ($\rho = CNR_L$).

$$\bar{\tau} B_L = \frac{\pi}{4} e^{2\rho} \qquad (10.4.9)$$

where $\bar{\tau}$ is the mean slip time and B_L is the loop noise bandwidth. Mean slip times can be used to predict the probability of a cycle slip occurring in a given time period τ. Studies [9] have confirmed that this probability $P_{cs}(\tau)$ is given quite accurately by the relation

$$P_{cs} \cong 1 - e^{-(\tau/\bar{\tau})} \qquad (10.4.10)$$

Equation (10.4.10) is plotted in Figure 10.11 for several values of ρ.

Oscillator Phase Noise

In Section 3.1 it was pointed out that oscillators and frequency generators invariably produce carriers with random phase noise. These carriers are then modulated, and later mixed with other noisy carriers, to produce received RF waveforms that appear at the tracking loop input with these combined random phase variations. These phase noise fluctuations must be tracked by the receiver VCO in order to maintain phase coherency. Phase noise on the received carrier superimposes a random phase variation on the carrier extraneous phase $\psi(t)$ in (10.1.1), which therefore enters the tracking loop as an input forcing function (Figure 10.6).

If we assume that the input RF carrier oscillator phase noise has spectral density $S_\psi(\omega)$, the mean squared loop phase error contribution that it produces is

$$\sigma_e^2 = \frac{1}{2\pi} \int_{-\infty}^{\infty} S_\psi(\omega)|1 - H(\omega)|^2 \, d\omega \qquad (10.4.11)$$

where $H(\omega)$ is again the tracking loop transfer function in (10.4.1). Oscillator phase noise typically has the spectral characteristics shown in Figure 10.12 and is often modeled as

$$S_\psi(\omega) = \frac{S_3}{\omega^3} + \frac{S_2}{\omega^2} + S_0 \qquad (10.4.12)$$

Here the first term is referred to as frequency *flicker noise*, the second term is referred to as *white frequency noise*, and S_0 is the *white phase noise* term, the latter assumed to exist to some upper frequency ω_u. For a second order loop with damping factor ζ, and loop natural frequency ω_n, $H(\omega)$ in (10.4.1) becomes

$$H(\omega) = \frac{1 + (2\zeta/\omega_n)j\omega}{1 + (2\zeta\omega_n)j\omega - (\omega/\omega_n)^2} \qquad (10.4.13)$$

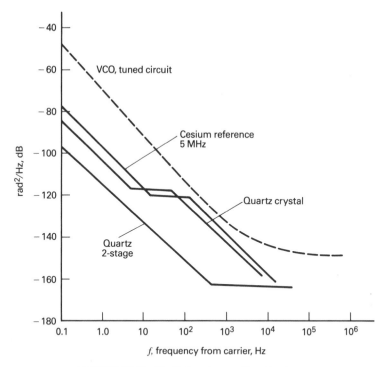

FIGURE 10.12. Oscillator phase noise spectrum.

When (10.4.12) and (10.4.13) are used in (10.4.11), the mean squared loop phase error contribution is

$$\sigma_e^2 = \frac{1}{2\pi} \int_{-\infty}^{\infty} \left[\frac{S_3}{\omega^3} + \frac{S_2}{\omega^2} + S_0 \right] \left[\frac{(\omega/\omega_n)^4}{[1 - (\omega/\omega_n)^2]^2 + 4\zeta^2(\omega/\omega_n)^2} \right] d\omega$$

(10.4.14)

For the case $\zeta = 1/\sqrt{2}$, the preceding integrates to (Section A.2, subsection on integrals),

$$\sigma_e^2 = \left(\frac{S_0}{2\pi} \right) \omega_u + \frac{0.94 S_2}{B_L} + \frac{0.03 S_3}{B_L^2}$$

(10.4.15)

where $B_L = 0.53 \omega_n$ is the loop noise bandwidth, obtained from Table 5.1. Hence oscillator phase noise effects are reduced by using as large a loop bandwidth as possible. Since the individual terms vary differently with B_L, it is important to characterize properly the phase noise spectrum, that is, determine whether the flat or reciprocal frequency portion is more significant. For more complicated phase noise spectra, (10.4.11) may have to be numerically integrated.

Phase noise on the loop VCO oscillator enters the equivalent system at a different point (Figure 10.6). If this local oscillator noise has spectrum $S_{os}(\omega)$, it produces the mean squared loop phase error

$$\sigma_{os}^2 = \left(\frac{\omega_c}{\omega_o}\right)^2 \frac{1}{2\pi} \int_{-\infty}^{\infty} S_{os}(\omega)|1 - H(\omega)|^2 \, d\omega \qquad (10.4.16)$$

Of obvious importance is the fact that the integral is multiplied by the frequency multiplication factor of the loop in determining the loop error. Thus VCO phase noise is extremely important in phase tracking loops with large frequency translations, and selection of "quiet" master oscillators is mandatory for good phase tracking.

Note that while oscillator phase noise effects are reduced with increasing B_L in (10.4.15), the receiver noise effect in (10.4.5) increases with B_L. Wider loop bandwidths allow better tracking of the oscillator phase variations, but more noise enters the loop. Hence a trade off always exists in determining loop bandwidth for balancing between the phase noise and receiver noise effects. Figure 10.13 plots σ_e^2 as a function of B_L for several values of P_c/N_0 ratios, and assuming white frequency noise on the carrier. The contribution of the receiver and phase noise combine to produce an obvious range of B_L values for minimal tracking error performance.

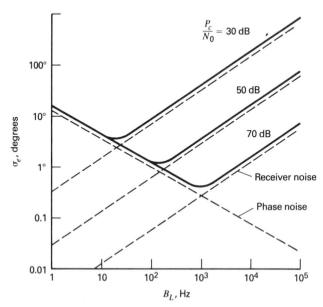

FIGURE 10.13. Phase error due to input noise and oscillator phase noise versus B_L.

Modulation

When communicating, modulation impressed on the carrier being tracked affects the transponder phase referencing operation. In addition to the suppression of the carrier power that it may cause, the modulation enters the tracking loop as an equivalent phase noise interference, as shown in Figure 10.6. Its mean squared error contribution can therefore be determined by replacing the noise spectrum in (10.4.2) with the corresponding phase modulation spectrum. Thus a phase modulation spectrum $S_m(\omega)$ produces the mean squared phase error

$$\sigma^2_{em} = \frac{1}{2\pi} \int_{-\infty}^{\infty} S_m(\omega)|H(\omega)|^2 \, d\omega \qquad (10.4.17)$$

If the modulation is on a subcarrier located outside the loop bandwidth, the integrated effect in (10.4.17) is negligible. If the modulation was applied directly to the RF carrier, the portion of it entering the loop bandwidth will cause phase tracking errors. Typically, the bandwidth of such modulation is much larger than that of the loop, and the carrier modulation spectrum can be modeled as a flat spectrum in (10.4.17). In this case, σ^2_{em} can be evaluated as in (10.4.2) with N_0/P_c replaced by the phase modulation spectral level. Also of particular concern is the presence of modulation spectral frequency lines that appear directly in the loop bandwidth. In the latter case it is necessary to insure that $|H(\omega)|$ is suitably small at the line frequency to reduce this component to a negligible phase error contribution.

In summary, we see that accurate receiver phase referencing is hindered by several primary causes. Which of these sources is involved generally dictates system implementation, since the minimization of each effect often requires contrasting design procedures for loop bandwidths. For example, dynamical and oscillator jitter phase errors are reduced by relatively fast loops, with large bandwidths and low damping factors, whereas noise and modulation effects require narrow bandwidths. Thus phase referencing loop design invariably reduces to careful tradeoff analyses of phase errors. In addition, phase referencing must be properly interfaced with the frequency acquisition operation that it initially performs. In Section 10.2 we found that frequency acquisition time, whether using quiescent or searching operations, is reduced with increasing bandwidths, which is contrary to the noise reduction requirements of phase referencing. Thus FAL design also involves careful tradeoff of the acquisition prerequisite as well as the required phase synchronization. In this regard several types of hybrid systems have been suggested, in which wideband RF loops are used for initial frequency acquisition, then switched to narrowband operaton for accurate phase referencing.

10.5 BPSK CARRIER REFERENCING

The previous analysis considered an acquisition and tracking loop that operated on the residual carrier component of the received RF carrier. When the received carrier is phase modulated as a BPSK carrier, there is no carrier component to be tracked (i.e., $\Delta_c = \pi/2$ and $A_c = 0$ in Table 10.1) and carrier recovery cannot be obtained via a standard phase lock loop. Instead, a modified system must be used, which first uses a nonlinearity to eliminate (wipeoff) the modulation while creating a carrier component having a phase variation proportional to that of the receiver carrier. Subsequent tracking of this residual carrier component then generates the desired carrier reference.

There are two standard methods for achieving phase referencing, both using modified tracking loops. One method is the use of a *squaring loop*, in which we precede a PLL with a nonlinear, memoryless, squaring device. The system is shown in Figure 10.14a. The received RF (or IF) signal is

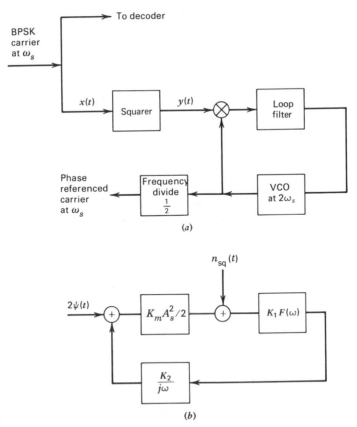

(a)

(b)

FIGURE 10.14. Squaring BPSK carrier phase referencing: (a) block diagram, (b) equivalent loop model.

squared and phase locked to provide the carrier reference for the BPSK decoder. The input to the squarer is the noisy BPSK signal

$$x(t) = Am(t) \cos (\omega_s t + \psi(t)) + n(t) \tag{10.5.1}$$

where $n(t)$ is the bandpass noise, $m(t) = \pm 1$ is the PCM baseband modulation, and $\psi(t)$ is the extraneous phase variation that must be tracked for coherent phase referencing. The output of the squarer is then

$$y(t) = x^2(t) = A_s^2 \sin^2 \left[\omega_s t + \frac{\pi}{2} m(t) + \psi(t) \right]$$

$$+ 2n(t) A_s \sin \left[\omega_s t + \frac{\pi}{2} m(t) + \psi(t) \right] + n^2(t)$$

$$= \frac{A_s^2}{2} + \frac{A_s^2}{2} \sin [2\omega_s t + 2\psi(t)] + n^2(t)$$

$$+ 2n(t) A_s \sin \left[\omega_s t + \frac{\pi}{2} m(t) + \psi(t) \right] \tag{10.5.2}$$

The output contains a constant term, a carrier component at $2\omega_s$ with no modulation and a phase variation multiplied by 2, and two interference terms involving the noise. A PLL with a VCO at $2\omega_s$ can now track the carrier component in (10.5.2). If it successfully performs the tracking, its VCO output is given by $\cos [2\omega_s t + \psi_2(t)]$, where $\psi_2(t)$ is a close replica of $2\psi(t)$. By now frequency dividing the VCO output by 2, we generate the reference carrier at ω_s

$$\cos \left[\omega_s t + \frac{\psi_2(t)}{2} \right] \tag{10.5.3}$$

where the phase $\psi_2(t)/2$ is now approximately phase locked to $\psi(t)$ in (10.5.1). Note that the role of the squarer is effectively to remove the digital modulation while frequency translating the carrier to $2\omega_s$ for phase locking.

The ability to achieve accurate phase referencing with squaring systems can be determined from the equivalent loop tracking diagram in Figure 10.14b. The loop is effectively driven by $2\psi(t)$ and therefore has twice the extraneous phase variation of the carrier itself. The equivalent noise $n_{sq}(t)$ entering the loop is simply the loop mixer output noise produced by the squarer noise terms [i.e., the last two terms in (10.5.2)]. These correspond to the squared carrier filtered noise and the signal–noise cross-product term. To evaluate the spectrum of $n_{sq}(t)$, we need only determine the spectrum of the squarer noise in the vicinity of $2\omega_s$. This can be determined by recalling that the squared Gaussian noise term has a spectrum given by twice the convolution of $S_n(\omega)$ with itself, and the spectrum of the cross-product term is given by the convolution of $S_n(\omega)$ with $2S_s(\omega)$, where $S_n(\omega)$, $S_s(\omega)$ are the spectrum of the filtered noise and carrier signals in (10.5.1). These spectra are sketched in Figure 10.15. The baseband noise is shown in part a, and part b shows the spectrum of its square. Figure 10.15c shows a typical carrier spectrum, and its convolution with the noise illustrated in part a is shown in part d. Figure 10.15e then shows the spectrum of $n_{sq}(t)$, obtained by shifting the superimposing the portion of parts b and d around $2\omega_s$. Note that the spectra are multiplied and spread by the squaring (convolution) operation. In the vicinity of zero, $n_{sq}(t)$ has approximately a flat spectrum with level

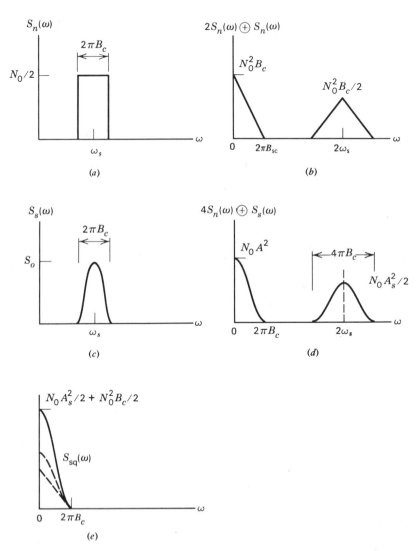

FIGURE 10.15. Noise spectral densities due to squaring: (a) input noise, (b) squared noise, (c) carrier, (d) carrier convolved with noise, (e) $n_{sq}(t)$ spectrum.

$$S_{sq}(0) \approx \frac{N_0^2 B_c}{2} + \frac{N_0 A_s^2}{2} \qquad (10.5.4)$$

Since the noise bandwidth B_L of the tracking loop is generally much smaller than that of the carrier filter bandwidth B_c, the mean squared noise value entering the loop bandwidth is therefore approximately $2S_{sq}(0)B_L$. The resulting carrier reference phase variance, after phase dividing by 2, is then

$$
\begin{aligned}
\sigma_e^2 &= \frac{1}{4} \left[\frac{2S_{sq}(0)B_L}{A_s^4/8} \right] \\
&= \frac{2N_0 B_L}{A_s^2} + \frac{2N_0^2 B_c B_L}{A_s^4} \\
&= \frac{N_0 B_L}{P_c} \left[1 + \frac{N_0 B_c}{P_c} \right] \qquad (10.5.5)
\end{aligned}
$$

where $P_c = A^2/2$ is again the carrier power. The bracket is called the *loop squaring factor* and accounts for the increases in error caused by the squaring operation. Since the bracket is greater than one, squaring loop errors are larger than those for standard tracking loops, the amount of the increase dependent on the carrier CNR. In effect, the squaring factor exhibits the fact that the loop noise changes from a squared noise term to a signal–noise cross-product as the carrier CNR increases. It is convenient to again denote

$$\frac{E_b}{N_0} = \frac{P_c T_b}{N_0} \qquad (10.5.6)$$

and write (10.5.5) as

$$\sigma_e^2 = \frac{B_L T_b}{(E_b/N_0)} \left[1 + \frac{B_c T_b}{E_b/N_0} \right] \qquad (10.5.7)$$

This relates the carrier referencing accuracy to the carrier E_b/N_0 used for the decoder. Since the reciprocal of the bracket in (10.5.7) effectively multiplies the E_b/N_0 factors, the former is often called a *squaring loss*, or a *squaring suppression factor*. Note that this factor depends on the presquaring bandwidth B_c. If this filter is too wide, excess noise will enter the loop, while if it is too narrow, the BPSK carrier may be distorted. Lindsey and Simon [10] have shown that for a particular filter type and bit waveform, an optimal prefilter B_c bandwidth relative to the bit rate exists for each E_b/N_0. For a typical operation ($E_b/N_0 \approx 8 = 9$ dB), the required bandwidth is about $3/T_b$ Hz. Figure 10.16 plots rms phase error versus E_b/N_0 for this prefilter bandwidth with $B_L T_b$ as a parameter. Note that rms tracking accuracies to within 5° can be maintained at $E_b/N_0 = 10$ with $B_L T_b \leq 0.01$; that is, a loop

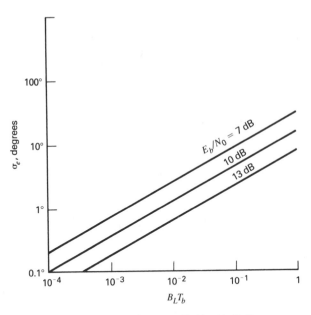

FIGURE 10.16. Phase error (rms) versus E_b/N_0 with $B_L T_b$ as a parameter.

bandwidth less than 1% of the bit rate. We can also see that the dc term in (10.5.2) is given by the carrier power $A_s^2/2$. Hence a low pass filter that extracts this term at the squarer output provides a convenient measure of carrier power for signal monitoring or automatic gain control.

A problem that occurs in phase referencing with a squaring loop is the phase ambiguity that must be resolved before the phase reference can be used for BPSK decoding. This ambiguity is caused by the one-half frequency division of the VCO output. The resulting divided down reference may be 180° out of phase with the desired reference (i.e., both the desired reference and its negative correspond to the same double frequency waveform at the VCO). This phase ambiguity must be resolved prior to BPSK decoding, since the negative reference will decode all bits exactly opposite to their true polarity. Resolving this ambiguity is usually accomplished by first decoding a known binary word.

A second way to achieve suppressed carrier synchronization is by the use of the Costas [11], or *quadrature*, loop shown in Figure 10.17a. The system involves two parallel tracking loops operating simultaneously from the same VCO. One loop, called the *in-phase* loop, uses the VCO directly for tracking, and the second (quadrature loop) uses a 90° shifted VCO. The mixer outputs are each multiplied, filtered, and used to control the VCO. The low pass filters in each arm must be wide enough to pass the carrier modulation without distortion. If the input to the Costas loop is the carrier signal in (10.5.1), the in-phase mixer generates the signal

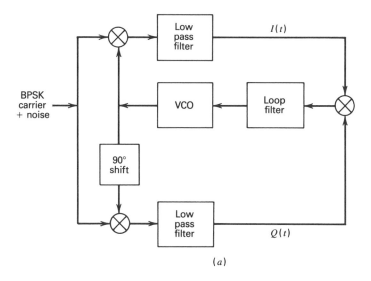

(a)

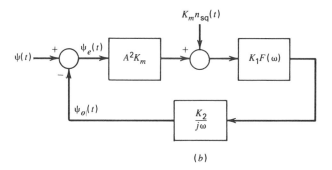

(b)

FIGURE 10.17. Costas BPSK phase referencing loop: (a) block diagram, (b) equivalent loop model.

$$I(t) = A_s \cos \left[\frac{\pi}{2} m(t) + \psi(t) - \psi_o(t) \right] + n_{mc}(t)$$

$$= A_s m(t) \cos \left[\psi_e(t) \right] + n_{mc}(t) \qquad (10.5.8)$$

while the quadrature mixer simultaneously generates

$$Q(t) = A_s m(t) \sin \left[\psi_e(t) \right] + n_{ms}(t) \qquad (10.5.9)$$

where $\psi_e(t) = \psi(t) - \psi_o(t)$, and $\psi_o(t)$ is the phase of the VCO in Figure 10.17a. The output of the multiplier is then

$$I(t)Q(t) = A_s^2 m^2(t) \sin \psi_e(t) \cos \psi_e(t) + n_{sq}(t)$$

$$= \frac{A_s^2}{2} \sin [2\psi_e(t)] + n_{sq}(t) \qquad (10.5.10)$$

where

$$n_{sq}(t) = n_{mc}(t)n_{ms}(t) + n_{mc}(t) \sin \left[\frac{\pi}{2} m(t) + \psi_e(t) \right]$$

$$+ n_{ms}(t) \cos \left[\frac{\pi}{2} m(t) + \psi_e(t) \right] \qquad (10.5.11)$$

The second equality in (10.5.10) follows, since $m(t) = \pm 1$. When the phase error $[\psi(t) - \psi_o(t)]$ is small, the loop has the equivalent linear model shown in Figure 10.17b. The Costas loop therefore tracks the phase variation $\psi(t)$ with the VCO without interference from the carrier modulation $m(t)$. The quadrature tracking loops have effectively removed the modulation, allowing the loop filter and VCO to accomplish the carrier synchronization. The mixer noises $n_{mc}(t)$ and $n_{ms}(t)$ in (10.5.11) are baseband versions of the carrier noise in (10.5.1). When this latter noise has the spectrum in Figure 10.15a, it can be shown (Problem 10.13) that $n_{sq}(t)$ in (10.5.11) has the identical spectrum as in Figure 10.15e. The frequency division by the factor of 2 can be accomplished directly in the loop gain, since the error term $\psi_e(t)$ must be derived from the argument of the sine term in (10.5.10). Thus the equivalent Costas quadrature loop is identical to the equivalent squaring loop system shown in Figure 10.14b and generates the same mean squared carrier phase error given in (10.5.5).

An advantage of the Costas loop is that the PSK decoding is partially accomplished right within the loop. If the loop is tracking well so that $\psi_o(t) \approx \psi(t)$, we see the quadrature mixer output in (10.5.9) is

$$I(t) \approx A_s m(t) + n_{mc}(t) \qquad (10.5.12)$$

This result is identical to that produced at the multiplier output in the coherent BPSK decoder in Figure 7.9. The addition of a bit integrator (Figure 10.18a) therefore completes the PSK decoder. Hence both data demodulation and phase referencing can be accomplished directly from the Costas loop. For this reason the quadrature arm is often called the *decisioning* arm and the in-phase arm, the *tracking* arm. One can think of the multiplier of the system as allowing the bit polarity of the decisioning loop to correct the phase error orientation of the tracking loop, thereby removing the modulation. Often Costas loops are designed with the decisioning arm filter followed by a limiter (Figure 10.18). At high SNR, the limiter output will have a sign during each bit interval that is identical to the present data bit polarity. This limiter output then multiplies the loop error and removes

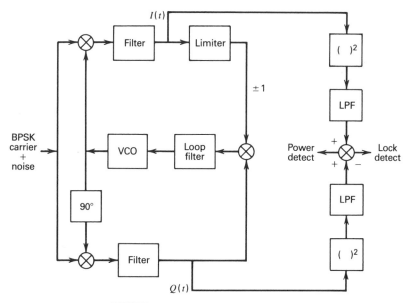

FIGURE 10.18. Modified Costas loop.

the loop modulation. In effect, the data bit sign is used to aid the tracking loop, and such modified systems are often called *data aided carrier extraction loops.*

An interesting property of Costas loops is that the system generates signals that can be used for other auxiliary purposes as well. This can be seen by reexamining (10.5.9) and noting the following (the overbar denotes low pass filtering):

$$\overline{I^2(t) - Q^2(t)} = A_s^2 \overline{m^2(t)} \cos(2\psi_e) \qquad (10.5.13)$$

Squaring, low pass filtering, and subtracting the arm voltages (Figure 10.18) produce an output that indicates phase lock [$\cos(2\psi_e) \to 1$ as $\psi_e \to 0$] and can therefore serve as a lock detector. When $\psi_e = 0$, this generates an output proportional to the average signal power, which can also be used for automatic gain control (AGC). We also note

$$\overline{I^2(t)} + \overline{Q^2(t)} = A^2 \overline{m^2(t)} + \overline{n_{ms}^2(t)} + \overline{n_{mc}^2(t)}$$
$$= \text{total power of } x(t) \qquad (10.5.14)$$

This produces a measurement of the total RF input power and therefore can be used for RF power control. Finally,

$$\overline{I(t)\frac{dQ}{dt}} - \overline{Q(t)\frac{dI}{dt}} \cong [A^2 \overline{m^2(t)} \cos 2\psi_e]\frac{d\psi_e}{dt} \qquad (10.5.15)$$

By differentiating in each arm and cross-multiplying, a signal proportional to the frequency error, $d\psi_e/dt$, is generated. This can be used for frequency offset measurements, or for aiding in frequency acquisition.

10.6 QPSK CARRIER REFERENCING

When digital data is modulated on the carrier with QPSK (quadraphase) encoding, carrier extraction must be accomplished by a slightly different means than with BPSK. This is due to the fact that the QPSK carrier, although still containing a suppressed carrier component, contains one of four possible phases, rather than two, during each bit interval. To extract the QPSK carrier, a fourth power, instead of a squaring, system must be used, as shown in Figure 10.19a. The QPSK carrier is passed into the fourth power device, and a phase lock loop can be locked to the fourth harmonic. To see this, we write the QPSK signal as

$$x(t) = A \cos \left[\omega_c t + \theta(t) + \psi(t) \right] \qquad (10.6.1)$$

where $\theta(t)$ is one of the phase angles $n\pi/2$, $n = 1, 2, 3, 4$, during each bit time. When raised to the fourth power, we obtain the following terms in the vicinity of frequency $4\omega_c$:

$$y(t) = \left(\frac{A^4}{4} \right) \cos \left[4\omega_c t + 4\theta(t) + 4\psi(t) \right] + n_4(t)$$

$$= \left(\frac{A^4}{4} \right) \cos \left[4\omega_c t + 4\psi(t) \right] + n_4(t) \qquad (10.6.2)$$

The first term is the desired harmonic to be tracked, with the modulation removed. The term $n_4(t)$ represents all the carrier–noise cross-products in the vicinity of $4\omega_c$. The subsequent tracking loop at $4\omega_c$ in Figure 10.19a tracks the phase variation $4\psi(t)$. Frequency division of the loop VCO reference output by a factor of 4 produces a phase locked carrier reference that is phase coherent with the suppressed QPSK carrier at ω_c. A phase ambiguity will again occur from this frequency division. In this case we have a fourth order ambiguity since frequency division by a factor of 4 can produce any of the four phase angles $i\pi/4$, $i = 1, 2, 3, 4$ on the divided reference frequency.

The contribution of the fourth order noise terms to the loop error can be computed as in the squaring loop analysis for BPSK. However, the procedure is now more complicated than it was for the squaring loop since the multiplication terms involving the QPSK carrier–noise cross-products are present in all orders up to the fourth power. It has been shown [11] that if the carrier bandpass filter in Figure 10.19 has the value $B_c = 1/T_b$, where T_b

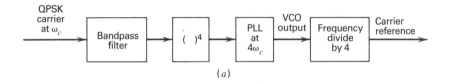

(a)

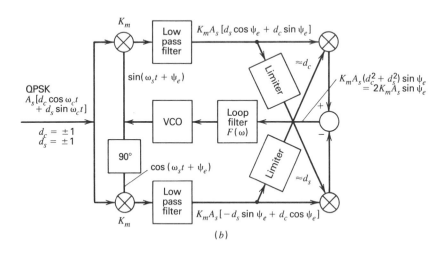

(b)

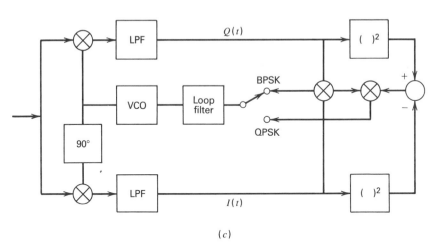

(c)

FIGURE 10.19. QPSK carrier referencing systems: (a) fourth power loop, (b) Costas crossover loop.

is the QPSK bit time, then the fourth-power loop generates the mean squared phase error at the carrier frequency of approximately

$$\sigma_e^2 \cong \frac{N_0 B_L}{P_c} \left[1 + \frac{3.8}{E_b/N_0} + \frac{4.2}{(E_b/N_0)^2} + \frac{1}{(E_b/N_0)^3} \right] \qquad (10.6.3)$$

The bracketed term now plays the same role for the fourth-power loop as the squaring factor in (10.5.5) for the squaring loop. We emphasize that (10.6.3) refers to the phase error between the received QPSK carrier and the divided-down reference carrier at ω_c. The phase error in the loop at $4\omega_c$ has a mean squared value that is $4^2 = 16$ times as large. This must be taken into account when analyzing the PLL tracking stability in Figure 10.19a.

Carrier extraction for QPSK signals can also be derived from a modified form of the Costas loop involving loop crossover arms, as shown in Figure 10.19b. The signal waveforms are shown throughout the figure. Note that each filter output contains data bits from both quadrature carriers that compose the QPSK subcarrier waveform. However, the sign of the output of the arm filters produced by the limiters is used to cross over and mix with the opposite arm signal. If the loop phase error is small, the output of the limiters corresponds to the cosine term of the arm filter outputs, and therefore has the sign of the data bits of the quadrature components as shown. Hence, the limiters effectively demodulate the QPSK quadrature bits, and the crossover produces a common phase error term that is cancelled after subtraction, leaving a remainder error term proportional to $\sin \psi_e$. The latter signal can then be used to generate an error signal to phase control the loop VCO, thereby closing the QPSK Costas loop. The crossover connections of the QPSK loop can be redrawn. For example, the ×4 tracking loop in Figure 10.19 will generate a tracking error voltage proportional to $\sin(4\psi_e)$. This can be expanded trigonometrically as

$$\sin(4\psi_e) = 4\cos(\psi_e)\sin(\psi_e)[\sin^2(\psi_e) - \cos^2(\psi_e)] \qquad (10.6.4)$$

Neglecting noise and assuming that $m(t)$ contains ideal pulses, the right side can be related to the Costas signals in (10.5.9) as

$$\sin(4\psi_e) = 4Q(t)I(t)[I^2(t) - Q^2(t)] \qquad (10.6.5)$$

This suggests a modified version of the BPSK Costas loop with the connections shown in Figure 10.19c for QPSK carrier recovery. Note that the system is basically an extension of the BPSK system by adding on the nonlinear section. We can easily connect from one to the other by simply reconnecting the VCO.

A more recent system for QPSK phase referencing is the demodulation–remodulation (Demod–Remod) [12–14] loop in Figure 10.20. The system obtains its phase referencing by attempting to reconstruct a local version of the received QPSK carrier. It does this by using the reference VCO to demodulate the input via the quadrature mixer to produce the baseband. It then remodulates the demodulated baseband back as to the same VCO reference carrier, thereby creating the local QPSK carrier. If the bits are demodulated correctly, the reconstructed carrier has the same modulation phase as the received carrier but differs by the phase reference error. Hence

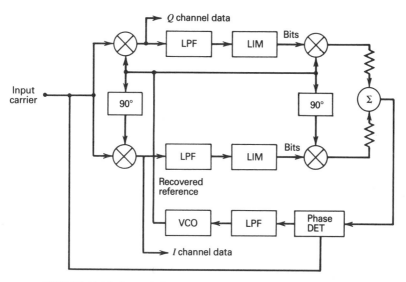

FIGURE 10.20. Demod–Remod loop for QPSK carrier referencing.

a phase comparison between the received waveform and the local reconstructed QPSK carrier generates a voltage error that corrects the local reference phase, bringing it into synchronism. Demod–Remod systems have the advantage that tracking is accomplished directly at the carrier frequency, so no divide-down ambiguities occur. However, the system requires more hardware (two more mixers than the Costas loop) and depends strongly on the accurate reconstruction of the local QPSK carrier. This means reliable bit decoding, and exacting balancing of the quadrature arms (as indicated by the gain adjustment in the output arm in Figure 10.20). For this reason the Demod–Remod system requires high input SNR and, since it primarily phase adjusts, is predominantly a phase tracking subsystem. It has been shown that if the bits are decoded correctly, the tracking error voltage in the Demod–Remod system evolves identically to that of a Costas loop error voltage, and the performance of the two systems become identical at high SNR values.

Unbalanced QPSK (UQPSK) carriers in (9.6.19) cannot be phase referenced by the standard fourth-power loops, since the modulation phase states are not equally spaced. Hence UQPSK carriers require modified Costas crossover loops or modified Demod–Remod loops for phase referencing. The modification occurs in the attenuation settings of each arm to account for the quadrature power imbalance. Since the crossover loop depends on interference subtraction from the crossover arms, the UQPSK loops are extremely sensitive to gain adjustments for exact cancellation. Likewise, the Demod–Remod loop must reconstruct the UQPSK carrier with the data arms having the exact power relation and is also sensitive to gain settings.

If the UQPSK imbalance is high, the carrier closely resembles a BPSK carrier, and phase referencing can be achieved by a standard squaring system in Figure 10.14. Denote the UQPSK carrier as

$$c(t) = \sqrt{2P_c} m_c(t) \cos(\omega_c t + \psi) + \sqrt{2P_s} m_s(t) \sin(\omega_c t + \psi) \quad (10.6.6)$$

where $P_c \gg P_s$. If the input to the squarer is $c(t) + n(t)$, the squarer output (neglecting dc values) is

$$[c(t) + n(t)]^2 = (P_c - P_s) \cos(2\omega_c t + 2\psi)$$
$$+ 2\sqrt{P_c P_s} m_c(t) m_s(t) \sin(2\omega_c t + 2\psi) + n_{sq}(t) \quad (10.6.7)$$

where $n_{sq}(t)$ is again the sum of the signal–noise and noise–noise cross-product terms in (10.5.2). This noise has the spectral level in (10.5.4)

$$S_{sq}(0) = \frac{N_0^2 B_c}{2} + N_0 P_T \quad (10.6.8)$$

with $P_T = P_c + P_s$ is the total UQPSK carrier power. The tracking term is given by the first term in (10.6.7) and has amplitude $(P_c - P_s)$. The second term represents interference due to the quadrature modulation and involves the cross-product of the two data waveforms. The power of this term, in the vicinity of $2\omega_c$, in the tracking loop bandwidth B_L, is then

$$\text{Quadrature noise} = [P_c P_s S_{cs}(0)] 2 B_L \quad (10.6.9)$$

where $S_{cs}(\omega)$ is the convolution of the spectral densities $S_c(\omega)$ and $S_s(\omega)$ of the two data waveforms. This means

$$S_{cs}(0) = \frac{1}{2\pi} \int_{-\infty}^{\infty} S_c(\omega) S_s(\omega) \, d\omega \quad (10.6.10)$$

The effective tracking loop SNR is then

$$\rho = \frac{(P_c - P_s)^2 / 2}{[N_0 P_T + (N_0^2 B_c / 2) + P_c P_s S_{sc}(0)] 2 B_L} \quad (10.6.11)$$

Let $P_c = q P_T$, $P_s = (1 - q) P_T$, and $P_T = P_c + P_s$, and write

$$\rho = \frac{P_T^2 [q - (1 - q)]^2 / 2}{\{N_0 P_T [1 + (N_0 B_c / 2 P_T)] + P_T^2 q (1 - q) S_{cs}(0)\} 2 B_L}$$

$$= \frac{1}{4} \left(\frac{P_T}{N_0 B_L} \right) \left\{ \frac{(2q - 1)^2}{[1 + (N_0 B_c / 2 P_T)] + (P_T / N_0) S_{cs}(0) q (1 - q)} \right\} \quad (10.6.12)$$

The resulting UQPSK phase referencing error, after dividing by 2, is then

$$\sigma_e^2 = \frac{1}{4\rho} \qquad (10.6.13)$$

Note that the bracket in (10.6.12) represents a UQPSK squaring loss and is due to both the squaring of the noise and the cross-product self-noise of the data itself. This self-noise depends on the spectral densities of the baseband waveforms and the type of digital modulation, while the amount of spectral overlap will depend on the individual data rates. Table 10.3 summarizes some basic formulae [15] for evaluating the parameter $S_{cs}(0)$ in (10.6.10) for some popular combinations of modulation formats and data rates used on each arm. For situations in which the self-noise dominates in (10.6.12), these results become important for determining and controlling the loop CNR. It may be necessary to actually place one of the data streams onto a digital subcarrier so as to further separate the spectra in (10.6.10) for increased self-noise reduction.

TABLE 10.3 Self-Noise Spectral Level for UQPSK Modulation

High rate (R_2 bps) modulation format	Low rate (R_1 bps) modulation format	$S_{cs}(0)$
NRZ	NRZ	$\dfrac{1}{R_2}\left(1 - \dfrac{R_1}{R_2}\right)$
Manchester	NRZ	$\dfrac{1}{6}\left[\dfrac{R_1}{R_2^2}\right]$
NRZ	Manchester	$\dfrac{1}{R_2}\left[1 - \dfrac{R_1}{R_2}\right], \quad R_2 \geq 2R_1$
		$\dfrac{1}{R_2}\left[\dfrac{R_2}{R_1} - \dfrac{1}{6}\left(\dfrac{R_2}{R_1}\right)^2 - 1 + \dfrac{1}{3}\left(\dfrac{R_1}{R_2}\right)\right],$ $\quad R_2 < 2R_1$
Manchester	Manchester	$\dfrac{1}{2}\left(\dfrac{R_1}{R_2^2}\right), \quad R_2 \geq 2R_1$
		$\dfrac{1}{R_2}\left[2 - \left(\dfrac{R_2}{R_1}\right) + \dfrac{1}{6}\left(\dfrac{R_2}{R_1}\right)^2 - \dfrac{5}{6}\left(\dfrac{R_1}{R_2}\right)\right],$ $\quad R_2 < 2R_1$

10.7 BIT TIMING

After the data modulated carrier has been demodulated to baseband via the coherent carrier reference, bit timing must be established to clock the bit or word decoding. Bit timing therefore corresponds to the operation of extracting from the demodulated baseband waveform a time coherent clock at the bit rate or word rate of the data. Bit timing subsystems generally operate in conjunction with the decoder (Figure 10.21) and use the same demodulated baseband waveform to extract the bit timing clock. Timing markers provided from this clock then can be used to synchronize the decoder.

One of the simplest ways to obtain a timing clock is by a separate subcarrier or pilot tone transmitted in the baseband with the data. The transmitter simply separates the digital data and timing waveforms in the baseband format (e.g., by FDM), and the receiver baseband subsystem processes the waveform, as in Figure 10.10. It merely separates the data and the timing information, carries out the timing acquisition and clocking operations by PLL tracking, and uses the generated timing markers to time the digital decoders. Since no data can be decoded until the acquisition is completed and the clock is locked in, timing parameters such as acquisition time become important in assessing the overall digital performance. Similarly, timing error variances will ultimately limit the decoding capability, as we saw in Section 7.7.

In many systems, however, the use of a separate subcarrier for the sole purpose of bit timing is avoided because of the requirement of excess baseband power and bandwidth. The preference is to derive the bit timing directly from existing hardware, if possible, or to generate it using modulation derived timing. For example, in a PCM format, the timing markers are required at the beginning of each bit to start integration. If the receiver achieved RF phase reference, then by making the transmitted bit rate time synchronous with the RF frequency (i.e., the bit interval begins at an RF positive going zero crossing), bit clocking can be maintained from the

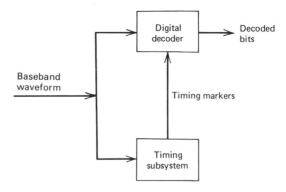

FIGURE 10.21. Bit timing subsystem and digital decoder.

receiver RF reference. We need only translate the RF, IF, or master oscillator frequency to the bit rate frequency and use the latter zero crossing for bit timing marker generation. This procedure, though theoretically feasible, is not ordinarily popular, since it requires exact synchronization between the RF frequency and the modulating data bits at the transmitter.

By far the most common method for obtaining bit timing information is to derive it directly from the noisy baseband waveform itself. In BPSK systems this would occur after the phase coherent demodulation shown in Figure 7.12d. In QPSK systems this would occur on either arm of the phase coherent quadrature correlator shown in Figure 10.14. If the arms have synchronized bits, only one bit synchronization system is required. If the data bits are unsynchronized, as in unbalanced QPSK, then a separate bit synchronization is needed for each arm. Subsystems for performing bit timing from the data can be classified into two basic types. One involves squaring of the data waveform to generate timing waveforms that can be used for time synchronization. The other uses some form of transition tracking of the bit edges to generate the timing. Both types of system inherently involve modulation removal and harmonic generation, but the mechanics of each is accomplished in different ways.

Squaring systems (Figure 10.22) contain some form of low pass waveform filtering followed by rectification or squaring to remove the modulation and generate a timing subsystem that follows. This can be seen from the diagram in Figure 10.23. Figure 10.23a shows a binary digital NRZ waveform composed of ideal pulses with random data bits. Figure 10.23b shows a low pass filtered version of this waveform, producing rounded pulses as shown. Squaring (or rectification) of this waveform produces the sequence in Figure 10.23c, in which the presence of a repetitive component is obvious. This means that the squared waveform contains harmonic power at the bit rate frequency. This harmonic is in exact time synchronism with the zero crossings of the original bit waveform. Thus tracking of the bit rate harmonic in a PLL following the squarer will produce timing markers at the zero crossings that can be used for bit timing.

The squared waveform in Figure 10.23c will have a spectrum composed of a continuous portion plus discrete harmonics of the bit rate frequency $1/T_b$. If $H_L(\omega)$ is the low pass prefilter function in Figure 10.22 and $P(\omega)$ the bit pulse transform, then for equal likely data bits, the first harmonic at $1/T_b$ at the input to the PLL will have power

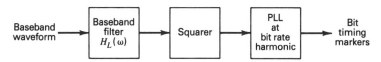

FIGURE 10.22. Filter–squarer bit timing subsystem.

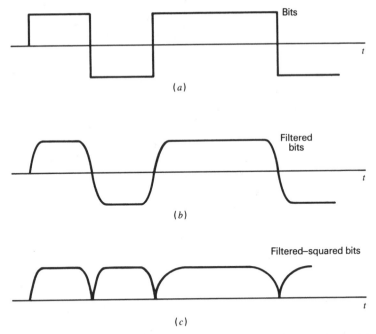

FIGURE 10.23. Bit clock generation from squared data: (*a*) digital waveform, (*b*) filtered waveform, (*c*) squared–filtered waveform.

$$P_1 = \frac{1}{T_b} \left| S_2 \left(\frac{2\pi}{T_b} \right) \right|^2 \qquad (10.7.1)$$

where $S_2(\omega) = [H_L(\omega)P(\omega)] \oplus [H_L(\omega)P(\omega)]$; that is, $S_2(\omega)$ is the convolution of $H_L(\omega)P(\omega)$ with itself. This component is to be tracked by the loop, while the continuous part adds spectral interference (called *self-noise*, or *data dependent noise*) to the loop input. This self-noise can generally be neglected, relative to the receiver noise entering the loop. Note that the harmonic tracking power depends on the filtering $H_L(\omega)$.

The baseband waveform is actually the sum of the data sequence in Figure 10.23*a* plus the baseband white noise (spectral level N_0). The filtering–squaring operation produces the harmonic in (10.7.1) plus the signal filtered noise cross-product and the squared noise component. The PLL tracks the harmonic with these cross-product and squared noise waveforms appearing as interference. The timing accuracy is therefore dependent on the phase tracking accuracy of the loop. A timing error τ in seconds is related to a loop phase error ψ_e in radians at $f = 1/T_b$ by $\tau = T_b(\psi_e/2\pi)$. Thus the timing error variance and phase error variance are related by

$$\sigma_\tau^2 = \sigma_e^2 \left(\frac{T_b}{2\pi} \right)^2 \qquad (10.7.2)$$

The phase error variance can be obtained by standard tracking loop analysis, with an input carrier tone having the power in (10.7.1) and considering the interference in the vicinity of $1/T_b$ entering the loop as noise. Holmes [16] has shown that for an RC presquaring filter the squared noise term produces a timing error variance in (10.7.2) of

$$\sigma_n^2 = C_n \left(\frac{N_0^2 T_b^3 B_L}{E_b^2} \right) \tag{10.7.3}$$

where C_n is a coefficient dependent on the $H_L(\omega)$ filter bandwidth and E_b/N_0 is the baseband bit energy-to-noise level at the timing system input. For the first order RC prefilter with 3 dB frequency f_3 and NRZ pulses, C_n has the value

$$C_n = \frac{r^3 \pi (1 + r^{-2})}{4(1 - e^{-2\pi r})} \tag{10.7.4}$$

where $r = f_3 T_b$. The signal–noise cross-product contributes a timing error variance of

$$\sigma_{sn}^2 = C_{sn} \left(\frac{B_L T_b^3}{E_b/N_0} \right) \tag{10.7.5}$$

where, for the RC prefilter,

$$C_{sn} = \frac{3r(4r^2 + 13r^4 - r^2 - 1)}{2\pi(1 - e^{-r})(r^2 + 1)(4r^2 + 1)(r^2 + 4)} \tag{10.7.6}$$

The total mean squared timing error for the filter–square bit synchronizer is then

$$
\begin{aligned}
\sigma_\tau^2 &= \sigma_n^2 + \sigma_{sn}^2 \\
&= B_L T_b^3 \left[\frac{C_n}{(E_b/N_0)^2} + \frac{C_{sn}}{(E_b/N_0)} \right]
\end{aligned}
\tag{10.7.7}
$$

Normalizing σ_τ^2 to the bit time, we can rewrite as

$$\left(\frac{\sigma_\tau}{T_b} \right)^2 = \frac{B_L T_b}{(E_b/N_0) \mathcal{S}_q} \tag{10.7.8}$$

where we have introduced the bit timing squaring loss

$$\mathcal{S}_q = \left[C_{sn} + \frac{C_n}{(E_b/N_0)} \right]^{-1} \tag{10.7.9}$$

Note that at low E_b/N_0, (10.7.8) behaves as $(E_b/N_0)^{-2}$, and as $(E_b/N_0)^{-1}$ at high E_b/N_0. It has been shown [16] that at $E_b/N_0 \approx 10 \text{ dB}$, the squaring loss in (10.7.9) is minimized with a value of $r \approx \frac{3}{16}$—that is, the prefilter bandwidth should be set to about $\frac{3}{16}$ of the bit rate frequency $1/T_b$. For this case

$$C_n \approx 0.318$$
$$C_{sn} \approx 0.545 \tag{10.7.10}$$

When these coefficients are used in (10.7.9), the normalized rms timing error in (10.7.8) plots as shown in Figure 10.24 for several values of $B_L T_b$. In Section 7.6 it was shown that bit timing must be maintained to within about 1% of a bit time to prevent serious degradation to the PE decoding performance. We see from Figure 10.24 that bit timing to within 1% of a bit time is feasible at $E_b/N_0 \approx 10 \text{ dB}$, with moderate values of loop bandwidth ($B_L T_b \leq 10^{-2}$). This means, for example, that with a bit clock of 10 Mbps, a timing accuracy on the order of nanoseconds can be retained with this type of timing system.

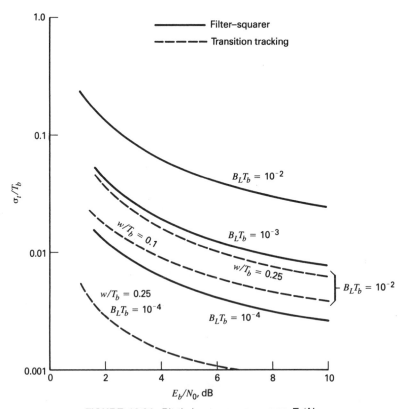

FIGURE 10.24. Bit timing rms error versus E_b/N_0.

A second way to achieve bit timing from a modulated NRZ waveform is to utilize *transition tracking* on the bit edges, in conjunction with bit decisioning. Transition tracking is obtained by integrating over the bit transitions (bit edges) to generate a timing error voltage for feedback control. To see this, consider Figure 10.25a, which shows a periodic pulse sequence with a transition being integrated over a w sec integration interval. We see that if the integration is centered exactly on the pulse transition, a zero integration value is produced, because of the equal positive and negative areas. If the integration is offset by δ sec, the integrator output generates the function $\pm g(\delta)$ shown, where the $\pm$ sign depends on the particular data bits (i.e., whether the transition changes from a positive to negative value or vice versa). To rectify this bit sign, a bit decision is made in a separate decoding arm and used to multiply the integrator output. By inserting a one-bit delay in the loop input and using the present bit decision, the bit polarity will always rectify the sign of the loop error voltage, producing a timing correction voltage of the proper sign independent of the data bit. This correction voltage is filtered and fed back to control the timing location of the integration for the next transition. Thus the previous bit

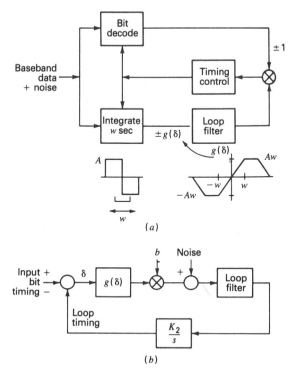

FIGURE 10.25. Transition tracking bit timing loop: (*a*) block diagram, (*b*) equivalent loop model.

transitions are effectively used to remove the modulation and the loop operates to hold the integrator centered on the transitions. Bit timing markers can then be generated in synchronism with the integration for timing control. The latter system is referred to as *data aided transition tracking* and is the bit timing equivalent of the suppressed carrier phase referencing system of the same type. The bit delay can also be eliminated in modified versions of the system by simply hard-limiting the baseband waveform in the upper arm and using the limiter output to invert the error. As long as the noise does not cause the bit waveform polarity to change sign during a bit, the tracking error will be properly rectified.

In transition tracking a timing error voltage is generated each pulse time when a transition occurs. (In NRZ data, this is once per bit if the bit changes sign but is twice per bit for Manchester data, with a transition guaranteed every midbit time.) The loop filter acts to smooth these error samples into a continuous loop control. In this sense the transition tracker appears as a sampling (discrete) control loop with error samples generated at regular intervals. The loop therefore has the equivalent timing model shown in Figure 10.25b. The input is the timing offset of the received data bits, and the feedback variable is the relative timing position of the integrator. The function $bg(\delta)$ converts timing errors to voltages, and the loop integrates to adjust integrator timing. The system is identical to the nonlinear sine wave phase tracking loop in Figure 10.2, except the nonlinear sine function is replaced by the nonlinear $g(\delta)$ function. Thus Figure 10.25b is the bit timing equivalent to the phase locking operation discussed earlier.

The multiplication by b accounts for the bit decision to rectify the error, with $b = +1$ if the bit is correctly detected and $b = -1$ if the bit is incorrectly decoded. The noise level entering the loop is the integrated noise variance per sample ($N_0 w$) divided by the sampling bandwidth $1/T_b$, producing the effective noise level $N_0 w T_b$. Note that the loop is linear for $|\delta| \leq w/2$, and therefore the transition tracking loop has an equivalent linear model with gain given by the mean slope of $bg(\delta)$ at $\delta = 0$. Hence

$$\text{Loop gain} = \frac{A\omega}{w/2} \, \mathscr{E}(b)$$

$$= 2A\mathscr{E}(b) \qquad (10.7.11)$$

where A is the data bit amplitude. The mean of the random variable b is given by $\mathscr{E}(b) = (1)(1 - \text{PE}) + (-1)\text{PE} = 1 - 2\text{PE}$, where PE is the bit decoding probability. Using the binary BPSK PE, (10.7.11) becomes

$$\text{Loop gain} = 2A\left[1 - 2Q\left[\left(\frac{2E_b}{N_0}\right)^{1/2}\right]\right] \qquad (10.7.12)$$

With this gain inserted into the linear model, the loop can be analyzed as a standard phase lock loop with noise bandwidth B_L. The timing error variance is then

$$\sigma_\tau^2 = \frac{(N_0 w T_b) 2 B_L}{4 A^2 [\text{Erf}\,(E_b/N_0)^{1/2}]^2} \qquad (10.7.13)$$

The variance normalized to the bit time T_b is then

$$\left(\frac{\sigma_\tau}{T_b}\right)^2 = \frac{N_0 2 B_L w/T_b}{4 A^2 [\text{Erf}\,(E_b/N_0)^{1/2}]^2} \qquad (10.7.14)$$

Note the direct dependence of the timing error variance on the integration width w. When a reduced integration time is used, the timing accuracy improves. However, it must be remembered that w is made smaller, the linear tracking range in Figure 10.25 is also decreased, making linear operation more difficult to retain. Again defining

$$\rho = \frac{N_0 2 B_L}{A^2} = \frac{B_L T_b}{E_b/N_0} \qquad (10.7.15)$$

we can write (10.7.14) as

$$\left(\frac{\sigma_\tau}{T_b}\right)^2 = \frac{1}{\rho \mathscr{S}_q} \qquad (10.7.16)$$

where $\mathscr{S}_q$ is now the timing squaring loss

$$\mathscr{S}_q = \left[\frac{4 \,\text{Erf}^2 \,(E_b/N_0)^{1/2}}{w/T_b}\right]^{-1} \qquad (10.7.17)$$

Equation (10.7.16) is also plotted in Figure 10.24 for several values of $B_L T_b$ and w/T_b. Timing performance is comparable to the filter–squarer system and can be improved by decreasing w. The transition loop with NRZ data also suffers from the fact that long sequences of identical bits produce no edge transitions for timing update. Insertion of intentional transitions may be necessary to guarantee the presence of a transition within a prescribed time interval. By using Manchester data, a transition is guaranteed at the center of every bit, thereby aiding the bit timing operation.

Transition tracking performance can be improved by accumulating a sequence of m error samples prior to timing correction. (This requires replacing the filter in Figure 10.25 by a digital accumulator.) Such an accumulation multiplies up the noise voltage variance by m but also increases the loop gain in (10.7.12) by m. The normalized timing error variance, assuming perfect bit decisioning, is then

$$\left(\frac{\sigma_\tau}{T_b}\right)^2 = \frac{m N_0 w}{m^2 4 A^2 T_b^2}$$

$$= \frac{(w/T_b) N_0/4}{m E_b/N_0} \qquad (10.7.18)$$

where w/T_b is the integration window expressed as a fraction of the bit time. Note the direct improvement as we accumulate over more samples. However, this improvement occurs only if the signal–error samples remain the same throughout the accumulation time. Hence m is restricted to the number of bits over which the received timing variation does not change appreciably.

The modern trend is toward all-digital bit timing, in which the clocked integrators are replaced by samplers and the error control is generated in software. Such a device is called a *digital transition tracking loop* (DTTL) [17, 18]. The block diagram of a DTTL is shown in Figure 10.26a, and its operation can be described by the time line in Figure 10.26b. The input is the baseband bit waveform (noise is not shown). This input is sampled at the rate of twice the bit rate, one sample approximately at the middle of the bit and one sample at bit edge. The samples are A-D converted and processed in software to generate the timing error voltage. If successive midbit samples are pairwise subtracted, they will indicate whether a bit transition has occurred, as well as the direction of the transition. The edge samples (even samples) will be zero if exactly at a bit edge crossover, but will generate

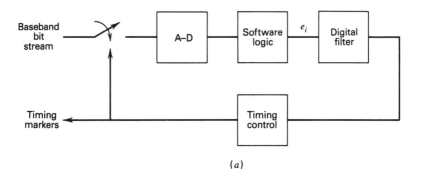

(a)

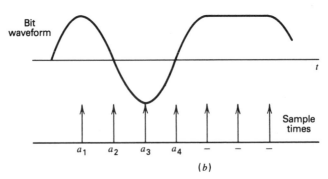

(b)

FIGURE 10.26. The digital transition tracking loop (DTTL): (a) block diagram, (b) time line.

positive or negative voltages if it is offset in one direction or the other, or if no transition has occurred. This latter information is provided by the midbit samples, and the difference value is used to multiply the edge sample value to achieve the correct voltage sign.

The multiplication is done in software using three-state logic $(+1, -1, 0)$, so the error sample is generated as

$$e_i = \text{sign}\left[(a_i - a_{i+2})\right] \text{sign}\left[a_{i+1}\right], \qquad i = 1, 3, 5, \ldots \quad (10.7.19)$$

where $\{a_i\}$ are the samples in Figure 10.26b. The error samples are then digitally accumulated and the result used to step the clocking of the sampler. When in synchronization, the edge samples will be at bit crossover, and zero correction voltages are generated. Bit timing can then be extracted from all odd (or all even) sample times. As the timing slides off, the voltages in (10.7.19) will be collected and the clock stepped in the proper direction to pull back into synchronization.

A DTTL has the advantage of high speed, digital processing and being chip implementable. At high SNR its accuracy can be obtained from (10.7.17), but performance degrades rapidly at low SNR. Because of its discrete nature, acquisition and pull-in times may be slow, and rigorous analysis requires state transition formulations [16–18]. Preloop filtering is important for properly shaping the bit waveforms in Figure 10.26 for linear zero crossing response.

REFERENCES

1. Lindsey, W. *Synchronization Systems in Communications and Control*, Prentice-Hall, Englewood Cliffs, N.J., 1972.

2. Cunningham, J. *Introduction to Nonlinear Analysis*, McGraw-Hill, New York, 1958.

3. Hayashi, C. *Nonlinear Oscillations in Physical Systems*, McGraw-Hill, New York, 1964.

4. Frazier, J. and Page, J. "Phase Lock Loop Frequency Acquisition," *IRE Trans. Space Electron. Telem.*, vol. SET-8, September 1962, pp. 210–227.

5. Viterbi, A. *Principles of Coherent Communicatons*, McGraw-Hill, New York, 1966.

6. Holmes, J. "First Slip Times is Static Offset Phase Error for the First and Second Order Phase Locked Loop," *IEEE Trans. Comm. Tech.*, vol. COM-19, April 1871.

7. Schuckman, L. "Time to Cycle Slip in First and Second Order Phase Lock Loops," *Proc. Int. Conf. Comm.*, San Francisco, June 1970, pp. 34–39.

8. Meyr, H. "Nonlinear Analysis of Corrective Tracking Systems Using Renewal Process Theory," *IEEE Trans. Comm. Tech.*, vol. COM-23, February 1975, pp. 192–203.

9. Lindsey, W. and Meyr, H. "Complete Statistical Description of Phase Error Process Generated by a Corrective Tracking System," *IEEE Trans. Inf. Theory*, vol. IT-23, March 1977, pp. 194–203.

10. Lindsey, M. and Simon, M. *Telecommunications Systems Engineering*, Prentice-Hall, Englewood Cliffs, N.J., 1974.

11. Costas, J. "Synchronous Communications," *Proc. IRE*, vol. 44, Sept. 1956, pp. 1713–1722.

12. Weber, C. and Alem, W. "Performance Analysis of Demod–Remod Coherent Receiver for QPSK," *IEEE Trans. Comm.*, vol. COM-28, December 1980.

13. Braun, W. and Lindsey, W. "Carrier Synchronization for Unbalanced QPSK," *IEEE Trans. Comm.*, vol. COM-26, September 1978.

14. Simon, M. and Alem, W. "Tracking Performance of Unbalanced QPSK Demodulators—Biphase Costas Loop," *IEEE Trans. Comm.*, vol. COM-26, August 1978.

15. Yuen, J. "Deep Space Telecommunications System Engineering," *JPL Publ.*, 82-76, 1982, Chap. 5.

16. Holmes, J.K. *Coherent Spread Spectrum Systems*, Wiley, New York, 1982.

17. Lindsey, W. and Anderson, T. "Digital Data Transition Tracking Loops," *Proc. Int. Tele. Conf.*, Los Angeles, October 1968, pp. 259–271.

18. Simon, M. "Optimization of the Performance of Digital Data Tracking Loop," *IEEE Trans. Comm. Tech.*, vol. COM-18, October 1970, pp. 686–690.

PROBLEMS

10.1   (10.1)   Derive the second order differential equation that describes the system in (10.1.13) when the loop filter is a pure integrator [i.e., $F(\omega) = 1/j\omega$] and the received frequency is linearly increasing at rate β Hz/sec relative to the VCO frequency of the loop.

10.2   (10.1)   Consider a tracking loop that replaces the frequency mixer by a general error detecting device $g(\psi_e)$. Derive the resulting system differential equation in the absence of modulation. Assume that the remaining part of the system (filter and VCO) remain the same.

10.3   (10.2)   Solve the first order system differential equation in (10.2.3). [You will not get an explicit solution for $\psi_e(t)$.] Consider the case where $\Omega/A_c k < 1$ and $\Omega/A_c k > 1$. Sketch the solution for $\psi_e(t)$ for both cases, paying careful attention to periodicity.

10.4   (10.2)   Assume that a frequency acquisition employs the lock detector subsystem in Figure P10.4. The output of the quadrature detector is $A \cos \psi_e(t) + n(t)$, where $n(t)$ is the RF mixer noise. This detector signal is low pass filtered and continually compared to a signal threshold. When threshold is crossed, acquisition lock-up is signaled. (a) Explain how the detector works, considering both the case where $\psi_e(t) = \Omega t$, $\Omega \gg 1$ and where $\psi_e(t) \approx 0$. Explain the role of the low pass filter and how it should be designed. (b) Assuming that the noise $n(t)$ is Gaussian, determine the expressions for probabilities of correct lock detection and probability of false lock indication for a particular low pass filter $H(\omega)$.

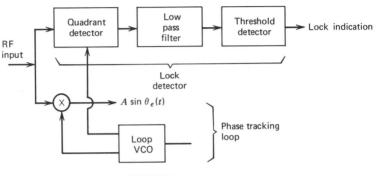

FIGURE P10.4.

10.5   (10.3)   A second order RF frequency acquisition system must acquire a 5 GHz carrier in an uncertainty bandwidth of 100 MHz. The FAL uses a master oscillator at 10 MHz and a loop bandwidth of 10 Hz at the master oscillator frequency. (a) How long will the acquisition operation take, assuming that we slew at the maximum rate allowed by (10.3.5) with an RF loop damping factor of 0.707? (b) How long will it take if we try to acquire at the master oscillator frequency?

10.6   (10.4)   A 2 GHz carrier is received with an unknown frequency offset of up to ± 1 ppm of the carrier frequency. It must be acquired in 1 msec by a slewed acquisition loop using the maximum slewing rate in (10.3.5). How much RF carrier component power is required to maintain a mean squared RF tracking error of 0.01 radian^2? The RF noise has a spectral level of $N_0 = -190$ dBW.

10.7   (10.4)   It is known that the steady state probability density of the first order nonlinear tracking loop in Figure 10.2 must satisfy the differential equation

$$\frac{dp(\psi_e)}{d\psi_e} + (\rho \sin \psi_e) p(\psi_e) = C$$

where C is an unknown constant to be determined. Solve the equation to derive the density of (10.4.8).

10.8   (10.4)   A carrier signal (amplitude A, frequency ω_c) is phase modulated with a sine wave at frequency ω_m and modulation index Δ. The modulated signal is filtered in a narrow bandpass filter about ω_c (bandwidth $\ll 2\omega_m$). The filtered output then has white noise (level N_0 one-sided) added to it. The combined output, filtered signal plus noise, is then used as the input to a linear carrier tracking phase lock loop having a noise bandwidth B_L Hz. Determine the mean square value of the tracking phase error in the loop.

10.9   (10.4)   A first order PLL is designed with a loop noise bandwidth of 50 Hz. It is to track a carrier in noise with a Doppler shift of 20 rps. Using linear loop theory, determine the received carrier power needed to guarantee that the tracking error (steady state plus random component) has a total mean squared phase error less than 0.02 rad (assume $N_0 = 10^{-9}$).

10.10   (10.4)   A first order carrier tracking loop, using an oscillator at 1 MHz, is used to track an RF carrier at 1 GHz with a loop noise bandwidth of 10 kHz. (a) If the RF frequency is received at 1.000001 GHz, what frequency will the loop oscillator operate at in the steady state? (b) If the RF carrier is divided down to the oscillator carrier frequency, what is the steady state phase error between the two carriers? Neglect noise.

10.11   (10.4)   Consider the probability density in (10.4.8). (a) By approximating the $I_0(\rho)$ term for large ρ (Appendix A) and the cosine term for small ψ_e, show that the density approaches a Gaussian density with variance $1/\rho$. (b) By using the Jacobi–Anger expansion equation, show that its mean squared value is given by

$$\sigma_e^2 = \frac{\pi^2}{3} + 4 \sum_{n=1}^{\infty} \frac{(-1)^n I_n(\rho)}{n^2 I_0(\rho)}$$

10.12   (10.5)   Show that the squaring factor in (10.5.5) can be written in general form as

$$\mu = \left[1 + \frac{2}{A_s^2 N_0} \int_{-\infty}^{\infty} R_n^2(\tau)\, d\tau \right]$$

where $R_n(\tau)$ is the autocorrelation function of the noise at the input to the squarer. (b) Evaluate this for the autocorrelation function $R_n(\tau) = (N_0/4)\, B \exp(-B|\tau|)$.

10.13   (10.5)   Determine the power spectrum of $n_{sq}(t)$ in (10.5.11). The processes $n_{mc}(t)$ and $n_{ms}(t)$ are uncorrelated Gaussian processes. Use autocorrelation analysis and apply convolution, using the spectral shapes in (Fig. 10.15) for the mixer noise. The autocorrelation of $n_{mc}(t)$, $n_{ms}(t)$ is one-half that of the mixer noise.

10.14 (10.5) Consider that the carrier signal in Figure 10.14 is passed through a hard-limiter prior to carrier tracking. Rederive the expression for ρ in the tracking loop by taking into account the effect of the limiter on the original carrier waveform. (*Hint:* Follow the analysis in Section 4.9.)

10.15 (10.7) Given a periodic sequence of impulses in which there is a probability p that an impulse will occur in a particular periodic position (i.e., there is probability $1 - p$ that a given impulse is missing). Show that the power at the impulse harmonic at the repetition rate frequency will depend on p. (*Hint:* Follow the spectral density development in Section 2.1.)

11

SATELLITE COMMUNICATIONS

The use of orbiting satellites has now become an integral part of our everyday communications. Each time we watch a TV cable program, make a long distance telephone call, or send teletype or facsimile data, we are most likely making use of a communication satellite. As the technology and hardware of such systems continues to improve, it is expected that satellites will continue to play an ever-increasing role in world wide communications.

To a communication engineer, a satellite channel represents a special type of communication link, complete with its own characteristics, limitations, and constraints. In order to properly design subsystems, and carry out tradeoff studies of the various design alternatives, one must be aware of the overall channel properties and anomalies. In this chapter we review some of the salient features of satellite communications.

11.1 DEVELOPMENT OF COMMUNICATION SATELLITES

Microwave propagation for long range communications has been readily used since the 1920s. However, these systems were, for the most part, restricted to line-of-sight links. This meant that two stations on Earth, located over the horizon from each other, could not communicate directly, unless by ground transmission relay methods. The use of tropospheric and ionospheric scatter to generate reflected skywaves for the horizon links tend to be far too unreliable for establishing a continuous system.

In the 1950s a concept was proposed for using orbiting space vehicles for relaying carrier waveforms to maintain long range, over-the-horizon communications. The first version of this idea appeared in 1956 as the Echo satellite—a metallic reflecting balloon placed in orbit to act as a passive reflector of ground transmissions to complete long rangelinks. Although the resulting communications were somewhat weak, the concept was nevertheless proved, and communications across the United States and across the Atlantic Ocean were successfully demonstrated this way. In the late 1950s

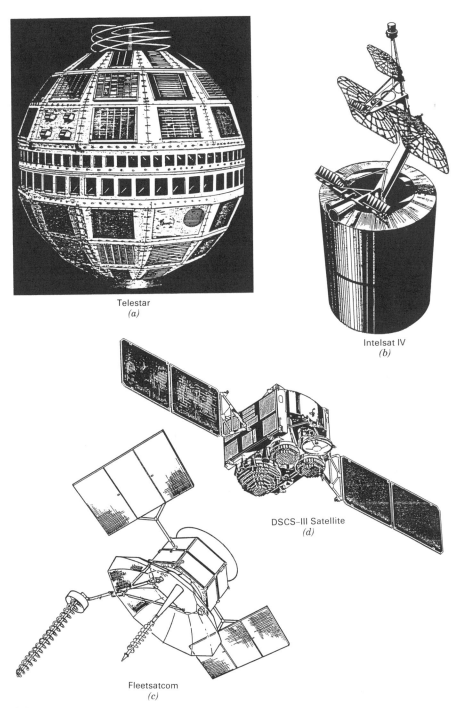

Telestar
(a)

Intelsat IV
(b)

DSCS–III Satellite
(d)

Fleetsatcom
(c)

FIGURE 11.1. Communication satellites: (*a*) Telstar, (*b*) Comstar or Intelsat, (*c*) Fleetsatcom, (*d*) DSCS III.

TABLE 11.1 Communications Satellites

Satellite	Launch Date	Launch Weight	Number of Transmitters	Total RF, MHz
Echo	1956	100	0	—
Score	1958	90	1	4
Courier	1960	500	2	13.2
Telstar	1962	170	2	50
Syncom	1963	86	2	10
Intelsat I	1965	76	2	50
Intelsat II	1966	190	2	130
Intelsat III	1968	270	4	450
Intelsat IV	1971	1400	12	500
ATS A, B, C, D	1966–1969	700	2	50
Telesat	1972–1975	1200	12	500
Westar	1974	120	12	500
Globecom	1975	1400	24	1000
DSCS I, II, III	1980–1981	2300	6	500
TDRSS	1980	1600	30	1200
Intelsat V	1981–1983	2000	35	2300
Intelsat VI	1986	3600	77	3366

new proposals were presented for using *active* satellites (satellites with power amplification) to aid in relaying long range transmissions. Early satellites such as Score, Telstar, and Relay verified these concepts. The successful implementation of the early Syncom vehicles proved further that these relays could be placed in extremely high orbits. These initial vehicle launchings were then followed by a succession of new generation vehicles, each bigger and more improved than its predecessors [1–3]. Table 11.1 lists some of the important early satellites, their launch dates, and some of their basic electronic characteristics. We note the continual increase in satellite size and communication capability as each new class of satellites was developed and placed in operation. Figure 11.1 depicts several of these satellites, and one can observe the increase in sophistication from the extremely simple early vehicles to the more recent models, with their multiple antennas and solar panels. In addition, each generation of communication satellites has been based on increasingly more refined and sophisticated technology, and this progress is expected to continue into the future.

11.2 SATELLITE LINKS

The basic satellite communication link is shown in Figure 11.2*a*. A transmitting Earth station sends a modulated electromagnetic field to an orbiting satellite. The satellite collects the impinging electromagnetic field and retransmits the modulated carrier as a downlink to specified earth stations.

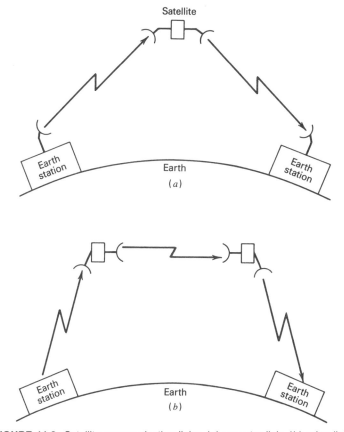

FIGURE 11.2. Satellite communication links: (*a*) repeater link, (*b*) relay link.

This allows communications to take place between stations that may be separated by mountainous terrain, be over-the-horizon, or even be continents apart.

A satellite that merely relays the uplink carrier as a downlink is referred to as a *relay satellite*, or *repeater satellite*. More commonly, since the satellite *transmits* the downlink by *responding* to the uplink, it is also called a *transponder*. A satellite that electronically operates on the received uplink to reformat it in some way prior to retransmission is called a *processing satellite*.

Figure 11.2*b* shows a satellite crosslink between two satellites prior to downlink transmission. Such systems allow communication between Earth stations not visible to the same satellite. By spacing multiple satellites in proper orbits around the Earth, worldwide communications between remote Earth stations in different hemispheres can be performed via such crosslinks.

Although the Earth stations and satellites each perform the basic communication operations—transmitting and receiving—the overall system is hindered by the inherent characteristics of the links involved. Satellites are placed above the Earth's atmosphere, so propagation path lengths between Earth and satellites will generally be 400 mi or more, sometimes even thousands of miles. It is therefore evident that satellite links involve extremely long transmission paths, having considerable propagation loss. In addition, satellite links involve space platforms in motion, which may require additional antenna pointing and tracking operations while communicating. Since the satellite is spaceborne, there are limits to the size, weight, and power that can be used, and all electronic equipment must be properly spaced qualified to assure operation in the space environment for almost a decade. These conditions place additional constraints on the link design, and often limit the choice of candidate design alternatives.

Earth stations form a vital part of the overall satellite system, and their cost and implementation restrictions must be integrated into system design. Basically, an Earth station is simply a transmitting and/or receiving power station operating in conjunction with an antenna subsystem. Earth stations are usually categorized into large and small stations by the size of their radiated power and antennas. Larger stations may use antenna dishes as large as 30 m in diameter. Smaller stations may use antennas of only 3–10 m and may, in fact, be constructed so as to be mounted on a mobile vehicle. Larger stations may often require antenna tracking and pointing subsystems to continually point at the satellite during its orbit, thereby ensuring maximum power transmission and reception. A given Earth station may be designed to operate as a transmitting station only, a receiving station only, or as both.

The internal electronics of an earth station is generally conceptually quite simple. In a transmitting station, the baseband information signals (telephone, television, data, etc.) are brought in on cable or microwave link from the various sources. The baseband information is then multiplexed and modulated onto IF carriers to form the station transmissions, either as a single carrier or perhaps as a multiple of contiguous carriers. The entire set of station carriers is then translated to radio frequencies for power amplification and transmission. A receiving earth station corresponds to a low noise, wideband RF front end followed by a translator to IF. At the IF, the specific uplink carriers wishing to be received are first separated, and then demodulated to baseband. The baseband is then demultiplexed (if necessary) and transferred on to the destination.

The satellite itself is merely a communication package placed in orbit around the Earth. A satellite transponder must relay an uplink or a crosslink electromagnetic field into a downlink or a crosslink. If the satellite was simply a passive reflector, the power levels of the downlink will be extremely low, because of the two-way propagation loss (plus the reflection loss). An active satellite repeater aids the relay operation by being able to add power amplification at the satellite prior to the downlink transmission. Hence an ideal active repeater would be simply an electronic amplifier in orbit, as sketched in Figure 11.3*a*. Ideally, it would receive the uplink carrier, amplify to the desired power level, and retransmit in the downlink. Practically, however, trying to receive and retransmit an amplified version of the same uplink waveform at the same satellite will cause unwanted feedback, or *ringaround*, from the downlink antenna back into the receiver. For this reason satellite repeaters must involve some form of frequency translation prior to the power amplification. The translation shifts the uplink frequencies to a different set of downlink frequencies so that some separation exists between the frequency bands. This separation allows frequency filtering at the satellite uplink antenna to prevent ringaround from the transmitting (downlink) frequency band. In more sophisticated processing satellites the uplink carrier waveforms are actually reformatted or restructured, rather than merely frequency translated, to form the downlink. Frequency band separation also allows the same antenna to be used for both receiving and transmitting, simplifying the satellite hardware.

The frequency translation requirements in satellites means that the ideal amplifier in Figure 11.3*a* should instead be reconstructed as in the diagram in Figure 11.3*b*. The satellite contains a receiving front end that first collects and filters the uplink. The collected uplink is then processed so as to translate or reformat to the downlink frequencies. The downlink carrier is then power amplified to provide the retransmitted carrier. The details of the hardware circuitry of a transponder of this type are discussed in Section 11.4. As more sophisticated satellites have evolved, the generic transponder model shown in Figure 11.3*b* has been modified and extended to more complicated forms.

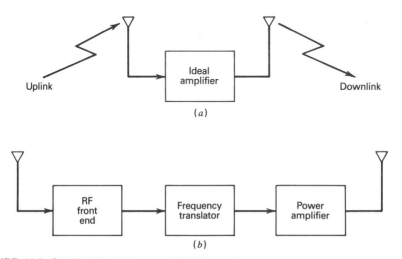

FIGURE 11.3. Satellite block diagram: (*a*) ideal amplifier, (*b*) frequency translation model.

In addition to the uplink repeating operation, a communication satellite will usually involve other subsystems as well. These include, for example, uplink command systems (that permit control of satellite position), downlink telemetry (for sending satellite "housekeeping" data), and ranging systems (for position monitoring from ground stations). These uplink and downlink subsystems used for tracking, telemetry, and command (TT&C) are usually combined with the uplink processing channels in some manner. This means that although they are not part of the mainline communication link, their design and performance does impact on the overall communication capability of the entire system.

The details of the fabrication of a satellite space vehicle is beyond the scope of the text. However, some aspects of satellite construction indirectly affect communication performance and therefore should be accounted for in initial system design. Primary power supply for all the communication electronics is generally provided by solar panels and storage batteries. The amount of primary power determines the usable satellite power levels for processing and transmission through the conversion efficiency of the electronic devices. The higher the primary power, the more power available for the downlink retransmissions. However, increased solar panel and battery size adds additional weight to the space vehicle. Thus there is an inherent limit to the power capability of the communication system.

Another important requirement in any orbiting satellite is attitude stabilization. A satellite must be fabricated so as to be stabilized (oriented) in space with its antennas pointed in the proper uplink and downlink direction. Satellite stabilization is achieved in one of two basic ways. Early satellites were stabilized by physically spinning the entire satellite (spin-stabilized) in order to maintain a fixed attitude axis. The second stabilization method is carried out via internal gyros, through which changes in orientation with respect to three different axes can be sensed and corrected by jet thrusters (three-axis stabilization). As requirements on satellite antennas and solar panels increased in size, it became correspondingly more difficult to despin, and three-axis stabilization became the preferred method.

Spin stabilization has the advantage of being simpler and providing better attitude stiffness. However, spinning is vulnerable to bearing failures, cannot be made redundant, and favors wide diameter vehicles, which may be precluded by launch vehicle size. Also, when despinning multiple-element antennas and solar panels, only a fraction of each can be used at any one time, thus reducing power efficiency. Three-axis stabilization tends to favor vehicles with larger antennas and panels and favors operation where stabilization redundancy is important.

Attitude stabilization determines the degree of orientation control and therefore the amount of error in the ability of the satellite to point in a given direction. Satellite downlink pointing errors are therefore determined by the stabilization method used. Both methods previously described can be made to produce about the same pointing accuracy, generally about a fraction of a degree. We shall see later that pointing errors directly affect antenna design and system performance, especially in the more sophisticated satellite models being developed.

Orbits of a satellite are classified as equatorial, polar, or inclined, defined by the path on Earth traced out by the subsatellite point during the orbit (Figure 11.4). The altitude selected for the satellite above the Earth directly determines both the period of the orbit (time to complete one cycle around the Earth) and the velocity that must be imparted to the vehicle to achieve that orbit. If a satellite in equatorial orbit has an orbital period of exactly one Earth-day, the satellite will orbit the Earth at exactly the same rate at which the Earth itself rotates. This means the satellite will appear stationary in the sky when observed from a point on Earth. This is referred to as a *synchronous*, or *geostationary*, orbit and occurs with a satellite at an altitude of 22,300 mi (35,784 km) above the Earth.

Geostationary satellites require simpler Earth stations, since the satellite location is always known, precluding the need for antenna tracking and pointing from the Earth station. (In actuality, however, even with station-keeping, a geostationary satellite may have a variation of about ±0.1°, simply as a result of orbit ellipticity. This means that an uncertainty in true satellite location of about ±40 km always exists in a synchronous satellite link.) In addition, countries relatively close to the equator observe the satellite at high elevation angles, avoiding much of the low angle degradation discussed in Chapter 4.

A disadvantage of geostationary satellites is that points on Earth beyond

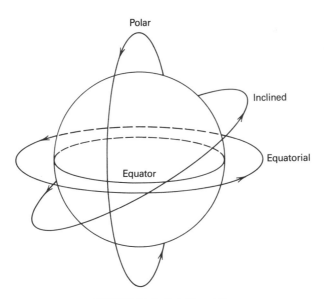

FIGURE 11.4. Satellite orbits.

about 80° latitude are not visible from the satellite. Inclined orbits, on the other hand, can provide visibility to the higher northern and southern latitudes, although they require Earth stations to continually track the satellite. This often necessitates an acquisition operation and sometimes involves handover from an orbiting satellite leaving the area to a new satellite entering the area. In addition, inclined orbits usually require multiple satellites to be spaced along the orbit in order to provide continuous coverage to a particular earth station.

The popularity of the geostationary orbit is evidenced by the number of existing satellites placed in that orbit. Figure 11.5 shows a sketch of the orbital location of some of the synchronous communication satellites currently in existence. Satellites are assigned specific orbit slots, with the spacing between satellites (when observed from the subsatellite point) maintained at about 2–3°. The continual demand for favorable orbit slots is expected to become a prime allocation problem in the near future.

Carrier frequencies used for satellite links are selected from bands that are most favorable in terms of power efficiencies, minimal propagation distortions, and reduction in noise and interference effects. These conditions tend to force operation into particular frequency regions that provide the best tradeoffs of these factors. Unfortunately, terrestrial systems (ground–ground) tend to favor these same bands. Hence there must be concern for interference effects between satellite and terrestrial systems. In addition, space itself is an international domain, just as are airline airways and the oceans, and satellite use from space must be shared and regulated on a worldwide basis. For this reason, frequencies to be used by satellites are established by a world body known as the International Telecommunications Union (ITU), with broadcast regulations controlled by a subgroup known as the *World Administrative Radio Conference* (WARC). The basic objective of these agencies is to allocate particular frequency bands for different types of

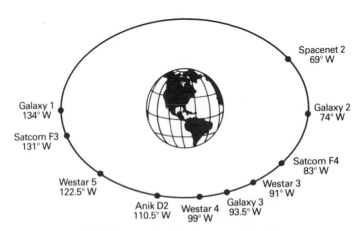

FIGURE 11.5. Satellites in equatorial orbits.

satellite service, and also to provide international regulations in the areas of maximum radiation levels from space, coordination with terrestrial systems, and the use of specific satellite locations in a given orbit. Within these allotments and regulations, an individual country operating its own domestic satellite system, or perhaps a consortium of countries operating a common international satellite system can make its own specific frequency selections based on intended uses and desired satellite services.

The frequency bands allocated by WARC for satellite communications is summarized in Figure 11.6. The tabulation in Table 11.2 shows the major use of these frequency bands in the United States and the available bandwidth within each band. Use of frequencies has been separated into military and nonmilitary, and services have been designated as *fixed point* (between ground stations located at fixed points on Earth), *broadcast* (wide area coverage), and *mobile* (aircraft, ships, land vehicles). Intersatellite refers to satellite crosslinks.

Most of the early satellite technology was developed for UHF, C-band, and X-band, which required the minimal conversion from existing microwave hardware. Major problems have been projected in these areas, however, because of the worldwide proliferation of satellite systems in these bands. The foremost problem is the fact that the available bandwidth in

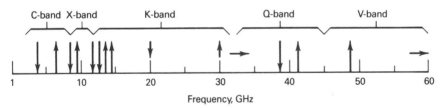

FIGURE 11.6. Frequency allocations for satellites.

TABLE 11.2 Allocated Satellite Bands for the United States

	Frequency Band, GHz		Major Uses in United States	Bandwidth
	Uplink	Downlink		
C-band	5.9–6.4	3.7–4.2	Fixed, point-to-point ground stations; nonmilitary	500 MHz
X-band	7.9–8.4	7.25–7.75	Mobile (ships, aircraft), radio relay; military only	500 MHz
Ku-band	14–14.5	11.7–12.2	Broadcast and fixed-point service; nonmilitary	500 MHz
Ka-band	27–30	17–20	(Unassigned)	—
	30–31	20–21		
V-band	50–51	40–41	Fixed-point, nonmilitary	1 GHz
Q-band		41–43	Broadcast, nonmilitary	2 GHz
V-band	54–58		Intersatellite	3.9 GHz
	59–64		Intersatellite	5 GHz

these bands will soon be inadequate to meet present and future traffic demands. Further, interference among various independent satellite systems and between satellite and existing terrestrial systems, will become more severe as additional satellites are put into use. Coordination among independent systems will be difficult to maintain. There can also be serious orbital congestion in the most favorable orbits for systems operating at C- and X-bands. For these reasons there is continued interest in extending operation to the higher K- and V-band frequencies. In most cases this means further development of technology and hardware but will allow the advantages of more spectral bandwidth and less terrestrial interference.

Although Figure 11.2 shows a single satellite link, a communication satellite will invariably be designed to handle many simultaneous uplinks and downlinks, as depicted in Figure 11.7. Here, separate Earth stations each transmit their individual carrier waveforms to the satellite, and all are relayed simultaneously to a similar group of separate receiving stations. A given transmitting station may wish to communicate its waveform to one or several different receiving stations. Similarly, a receiving station may wish to receive the transmissions of several different transmitting stations. Since all the uplink carriers must access through a common satellite to complete their downlink transmissions, the overall system operation has been referred to as *multiple-access communications*. In general, all receivers observe the same satellite transmissions, and therefore a multiple-access satellite system must allow for separability in the downlink waveforms—that is, multiple accessing must allow a downlink receiver to separate out a desired uplink transmission while tuning out undesired ones. This separability is achieved by requiring the uplink carrier to conform to a specific multiple-access format. The multiple-access format is simply a form of carrier wave multiplexing that allows many uplink carriers, even when emitted from remotely located

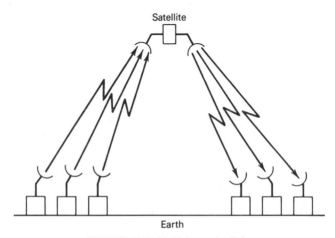

FIGURE 11.7. Multiple carrier links.

ground stations, to pass their waveforms through the satellite electromagnetically so as to be separable during the downlink transmission.

The three most common forms of multiple-accessing formats are summarized in Table 11.3. In *frequency division multiple access* (FDMA), Earth stations using the satellite are assigned specific uplink and downlink carrier frequency bands within the allotted satellite bandwidth. Station separability is therefore achieved by separaton in frequency. After retransmission through the satellite, a receiving station can receive the transmitted waveform of an uplink station by simply tuning to the proper frequency band. FDMA is the simplest and most basic format to implement since it requires Earth station configurations most compatible with existing hardware. FDMA formats are also the most popular and were used almost exclusively in all early satellite systems. The primary disadvantage of FDMA is its susceptibility to station crosstalk and intercarrier interference from nearby carriers while all are passing through the satellite.

In *time division multiple access* (TDMA), each uplink station is assigned a specific time slot in which to use the satellite. Each station must carefully ensure that its waveform passes through the satellite during its prescribed interval only. Receiving stations receive an uplink station by receiving the downlink only at the proper time period. Clearly, TDMA involves most complicated station operations, including some form of precise time synchronization among all users. Frequency crosstalk between users is no

TABLE 11.3 Summary of Multiple-Access Formats

Multiple-access Format	Designation	Characteristic
Frequency-division multiple access	FDMA	Frequency separation Carrier bands → Frequency
Time-division multiple access	TDMA	Time separation Carrier time slots → Time
Code-division multiple access (spread-spectrum multiple access)	CDMA (SSMA)	Waveform separation Coded carrier spectra → Frequency

longer a problem, since theoretically only one station uses the satellite at a time. Since each station uses the satellite for intermittent time periods, TDMA systems require short burst communications. This type of communications allows each station to transmit a burst of information on its carrier waveform, during its allotted time interval. An operation like this makes TDMA primarily applicable to special purpose systems involving relatively few Earth stations.

In *code division multiple access* (CDMA), carriers are separated by assigning a specific coded address waveform to each. Information is transmitted by superimposing it onto the addressing waveform, and modulating the combined waveform onto the station carrier. A station can use the entire satellite bandwidth and transmit at any desired time. All stations transmitting simultaneously therefore overlap their carrier waveforms on top of each other. Receiving the entire satellite transmission, and demodulating with the proper address waveform, allows reception of only the appropriate uplink carrier. Accurate frequency and time interval separation are no longer needed, but station receiver equipment tends to be more complicated in order to carry out the address selection required.

Since addressing waveforms tend to produce carrier spectra over a relatively wide bandwidth, CDMA signals are often called *spread-spectrum signals*, and CDMA is alternatively referred to as *spread-spectrum multiple access* (SSMA).

The specific implementation and attainable performance associated with these multiple-access formats are examined in detail in Sections 11.6–11.8. It is first necessary, however, to develop reliable communication models of the various types of satellite links. This topic is covered in the next section.

11.3   SATELLITE TRANSPONDERS

A satellite transponder receives and retransmits the RF carrier transmitted from Earth stations. A block diagram of a generic transponder is shown in Figure 11.8. The RF front end receives and amplifies the uplink carrier while filtering off as much receiver noise as possible. The received carrier is then processed so as to prepare the retransmitted waveform for the return link. Carrier processing involves either some form of direct spectral translation or some form of remodulation. In spectral translation, the entire uplink spectrum is simply shifted in frequency to the desired downlink frequency. In remodulation processors, the uplink waveforms are demodulated at the satellite, and then remodulated onto the downlink carrier. Remodulation processors involve more complex circuitry but provide for changeover of the modulation format between uplink and downlink. This restructuring of the downlink can provide advantages in decoding, power concentration, and interference rejection.

While the earlier transponder diagram showed a separate antenna for

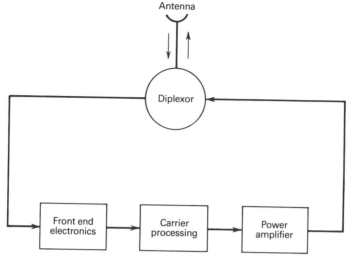

FIGURE 11.8. Transponder block diagram.

uplink receiving the downlink transmission, the same antenna can actually be used for both. This is possible since the uplink and downlink frequency bands are separated. A *diplexor* is used in the front end to allow simultaneous transmission and reception. The diplexor is a two-way microwave gate that permits received carrier signals from the antenna and transmitted carrier signals to the antenna to be independently coupled into and out of the antenna cabling. The carriers, being at different frequency bands, can flow in the same cabling and antenna feeds without interfering.

After front end filtering and signal processing, the downlink carrier is power amplified to provide the required power level for the downlink receiver. This power amplification is generally provided by traveling wave tube amplifiers (TWTA). These amplifiers may be preceded by stages of preamplification that set operating points for the TWTA. Since uplink carrier power levels will be on the order of fractions of a microwatt (see Section 4.5), and since downlink power values of watts are needed, a typical transponder must provide a composite power gain of around 80–100 dB. Taking into account the power losses of filters and cabling, the required gain is generally far above that that can be provided by a TWTA alone. Hence transponders require intermediate amplification to achieve the power levels suitably matched to the capability of the high power downlink amplifier. These intermediate gain stages are usually provided by low weight semiconductor amplifiers. Note that an amplifier (TWTA) requires both high gain and high output power levels.

Satellite front end electronics are similar to the front end type discussed in Section 4.2. The receiving antenna is coupled through cabling, usually to a diplexor or power splitter and followed by stages of RF filtering and amplifiers. The cabling, diplexors, and power splitters are lossy elements, while the filters and amplifiers determine bandwidth and power gain values. The satellite front end is therefore characterized by an effective front end gain, an overall filtering characteristic, and a noise figure F. The latter sets the satellite input noise level at

$$N_0 = \kappa T^\circ_{eq} \tag{11.3.1}$$

where κ is Boltzmann's constant and

$$T^\circ_{eq} = T^\circ_b + (F - 1)290^\circ \tag{11.3.2}$$

Background noise temperature T°_b was discussed in Section 4.4. An Earth-based antenna sees the sky and galactic background from all directions, producing the approximate T°_b values shown in Figure 4.8. A satellite looking down at the Earth sees the illuminated Earth at a uniform reradiation temperature of about 300 K, surrounded by the galactic noise of outer space. A spacecraft antenna will see the Earth in its mainlobe, and the lower level galactic noise in its sidelobes. The antenna averaging will then tend to lower the combined background temperature. A satellite spot beam sees the illuminated Earth in its entire pattern, and therefore will have a background temperature closer to 300 K.

Front end noise figures depend on the electronic components used, and on their specific arrangement, as was discussed in Section 4.4. A basic requirement is for low loss coupling and low noise amplifiers, the latter usually of the solid state type. Satellite transponder noise figures can generally be designed in the range 3–7 dB.

Background and noise figures of the satellite sets the noise spectral level in (11.3.1). The amount of uplink noise power entering the satellite also depends on the noise bandwidth of the front end filtering. The latter depends on the type of filter function used, its spectral width, and its actual shape. However, the noise bandwidth of the RF filter must be balanced against carrier distortion. If the front end filter is made too narrow, the carrier–signal spectrum will be distorted as it passes through. Distortion is caused by both the nonflat amplitude responses over the carrier bandwidth and by group-delay distortion (different frequencies being delayed by different amounts). Delay variation occurs primarily at band edge, where the amplitude characteristic is rapidly decreasing. The amount of tolerable distortion caused by amplitude and group delay depends on the properties of the signal spectrum. If the uplink spectrum is that of a single wideband modulated carrier, with center frequency in the middle of the RF bandwidth, band edge effects may not be that significant and some degree of bandlimiting may be tolerable. However, when the uplink contains multiple carriers spread over the bandwidth, band edge effects become extremely important to the outer carriers.

RF carrier spectra in satellite systems often have designated *masks* inside of which the downlink spectrum must be contained (Figure 11.9). These masks control the out-of-band spectral content of the carrier. If the uplink carrier spectrum does not satisfy the mask at the satellite, RF filtering must be applied either at the transmitter or at the front end to accomplish the desired filtering. As the order of the filter is increased for a given 3 dB bandwidth, the out-of-band attenuation increases, along with the in-band band edge group delay. This means that if the mask attenuation is to be held constant, the 3 dB frequency of the filter can be increased, providing a reduction in group delay distortion. Hence, implementing higher order RF filters can be traded directly for reductions in delay distortion. Group delay distortion can also be partially compensated with delay *equalization* networks. These networks are designed to have no attenuation distortion (flat gain curves across the RF bandwidth) but have delay variations that tend to cancel the delay distortion of the RF filters. Usually a filter is first constructed to satisfy the mask, and then an equalizer is added to correct the delay distortion.

Satellite filters must be designed to be lightweight while achieving the desired mask, noise rejection, and equalization. With the high satellite frequencies involved, RF filters are typically constructed as microwave waveguide filters [4–11]. Such integrated circuits allow better packaging and lower weight and require less power. Increased use is also being made of dual-mode filters using cavity coupling. Such filters have center frequency–bandwidth ratios of about 10^4 at C-band and about 10^3 at K-bands. For example, satellites at C-band typically have carrier filters with RF bandwidths of about 36 MHz.

Satellite signal processing generates the microwave carrier for the return link. This carrier can be achieved by direct frequency translation or by carrier remodulation. Subsystems that accomplish these operations are shown in Figure 11.10. Figure 11.10*a* shows a direct RF-to-RF conversion

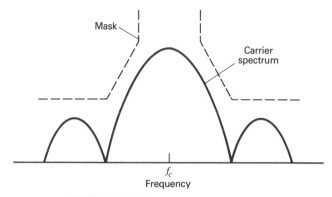

FIGURE 11.9. Spectral mask for RF uplink.

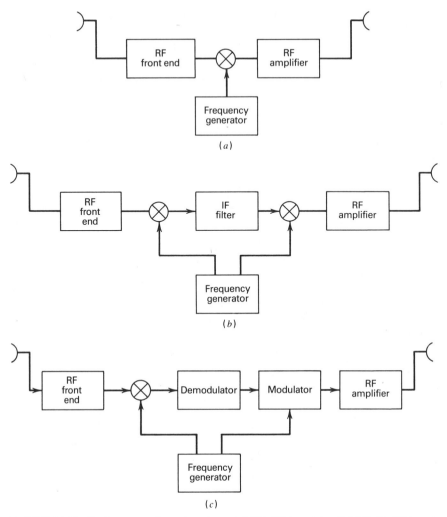

FIGURE 11.10. Carrier processing subsystems: (*a*) RF–RF translator, (*b*) RF–IF–RF translation, (*c*) Demod–Remod processor.

using a single mixer system. Figure 11.10*b* shows a double conversion from RF to an intermediate lower IF, then back up to RF using a single mixing oscillator. This system has the advantage of allowing carrier filtering and amplification to be performed at the lower IF frequency band rather than at the higher RF band. In addition, RF–IF–RF systems allow uplink command carriers to be more easily removed (or new telemetry carriers inserted) at the satellite before return retransmission. Note that frequency translators of either type always cause all uplink noise to be similarly translated, and, therefore, directly superimposed onto the downlink spectra. When the transponder merely frequency translates the uplink carrier to the downlink, the satellite appears to be simply "bending" the Earth-station uplink into the receiving downlink. For this reason, frequency translating satellite links are often referred to as *bent pipes*.

Remodulating processes involve a stage of modulation that generates the return carrier. This is accomplished by directly demodulating the uplink carrier, as in Figure 11.10*c*. With digitally modulated carriers, baseband demodulation can involve baseband decoding of the information bits, followed by digital remodulation. Remodulation removes uplink noise and interference from the return modulation, while facilitating uplink on-board digital processing and return link bit insertion.

Satellite power amplifiers provide the primary amplification for the retransmitted carrier and are obviously one of the key elements in a communication satellite. Power amplifiers, besides having to generate sufficient power levels and amplification gain, have additional requirements for reliability, long life, stability, high efficiency, and suitability for the space (orbiting) environment. These requirements have been sufficiently met by the use of the TWTA, either of the cavity-coupled or helix type. TWTA have been extensively developed, their theory of operation is well understood, and they have been successfully implemented in all types of space missions. For this reason TWTA have emerged as the universal form of both Earth station and satellite power amplifier. Even as increased demands on power amplifiers will push them to higher power levels and higher frequencies, the TWTA will undoubtedly continue to be the dominant amplification device. Their continual development has already produced sufficient power levels well into the 30 GHz frequency range.

It is expected that there will be continued effort to develop smaller, lighter-weight solid state amplifiers, such as galium arsenide *field-effect transistors* (GA FET) for future satellite operations [12, 13]. These devices, however, have not been established in high power operating modes with reasonable size bandwidths. They most likely will have future applications with appropriate power combining, or lower power, multiple-source operations. FET operation is generally confined to upper frequencies of about 30 GHz. For projected amplification above 30 GHz, *impact avalanche transit time* (IMPATT) diode amplifiers are rapidly developing as a capable medium power amplifier. Such diodes have been developed at frequencies up to about 40 GHz, and it appears that they will become extendable to 60 GHz operation in the near future.

TWTAs achieve their power gain by using the input microwave carrier to phase control resonant waves in cavity so as to produce wave reinforcement. An output from the resonant cavity or tube, then, is an amplified, phase replica of the input carrier. Such mechanisms achieve the significantly large power gains needed for the satellite transponder links. Since TWTAs amplify phase modulated carriers, they are intended primarily for constant amplitude carriers. Variations in input carrier amplitudes during amplification produce an additional unintentional phase modulation that appears as phase interference on the amplified carrier. The conversion of amplitude fluctuations to phase variations is referred to as the *AM/PM effect*.

For constant amplitude input carriers, there is a direct gain conversion to the output amplitude. As the amplitude of the input carrier is increased, the output amplitude is also increased until a saturation effect occurs within the cavity. This is exhibited by the input–output power curve of the TWTA, which is typically of the form shown earlier (Figure 3.7). As the input power is increased, a direct linear gain occurs in output power until the output power saturates, and further increase in input power no longer produces larger outputs. Achievement of this maximum output power is therefore accompanied by a nonlinear amplification within the amplifier as the saturation condition is approached. When only a single input carrier is involved, this saturation causes no carrier distortion, but only a limitation to its output power. When the TWTA input corresponds to multiple carriers, this nonlinearity of the saturation effect becomes extremely important, as is discussed in Section 11.6.

The drive power at which the output power saturation occurs is called the *input saturation power*. The ratio of input saturation power to desired drive power is called the amplifier *input backoff*. Increasing input backoff (decreasing input drive power) produces less output power but improves the linearity of the device, since the degree of nonlinearity is reduced. The output saturation power of the amplifier is the maximum total power available from the amplifier. Output backoff is the ratio of the maximum output (saturation) power to actual output power. Output backoff obviously depends on input backoff, that is, where the drive power is operated. Increasing input backoff lowers the output power and increases the output backoff.

Proper input control of the operating drive power of an amplifier is important in TWTA operation. Power control for the TWTA is often obtained by a BPL-amplifier combination, similar to that shown in Figure 4.23. Since the limiter produces fixed-output power levels, proper gain adjustment of the drive amplifier can carefully set the TWTA input to desired input backoff. This backoff setting can be extremely critical in satellite operation for achieving satisfactory downlink performance.

11.4 SATELLITE LINK ANALYSIS

As a communication channel, a satellite system can be analyzed by the link equations presented in Chapter 4. The satellite channel can be separated into an uplink, a downlink, a repeater link (combined uplink and downlink), and a crosslink. Each link is composed of an electromagnetic field sent from a transmitter to a receiver. As such, CNR (or E_b/N_0) can be determined for each and the performance related to the key parameters of both the Earth station and the satellite. In this section we examine these satellite links.

Uplink

Figure 11.11 shows a simplified satellite uplink, with a transmitting Earth station communicating with an orbiting satellite with possible adjacent satellites in the vicinity. The prime requirement is to transmit only to the desired satellite, while avoiding interference to the adjacent satellites. Since the latter may be only a few degrees away (when viewed from Earth), the uplink beam pattern may often be of more concern than the actual uplink effective isotopic radiated power (EIRP). While the latter determines the power to the desired satellite, the shape of the pattern determines the amount of off-axis (sidelobe) interference power impinging on nearby satellites. The narrower the Earth-station beam, the closer an adjacent satellite can be placed without receiving significant interference. On the other hand, an extremely narrow beam may incur significant pointing losses due to uncertainties in exact satellite location. For example, if a satellite location is known only to within $\pm 0.2°$, a minimum Earth-station half-power beamwidth of about $0.6°$ is necessary. This sets the transmit antenna gain at about 55 dB. For the parabolic ground antennas, this produces an off-axis gain curve such that the peak power is about 20 dB down to an adjacent satellite $3°$ away. Thus the uplink beamwidth is set by both the pointing accuracy of the Earth station and the satellite orbit separation satisfactory for an acceptable sidelobe interference. If satellite pointing is improved, the uplink beamwidth can be narrowed, producing less interference. Alternatively, it would permit a closer satellite spacing in a given orbit, thereby increasing the number of satellites that can be inserted in a common orbit

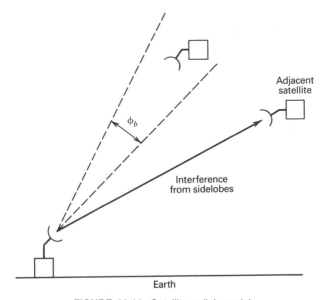

FIGURE 11.11. Satellite uplink model.

(such as the synchronous orbit). Thus there is a direct relationship between beam patterns assigned to Earth stations and the spacing of satellites in a given orbit.

With the half-power beamwidth set, the carrier frequency will determine the required Earth-station antenna size. Figure 11.12 shows the relation between Earth-station antenna diameter and frequency in producing a given uplink beamwidth and gain, using equations (3.5.7). Note that while increase of carrier frequency does not directly aid receiver power, we see here that an advantage does accrue in reducing Earth-station size, and possibly in improving satellite trafficking.

Transmitter power for Earth stations is generally provided by high powered amplifiers, such as TWTs and Klystrons. Since the amplifier and transmitting antenna are located on the ground, size and weight are not prime considerations, and fairly high transmitter EIRP levels can be achiev-

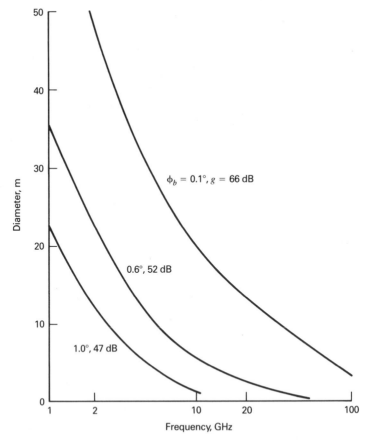

FIGURE 11.12. Uplink ground station antenna diameter versus frequency for different half-power beamwidths and gain.

ed. Earth-based power outputs of 40–60 dBW are readily available at frequency bands up through K-band, using cavity-coupled TWTA or Klystrons. With a 0.6° uplink beamwidth (gain ≈ 52 dB), Earth-station EIRP values of about 80–90 dBW are readily available. Using the synchronous orbit space losses from Figure 4.5, Figure 11.13a shows the achievable CNR in a 10 MHz bandwidth at the spacecraft as a function of the $g/T°$ of the

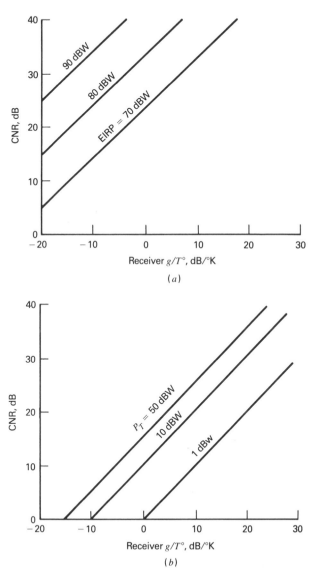

FIGURE 11.13. Carrier to noise ratio (CNR) in 10 MHz bandwidth versus receiver $g/T°$ for: (a) uplink wth stated EIRP, (b) downlink with stated satellite power.

satellite. Even with significant range losses (≈ 200 dB) and relatively low $g/T°$ values, an acceptable uplink communication link can usually be established.

Satellite Downlink

A satellite downlink (Figure 11.14a) is constrained by the fact that the power amplifier and transmitting antenna must be spaceborne. This limits the power amplifiers to efficient, lightweight devices, with limited output

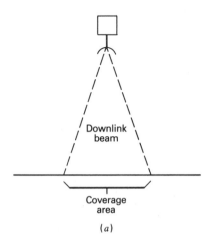

Downlink beam

Coverage area

(a)

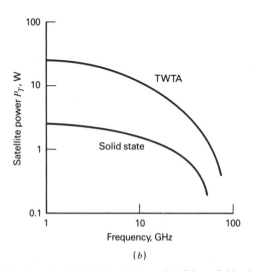

(b)

FIGURE 11.14. Satellite downlink: (a) system model, (b) available downlink power levels.

power capabilities that are dependent on the carrier frequency (Figure 11.14*b*). The spacecraft antenna, while similarly limited in size, must use beam patterns that provide the required area on Earth. A satellite may be required to illuminate the entire Earth (global beam), or only a particular country or subregion (spot beam). The smaller the coverage area, the larger the required satellite antenna and the higher the antenna gain. By operation at higher downlink frequencies, a spot beam can be achieved with a smaller spacecraft antenna. This is a primary incentive in developing K-band satellite technology. Satellite antennas generally run from 3 to 6 ft, and at C-band provide gain levels from 25 to 35 dB.

With satellite power level and antenna gain established, the carrier power collected at the Earth station depends only on the $g/T°$ factor, just as for the uplink. Figure 11.13*b* includes the required values of $g/T°$ needed in the downlink for achieving the same CNR values, assuming synchronous orbit and global satellite antennas. It is evident that relatively large Earth-station $g/T°$ is needed to overcome the smaller EIRP of the satellite. This means that small Earth stations will be severely limited in their ability to receive large bandwidth carriers.

Although use of higher carrier frequencies allows smaller satellite antennas, care must again be used in accounting for its effect in downlink analysis, as discussed in Section 4.5. It will produce higher Earth-station $g/T°$ values, but it will not increase CNR because of the increased downlink space loss. The choice of downlink frequency band is, of course, important in determining spot beam size (for a given antenna size) and in determining atmospheric losses and available transmitter power.

Repeater Link

In transponding satellites the primary function of the spacecraft is to relay the uplink carrier into a downlink. For simplicity, we neglect the frequency translation between the uplink and downlink and model the satellite as an ideal linear amplifier that converts the uplink waveform to a downlink waveform through a power gain G, as shown in Figure 11.15. This represents the most basic, idealized, repeater link that can be constructed. The uplink power is composed of a signal term from the uplink Earth station P_{us} and the noise power collected at the satellite front end P_{un}. The downlink power P_T is composed of an amplified signal and noise power term

$$P_T = G(P_{us} + P_{un}) \qquad (11.4.1)$$

Let L represent the combined total power gain (or loss) in the downlink, including antenna gains and channel losses,

$$L = g_t L_a L_p g_r \qquad (11.4.2)$$

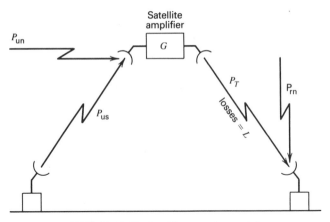

FIGURE 11.15. Satellite repeater model.

The received downlink carrier power is then

$$P_{rs} = GP_{us}L \tag{11.4.3}$$

The retransmitted noise appearing at the downlink receiver is

$$P_{rn} = GP_{un}L \tag{11.4.4}$$

In addition, a noise power $P_{rd} = \kappa T_d^\circ B_d$ appears at the downlink receiver due to its noise temperature T_d° and bandwidth B_d. Hence the combined CNR at the downlink receiver is

$$
\begin{aligned}
\mathrm{CNR}_d &= \frac{P_{rs}}{P_{rn} + P_{rd}} \\
&= \frac{P_{us}}{P_{un} + (P_{rd}/GL)}
\end{aligned}
\tag{11.4.5}
$$

Dividing by P_{un}, we have

$$
\begin{aligned}
\mathrm{CNR}_d &= \frac{P_{us}/P_{un}}{1 + (P_{rd}/P_{rn})} \\
&= \frac{(\mathrm{CNR}_u)(\mathrm{CNR}_r)}{\mathrm{CNR}_u + \mathrm{CNR}_r}
\end{aligned}
\tag{11.4.6}
$$

where we have denoted

$$\mathrm{CNR}_u = \frac{P_{us}}{P_{un}} \tag{11.4.7}$$

as the uplink CNR at the satellite, and

$$\text{CNR}_r = \frac{P_{us}GL}{P_{rd}} \qquad (11.4.8)$$

as the downlink receiver CNR. This last CNR equation is based on satellite transmitted carrier power and receiver noise only, that is, as if there were no uplink noise. Thus, even with a relatively simple and ideal satellite repeater model, we establish a basic property of a repeater system. The downlink CNR depends on *both* the uplink CNR and the receiver CNR and can never exceed either one. Thus the weakest of the uplink and downlink channels will determine the performance level of a repeater system. Even with perfect repeater amplifiers, design of the uplink, as well as the satellite downlink, must be taken into account. We point out that by inverting CNR_d, we can rewrite (11.4.6) as

$$(\text{CNR}_d)^{-1} = (\text{CNR}_u)^{-1} + (\text{CNR}_r)^{-1} \qquad (11.4.9)$$

This is sometimes more convenient to use in computing CNR_d in repeater analyses.

For a transponded digital link, the downlink CNR_d can be converted to E_b/N_0 to determine bit error probability for the linear amplifier satellite. This requires replacing B_d by $1/T_b$, and writing

$$\left(\frac{E_b}{N_0}\right)_d = \frac{(E_b/N_0)_u (E_b/N_0)_r}{(E_b/N_0)_u + (E_b/N_0)_r} \qquad (11.4.10)$$

where $(E_b/N_0)_u$ and $(E_b/N_0)_r$ are obtained from (11.4.7) and (11.4.8) by taking the noise bandwidths as $1/T_b$ Hz. For example, the resulting bit error probability, PE, for a phase coherent BPSK repeater link is then

$$\text{PE} = Q\left[\left(\frac{2E_b}{N_0}\right)_d^{1/2}\right] \qquad (11.4.11)$$

where $Q(x)$ is the Gaussian tail integration. Note again that digital performance depends on both uplink and downlink CNR.

In satellite processing using demodulation and remodulation, the uplink is demodulated to baseband, and the entire baseband is remodulated onto a downlink carrier. When digital modulation is used, the transponder demodulation can reinterpreted as bit decoding. Remodulation corresponds to encoding these decoded bits back onto the RF return carrier. This means a given source bit, in traveling through the transponder to the specific Earth station, undergoes two states of decoding in cascade. This may be diagrammed as shown in Figure 11.16. A given bit will be decoded correctly at the end if two correct decodings occurred, or if two incorrect decodings occurred. Let PE_u denote the probability of a bit error during the uplink decoding, and let PE_d be that for the downlink decoding. The average bit error probability is then

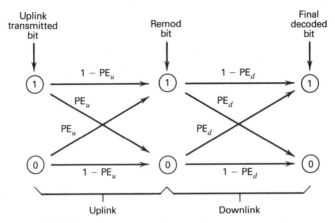

FIGURE 11.16. Demod–Remod bit decoding model.

$$PE = (1 - PE_u)PE_d + PE_u(1 - PE_d)$$
$$= PE_u + PE_d - 2PE_uPE_d \qquad (11.4.12)$$

If both links have been designed for small PE ($PE_u, PE_d \leq 10^{-1}$), the last term will be negligible compared to the sum. This means the overall PE for the transponder link is the sum of the individual link error probabilities. The weakest link (largest PE_i) will therefore determine the overall PE, and there is no advantage in having one digital link significantly better than the other in terms of error probability.

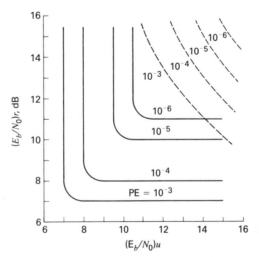

FIGURE 11.17. Uplink–downlink E_b/N_0 for given PE (solid line—Demod–Remod, dashed line—linear repeater).

As an example, consider a BPSK uplink and downlink, operating with an uplink $(E_b/N_0)_u$ and a downlink $(E_b/N_0)_r$. We wish to operate the overall transponder remodulation link with an error probability PE. Figure 11.17 shows the uplink–downlink E_b/N_0 combinations needed to produce a specified PE of 10^{-3}–10^{-6} for the cascaded Demod–Remod link. Also included is the required E_b/N_0 values for the same PE in (11.4.11) using a linear amplifying up–down link. The result shows clearly the advantage of inserting decoding–encoding hardware at the satellite and why on-board processing on the satellite is increasing in importance. Curves of this type are useful for performing initial system design and sizing the hardware for satellite and Earth station.

Satellite Crosslinks

Satellite systems often require communication between two satellites via a crosslink. A crosslink between two orbiting satellites is referred to as an *intersatellite link* (ISL). As a communication link, an intersatellite link has the disadvantage that both transmitter and receiver are spaceborne, limiting operation to both low P_T and low $g/T°$ values. To compensate in long links, it is necessary to increase EIRP by resorting to narrow transmit beams for higher power concentration. With satellite antenna size constrained, the narrowbeams are usually achieved by resorting to higher carrier frequencies. Hence satellite crosslinks are designed for EHF (60 GHz) frequencies. Satellite crosslink distances may become quite large for high orbits and, in fact, can approach approximately twice the synchronous altitude for geo-stationary satellites.

We assume first that both satellite locations are known exactly by each, and each is perfectly stabilized. Each satellite uses an antenna of diameter d with a beamwidth ϕ_b pointed at each other (Figure 11.18). The CNR delivered to the receiving satellite over the crosslink, using (4.5.6), is then

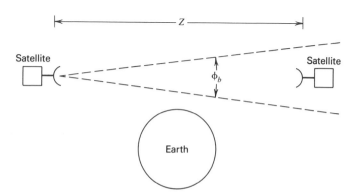

FIGURE 11.18. Crosslink model.

$$\text{CNR} = \frac{P_T(\pi d^2/4)}{\phi_b^2 Z^2 (\kappa T_{eq}^\circ) B_c} \tag{11.4.13}$$

Assuming $\rho_a = 1$, and letting $\phi_b = \lambda/d$, this can be simplified to

$$\text{CNR} = \left[\frac{0.71 P_T}{Z^2(\kappa T_{eq}^\circ)}\right]\left[\frac{d^4}{\lambda^2}\right] \tag{11.4.14}$$

We see that CNR increases as the fourth power of the antenna diameters. Hence the more efficient way to improve a crosslink is to increase antenna size.

With antenna diameter selected, CNR can be increased only by lowering front end temperature, increasing transmitter power, or increasing carrier frequency. This is why crosslink frequencies have been assigned at the Q-band and the V-band.

Equation (11.4.14) shows that the system improves by using higher frequency bands. This is a direct result of a more concentrated beamwidth and the absence of atmospheric effects. The limitation of extremely narrow crosslink beams, however, is the pointing error that exists as a result of the relative uncertainty of the location of each satellite with respect to each other, and to the satellite attitude error (the inability of a satellite to properly orient itself so as to point exactly in a desired direction). These errors are invariably much larger than those encountered in Earth-based links where ground control of tracking and pointing is feasible.

Typically satellite uncertainty is about 100 km, which produces a pointing error of about 0.0014 radians at maximal geostationary crossranges, (45,000 mi). Attitude errors however are usually in the range 0.01–0.002 radian and therefore are usually the prime contributor to total pointing error and eventually limit the available CNR. Although this is basically true for any communication link, it becomes particularly important in the crosslink, where pointing errors tend to be compounded.

11.5   FREQUENCY DIVISION MULTIPLE ACCESS (FDMA)

In an FDMA satellite system each uplink carrier is assigned a frequency band within the availble RF bandwidth of the satellite (Figure 11.19). In the satellite transponder, the entire RF frequency spectrum appearing at the satellite imput is frequency translated to form the downlink. A receiving station receives a particular uplink station by tuning to, and filtering off, the proper band in the downlink spectrum. Each carrier can be independently modulated, either analog or digital, from all others. FDMA represents the simplest form of multiple accessing, and the required system technology and hardware are almost all readily available in today's communication market.

Each uplink carrier may originate from a separate Earth station, or

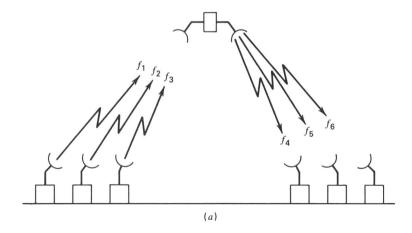

(a)

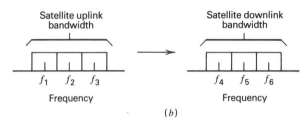

(b)

FIGURE 11.19. FDMA model: (a) transmission diagram, (b) frequency plan.

several carriers may be transmitted from a particular station. Frequency band selection may be fixed or assigned. In *fixed frequency* operation each carrier is assigned a dedicated frequency band for the uplink, and no other carrier utilizes that band. In *demand-assignment multiple access* (DAMA) frequency bands are shared by several carriers, with a particular band assigned at time of need, depending on availability. DAMA systems can serve a greater number of carriers if the usage time of each is relatively low, but may require more complex ground routing hardware [14].

Individual RF carrier spectra in an FDMA system must be sufficiently separated from each other to both allow the carriers to filter off at the downlink stations and to prevent carrier crosstalk (frequencies of one carrier spectrum falling into the band of another carrier). This is why spectral tails associated with digital carriers was discussed in detail in Chapter 2. However, excessive separation causes needless waste of satellite bandwidth. To determine the proper spacing between FDMA carrier spectra, crosstalk power must be calculated. Spacings can then be selected for any acceptable crosstalk level desired.

When an FDMA format is used with a linear transponding channel, the downlink performance can be determined by extending the single channel analysis given in Section 11.4. Let an RF satellite bandwidth B_{RF} be available to the Earth-station carriers, each carrier using an individual bandwidth of B Hz (including spectral spacing). The number of FDMA carriers allowed by the satellite bandwidth is then

$$K = \frac{B_{RF}}{B} \tag{11.5.1}$$

Let P_{ui} be the ith carrier uplink power at the satellite amplifier input, and let P_{un} be the corresponding uplink noise in the same carrier bandwidth. The total amplifier input power is then

$$P_u = \sum_{i=1}^{K} P_{ui} + KP_{un} \tag{11.5.2}$$

For a linear amplifier transponder, the satellite RF bandwidth is frequency translated and amplified by the power gain G defined by

$$G = \frac{P_T}{P_u} = \frac{P_T}{\Sigma_i^K P_{ui} + KP_{un}} \tag{11.5.3}$$

The downlink receiver power of the ith carrier after amplifying and downlink transmission is then

$$P_{di} = LGP_{ui} = LP_T\left[\frac{P_{ui}}{\Sigma_i^K P_{ui} + KP_{un}}\right] \tag{11.5.4}$$

where L is again the combined downlink power losses and gains from satellite amplifier output to Earth-station receiver input. The downlink power is therefore aportioned by using the fraction shown of the available satellite power P_T. If all uplink carrier powers were the same, and if there were no uplink noise, then each downlink carrier would have exactly $1/K$th of P_T. The presence of the uplink noise reduces this distribution, effectively taking power away from individual carriers. This is referred to as "power robbing" caused by the noise. Note that if the uplink carrier powers are not equal, the downlink power will be unequally redistributed, with the stronger uplinks receiving the larger portion of the available satellite power P_T.

The total receiver noise is the sum of the transponded uplink noise and the receiver noise. Hence the downlink CNR of a single carrier in its own bandwidth B is then

$$CNR_d = \frac{P_{di}}{LGP_{un} + N_{0d}B} \tag{11.5.5}$$

where N_{0d} is the receiver noise spectral level. We again rewrite this as

$$(\text{CNR}_d)^{-1} = (\text{CNR}_u)^{-1} + (\text{CNR}_r)^{-1} \tag{11.5.6}$$

where now

$$\text{CNR}_u = \frac{P_{ui}}{P_{un}} = \frac{P_{di}}{LGP_{un}} \tag{11.5.7}$$

$$\text{CNR}_r = \frac{P_{di}}{N_{0d}B} = \frac{P_T L}{N_{0d}B}\left(\frac{P_{ui}}{P_u}\right) \tag{11.5.8}$$

Here CNR_u is the uplink CNR of a single carrier. The receiver CNR_r is that which the satellite power P_T can produce at the receiver, reduced by the power robbing loss of the uplink. If the system is power balanced (all carriers have some uplink power) and if the uplink CNR_u is high ($\text{CNR}_u \gg 1$), then $\text{CNR}_d \approx P_T L / K N_{0d} B$; that is, the available satellite P_T is equally divided among the FDMA downlink carriers.

If we solve (11.5.6) for K, we can determine the number of uplink carriers that can be supported with a satellite of power P_T and a desired CNR_d. This yields

$$K = \left[\frac{P_T L}{N_{0d}B}\right]\left[\frac{1 - (\text{CNR}_d/\text{CNR}_u)}{\text{CNR}_d}\right] \tag{11.5.9}$$

With a specified value of CNR_d and CNR_u, we see that K increases linearly with P_T. This equation can be used to compare the number of carriers allowed by power constraints with the number allowed by the bandwidth in (11.5.1). Hence a linear FDMA system can be either power or bandwidth limited, depending on which equation yields the lower number of allowable carriers.

When satellite nonlinearities are present, the previous equations must be modified. The most important nonlinearity is the power amplifier when operated near saturation for maximum output power. When multiple FDMA carriers pass through a common nonlinearity two basic effects occur: (1) the carriers mix together to produce unwanted intermodulation products that fall into the FDMA bands as interference or noise, and (2) available output power decreases as a result of conversion of useful satellite power to intermodulation noise. Both of these effects depend on the type of non-linearity and the number of simultaneous FDMA carriers present, as well as their power levels and spectral distributions.

An exact intermodulation analysis would be required to determine the actual spectral distribution that occurs as a result of the mixing of the FDMA carriers. The commonly used apporaches either involve direct simulation and spectral analysis, or direct computation. The intermodulation spectrum is determined by analytically computing the third order beat products, then the fifth order beat products, and so forth. Figure 11.20 shows the result of an analysis of a soft-limiting power amplifier computed in

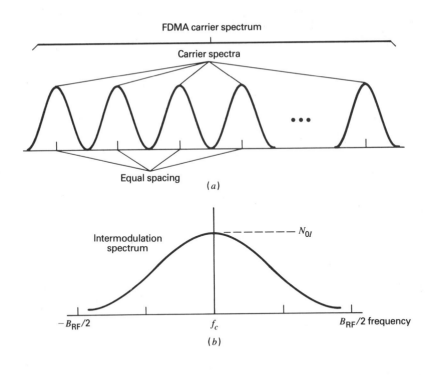

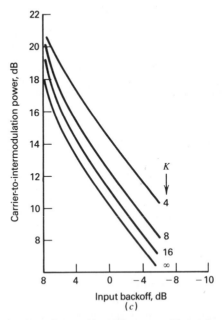

FIGURE 11.20. Intermodulation effects with nonlinear amplifier: (a) uplink FDMA spectrum, (b) resulting intermodulation spectrum, (c) downlink carrier power to intermodulation noise in 36 MHz bandwidth versus amplifier backoff (K = number FDMA carriers), (d) normalized available satellite power versus amplifier backoff (K = number FDMA carriers).

528

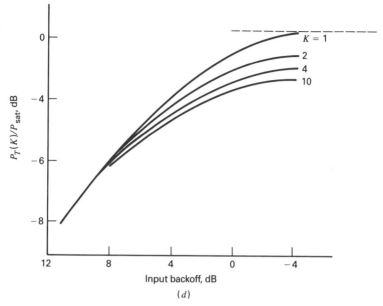

FIGURE 11.20 (*Continued*).

this way [2]. Figure 11.20a shows the received FDMA spectrum, composed of K equal power, identical modulated carriers. Figure 11.20b shows the resulting intermodulation spectrum caused by the mixing of the received spectra. The intermodulation tends to be concentrated in the center of the RF bandwidth, so that the center carriers receive the most interference. Figure 11.20c shows the ratio of carrier power P_i to peak intermodulation spectral level N_{0I}, as a function of the number of carriers K and the degree of amplifier backoff. The curve indicates the decrease in intermodulation level as backoff is increased, and the amplifier is operated more as a linear amplifier. Figure 11.20d plots the ratio of the total carrier power available for the K carriers $[P_T(K)]$ to the full saturation power of the amplifier, as a function of the amplifier backoff and number of carriers.

With the intermodulation and power suppression included, the CNR_d in (11.5.5) now becomes

$$\mathrm{CNR}_d = \frac{P_T(K)L\alpha_s^2/K}{N_{0d}B + P_T(K)\alpha_n^2 L + N_{0I}BL} \tag{11.5.10}$$

where we have assumed a flat intermodulation spectrum over the band, and let α_s, α_n be the limiter suppression factors in (4.10.21),

$$\alpha_s^2 = \frac{P_u}{P_u + P_{un}}$$

$$\alpha_n^2 = \frac{P_{un}}{P_u + P_{un}}$$

This represents the downlink receiver CNR_d of a *particular* FDMA carrier. As such it represents the extension of the single carrier result in (11.5.6) to the general FDMA carrier, nonlinear transponder case. To emphasize this, we divide through by the numerator to obtain

$$(CNR_d)^{-1} = \left(\frac{\alpha_s^2}{\alpha_n^2} CNR_u\right)^{-1} + (CNR_r)^{-1} + (CNR_I)^{-1} \quad (11.5.11)$$

where $CNR_u = P_{ui}/P_{ni}$ = uplink carrier CNR at the satellite limiter input

$CNR_r = P_{di}/N_{0d}B$ = downlink carrier CNR due to available satellite power

$CNR_I = (P_T(K)/K)/N_{0I}B$ = carrier-to-intermodulation ratio

Equation (11.5.11) extends (11.5.6) by inserting the backoff suppression [via $P_T(K)$], and intermodulation (via N_{0I}), effects for multiple carriers. The first two effects are both dependent on the input backoff (operating drive power) of the nonlinear power amplifier. When uplink and downlink power levels are sufficient, downlink CNR is determined primarily by the intermodulation, and we can approximate:

$$CNR_d \approx CNR_I = \frac{P_T(K)/K}{N_{0I}B} \quad (11.5.12)$$

The numerator term of CNR_I decreases with input backoff (see Figure 11.20d), while the intermodulation level N_{0I} is likewise reduced with backoff (Figure 11.20c). When these results are superimposed, the ratio of the two behaves as shown in Figure 11.21. As backoff increases, the intermodulation

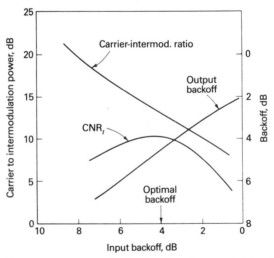

FIGURE 11.21. Input backoff selection, based on maximizing CNR_I.

decreases, increasing CNR_I; eventually, however, the available satellite power will decrease, producing a backoff point at which CNR_I is maximized. Plots of this type are most convenient for locating optimal backoff operating points in FDMA systems. We emphasize that the desired operating backoff is strongly dependent on the shape of the power curves (in Figure 11.20d) and therefore will vary with the particular power amplifier involved. Note that the satellite power P_T and the number of FDMA carriers K in (11.5.12) are no longer simply linearly related, as in (11.5.9). In general, when operating with a fixed satellite power, fewer FDMA carriers can be supported than when a linear amplifier is used. Again the actual number of carriers will be the smaller of the number allowed by the transponder power and the number allowed by the satellite bandwidth.

The FDMA analysis in (11.5.10) assumed a single transponder, and all the available satellite power in the downlink must be divided among all carriers. Furthermore, strong carriers tend to suppress weak carriers in the downlink. This means that when a mixture of both strong and weak carriers are to use the satellite simultaneously, we must ensure that the weaker carriers can maintain a communication link especially if it is to be transmitted to a relatively small (small $g/T°$) receiving station. One way to alleviate these effects, and improve the downlink performance, is by the use of satellite channelization. In channelization, the FDMA uplinks are grouped in separate transponder channels, each with its own bandpass filter and power amplifier (Figure 11.22). Each satellite channel then becomes an independent transponder. Only carriers of the same power are used in the same channel. The power of each amplifier is therefore divided only among the carriers in its own bandwidth. In addition, the intermodulation and

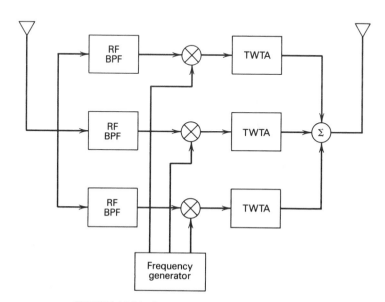

FIGURE 11.22. Channelized FDMA transponder.

power suppression effects are reduced since there are fewer carriers in each transponder. The limit, of course, is when each uplink carrier is assigned its own transponder channel, which is the so-called SCPC (*single channel per carrier*) format and all nonlinear effects are removed. The advantages of channelization are achieved, of course, at the expense of a more complex satellite, since the weight of not only the additional power amplifiers and filters must be included but also that of the supporting auxiliary primary power. The advantages in performance of the increased number of independent transponders must be carefully weighed against the additional satellite cost.

11.6   TIME DIVISION MULTIPLE ACCESS (TDMA)

In TDMA through a satellite, uplink carriers are separated in time rather than frequency. Instead of assigning a frequency band, each uplink carrier is assigned a prescribed time interval in which to relay through the satellite. During its interval, a particular station has exclusive use of the satellite, and its uplink transmission alone is processed by the satellite for the downlink. This means each carrier can use the same carrier frequency and make use of the entire satellite bandwidth during its interval. Since no other carrier uses the satellite during this time interval, no intermodulation or carrier suppression occurs, and the satellite amplifier can be operated in saturation so as to achieve maximal output power. Thus TDMA downlinks always operate at full saturation power of the satellite. However, the entire TDMA system must have all Earth stations properly synchronized in time so that each can transmit through the satellite only during its prescribed interval, without interfering with the intervals of other stations. This time synchronization between satellite and all Earth stations is called *network synchronization*. A downlink Earth station, wishing to receive the transmissions from a particular uplink, must gate in to the satellite signal during the proper time interval. This means that all Earth stations, whether transmitting or receiving, must be part of the synchronized network.

Since there may be many users of the TDMA satellite, each wishing to establish a communication link in approximately real time, the total transmission time must be shared by all users. Thus the time intervals of each station must be relatively short, and repeated at regular epochs. This type of short-burst, periodic operation is most conducive to digital operation, where each station transmits bursts of data bits during its intervals. However, TDMA systems using digital transmissions require that all receiving stations must obtain decoder synchronization for slot timing. For phase-coherent decoding, decoder synchronization requires establishing both a coherent phase reference and a coherent bit timing clock before any bits can be decoded within a slot. Also, word synchronization may be needed to separate the digital words occurring during a slot. This hierarchy of decoding synchronization must be established at the very beginning of each slot if the subsequent slot bits are to be decoded. Furthermore, since each slot contains data from a different carrier, synchronization must be separately established for each slot being received. In fact, even when receiving from the same station, synchronization must generally be reestablished from one periodic burst to the next. Hence digital communications with TDMA has an inherent requirement for rapid synchronization in order to perform successfully. The technology for short-burst communications is rather new and, of course, will be closely linked to the development of high speed digital processing hardware.

A TDMA satellite system is shown in Figure 11.23. Each uplink station is assigned one of a contiguous set of time slots, and the group of all such time slots forms a transponder time frame. We assume a single transponder operating at the satellite. During each successive frame, each transmitting station has its own specific slot. Hence, if slots are τ sec, a given station uses the satellite τ sec during each T_f sec frame. The station transmits bursts of data bits during its τ-sec transmission time while using the remaining frame time to generate the next set of data bits.

Ideally, all uplink carriers should operate at approximately the same data rate. Accommodation of stations with widely varied transmission bit rates in a common frame requires station buffering and storage. It may therefore be

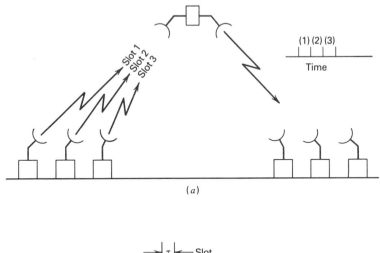

(a)

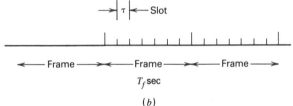

(b)

FIGURE 11.23. TDMA model: (a) transmission diagram, (b) slot-frame alignment.

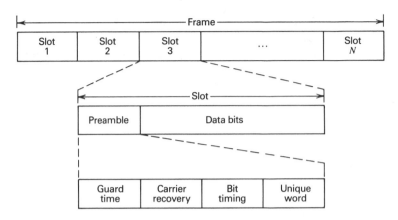

FIGURE 11.24. Slot formatting in TDMA.

more advantageous to provide several separate TDMA transponder channels with stations of approximately the same rates grouped in a common frame.

A transponder TDMA frame is typically formatted as in Figure 11.24. The frame is divided into slots, each assigned to an uplink station. Each slot interval is then divided into a preamble time and a data transmission time. The preamble time is used to send a synchronization waveform so that a receiving station gated to the slot can lock up its receiver decoder. The preamble generally contains guard time (to allow for some errors in slot timing), a phase referencing and bit timing interval (to allow a phase coherent decoder to establish carrier and bit synchronization), and a unique code word (to establish word synchronization). Observation of this unique word can be used to set the word markers for the data transmission during the remainder of that slot interval. Each frame slot can be formatted in this way. Any station can be received by gating in at the proper slot time, referred to the Earth-station time axis. The latter must be separately referenced by each Earth station to the satellite time axis through network synchronization. All transmitting stations use the same carrier frequency, so that every receiving station has (approximately) frequency synchronization throughout a frame. Thus the preamble time is needed primarily to adjust to the phase and bit timing of a given station, which can possibly drift from one frame to the next.

Preamble time should be long enough to establish reliable synchronization but should be short compared to the data transmission time. If a slot contains P preamble bits and D data bits, then

$$\eta = \frac{P}{D + P} \qquad (11.6.1)$$

is the slot preamble overhead. Systems are typically designed with no more than a 10% overhead factor. For a satellite to transmit each slot, it must operate at a bit rate of

$$R_s = \frac{D + P}{\tau} \quad \text{bps} \tag{11.6.2}$$

Each uplink carrier utilizing one slot per frame therefore sends D bits every frame time, or at a data rate of

$$R_b = \frac{D}{T_f} \quad \text{bps} \tag{11.6.3}$$

Thus the number of slots per frame (number of separate uplink carriers) is then

$$K = \frac{T_f}{\tau} = \frac{D/R_b}{(D + P)/R_s}$$

$$= (1 - \eta) \frac{R_s}{R_b} \tag{11.6.4}$$

The desired overhead and the bit rates therefore limit the number of TDMA users. Note carefully the difference between the satellite bit rate (rate at which bits are generated and transmitted through the satellite) and the source data rate (bit rate at which each uplink user is actually transmitting data bits). The higher the ratio of these bit rates, the larger the number of uplink carriers that can be accommodated by the TDMA system.

As an example, suppose that each uplink carrier is to transmit digital voice at 64 kbps, and assume that each slot requires 40 bits of preamble to establish the required synchronization. Then for a 10% preamble overhead, each slot will contain 400 bits, 360 of which are data bits. The TDMA frame must therefore be $360/64 \times 10^3 = 5.625$ msec long. If the satellite transponder has a bit rate of $R_s = 10$ Mbps, then each slot can be $400/10^7 = 40 \ \mu\text{sec}$ long, and a total of 5.625 msec$/40 = 140$ uplink carriers can be sustained.

Frame times in TDMA voice formats are usually relatively short (on the order of milliseconds), which of course complicates the slot synchronization operation. Increasing the number of bits per burst lengthens the frame time, but as the frame is lengthened, more time elapses between bursts from a given station, allowing more drift time to produce larger station oscillator phase shifts. Hence shorter frames tend to produce better burst-to-burst synchronization coherency.

Decoding performance of the slot bits at a ground receiver depends on the transponded E_b/N_0. Since there is no intermodulation noise or crosstalk, the receiver decoder E_b/N_0, after transponding through the satellite to the ground receiver, is given by (11.4.6) as

$$\frac{E_b}{N_0} = \left[(\text{CNR}_u)^{-1} + \left(\frac{P_T L}{N_{0d} R_s} \right)^{-1} \right]^{-1} \tag{11.6.5}$$

where CNR_u is the uplink CNR in the satellite bit rate bandwidth R_s and P_T is the maximum saturation power of the satellite amplifier. For a sufficiently strong uplink, the power P_T must be satisfactory to support a bit rate R_s with the specified PE. If an $E_b/N_0 = Y$ is required to decode the slot bits with PE, then satellite power and slot bit rate are related by

$$P_T = \frac{Y R_s N_{0d}}{L} \tag{11.6.6}$$

Thus a TDMA system may be satellite power limited (in producing a slot decoding PE with a satellite bit rate R_s) or may be satellite bit rate limited in supporting a required number of carriers in (11.6.4).

A key operation in TDMA is the network synchronization. Network synchronization is achieved by clocking together all the transmitting and receiving stations of the TDMA system. Theoretically, if each transmitting station knew its range to the satellite precisely, network synchronization could be achieved by a single ground master clock used to time all Earth stations. Accurate timing could be obtained from the master clock to each station using terrestrial links, and each station could be assigned a satellite time slot beginning at a fixed-time epoch relative to the master clock. Each station then need only adjust its uplink transmission so as to arrive at the satellite at exactly the correct time interval. The Earth station would merely compensate in the uplink for the time delay due to its range to the satellite. This is referred to as *open loop timing*. In practice, however, range values to a satellite cannot be determined precisely due to inherent uncertainty in satellite location. For example, a 40 km uncertainty in satellite range will cause a timing uncertainty of hundreds of microseconds in transmission time. However, timing accuracy for network synchronization must be maintained to within a small fraction of a slot time, which, from the previous examples, may be on the order of several microseconds. In addition, maintaining common clocks with microsecond accuracy over remote Earth stations for long periods of time may be difficult. It is therefore necessary to combine a common, simultaneous range measurement for each station with a timing marker transmission from the satellite in order to achieve the desired timing. The synchronization markers are sent in the downlink, and network synchronization is initiated directly from the satellite.

The timing diagram in Figure 11.25 illustrates how this can be easily accomplished. The satellite transmits continually a periodic sequence of timing markers (in the form of some convenient waveform). These timing markers must be sent over a separate satellite bandwidth and cannot interfere with the TDMA channels. The markers can be self-generated on board by a stable satellite clock, or can be relayed through the satellite from an Earth-station clock. Each transmitting station is assigned a time slot at the satellite with respect to the marker points; that is, a station is assigned a time length t_s after each marker initiation, indicating where its time slot

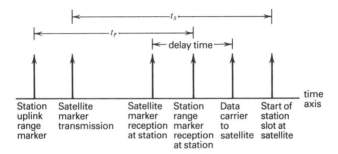

FIGURE 11.25. Timing diagram for TDMA network synchronization.

begins at the satellite. A station wishing to transmit first sends up its own ranging markers, which are retransmitted by the satellite and received back at the transmitting station, along with the satellite markers. By measuring the total two-way transmit time t_r of its range signal, a transmitting station can then adjust its own uplink transmission so it arrives exactly in its own time slot. This now allows the Earth station to "acquire" the correct slot, and transmission of the slot burst can begin. By monitoring the return time of the unique word of its own burst preamble (relative to the received timing marker of the satellite), closed loop control of the slot timing can be maintained.

In effect, the Earth station replaces the ranging signal by the preamble word and adjusts uplink burst transmission so that the returned synchronization word falls exactly in the proper slot position t_s sec after marker reception. Since the satellite may drift slowly, the return preamble measurement must be continually updated to provide stationkeeping and maintain the transmitting station synchronized within the network. This discussion points out a basic disadvantage of this type of TDMA operation. The necessity for continual network synchronization requires that each transmitting station have the capability of receiving its own transmissions. This can occur only with wide beam downlink antennas, and requires all transmitting stations to have reception capability also. Note however that a receiving Earth station does not require a ranging operation, but only the ability to receive the satellite markers. If it wishes to recover the transmission from a station using a time slot t_s sec after marker generation, it need only adjust to the time slot beginning t_s sec after marker reception.

The role of the ranging signal is only to aid in the initial acquisition of the slot. While the preamble unique word can be used to maintain slot timing, it cannot be used for the initial acquisition. This is due to the fact that without some a priori indication of slot location, transmission of the preamble burst from an earth station would directly interfere with other slots. For this reason, the ranging operation is generally performed in an adjacent satellite band to avoid interference with the TDMA channels. An alternative that avoids use of other bands is use of low level range codes that are spread over the entire satellite bandwidth. If the power level of the range code is low enough, it will not interfere with any TDMA slot, and its length can be correlated up to provide accurate range markers. However, each station would then have to have its own recognizable range code.

11.7   CODE DIVISION MULTIPLE ACCESS (CDMA)

In CDMA satellite systems uplink carriers are identified by unique separable address codes embedded within the carrier waveform. Each uplink station uses the entire satellite bandwidth and transmits through the satellite whenever desired, with all active stations superimposing their waveforms on the downlink. Thus no frequency or time separation is required. Carrier separation is achieved at an Earth station by identifying the carrier with the proper address. These addresses are usually in the form of periodic binary sequences that either modulate the carrier directly or change the frequency state of the carrier. Address identification is accomplished by carrier correlation operations. CDMA carrier crosstalk occurs only in the inability to correlate out the undesired addresses while properly synchronizing to the correct address for decoding. As in TDMA, CDMA carriers have the use of the entire satellite bandwidth for their total activity period, and CDMA has the advantage that no controlled uplink transmission time is required and no uniformity over station bit rates is imposed. However, system performance depends quite heavily on the ability to recognize addresses, which often becomes difficult if the number of stations in the system is large.

Digital addresses are obtained from code generators that produce periodic sequences of binary symbols. A station's address generator continually cycles through its address sequence, which is superimposed on the carrier along with the data. If the address is modulated directly on the carrier, the format is referred to as *direct sequence* CDMA (DS-CDMA). Superimposing addresses on modulated uplink carriers generally produces a larger carrier bandwidth than that that will be generated by the modulation alone. This spreading of the carrier spectrum has caused CDMA systems to be referred to also as *spread-spectrum multiple access* (SSMA) systems.

A digital version of a direct sequence CDMA link is shown in Figure 11.26a. The ith uplink station has assigned to it a digital address code $q_i(t)$, the latter a periodic binary sequence with binary symbols (chips) of width w sec and code length k (i.e., k chips per period). Each Earth station has its own such address code. Digital information bits are transmitted by superimposing the bits onto the address code. If the ith station is to transmit the binary data waveform $d_i(t)$ [$d_i(t) = \pm 1$], it forms the binary sequence

$$m_i(t) = d_i(t)q_i(t) \qquad (11.7.1)$$

If T_b is the bit time of $d_i(t)$, then T_b may correspond to either a full period

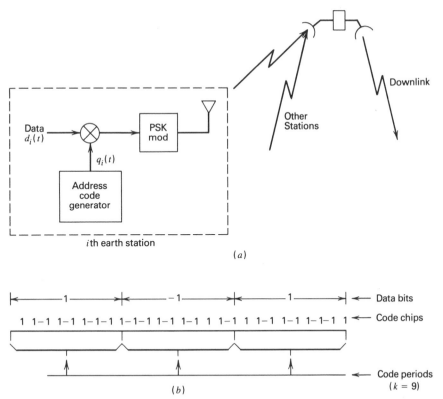

FIGURE 11.26. CDMA model: (*a*) transmission diagram, (*b*) addressed bit sequence, data bits on code.

of $q_i(t)$, or to a fraction of a period. If each T_b is exactly one address code period, then the data bits are modulating the polarity of a portion of a code period. Figure 11.26*b* shows an example of these situations for a particular address code and data sequence. In either case the address code $q_i(t)$ is serving as a subcarrier for the source data. The binary sequence in (11.7.1) is then PSK directly onto the station RF carrier located at the center frequency of the satellite uplink RF bandwidth. Since each station can use the entire satellite bandwidth, and since (11.7.1) has a code chip rate of $1/w$ chips per second, each BPSK carrier will utilize a RF bandwidth of approximately

$$B_{\mathrm{RF}} = 2/w \quad \mathrm{Hz} \tag{11.7.2}$$

Conversely, the available satellite RF bandwidth determines the minimum chip width, while the code period determines its relation to the bit times. That is, the number of code chips per bit is given by

$$k = \frac{T_b}{w} = \frac{B_{RF}}{2R_b} \tag{11.7.3}$$

The satellite RF bandwidth therefore determines the code length per bit. Each station forms its PSK carrier in exactly the same way, each using the same RF carrier frequency and RF bandwidth, but each with its own address code $q_i(t)$. At the satellite, the frequency spectra of all active carriers are superimposed in the RF bandwidth.

The satellite repeater retransmits the entire uplink RF spectrum in the downlink, using straightforward RF–RF or RF–IF–RF conversion. Since all active carriers pass through the satellite simultaneously, limiting driver stages for the satellite amplifier will produce power robbing due to uplink noise, and weak carrier suppression by strong carriers, just as in FDMA. Hence uplink power control is usually required with CDMA. In addition, if nonlinear amplification is used, intermodulation interference will be produced in the downlink, and the available satellite power must be shared by all stations. The amount of intermodulation can be controlled by adjustment of the satellite amplifier backoff.

In CDMA no attempt is made to synchronize or align the bit intervals, and the various uplink carriers operate independently with no overall network timing. Each active station simply transmits its modulated addressed carrier through the satellite into the downlink. A ground receiver must again obtain phase, bit, and code coherency with the desired uplink transmission in order to detect coherently, in the presence of the undesired carriers, the bits of the desired addressed carrier.

Temporarily ignoring all noise and intermodulation, the satellite combined BPSK carrier signals in the downlink is

$$x(t) = \sum_{i=1}^{K} A_i \sin\left[\omega_c t + \frac{\pi}{2} m_i(t - \tau_i) + \psi_i\right]$$

$$= \sum_{i=1}^{K} A_i m_i(t - \tau_i) \cos(\omega_c t + \psi_i) \tag{11.7.4}$$

where the sum is over all active carriers and where $\{\tau_i\}$ and $\{\psi_i\}$ account for the different time shifts and phase shifts in passing through the satellite. At a particular Earth receiving station, a correlating decoder is used to recover the message bits of a particular uplink station, as shown in Figure 11.27. A receiver decoding the jth uplink therefore would generate the coherent coded reference

$$r_j(t) = 2q_j(t - \tau_j) \cos(\omega_c t + \psi_j) \tag{11.7.5}$$

Note that the generation of this receiver reference signal requires a time referenced code and a phase coherent RF carrier. Time and phase coheren-

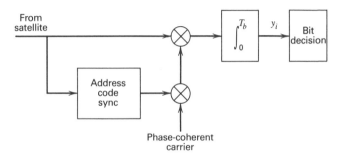

FIGURE 11.27. CDMA receiver.

cy of this type requires a combined phase locking–code locking loop, as will be discussed shortly. In order to maintain the required address code synchronism, the code locking subsystem must first acquire, then accurately track, the received address code. It does this by operating its own address code generator (identical to that of the transmitting station to be decoded) in time synchronization with the arriving address.

Assuming perfect code acquisition and lock-up, a decoder for the jth uplink correlates over a bit period the received waveform in (10.7.4) with the coded reference in (10.7.5), generating

$$y_j = \frac{1}{T_b} \int_{\tau_j}^{\tau_j + T_b} r_j(t)x(t)\, dt$$
$$= d_j A_j + C_j \tag{11.7.6}$$

where $d_j = \pm 1$ is the desired data bit being received during the jth decoding bit time and

$$C_j = \sum_{\substack{i=1 \\ i \neq j}}^{K} A_i \cos(\psi_i - \psi_j)c_{ij} \tag{11.7.7}$$

$$c_{ij} = \frac{1}{T_b} \int_0^{T_b} m_i(t - \hat{\tau}_i)q_j(t)\, dt$$
$$\hat{\tau}_i = \tau_i - \tau_j \tag{11.7.8}$$

The parameter C_j is the carrier crosstalk term arising in the jth decoder during the bit interval $(0, T_b)$ and is composed of a crosstalk contribution due to each active carrier in the downlink. The effect of each depends on the modulated address code cross-correlation c_{ij} in (11.7.8). The crosstalk C_j appears as an effective data dependent interference term that adds to, or subtracts from, the desired correlation value $d_j A_j$ in (11.7.6).

Crosstalk depends on the data bits, phase differences, and time shifts, all of which must be considered a set of random, independent variables occurring between the various uplink carriers as they arrive at the satellite. This means that the crosstalk parameter C_j evolves as a random disturbance. We proceed by assuming all carriers use the same bit time T_b. In this case, during a decoding interval of the jth carrier, the ith carrier arives with relative delay $\hat{\tau}_i$. This means we can write

$$d_i(t - \hat{\tau}_i) = \begin{cases} d_{i_1}, & 0 \le t \le \hat{\tau}_i \\ d_{i_0}, & \hat{\tau}_i \le t \le T_b \end{cases} \tag{11.7.9}$$

where d_{i_1} and d_{i_0} are the previous and present bits, respectively, of the ith carrier during $(0, T_b)$. Because of the equal bit rate assumption, only two adjacent bits can cause crosstalk interference. This means that the crosstalk parameter C_j, conditioned on the data bits and the set of random shifts $\hat{\tau}_i$, is again given by (11.7.8) only now with

$$c_{ij} = d_{i_1}\zeta_{ij}(\hat{\tau}_i) + d_{i_0}\zeta'_{ij}(\hat{\tau}_i) \tag{11.7.10}$$

where

$$\zeta_{ij}(\hat{\tau}_i) = \frac{1}{T_b} \int_0^{\hat{\tau}_i} q_i(t - \hat{\tau}_i) q_j(t)\, dt \tag{11.7.11a}$$

$$\zeta'_{ij}(\hat{\tau}_i) = \frac{1}{T_b} \int_{\hat{\tau}_i}^{T_b} q_i(t - \hat{\tau}_i) q_j(t)\, dt \tag{11.7.11b}$$

The terms ζ_{ij} and ζ'_{ij} are partial cross-correlations of the address codes, each involving integrations over only a portion of the T_b sec interval. Averaging over equal likely data bits and uniform phase angles shows that the mean of $C_j = 0$ for all $\{\hat{\tau}_i\}$, while the mean squared value follows as

$$\overline{C_j^2} = \sum_{\substack{i=1 \\ i \ne j}}^{K} \frac{A_i^2}{2} \mathscr{E}[c_{ij}^2] \tag{11.7.12}$$

Substituting from (11.7.10), and averaging over the data bits yields

$$\mathscr{E}[c_{ij}^2] = \int_{-\infty}^{\infty} [\zeta_{ij}^2(\tau) + \zeta'^{2}_{ij}(\tau)] p(\tau)\, d\tau \tag{11.7.13}$$

where $p(\tau)$ is the probability density of the delay $\hat{\tau}_i$. If each delay is uniformly distributed over $(0, T_b)$, implying that the relative carrier delays are independent and completely random, we then have

$$\overline{C_j^2} = \sum_{\substack{i=1 \\ i \ne j}}^{K} \frac{A_i^2}{2} \gamma_{ij}^2 \tag{11.7.14}$$

where now

$$\gamma_{ij}^2 = \frac{1}{T_b} \int_0^{T_b} [\zeta_{ij}^2(\tau) + \zeta_{ij}'^2(\tau)] \, d\tau \qquad (11.7.15)$$

We see that γ_{ij}^2 plays the role of a code cross-correlation parameter and the mean squared value of the address code interference $\overline{C_j^2}$ increases directly with the set of address code cross-correlation $\{\gamma_{ij}^2\}$. Note that performance depends on the squared integrated cross-correlation functions in (11.7.15). It is this complexity that makes the CDMA case difficult to analyze. Investigation of maximum correlation values alone can be extremely misleading here, since it is the integrated squared value that is of prime importance. The interested reader is referred to References 15–18 for further discussion of the properties of γ_{ij}^2, including some useful bounds for estimating (11.7.15).

An interesting result is that of Welch [19], who showed that when cross-correlations involve k code symbols and K separate codes the maximum value of γ_{ij} in (11.7.15) must be greater than $[(K-1)/kK-1]^{1/2}$. For large K, this implies that the worst case effect decreases as $1/\sqrt{k}$, where k is the number of code chips per bit. Hence cross-correlation γ_{ij} can only be reduced by increasing the code length per bit. From (11.7.3), this can be restated as

$$\gamma_{ij}^2 \geq \frac{2R_b}{B_{RF}} \qquad (11.7.16)$$

and requires an increase in the satellite bandwidth to reduce the cross-correlation. Gold codes [20] and other classes of code [21–29] have been found that produce cross-correlations close to this bound.

The effect of code cross-correlation during CDMA decoding is to add an interference term to the desired correlator output during bit decoding. The bit correlator output used for the receiver decisioning in Figure 11.27 is then

$$y_j = d_j A_j + n_j + C_j \qquad (11.7.17)$$

where n_j is a zero mean Gaussian noise variable with variance $N_{0T}/2T_b$. Here N_{0T} is the total one-sided noise spectral level due to both receiver input noise (downlink and retransmitted uplink noise) and any intermodulation noise produced during the satellite transponding. If there were no cross-correlation interference (orthogonal system), the decoder would operate with

$$\left(\frac{E_b}{N_0}\right)_o = \frac{P_j T_b}{N_{0T}} \qquad (11.7.18)$$

To determine performance with C_j present in (11.7.17), we must proceed formally by treating receiver correlations as intersymbol interference. The bit error probability for the jth receiver must then be obained by averaging over all possible bit sequences and code delays of all interfering stations. Although the resulting PE_j will then represent the exact bit error probability for a particular receiver, its computation may become quite lengthy if the number of interfering carriers is large. This is due to the fact that the number of possible data vectors that must be averaged grows exponentially with the number of carriers K. This computation can be circumvented, however, when K is large, since we can model C_j as an additive Gaussian crosstalk variable that simply adds to the Gaussian interference of the downlink carrier. Justification and conditions for the validity of this Gaussian assumption were studied in depth in References 29–31. Using (11.7.14), we can establish that C_j has a zero mean and a variance given by

$$\overline{C_j^2} = \sum_{\substack{i=1 \\ i \neq j}}^{K} P_i \gamma_{ij}^2 \qquad (11.7.19)$$

We can interpret (11.7.19) as an interference power, spread over the data bandwidth. Hence the decoder E_b/N_0 is now modified from (11.7.18) to

$$\frac{E_b}{N_0} = \frac{P_j T_b}{N_{0T} + (\overline{C_j^2}/R_b)} \qquad (11.7.20)$$

This can be rewritten in the form of (11.7.18) as

$$\left(\frac{E_b}{N_0}\right) = \frac{(E_b/N_0)_o}{1 + (E_b/N_0)_o[\overline{C_j^2}/P_j]} \qquad (11.7.21)$$

This relates the (E_b/N_0) of CDMA system to that that would occur if the system were truly orthogonal (i.e., no station cross-correlation). The latter system depends only on the intermodulation and power division that occurs in passing through the satellite with the addressed carriers. Note that the operating E_b/N_0 of a CDMA receiver is always less than that with the same number of orthogonal carriers, because of the denominator in (11.7.21). Thus a CDMA system must be designed with higher power levels than an orthogonal system if it is to have the same performance.

To further examine the cross-correlation effect, consider the simplified case where $P_i = P = LP_T/K$ (equal carrier power) and $\gamma_{ij} = \gamma$ (equal address correlation). Under these conditions, $\overline{C_j^2} = (K-1)\gamma^2 P$ for all j, and

$$\left(\frac{E_b}{N_0}\right) = \frac{(E_b/N_0)_o}{1 + (E_b/N_0)_o[(K-1)\gamma^2]} \qquad (11.7.22)$$

where $(E_b/N_0)_o = P_T L/KN_{0T}$. This represents the operating E_b/N_0 (from which PE can be determined) for a CDMA system of K simultaneous carriers and fixed cross-correlation γ^2.

Equation (11.7.22) can be solved for the required satellite power P_T needed to achieve a desired (E_b/N_0) of Y. Hence

$$P_T = \frac{1}{L} \left[\frac{Y(N_{0T}/T_b)K}{1 - Y[(K-1)\gamma^2]} \right] \qquad (11.7.23)$$

When $\gamma = 0$ (perfectly orthogonal system), P_T is linearly related to K. However, for K binary codes to be orthogonal they must have at least length K; that is, there must be K chips per bit. From (11.7.3) we see, therefore, that again the system may be either power limited or bandwidth limited.

When $\gamma^2 \neq 0$ buildup of interference with γ^2 will further limit the number of CDMA carriers. In fact, an infinite satellite power is required in (11.7.23) if K is larger than

$$K_{\max} = \frac{1}{Y\gamma^2} \qquad (11.7.24)$$

If we insert the lower bound in (11.7.16), this is approximately

$$K_{\max} \approx \frac{B_{RF}}{Y(2R_b)} \qquad (11.7.25)$$

With $Y = 10\,dB$, the maximum number of CDMA carriers is about one-tenth the number of potential orthogonal carriers with the same bandwidth. Hence CDMA system, although having the advantages of simplicity, non-synchronism, and independent operation, is less efficient in bandwidth usage. However, it should be emphasized that (11.7.25) refers to the actual number of CDMA carriers passing through the satellite at one time, which may be considerably fewer than the total CDMA carriers in the system, if each operates only a small fraction of the time.

11.8 CODE SYNCHRONIZATION IN DS-CDMA

It was stated that DS-CDMA operation required a time-coherent replica of the received code and carrier in order to decode the data. This code synchronization is obtained by a code locking subsystem operating in parallel with the data decoding channel in Figure 11.27. A local version of the code is generated at the receiver, using an identical code generator. Initially, however, the local code is out of alignment with the received code from the transmitter, and it is necessary to bring the two codes into synchronization before data can be decoded. This synchronization is accomplished by an acquisition operation that basically applies a search-and-test aligning procedure. When an indication appears that the codes are nearly aligned, a tracking subsystem is activated to maintain the code synchronization throughout the subsequent data transmissions.

The code acquisition subsystem appears as in Figure 11.28. The output code sequence from the receiver code generator is multiplied with the received modulated carrier. The mixer output is then bandpass filtered, envelope detected, and integrated for a fixed interval T_{in}. The voltage value of the integrator output serves as an indication of whether the two codes are in alignment. If it is concluded that the codes are not aligned, the local code sequence can be delayed or advanced, and another correlation measurement is performed. This testing can be repeated until code alignment is indicated.

The fact that the integrated envelope voltages serve as an indicator of code alignment can be shown as follows. Assume the received addressed carrier $c(t)$ arrives with a δ sec code offset relative to the local code $q(t)$. Thus, neglecting interfering carriers,

$$c(t) = Ad(t + \delta)q(t + \delta) \cos(\omega_c t + \psi) \qquad (11.8.1)$$

where $d(t)$ is the data waveform. The multiplier in Figure 11.28 multiplies $c(t)$ by the local code. The bandpass filter is centered at ω_c and is wide enough to pass the data $d(t)$, but not the code $q(t)$. Hence the filter output is

$$f(t) = A\overline{q(t)q(t + \delta)}d(t + \delta) \cos(\omega_c t + \psi) \qquad (11.8.2)$$

where the overbar denotes the averaging effect of the BPF on the code. The output of the envelope detector is then

$$\text{Envelope detector output} = |f(t)|$$

$$= |A\overline{q(t)q(t + \delta)}| \qquad (11.8.3)$$

where we have used the fact that the envelope of the PSK carrier is unity. The envelope detector output is then integrated to produce the voltage v_s. If we take the integration time to be an integer multiple of a code period T_c, we have $T_{in} = \eta T_c$, and

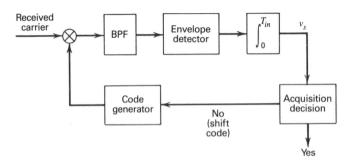

FIGURE 11.28. Code acquisition loop.

$$v_s = \left(\frac{AT_c}{2}\right)\eta R_q(\delta) \tag{11.8.4}$$

where we have written

$$R_q(\delta) = \overline{q(t)q(t+\delta)}$$

$$= \frac{1}{T_c}\int_0^{T_c} q(t)q(t+\delta)\,dt \tag{11.8.5}$$

The function $R_q(\delta)$ is the periodic autocorrelation function of the code. Thus the integrator voltage reading is directly related to the code correlation function evaluated at the offset δ. It is desirable to have this voltage v_s low when the codes are out of alignment and high when the codes are perfectly aligned. This requires that the code autocorrelation function be (ideally) zero for $\delta \neq 0$ and maximum when $\delta = 0$. Such a function would allow observations of the voltages v_s to directly indicate when the codes are aligned. Note that this autocorrelation condition places an additional requirement on the codes selected for addressing in the CDMA format. We have previously found that address codes should be long (many chips per data bit) and that sets of codes should exhibit low pairwise cross-correlation, or partial cross-correlation, with other address codes. We now see that the acquisition operation further constrains the autocorrelation properties of each address code as well. Note that the correlation values v_s are generated without knowledge of the carrier phase ψ and independent of the specific data $d(t)$ being sent. This is referred to as *noncoherent* code acquisition.

   The desired code autocorrelation for $R_q(\delta)$ can be adequately approximated by the correlation function of codes generated from digital feedback shift registers. Such a code generator is shown in Figure 11.29a. Digital sequences are shifted through the register one bit at a time at a preselected shifting rate. The register output bits form the address code. At any shifting time, the contents of the register are binary combined and fed back to form the next input bit to the register. Once started with an initial bit sequence stored in the register, the device continually regenerates its own inputs while shifting bits through the register to form the output code. The feedback logic and the initial bit sequence determine the structure of this output code. With properly designed logic, the output codes can be made to be periodic, and the register acts as a free running code oscillator. Codes generated in this manner are called *shift register codes* [32]. The class of periodic shift register codes having the desired binary waveform correlation of Figure 11.29b are called *pseudo-random noise* (PRN) codes. PRN codes are known to have the longest period associated with a given register size, and are therefore also called *maximal length* shift register codes. If the register length is l stages, the PRN code will have a period of $n_l = 2^l - 1$ chips. The underlying mathematical structure of shift register codes and the technique

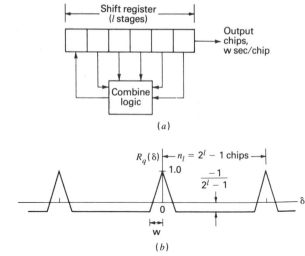

FIGURE 11.29. Shift register: (*a*) code generator, (*b*) PRN code correlation function.

for determining proper feedback logic are well documented in the literature [32–34] and are not considered here.

A property of shift register codes well suited to acquisition is that time shifting of the code can be easily implemented by externally changing the bits in the register at any time shift. Changing these bits causes the output to jump to a new position in the code period, which then continues periodic generation from that point on. Thus, within one shift time the output code can be forced to skip ahead in its cycle. Practical code generators are obtained by driving the shift register with a clock oscillator. Each cycle of the clock provides the register shifting rate, so that an oscillator frequency of f_c would produce a code rate of f_c chips/sec and a chip width of $w = 1/f_c$ sec. Hence code rates are determined by the maximal rate at which a shift register can be driven.

The code correlation voltages in (11.8.4) are observed in the presence of additive noise arriving with the addressed carrier. The code acquisition system must decide when true acquisition has been achieved. There are two basic procedures for deciding the correct code position from these noisy voltage observations. One method is called *maximum-likelihood* (ML) acquisition and involves an observation of the correlation of all code positions in a period, selecting the position with the maximum voltage for acquisition. The second method, called *threshold acquisition*, selects the first position whose voltage exceeds a fixed threshold value as the acquisition position. Note that in maximum-likelihood acquisition every code position is examined, and a decision is always made after completing the entire examination. In threshold acquisition, code positions are continually examined until a threshold crossing occurs.

In ML acquisition, incorrect acquisition occurs if an incorrect position voltage exceeds the correct one. As such, the effect is similar to that occurring in an M-ary noncoherent orthogonal decoder. In this case, word decisions are also based on maximal envelope samples, where all incorrect words have a zero signal voltage. (For shift register codes, it is assumed that the codes are long enough that the offset code correlation values in Figure 11.29b can be considered zero.) The probability of incorrect ML acquisition is therefore given by the word error probability of a corresponding non-coherent orthogonal M-ary system where M is equal to the number of possible code positions; that is, the code length n_l. Such curves were shown in Figure 9.25 as a function of the bit energy and noise spectral level. For the acquisition case, the energy corresponds to the integrated carrier envelope energy collected in the T_{in} sec integrator,

$$\frac{E}{N_0} = \frac{(A^2/2) T_{\text{in}}}{N_{0T}}$$

$$= \eta \left(\frac{E_c}{N_{0T}} \right) \tag{11.8.6}$$

where η is the number of code periods in T_{in} and E_c is the carrier energy in a code period.

With fixed carrier power (11.8.6) serves as a basic for determining the required envelope integration time. If E_c is not satisfactory for the desired acquisition probability, we must integrate over more periods in the correlation detector in order to integrate up the acquisition power to the desired energy value. This means we spend more time examining a code shift position. Since each position must be examined, the total acquisition time to perform the ML test is then

$$T_{\text{acq}} = n_l T_{\text{in}} = n_l (\eta T_c) \tag{11.8.7}$$

Hence acquisition time depends directly on the code length. We see the obvious tradeoff of long length address codes (many chips per bit) versus the disadvantage of increased acquisition time.

For threshold acquisition with a threshold value of Y volts, a false threshold crossing will occur with envelope samples with probability

$$\text{PFC} = Q(0, Y) \tag{11.8.8}$$

where $Q(a, b)$ is defined in (8.1.31). A correct threshold crossing occurs with probability

$$\text{PCC} = Q \left(\frac{E}{N_0}, Y \right) \tag{11.8.9}$$

with E/N_0 given in (11.8.6). If the jth code position is correct, and the acquisition code search begins with the first, correct acquisition will occur with probability

$$\text{PAC}_j = \text{PCC}(1 - \text{PFC})^{j-1} \qquad (11.8.10)$$

The average acquisition probability is then

$$\begin{aligned}
\text{PAC} &= \frac{1}{n_l} \sum_{j=1}^{n_l} \text{PAC}_j \\
&= \frac{\text{PCC}}{n_l} \left[\frac{1 - (1 - \text{PFC})^{n_l}}{\text{PFC}} \right]
\end{aligned} \qquad (11.8.11)$$

If $\text{PFC} \ll 1/n_l$, then $\text{PAC} \approx \text{PCC}$, and (11.8.9) can be used to adjust threshold and integration time T_{in} to the desired acquisition probability.

The length of code acquisition time for the threshold test depends on the number of code periods that are searched before acquiring. The probability that the test will go through a complete period without reporting an acquisition is

$$\text{PNC} = (1 - \text{PCC})(1 - \text{PFC})^{n_l - 1} \qquad (11.8.12)$$

The probability that it will take exactly i periods to successfully acquire the jth position is then $\text{PAC}_j (\text{PNC})^{i-1}$. The average acquisition time is then

$$T_{\text{acq}} = \frac{1}{n_l} \sum_{j=1}^{n_l} \sum_{i=1}^{\infty} T_{\text{in}}[(i-1)n_l + j] \, \text{PAC}_j \, (\text{PNC})^{i-1} \qquad (11.8.13)$$

When $\text{PFC} \ll 1/n_l$, $\text{PAC} \approx \text{PCC} \approx 1 - \text{PNC}$, and

$$\begin{aligned}
T_{\text{acq}} &\approx T_{\text{in}} \left[\frac{n_l(1 - \text{PAC})}{\text{PAC}} + \frac{n_l}{2} \right] \\
&\approx T_{\text{in}} \frac{n_l}{2}
\end{aligned} \qquad (11.8.14)$$

Thus, on the average, approximately one-half the code chips will be searched before acquisition. We see, therefore, that threshold testing also has an acquisition time dependent on the code length. However, it must be remembered that T_{acq} is an average time, and individual acquisition operations may run considerably longer.

Once code acquisition has been accomplished (local and received codes brought into alignment), the received code must be continually tracked to maintain the synchronization. A code-tracking loop for discrete sequences is shown in Figure 11.30. It contains two parallel branches of a multiplier, bandpass filter, and envelope detector. (One branch can be that used in the

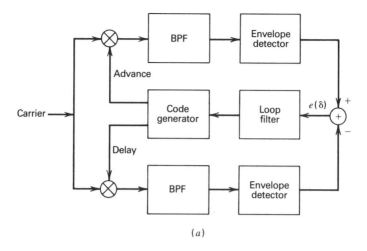

(a)

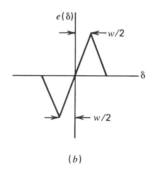

(b)

FIGURE 11.30. Code tracking loop: (a) block diagram, (b) loop error function.

acquisition operation, so that initiation of tracking requires only the insertion of the second branch.) The two-branch system in Figure 11.30 is referred to as a *delay-locked loop* [35]. A delay-locked tracking loop uses the local code to generate error voltages $e(t)$ that are fed back to correct any timing offsets of the local code generator as they arise. Timing error voltages are obtained by using the local code to generate two offset sequences advanced and delayed by one-half chip. With shift register codes, these are obtained from different taps within the same register. Thus, if $q(t)$ is the local code, the sequences $q(t + w/2)$ and $q(t - w/2)$ are simultaneously generated from the register. These offset codes are each separately multiplied with the input in the two channels of the delay-locked loop. The multiplied outputs are then bandpass filtered and envelope detected to produce output voltages that are subtracted to form the correction signal for code timing adjustment. Again, each BPF passes the data bits but averages the code product, which is then envelope detected. Following the discussion in (11.8.3), the advanced code channel output is

$$V_a = A \left| R_q \left(\delta - \frac{w}{2} \right) \right|$$ (11.8.15)

where $R_q(\delta)$ is defined in (11.8.5). Similarly, the delayed code channel produces

$$V_d = A \left| R_q \left(\delta + \frac{w}{2} \right) \right|$$ (11.8.16)

The output of the subtractor is therefore the correction voltage

$$e(\delta) = V_d - V_a$$

$$= A \left| R_q \left(\delta + \frac{w}{2} \right) \right| - A \left| R_q \left(\delta - \frac{w}{2} \right) \right|$$ (11.8.17)

for any offset δ.

This function is plotted in Figure 11.30b for the code correlation functions depicted in Figure 11.29b. We see that regardless of whether δ is positive or negative, a voltage proportional to δ is produced having the proper sign to adjust correctly the local code timing to reduce δ. This is

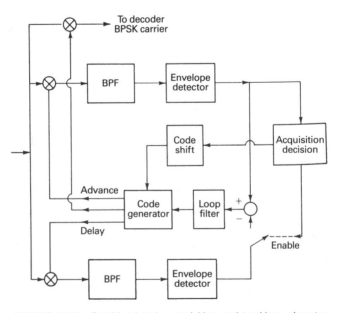

FIGURE 11.31. Combined code acquisition and tracking subsystem.

accomplished by using $e(\delta)$ to adjust the code clock that drives the code register (slow it down or speed it up). Note that a proportional correction voltage is generated only if $\delta \leq w/2$; that is, only if we are within a half chip time of code lock. This is why an initial acquisition procedure is necessary to first align the codes to within this accuracy. The delay-locked loop then operates to continually correct for subsequent timing errors. Note that, as a result of the envelope detection, the delay-locked loop achieves code tracking noncoherently, that is, without knowledge of the carrier phase or frequency (as long as the multiplier outputs pass through the BPF).

The acquisition system in Figure 11.29a and the delay-locked loop in Figure 11.30a can be combined to form the total DS-CDMA acquisition and tracking subsystem shown in Figure 11.31. When the tracking error is zero, the local code generator (without the offsets) is exactly in phase with the received code and can therefore be used to despread the received carrier. Thus, in the data channel in Figure 11.31, we use the fact that $q(t) \cdot q(t) = q^2(t) = 1$ to form the product

$$[d(t)q(t)\cos(\omega_c t + \psi)]q(t) = d(t)\cos(\omega_c t + \psi) \qquad (11.8.18)$$

The despread PSK carrier can now be processed in a standard PSK decoder for RF carrier phase referencing, bit timing, and data recovery.

REFERENCES

1. Martin, D. "Communication Satellites, 1958–1982," *Aerosp. Med. Res. Lab.* [*Tech. Rep.*], AMRL-TR (U.S.), AMRL-TR-79-078, September 1979.

2. Gagliardi, R. *Satellite Communications*, Van Nostrand-Reinhold, New York, 1984.

3. Gould, R. *Communication Satellite Systems—An Overview*, IEEE Press, New York, 1976.

4. Temes, G.C. and Mitra, S.K. *Modern Filter Theory and Design*, Wiley, New York, 1973.

5. Williams, A.E. and Atia, A.E. "Dual Mode Canonical Waveguide Filters," *IEEE Trans. Microwave Theory Tech.*, vol. MTT-25, December 1977, pp. 1021–1026.

6. Rhodes, J.D. "The Design and Synthesis of a Class of Microwave Band Pass Linear Phase Filters," *IEEE Trans. Microwave Theory Tech.*, vol. MTT-17, April 1969.

7. Atia, A.E., Williams, A.E., and Newcomb, R.W. "Narrow-Band Multiple-Coupled Cavity Synthesis," *IEEE Trans. Circuits Syst.*, vol. CAS-21, September 1974, pp. 649–654.

8. Atia, A.E. "Computer-Aided Design of Waveguide Multiplexers," *IEEE Trans. Microwave Theory Tech.*, vol. MTT-22, March 1976, pp. 332–336.

9. Kurzrok, R.M. "General Four-Resonator Filters in Waveguide," *IEEE Trans. Microwave Theory Tech.*, vol. MTT-16, May 1966, pp. 46–47.

10. Matthaei, G.L., Young, L. and Jones, E.M.T. *Microwave Filters, Impedance Matching Networks and Coupling Structures*, McGraw-Hill, New York, 1964.

11. Kudsia, C.M. and Donovan, V.O. *Microwave Filters for Communications Systems*, Artech House, Dedham, Mass., 1974.

12. Liechti, C. "Microwave Field Effect Transistors," *IEEE Trans. Microwave Theory*, vol. MT-24, June 1976, pp. 279–300.

13. Di Lorenzo, J. "GaAs FET Development," *Microwave J.*, vol. 21, February 1978, pp. 39–42.

14. "Demand Assignment," in *Satellite Communications*, H. Van Trees, ed., IEEE Press, New York, 1979, sect. 3.6.5.

15. Sarwate, D.V. and Pursley, M.B. "Crosscorrelation Properties of Pseudorandom and Related Sequences," *Proc. IEEE*, vol. 68, May 1980, pp. 593–619.

16. Pursley, M. "Performance Evaluation for Phase-Coded Spread Spectrum Multiple-Access. Part I. System Analysis," *IEEE Trans. Comm.*, vol. COM-25, August 1977, pp. 795–799.

17. Pursley, M.B. and Roefs, H.F.A. "Numerical Evaluation of Correlation Parameters for Optimal Phases of Binary Shift-Register Sequences," *IEEE Trans. Comm.*, vol. COM-27, October 1979, pp. 1597–1604.

18. See section on "Coded Division Multiple Access," in *IEEE Trans. Comm.*, vol. COM-30, May 1982.

19. Welch, L. "Lower Bounds on the Maximum-Cross Correlation of Signals," *IEEE Trans. Inf. Theory*, vol. IT-20, May 1974, pp. 397–399.

20. Gold, I.R. "Maximal Recursive Sequence with Multi-valued Crosscorrelation," *IEEE Trans. Inf. Theory*, vol. IT-14, January 1968, pp. 154–156.

21. Kasami, T. and Lin, L. "Coding for a Multiple Access Channel," *IEEE Trans. Inf. Theory*, vol. IT-22, March 1976, pp. 123–132.

22. Frank, R. and Zadoff, S. "Phase Shift Codes with Good Periodic Correlation Proportion Properties," *IEEE Trans. Inf. Theory*, vol. IT-8, October 1962, pp. 381–382.

23. Heimiller, R. "Codes with Good Period Correlations," *IEEE Trans. Inf. Theory*, vol. IT-7, October 1961, pp. 254–256.

24. Chu, D. "Polyphase Codes with Good Correlation Properties," *IEEE Trans. Inf. Theory*, vol. IT-18, July 1972, pp. 531–540.

25. Gold, R. "Optimal Binary Sequences for Spread Spectrum Mutiplexing," *IEEE Trans. Inf. Theory*, vol. IT-13, October 1967, pp. 619–621.

26. Mohanty, N. "Multiple Frank-Heimiller Signals for Multiple Access System," *IEEE Trans. Aerosp. Electron. Sys.*, vol. AES-11, July 1975.

27. Schneider, K. and Orr, R. "Aperiodic Correlation Constraints on Binary Sequences," *IEEE Trans. Inf. Theory*, vol. IT-12, January 1975.

28. Turyn, G. "Sequences with Small Correlation," in *Error Correcting Codes*, H. Mann, ed., Wiley, New York, 1968.

29. "Special Issue on Spread Spectrum Communications," *IEEE Trans. Comm.*, vol. COM-25, August 1977; vol. COM-30, May 1982.

30. Yao, K. "Error Probability of Asynchronous Spread Spectrum Multiple-Access Communication Systems," *IEEE Trans. Comm.*, vol. COM-25, August 1977, pp. 803–809.

31. Weber, C., Huth, G., and Batson, B. "Performance Considerations of CDMA Systems," *IEEE Trans. Veh. Technol.*, vol. VT-30, February 1981.

32. Golomb, S. *Shift Register Sequences*, Holden-Day, San Francisco, Calif., 1967.

33. Golomb, S. et al., *Digital Communications with Space Applications*, Prentice-Hall, Englewood Cliffs, N.J., 1964.

34. Lee, J. and Smith, D. "Families of Shift Register Sequences with Impulse Correlation Registers," *IEEE Trans. Inf. Theory*, vol. IT-20, March 1974.

35. Spilker, J. *Digital Communication by Satellite*, Prentice-Hall, Englewood Cliffs, N.J., 1977.

PROBLEMS

11.1   (11.2)   Assume that a satellite location is known to within a distance of δr at altitude h. Convert this to angular uncertainty as observed from Earth, and sketch a plot of angle uncertainty versus position uncertainty for various orbit altitudes.

11.2   (11.2)   Derive an equation for field propagation time in terms of propagation path length, using the fact that an electromagnetic wave travels at the speed of light ($c = 3 \times 10^8$ m/sec).

11.3   (11.4)   Show that the number of satellites that can be placed in geosynchronous orbit with no more than 20 dB sidelobe interference from the ground is $n \approx (20)fd$, where f is frequency in GHz and d is antenna diameter in meters.

11.4   (11.4)   Sketch a curve of required antenna size needed from geosynchronous orbit at frequency 10 GHz to provide a given spot size on Earth.

11.5   (11.4)   Given the uplink–downlink satellite system in Figure 11.15. The carrier is transmitted to the satellite, amplified, and retransmitted. Determine the received CNR at the ground receiver, using the following parameters: Uplink EIRP = 90 dB, Uplink loss = -200 dB, Satellite gain, $Gg_t = 100$ dB, Satellite NF, $F = 5$ dB, Uplink sky temperature = 300°, Satellite $g/T° = -40$ dB, Satellite bandwidth = 1 MHz, Downlink loss = -190 dB, Receiver $g/T° = 50$ dB, Receiver bandwidth = 1 MHz.

11.6   (11.4)   A crosslink is operated with an FM carrier and has a total carrier bandwidth of 2 GHz. The downlink requires a CNR of 40 dB. (a) What receiver CNR must the crosslink provide in an RF–RF conversion system? (b) An IF remodulation crosslink uses a

450 MHz carrier with a bandwidth of 100 MHz. What receiver CNR must this crosslink have? (c) Compare the results in parts a and b in terms of antenna sizes.

11.7   (11.5)   Given two adjacent BPSK carriers, each with power P, data rate R bps, and carrier frequency separation Δf Hz. Derive an expression for the crosstalk interference of one carrier spectrum onto the other's main-hump bandwidth in terms of the $\text{Si}(x)$ function:

$$\text{Si}(x) = \int_0^x \left(\frac{\sin u}{u} \right)^2 du$$

11.8   (11.5)   An FDMA system transmits carriers with a 50 MHz bandwidth and uses a linear satellite with a 500 MHz RF bandwidth. Each uplink carrier has a $\text{CNR}_u = 20$ dB. Let the satellite power be 5 W, the net downlink loss be 140 dB, and the downlink receiver have $N_0 = -200$ dB watts/Hz. (a) Determine if this FDMA system is power- or bandwidth-limited. (b) Repeat if the carrier uplink CNR_u is reduced to 11 dB. For both parts, use $\text{CNR}_d = 10$ dB.

11.9   (11.5)   An FDMA-linear satellite system is designed to accommodate a total of K carriers with a satellite power of P_T. (a) Show that if only Q of the K carriers are active at one time (and the satellite knows it), the satellite power can be reduced without lowering performance (i.e., all active downlinks will still have the required CNR_d). (b) Assuming that each carrier has a 60% activity time, determine the reduction in average satellite power P_T if Q can be continuously monitored.

11.10   (11.5)   An FDMA repeater is to be channelized as in Figure 11.22. Consider the following simple five-carrier system with a linear satellite amplifier:

Carrier	P_{uc}(mW)	P_{sd}(W)
1	10	10
2	10	5
3	20	10
4	20	2
5	20	4

Here P_{uc} is the uplink power, and P_{sd} is the required satellite downlink power for each carrier. (a) What is the minimum value of P_T that can satisfy all downlinks? (b) What P_T is needed for a single transponder (no channelization)? (c) Repeat part b for a two- and three-channel transponder. (d) How many channels are needed to achieve the minimum P_T in part a?

11.11 (11.5) In nonlinear FDMA, optimal backoff points for maximizing CNR_d depend on the parameter Q, the number of carrier signals passing through the satellite at any one time. Devise a satellite block diagram (onboard equipment) that will optimally adapt an FDMA satellite for optimal operation, as the number of active carriers change in time.

11.12 (11.6) A BPSK TDMA system is to transmit 1000 digital voice channels, each with 4 bits per sample, at a 64 kbps rate. The system must accommodate 1000 data bits/slot, at a frame efficiency of 10%. (a) What satellite bandwidth is needed? (b) How long is a TDMA frame? (c) How many slots in a frame? (d) How many preamble bits can be used?

11.13 (11.6) A TDMA system requires timing to occur within 5% of a slot time. Plot a curve showing the available slot length versus satellite location uncertainty. [*Hint:* Recall Problem 10.2].

11.14 (11.6) A TDMA synchronization system uses 1 msec slots, and the satellite transmits 10-bit marker words of bit length 5×10^{-4} seconds, with $(E_b/N_0)_s = 10$ dB. Assume no range timing error. What timing clock frequency is needed to ensure an rms timing error of no more than 5% of a slot time?

11.15 (11.6) A TDMA slot acquisition system uses a frequency burst of 20 μsec to determine its range. The satellite has a 500 MHz bandwidth and can produce a carrier downlink $CNR_{RF} = 10$ dB. How much lower in power can the frequency burst be, relative to a TDMA carrier, to produce the same CNR in its filter bandwidth?

11.16 (11.7) Given the two periodic code sequences:

$$\text{Code 1:} \quad 1 \quad -1 \quad 1 \quad 1 \quad 1 \quad -1 \quad 1 \quad 1$$
$$\text{Code 2:} \quad -1 \quad 1 \quad 1 \quad -1 \quad -1 \quad -1 \quad -1 \quad 1 \quad \text{(one period)}$$

(a) Compute the *periodic* cross-correlation of the sequences at each chip shift time. (b) Without shifting, compute the *partial* cross-correlations of length three.

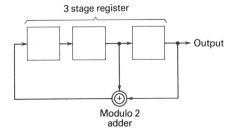

3 stage register

Output

Modulo 2
adder

FIGURE P11.17.

11.17 (11.8) Consider the shift register in Figure P11.17. Start with all zeros in the register, and shift in one bit from the left. Compute the output sequence from the register as the bits are clocked through.

11.18 (11.8) Show that the power spectrum of a PRN binary waveform of period n bits and pulse width w is given by

$$S(\omega) = \left(\frac{n+1}{n^2}\right)\left[\frac{\sin(\omega w/2)}{(\omega w/2)}\right]^2 \sum_{\substack{i=-\infty \\ i \neq 0}}^{\infty} \delta\left(\omega - \frac{2\pi i}{nw}\right) + \frac{2\pi}{n^2}\delta(\omega)$$

11.19 (11.8) In *pulse-addressed multiple accessing* (PAMA), a station pair uses a τ sec burst of the PSK carrier modulation placed in a specific location in a T sec data frame. The burst locations are selected randomly by each station pair, and any station transmits when desired, without knowledge of other station pairs. (a) Show that the mean square cross-correlation for any station pair is $(\tau/3T)$. (b) For equal amplitude carriers, show that the maximum number of users is $K \leq 3T/\tau$.

11.20 (11.8) In *slotted PAMA*, only specific τ sec slots can be used. Assume the T sec frame is divided into T/τ disjoints slots and each station pair selects its slot randomly. Show that the probability that r of k users will lie in the same slot is

$$P(r) = \binom{k}{r}\left(\frac{1}{\mu}\right)^r\left(\frac{\mu-1}{\mu}\right)^{k-r}$$

where $\mu = T/\tau$.

12

FIBER OPTIC COMMUNICATIONS

One of the more important guided communication links that has developed over the past several decades is the optical link. In one sense the fiber is simply another form of cable transmission system, similar to coaxial lines and metallic waveguides used at RF frequencies. On the other hand, the development of low loss glass fibers, and efficient light sources and detectors, has produced significant advantages that are causing fiber links to rapidly replace their metallic counterparts in high quality guided point-to-point communications. In this chapter we attempt to outline the basic communication characteristics of the fiber channel.

12.1   THE OPTICAL FIBER LINK

A fiber link operates on the same principle as a metallic cable link. Just as a microwave field inserted into a cable will appear at the output, a light field inserted into a fiber will shine out at the output. If the inserted light field is modulated, information can be transmitted by detecting the light modulation at the output of the fiber.

Figure 12.1 shows the basic fiber communication link. A data source produces information that modulates a light source that shines into the fiber. The light propagates in the fiber and appears at the output, where it is detected to regenerate the information signal. While microwave cables use electronic modulators and receivers at its terminals, a fiber system uses optical elements. Although the operational objectives are the same (modulation and detection), optical devices are constructed from quantum-mechanical principles rather than electronic. As a result, link models differ slightly and contain some new parameters that must be properly accounted for in communication analysis.

Fiber links are generally associated with digital sources, and fiber light sources typically operate in a pulsed format. This is due primarily to the simplicity of pulsed modulation (it is relatively easy to pulse a light source)

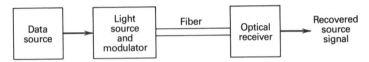

FIGURE 12.1. The optical fiber link.

and for advantages in power flow performance (as will be discussed later). Thus, as shown in Figure 12.1, the source bits are used to pulse the light source, which shines into the fiber. The most common format is simple on–off keying, in which a binary pulse waveform from the source is used to clock the light source on or off, as shown in Figure 12.2*a*. This corresponds to basic multiplication of the light field by a biased NRZ bit waveform and therefore is a form of amplitude, or intensity, modulation. Analog sources (voice, video, etc.) can be A-D converted into bit waveforms, which then blink the light source according to the data stream. Detection of the on–off light state in each bit interval at the receiver then decodes the bits.

Another popular pulsed encoding format is pulse position modulation (PPM) discussed earlier (Section 9.8). Here a block of k source bits is converted to an electronic pulse placed in one of a set of specific position locations. The pulse then triggers the light source, producing an optical

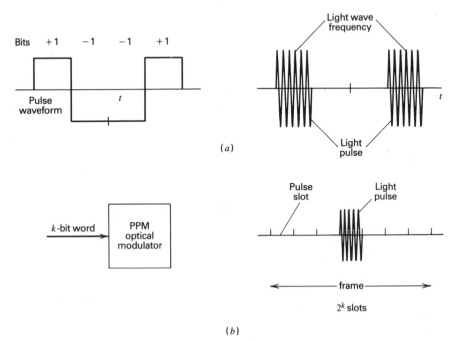

FIGURE 12.2. Pulse modulation of light waves: (*a*) on–off keying, (*b*) pulse position modulation.

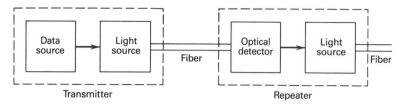

FIGURE 12.3. Fiber link with repeater.

pulse in one of a set of time slots. Hence PPM corresponds to a word encoding format in which k bits are encoded into an optical pulse placed in one of $M = 2^k$ possible slot positions (Figure 12.2b). Decoding is achieved at the fiber output by determining which of the possible slots in a word frame contain the light pulse. A pulsing format is still used, but each pulse now carries k bits of information. PPM becomes advantageous if narrow width, high peaked pulses can be used in the fiber [1].

As a light pulse shines through a fiber, it is attenuated and becomes weaker. If it becomes too dim, a repeater may be needed to boost its power level. This is accomplished by terminating the fiber, detecting the light pulse, and restrobing another high power light source into a subsequent fiber, as shown in Figure 12.3. The amplified pulse then flows in the second fiber to complete the line. Since the insertion of a repeater adds cost and complexity to the link, it is advantageous to have as long a fiber as possible before repeaters are necessary. This is why attenuation losses in any cable-type link are extremely important.

12.2 OPTICAL FIBERS

The optical fiber is the heart of the fiber link. A fiber is a thin "pipe" of glass through which one can shine light and transmit optical power to the output [2–6]. The basic construction of an optical fiber is shown in Figure 12.4. The core of the fiber is made of high quality glass over which the light field can propagate. Glass or a similar silicon compound material is used since it offers the minimal attenuation to the optical transmission. The glass core is supported by a shielding, or cladding, which provides mechanical strength, isolates the core from external interference and radiation, and aids

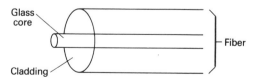

FIGURE 12.4. Fiber construction.

in confining the light propagation to only the internal glass paths. The cladding is opaque, usually constructed of a form of silicon plastic, and attempts to reflect any escaping light back into the core. Typically, the core diameter is on the order of micrometers, while the fiber cross section (core plus cladding) is on the order of several millimeters. Thus the fiber is literally a thread of transmission path over which the modulated light field can be propagated. The ability to pack large amounts of modulated data over an extremely small spatial area is an overwhelming advantage for optical fiber communications.

Attenuation in fibers is due to both absorption and scattering by the glass impurities, with the scattering effects decreasing at the higher wavelengths. Over the last decade the ability to improve the purification of silicon for fiber applications has led to significant reduction in attenuation losses. Figure 12.5a shows the attenuation loss in decibels per kilometers of glass fibers, as a function of the optical wavelength. Minimal attenuation occurs in the vicinity of 1 μm wavelength, where attenuation loss of less than 1 dB/km is possible. Lowest loss tends to favor operation at about 1.3 and 1.5 μm, producing the so-called optical windows for fiber links.

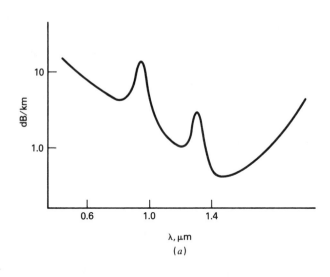

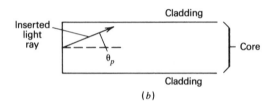

FIGURE 12.5. (a) Attenuation loss factors versus optical wavelength, (b) fiber ray lines.

To investigate further properties of the fiber as a guided light channel, it is necessary to review power flow of the light in the fiber [7]. Only light rays inserted into the fiber at narrow angles (θ_p in Figure 12.5b) will propagate down the fiber. If θ_p is too large, the light ray will be absorbed into the cladding and will not propagate. Propagation will occur down the fiber only if the incident ray angle at the cladding boundary is shallow enough to be reflected back into the core. The ray will then continue to propagate down the core as it continually reflects off the core walls. The maximum angle θ_p (critical angle) that will permit propagation is related to the relative refractive indices of the core and cladding and can therefore be accurately controlled. Critical angles in most fibers generally are about 10° or less.

The distribution of fiber power over propagating angles out to the critical angle depends on the core size. If the core diameter is extremely small (several times the optical wavelength), almost all the power flows at a small angle θ_p (i.e., down the center of the fiber). Such fibers are called *single-mode* fibers and have the attenuation loss characteristic given in Figure 12.5, due only to the glass impurities.

For wider diameter fiber cases (*multimode* fibers) the propagating power is distributed over all angles out to the critical angle. Wider core fibers are less expensive and easier to couple, splice, and interconnect. Their attenuation factors are slightly higher than for single-mode fibers, as a result of power lost at the cladding boundaries at the outermost angle propagation.

At a pulse communication channel, one is interested in not only the power loss, but the pulse distortion as well. The primary cause of pulse distortion is the fiber dispersion, which causes an inserted pulse width to be broadened during propagation as shown in Figure 12.6. That is, if a light pulse of width Δ sec is inserted, it will appear at the fiber output as a pulse of width ($\Delta + t_d$). The value of t_d is typically given as [7]

$$t_d = \begin{cases} \beta L , & L < 1 \text{ km} \\ \beta \sqrt{L} , & L > 1 \text{ km} \end{cases} \qquad (12.2.1)$$

where L is the fiber length in kilometers and β is the fiber *dispersion* coefficient, stated in nanoseconds per kilometer. The value of β depends on

FIGURE 12.6. Pulse spreading model: (*a*) fiber input pulse, (*b*) fiber output pulse after dispersion.

the fiber characteristics and is due to both the inherent dispersion of the glass itself (material dispersion) and the delay dispersion of the propagating ray angles (mode dispersion). The latter effect is caused by the fact that the ray propagation at the outer angles within the fiber take longer to reach the output than those at the center angles. A single-mode fiber has only zero-angle propagation and therefore has only material dispersion, which is only fractions of a nanosecond per kilometer. A multimode fiber is dominated by mode dispersion, which can be in the range from 1–50 nsec/km. Mode dispersion in a multimode fiber can often be reduced by radially varying the refractive index of the glass core (*graded index* fibers), but this tends to increase their cost.

The principal effect of pulse spreading is a limiting of the pulse rate. To avoid spreading of the light pulse into the next pulse interval, the pulse rate must be reduced to accommodate the expected spreading. As a rough rule of thumb, pulse intervals are set at approximately twice the pulse width, permitting a pulse to spread to twice its original width. This means that the pulse rate will be limited to

$$R_p = \frac{1}{\Delta + 2t_d} \text{ pulses per sec} \tag{12.2.2}$$

where t_d is given in (12.2.1). This implies that the allowable pulse rate into a fiber is limited by the fiber dispersion coefficient and fiber length. The longer the fiber, the greater will be the expected pulse spreading, and the lower the distortionless pulse rate.

12.3 LIGHT SOURCES

Light sources for fiber optic systems must produce focused light beams at a near-monochromatic (single frequency) wavelength conducive to the fiber propagation. The two important light sources used in fiber systems are the semiconductor light emitting diode (LED) and the injection laser [8–10]. The LED uses a junction between two semiconductors [an N- and P-type gallium aluminum arsenide (GaAlAs)] to emit light when stimulated by an external current (Figure 12.7). GaAlAs emits in the visible frequency range at wavelengths of about 0.8–0.9 μm, and, by modifying the material band structures, its emissions can be increased to the 1.3–1.5 μm range. This places its output light wavelengths at the desirable range for glass transmission, as was shown in Figure 12.5.

LEDs are inexpensive, easy to operate, rugged, and small (about the size of an aspirin tablet). Their efficiency is low (1%), their output power is typically limited to about 1–10 mW, and the emitted light field is not collimated but spread over a fairly wide angle (about 10–20°). In addition, the frequency spectrum of the LED light wave is not a pure single frequency, but contains "line broadening" that can be as large as several

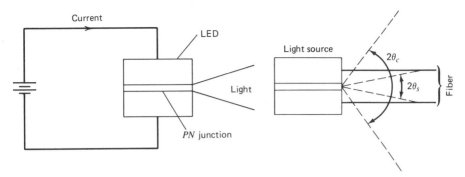

FIGURE 12.7. Light wave source: (a) light emitting diode or injection laser, (b) power coupling into fiber.

hundred megahertz around the desired optical wavelengths. This line broadening will appear as phase noise to any angle modulation that is attempted to be inserted on the LED frequency. For this reason, its primary use is with amplitude modulation, in which the power of the entire spectral content is available for carrier transmission. This is another reason why pulsed formats are commonly used with LED light sources.

Injection lasers are similar to LED junctions, but contain inserted mirrors at the junction ends. This causes emitted photons to become trapped, thereby stimulating further photon emissions. When enough photons are excited, the light penetrates one of the mirrors and appears at the output as focused laser light. Injection lasers are more efficient (10–15%), have higher output average powers (tens of milliwatts), and almost spectrally pure light emissions. In addition, the emitted light is almost completely focused (<1° beamspread).

An important communication property of both LED and injection lasers is that they can be rapidly switched on and off by control of the external current source. An LED can be pulsed at about 100 Mpps (megapulses per second), while injection lasers can be pulsed at rates approaching Gpps (gigapulses per second). This is again a reason why these light sources are especially conducive for operation in a pulsed format, such as on–off keying and PPM. In on–off keying (Figure 12.2a) the data bit waveform is used to drive the current source shown in Figure 12.7 into and out of the lasing region, and the light source is effectively switched on and off. This produces a light output data rate equal to the source bit rate. Since source rates are usually well below 100 Mbps, they can be easily accommodated by either an LED or laser light source. PPM also uses a pulsed source and achieves its principal advantage if the average power of the source can be converted to increased peak powers at lower pulse repetition rates.

The focusing angle of the source in Figure 12.7 is important for determining the amount of source power that will actually be coupled into the fiber. Figure 12.7b shows a light source with a wide angle, unfocused output light beam butted to a fiber having a critical angle $\pm\theta_c$. If the source has a uniform power distribution over its beam angle $\pm\theta_s$, the fraction of its output power coupled into the fiber for propagation is

$$L_t \cong \left(\frac{\theta_c}{\theta_s}\right)^2, \qquad \theta_s \geq \theta_c \qquad (12.3.1)$$

Since θ_c is typically less than θ_s, (12.3.1) represents a coupling loss from source to fiber, just as an impedance mismatch loss occurs for cables. An injection laser, with its more focused output beam, clearly produces improved power levels. For a single-mode fiber, where $\theta_c \approx 0$, it is almost mandatory to use an injection laser, where the focusing angle can be reduced to this same magnitude.

12.4 OPTICAL PHOTODETECTORS

The modulated light fields propagating in the fiber must be detected. This is accomplished by a photodetector placed at the fiber output. In RF systems an antenna or amplifier electronically detects an impinging field by converting the radiation into a current that varies wth the field itself. At optical wavelengths, the carrier is at such a high frequency it cannot be directly detected by any electronic elements. Instead, a photodetector responds quantum mechanically to the impinging field by using the light radiation to excite a photoemissive surface and produce a current flow [9–11]. The latter depends on the instantaneous power received over the detector surface. Hence a photodetector detects the time varying power variations of the optical field focused from the fiber. When the light field in the fiber is pulsed (power turned on and off), the photodetector responds by producing equivalent current pulses at its output.

Photodetectors used for fiber links are usually semiconductor photodiodes that respond to the optical wavelengths used in the fibers. The important photodiodes are the PIN diode and the avalanche photodetector (APD) [12]. The PIN diode converts input optical radiation directly to output current. The APD has an internal photomultiplication mechanism that produces an inherent power gain during detection.

The important parameters of any photodetector are its *efficiency, gain, reponsivity,* and *bandwidth*. The detector efficiency indicates the fraction of the incident light power on the photoemissive surface that is actually detected. Efficiency is wavelength dependent, (Figure 12.8a) and depends on the material used on the surface (usually silicon or germanium). Efficiency values are generally in the range (0.15–0.90) for fiber wavelengths. Detector efficiency η allows us to convert impinging field power P_r to effective detected power ηP_r.

As stated, the APD has an inherent photomultiplication mechanism that

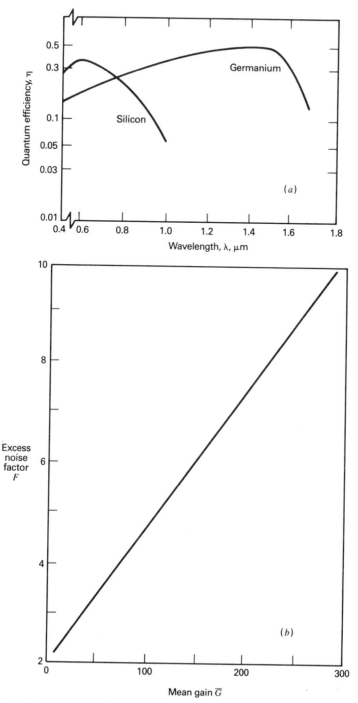

FIGURE 12.8. Parameter variations: (*a*) quantum efficiency versus optical wavelength, (*b*) excess noise factor of APD versus photomultiplication mean gain.

produces an internal power gain during detection. Unfortunately, the multiplication is often random in nature, producing a random gain distributed about same mean gain $\bar{G}$. This causes a statistical dependence of the output current on this gain and introduces an additional degree of randomness in the output current. The extent of this randomness is indicated by the APD *excess noise factor F*. Noise factor is the ratio of the gain variance to the square of the mean gain. These factors depend on the photoemissive material, and generally increase with mean gain $\bar{G}$. Figure 12.8*b* plots excess noise factor F as a function of the mean gain for a germanium based APD. Typically, an APD has mean gain values $\bar{G}$ of about 50–300 and F values in the range of 2–5. The PIN photodiode has no internal gain ($\bar{G} = 1$, $F = 1$).

Responsivity of a photodetector is the conversion coefficient from input incident light power to output current and simply indicates how much current will be produced for a given input power. Responsivity has units of amperes per watt and includes both the efficiency factor η and mean gain $\bar{G}$. Formally, the responsivity u of a photodetector is given by

$$u = \frac{e\eta\bar{G}}{hf_0} \quad \text{A/W} \tag{12.4.1}$$

where e is the charge of an electron ($e = 1.97 \times 10^{-19}$ C), h is Planck's constant ($h = 6.6 \times 10^{-34}$ W/Hz), and f_0 is the optical light frequency. At an efficiency of 70%, and operating at $\lambda = 1\ \mu$m, this is approximately $u \cong 0.7\bar{G}$.

Photodetector bandwidth determines the rate of input power variation that can be photodetected. For pulsed inputs, it determines the amount of filtering that will be applied to output current pulses during detection. Photodiode bandwidths are usually 1–10 GHz and therefore are much wider than the pulse rates typically occurring in fiber links. An anomaly in optical detection is that a photodetector does not respond perfectly to impinging field power. This can be demonstrated as in Figure 12.9. If a fixed power

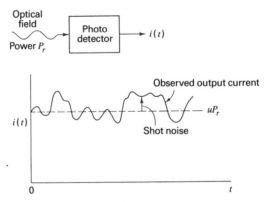

FIGURE 12.9. Photodetector output time function with constant input power.

field is focused onto a photodetector, the output current should be constant at a value proportional to the input power. In reality, however, the output current observed will appear as a random variation around that proportional value, as shown. This random variation around the true value, called detector *shot noise*, is caused by the statistical nature of the emission of photoelectrons during photodetection. As a noise process (i.e., the process obtained by subtracting the mean output value from the total output), shot noise has a flat spectral density with a level that depends on the input power value. Hence shot noise is stronger during the reception of high power levels and weaker during weak power reception. This means that shot noise is in reality a nonstationary process when the input field power is time varying. This fact obviously creates a degree of complication in any type of rigorous receiver study. The actual shot noise spectral level is known to be [1]

$$S_{\mathrm{sn}}(\omega) = \bar{G}^2 FeuP \quad \mathrm{A}^2/\mathrm{Hz} \tag{12.4.2}$$

where $\bar{G}$, F, and u are respectively the detector gain, noise factor, and responsivity parameters previously discussed; e is the electron charge; and P is the received power.

In addition to shot noise, a photodetector produces *dark current*. Dark current is output current that appears even with no imput radiation and is caused by the random thermal emission of photoelectrons due to the inherent temperature of the device. Dark current is a random process and must be considered an additional noise in the receiving operation. It also is described as a shot noise process, with a spectral level proportional to the average output dark current. Thus its spectral level is [13]:

$$S_{\mathrm{dc}}(\omega) = eI_{\mathrm{dc}} \quad \mathrm{A}^2/\mathrm{Hz} \tag{12.4.3}$$

where I_{dc} is the detector mean dark current. Typical detector dark current mean values are in the range 10^{-16}–10^{-12} A and can be further reduced by detector cooling.

The fiber photodetector can be summarized by the model shown in Figure 12.10. The input pulse power $P_r(t)$ is photodetected to the output current waveform as shown. To this signal current is added the signal

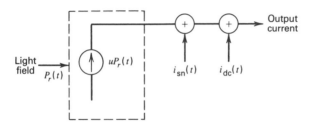

FIGURE 12.10. Photodetection model.

dependent shot noise current $i_{sn}(t)$, with the spectral level in (12.4.2), and the dark current noise $i_{dc}(t)$, with the spectral level in (12.4.3). The relative effects of these signal and noise currents at the photodetector output will depend on the circuitry connected to the detector output. This we consider in the next section.

12.5 FIBER LINK ANALYSIS

As in any communication system, fiber performance can be determined by a basic end-to-end link analysis. Consider the fiber system in Figure 12.11a. A light source produces pulsed light, according to an on–off keyed bit modulation, which is fed into the fiber with an average power P_a. The fiber has length L, attenuation α (decibels per length), and dispersion coefficient β (nanoseconds per length). The fiber output is detected with a photodiode having responsivity u, gain $\bar{G}$, excess noise factor F, and dark current I_{dc}. The photodiode current output is amplified and processed for bit decoding. We assume that on–off decoding is achieved by integrating over each T_b pulse time, and by comparing the integrated signal to a preselected threshold. A one-bit is decoded (pulse is on) if above threshold, and a zero-bit is decoded (pulse is off) if below threshold.

The source produces in the fiber a light pulse (or its absence) of peak power P_t, where $P_t = 2P_a$ for equal likely bits. The pulse has width T_b sec, where $R_b = 1/T_b$ is the transmitted bit rate. The light pulse appears at the fiber output with peak power

$$P_p = 2P_a[10^{-\alpha L/10}] \qquad (12.5.1)$$

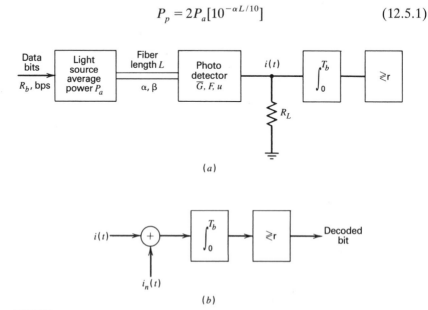

FIGURE 12.11. Fiber link model: (a) block diagram, (b) equivalent decoding model.

The pulse rate is limited by the fiber dispersion to be no greater than

$$R_b = \frac{1}{T_b} \le \frac{1}{\beta \sqrt{L}} \quad \text{bps} \tag{12.5.2}$$

The photodetected pulse current has added to it the noise processes of the detector output, including shot noise and dark current noise. In addition, the receiver electronics adds output thermal noise current having spectral level [14]

$$N_{oc} = \frac{4 \kappa T_{eq}^{\circ}}{R_L} \quad \text{A}^2/\text{Hz} \tag{12.5.3}$$

where κ is Boltzmann's constant, T_{eq}° is the receiver noise temperature (including the amplifier noise figure), and R_L is the input impedance of the amplifier. The noise currents combine to produce a total output noise $i_n(t)$ having a flat spectral level of

$$N_0 = \left[(\bar{G})^2 FeuP_r + eI_{dc} + \frac{4 \kappa T_{eq}^{\circ}}{R_L} \right] \quad \text{A}^2/\text{Hz} \tag{12.5.4}$$

Combination of Figures 12.10 and 12.11a produces the decoding model in Figure 12.11b. The integrator output after each bit time is therefore

$$y = \left[\int_0^{T_b} [uP_r(t) + i_n(t)] \, dt \right]$$

$$= u \int_0^{T_b} P_r(t) \, dt + \int_0^{T_b} i_n(t) \, dt \tag{12.5.5}$$

The mean value of y, given by the first term, is

$$\text{Mean } y = \begin{cases} uP_p T_b & \text{if pulse is present} \\ 0 & \text{if pulse is absent} \end{cases} \tag{12.5.6}$$

The variance of y is determined by the second term in (12.5.5) corresponding to the integration of white noise and is $N_0 T_b$. The value of N_0 is given in (12.5.4), but its value depends on whether the pulse is present, that is, whether there is detector shot noise. Thus the noise spectral level N_0 is either

$$N_{01} = \left[(\bar{G})^2 FeuP_p + eI_{dc} + \frac{4 \kappa T^{\circ}}{R_L} \right] \quad \text{if pulse present}$$

or

$$N_{00} = \left[eI_{dc} + \frac{4 \kappa T^{\circ}}{R_L} \right] \quad \text{if pulse absent} \tag{12.5.7}$$

If the decoding threshold is set at one-half the mean in (12.5.6), and if we assume Gaussian integrator noise [15], the resulting bit error probability follows as

$$PE = \tfrac{1}{2}Q(\sqrt{\tfrac{1}{4}SNR_1}) + \tfrac{1}{2}Q(\sqrt{\tfrac{1}{4}SNR_0}) \tag{12.5.8}$$

where

$$SNR_1 = \left[\frac{(uP_pT_b)^2}{N_{01}T_b}\right]$$

$$= \frac{(uP_pT_b)^2}{[\bar{G}^2FeuP_p + eI_{dc} + (4\kappa T^{\circ}_{eq}/R_L)]T_b} \tag{12.5.9a}$$

and

$$SNR_0 = \frac{(uP_pT_b)^2}{[N_{00}T_b]}$$

$$= \frac{(uP_pT_b)^2}{[eI_{dc} + (4\kappa T^{\circ}_{eq}/R_L)]T_b} \tag{12.5.9b}$$

The above SNRs are often called the *electronic*, or *postdetection*, SNRs. These SNR values can be rewritten by dividing through by $(e\bar{G})^2$ and combining as

$$SNR_1 = \frac{K_s^2}{FK_s + K_n}$$
$$SNR_0 = \frac{K_s^2}{K_n} \tag{12.5.10}$$

where we have defined

$$K_s = \frac{\eta P_pT_b}{hf} \tag{12.5.11a}$$

$$K_n = \frac{(eI_{dc} + 4\kappa T^{\circ}/R_L)T_b}{(e\bar{G})^2} \tag{12.5.11b}$$

The parameters K_s and K_n play the role of signal and noise energy variables and are often called the signal and noise bit *counts*. The latter arises from the fact that they can be related to the number, or count, of optical photons collected in a bit time at the photodetector output. We see from (12.5.10) that SNR depends on the square of the signal count K_s and on both K_s and K_n (not simply on their ratio). This means that front end optical losses (such as fiber coupling and detector quantum efficiency) must be properly accounted for in computing K_s and K_n. (In RF analysis, all front end gains and losses canceled out, since only their ratio is needed in SNR evaluation.) Note further that both the dark current and thermal noise are reduced by the square of the photodetector mean gain. Hence high gain photomultipliers aid in reducing the effect of postdetection noise, although the detector noise factor F also tends to increase with higher gain values.

Figure 12.12 plots PE in (11.5.8) as a function of K_s for $F = 5$ and different noise values K_n, for the on–off keyed system. With the fiber

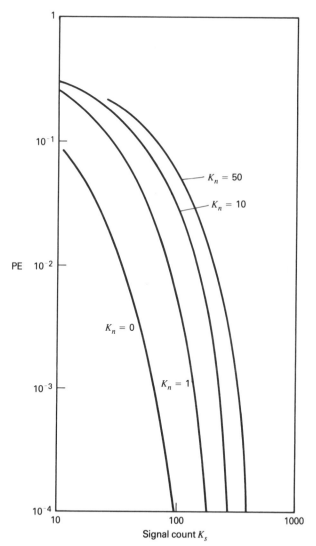

FIGURE 12.12. Bit error probability versus K_s, K_n for on–off keying ($F = 5$).

TABLE 12.1 An Example of Fiber PE Calculation

Fiber	
Transmitter parameters	
$R_b = 100$ Mbps	
Receiver parameters	
$I_{dc} = 10^{-12}$ A	
$T^\circ_{eq} = 300^\circ$	
$R_L = 10^6$ ohms	
$F = 5$	
$T_b = 10^{-8}$ sec	
$G = 100$	
Compute:	
eI_{dc},	1.97×10^{-31}
$4\kappa T^\circ_{eq}/R_L$,	1.64×10^{-26}
$e\bar{G}$,	1.97×10^{-17}
Noise count rate, cps,	$[(0.5 \times 10^3) + 0.87 \times 10^8]$
$K_n = $ noise count in T_b sec,	0.87
From Figure 12.12 at PE $= 10^{-4}$,	$K_s \cong 200$

receiver parameters specified, Figure 12.12 can be used to determine the required K_s that must be provided by the fiber link to achieve a desired PE. Table 12.1 shows an exemplary computation of how this is carried out. As a rough rule of thumb, if Y is the value of E_b/N_0 needed in a Gaussian link to achieve the desired PE, then $K_s \cong FY$ if $K_n \ll Y$, and $K_s \approx \sqrt{K_n Y}$ if $K_n \gg Y$.

In the PPM format of Figure 12.2b, word decoding is based on determining which of the M time slots in a frame containing the pulse. This decision is made by integrating the photodetector output over each slot time and selecting the largest integrator voltage as the signaling slot. Using the model of Figure 12.11, the correct slot integrates to a value

$$y_0 = uP_p \tau + n_0 \qquad (12.5.12)$$

where τ is the slot time and n_0 is noise variate with zero mean and variance $N_{01}\tau$, with N_{01} given in (12.5.7a). All incorrect slots integrate to independent variables y_i, $i = 1, 2, \ldots, M-1$, depending on only the noise, with each having the same variance, $N_{00}\tau$. The probability of correct word decoding in PPM is the probability that the correct variable y_0 exceeds all other y_i. For the Gaussian noise model with M slots per frame, the word error probability is given by an M-ary orthogonal PWE [1, 15, 16], similar to that in (9.5.3)

$$\text{PWE} = 1 - \int_{-\infty}^{\infty} \left[\int_{-\infty}^{x} G(y, 0, K_n)\, dy \right]^{M-1} G(x, K_s, FK_s + K_n)\, dx \qquad (12.5.13)$$

where we have now defined

$$K_s = \frac{\eta P_p \tau}{hf} \qquad (12.5.14a)$$

$$K_n = \frac{(eI_{dc} + 4\kappa T^{\circ}_{eq}/R_L)\tau}{(e\bar{G})^2} \qquad (12.5.14b)$$

Here K_s and K_n are the signal and noise counts in a slot time τ [as opposed to the bit time counts in (12.5.11)]. Figure 12.13 plots PWE in (12.5.13) as a function of K_s, for different K_n and M values, with $F = 5$. The PPM system has the advantage that no threshold is required and decoding is based solely on maximum slot integration voltages. Note that binary $(M = 2)$ PPM corresponds to bit signaling with a pulse in either the first or second half of the bit interval, similar to a Manchester encoding format.

With K_s determined, the required pulse power P_p needed at the photo-detector input to achieve K_s can be obtained from (12.5.11a). This latter equation shows that the parameter K_s is similar to an *effective* E_b/N_0 parameter, with a noise spectral level given by hf/η. The latter is often

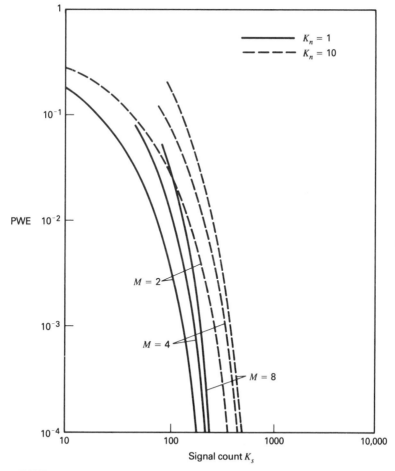

FIGURE 12.13. Word error probability versus K_s, K_n for PPM system $(F = 5)$.

referred to as the *quantum noise level*. The value of pulse power needed is therefore that required to achieve a given $E_b/N_0 = K_s$, with respect to the quantum noise level. Solving (12.5.11a), we obtain

$$P_p = \frac{K_s(hf)}{\eta T_b}$$

$$= \left(\frac{hf}{\eta}\right)(K_s R_b) \tag{12.5.15}$$

The fiber input average power in (12.5.1) must then satisfy

$$2P_a 10^{-\alpha L/10} = \left(\frac{hf}{\eta}\right)(K_s R_b) \tag{12.5.16}$$

When expressed in decibels, this becomes

$$(P_a)_{dB} - \alpha L = \left(\frac{hf}{\eta}\right)_{dB} + \left(\frac{K_s}{2}\right)_{dB} + (R_b)_{dB} \tag{12.5.17}$$

When expressed in this form, we see that the fiber power loss equation is identical to that of a cable (electronic) system attempting to achieve a specified E_b/N_0 with a given noise level N_0 (see Section 3.10). In an optical system, at $\lambda = 1\ \mu$m, the quantum spectral level is

$$hf = (6.6 \times 10^{-34})(3 \times 10^{14}) = 19.8 \times 10^{-20} = -177\ \text{dB/Hz} \tag{12.5.18}$$

This can be compared, for example, with an electronic receiver at $T_{eq}^\circ = 300^\circ$, having a noise level of $N_0 = kT_{eq}^\circ = (-228.6 + 25)\ \text{dB/Hz} = -203.6\ \text{dB/Hz}$. Hence the optical system involves an effective noise level that is about 16 dB higher than a 300° electronic system.

Equation (12.5.15) can also be interpreted as relating achievable fiber data rate R_b to the fiber length for a specified source power and attenuation loss. When solved for data rate this shows

$$R_b = \left[\frac{\eta^2 P_a}{hf K_s}\right][10^{-\alpha L/10}] \tag{12.5.19}$$

This should be contrasted with the data rate limitation caused by the fiber dispersion, as given in (12.5.2). Hence a fiber may be either attenuation (power) limited or dispersion (bandwidth) limited. Figure 12.14 plots the two equations, (12.5.2) and (12.5.19), as a function of fiber length for several different dispersion and attenuation values, assuming a desired $K_s = 100$, $\lambda = 1\ \mu$m, $\eta = 0.70$, and $P_a = 0.1$ mW. From a figure of this type, one can estimate the fiber length that can be used (without repeaters) for achieving a specified data rate. It can be seen that data rates of 100 kbps can

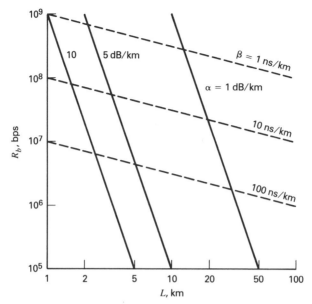

FIGURE 12.14. Achievable fiber bit rate versus fiber length (β = dispersion coefficient, α = attenuation loss factor).

be obtained over links of tens of kilometers, even with relatively low power sources (<1 mW) if high quality fibers are used. It is this fact that makes fiber communications a predominant favorite for point-to-point guided links.

REFERENCES

1. Gagliardi, R.M. and Karp, S. *Optical Communications*, Wiley, New York, 1976.
2. Howes, M.J. and Morgan, D.V. *Optical Fiber Communications*, Wiley, New York, 1980.
3. Sandbank, C.P. *Optical Fibre Communication Systems*, Wiley, New York, 1980.
4. Seippel, R. *Fiber Optics*, Reston Publishing, Virginia, 1984.
5. Jeunhomme, L. *Single Mode Fiber Optics*, Dekker, New York, 1983.
6. Marcuse, D. *Theory of Dielectric Optical Waveguides*, Academic, New York, 1974.
7. Gloge, D. "Impulse Response of Clad Optical Fibers," *Bell Syst. Tech. J.*, vol. 52, July 1973.
8. Pratt, W. *Laser Communication Systems*, Wiley, New York, 1969.
9. Ross, M. *Laser Receivers*, Wiley, New York, 1966.
10. Yariv, A. *Quantum Electronics*, Wiley, New York, 1975.
11. Spicer, W. and Wooten, F. "Photoemission and Photomultipliers," *Proc. IEEE*, August 1963.

12. McIntyre, R. "The Distribution of Gains in APD," *IEEE Trans. Electron. Devices*, vol. ED-19, June 1972.

13. Kruse, P. et al., *Elements of Infrared Technology*, Wiley, New York, 1982.

14. Van der Ziel, A. *Noise*, Prentice-Hall, Englewood Cliffs, N.J., 1954.

15. Sorensen, N. and Gagliardi, R. "Performance of Optical Receivers with APD," *Trans. IEEE Comm.*, vol. COM-27, September 1979.

16. Garrett, I. "PPM for Transmission Over Fibers," *IEEE Trans. Comm.*, vol. COM-33, April 1983.

PROBLEMS

12.1 (12.1) Engineers deal with frequency separation measured in hertz. Physicists use equivalent separation measured in wavelength λ. Determine the relation between a frequency difference df about a center frequency f_c and its equivalent wavelength difference $d\lambda$ about the corresponding λ_c.

12.2 (12.2) In a properly designed optical lensing system with focal length f_c, the field imaged on the detector $f_d(x, y)$ is the two-dimensional Fourier transform of the field passing through the lens aperture, $f(x, y)$. That is,

$$f_d(x, y) = \int_{\text{plane}} f(u, v) \exp\left[-j \frac{2\pi}{\lambda f_c} (xu + yv)\right] du\, dv$$

Assume an arriving plane wave (field constant over an infinite plane) at the fiber output and a square lens aperture of width d. Compute the imaged field on the photodetector.

12.3 (12.3) In certain optical detection models, the count detected over a time interval has a Poisson distribution, in which the probability that the count is n (an integer) is given by

$$P(n) = \frac{K^n}{n!} e^{-K}$$

where K is the mean interval count. Use this to write the Poisson error probability of an M-slot PPM system, assuming that the pulsed slot has mean count $K_s + K_b$ and all the other slots have mean count K_b. Pulse decoding is based on selecting the signal slot as that with the largest count.

12.4 (12.5) The *noise equivalent power* (NEP) of an optical receiver is defined as the value of received power P_r such that $\sqrt{\text{SNR}_s} = 1$ in a 1 Hz noise bandwidth. Use (12.5.4) and (12.5.9) to derive an equation for the NEP of an optical receiver with shot noise, dark current, and thermal noise.

12.5 (12.5) The count rate K_s defined in (12.5.11) actually refers to the number of *photons* collected from a quantum-mechanical field. (For this reason, an optical receiver is often called a *photon bucket*.) Prove that K_s is the number of field photons occurring in a time period T_b. (*Hint:* The energy of a photon of frequency f_0 is hf_0.)

12.6 (12.5) It is desired to send 10^5 bps with OOK encoding over a 5 km fiber with negligible dispersion and a loss of 1 dB/km, and achieve a $PE = 10^{-4}$. How many joules (watts-second) are required at 1 μm from a light source, assuming a photodetector with quantum efficiency of 0.5 and a noise level of 10^6 counts/sec?

12.7 (12.5) A 4-bit PPM direct detection optical communication system is desired with a word-error probability of 10^{-4} and has a noise count energy of $K_b = 3$ in each slot. Approximately how many joules per pulse must be received to operate the system. (1 J = 1 W-sec.)

12.8 (12.5) An optical PPM system uses a source of average power P and pulse repetition rate of r pulses per second. Derive the equation for the peak pulse power in a system sending R_b bits per second.

FOURIER TRANSFORMS
AND IDENTITIES

A.1 THE FOURIER TRANSFORM

$$X(\omega) = \int_{-\infty}^{\infty} x(t) e^{-j\omega t} \, dt$$

The Inverse Fourier Transform

$$x(t) = \frac{1}{2\pi} \int_{-\infty}^{\infty} X(\omega) e^{j\omega t} \, d\omega$$

Time Convolution

$$x(t) \oplus y(t) = \int_{-\infty}^{\infty} x(t - \rho) y(\rho) \, d\rho$$

Frequency Convolution

$$X(\omega) \oplus Y(\omega) = \frac{1}{2\pi} \int_{-\infty}^{\infty} X(\omega - \rho) Y(\rho) \, d\rho$$

Transform Properties

Linearity $\qquad\qquad \alpha x(t) + \beta y(t) \leftrightarrow \alpha X(\omega) + \beta Y(\omega)$

Delay $\qquad\qquad\quad x(t - \alpha) \leftrightarrow e^{-j\omega\alpha} X(\omega)$

Scale $\qquad\qquad\quad x(\alpha t) \leftrightarrow \dfrac{1}{|\alpha|} X\!\left(\dfrac{\omega}{\alpha}\right)$

Conjugate $\qquad\quad\; x^*(t) \leftrightarrow X^*(-\omega)$

Duality	$X(t) \leftrightarrow x(\omega)$
Shift	$e^{j\alpha t}x(t) \leftrightarrow X(\omega - \alpha)$
Differentiation	$\dfrac{d^n x(t)}{dt^n} \leftrightarrow (j\omega)^n X(\omega)$
Integration	$\displaystyle\int_{-\infty}^{t} x(\rho)\, d\rho \leftrightarrow \dfrac{X(\omega)}{j\omega}$
Multiplication	$x(t)y(t) \leftrightarrow X(\omega) \oplus Y(\omega)$
Convolution	$x(t) \oplus y(t) \leftrightarrow X(\omega)Y(\omega)$

Useful Transforms

$x(t)$	$X(\omega)$
$e^{-bt}, \quad t>0$	$\dfrac{1}{b + j\omega}$
$\cos(\omega_c t + \psi)$	$\pi[\delta(\omega - \omega_c)e^{j\psi} + \delta(\omega + \omega_c)e^{-j\psi}]$
$\sin(\omega_c t + \psi)$	$\dfrac{\pi}{j}[\delta(\omega - \omega_c)e^{j\psi} - \delta(\omega + \omega_c)e^{-j\psi}]$
1	$2\pi\delta(\omega)$
$\delta(t)$	1
$1, \quad t>0$	$\dfrac{1}{j\omega}$
$\dfrac{1}{\sqrt{2\pi}} e^{-t^2/2}$	$e^{-\omega^2/2}$
$t, \quad t>0$	$\left(\dfrac{1}{j\omega}\right)^2$
$t^2, \quad t>0$	$\dfrac{2}{(j\omega)^3}$

A.2 THEOREMS, IDENTITIES, AND EXPANSIONS

Parceval's Theorem

$$\int_{-\infty}^{\infty} x(t)y(t)\, dt = \frac{1}{2\pi} \int_{-\infty}^{\infty} X(\omega)Y^*(\omega)\, d\omega$$

$$\int_{-\infty}^{\infty} |x(t)|^2\, dt = \frac{1}{2\pi} \int_{-\infty}^{\infty} |X(\omega)|^2\, d\omega$$

Schwartz Inequality

$$\left| \text{Real} \int_{-\infty}^{\infty} x(t) y^*(t)\, dt \right| \leq \left[\int_{-\infty}^{\infty} |x^2(t)|\, dt \int_{-\infty}^{\infty} |y^2(t)|\, dt \right]^{1/2}$$

equality holds if $x(t) = Cy(t)$.

Holder Inequality

$$\left| \sum_{i=0}^{\infty} a_i b_i \right| \leq \left[\sum_{i=0}^{\infty} |a_i^2| \sum_{i=0}^{\infty} |b_i^2| \right]^{1/2}$$

equality if $a_i = Cb_i$.

Final Value

$$\text{Lim}_{t \to \infty} x(t) = \text{Lim}_{\omega \to 0} [\, j\omega X(\omega)]$$

Bessel Identities

$$J_n(x) = \left(\frac{x}{2} \right)^n \sum_{k=0}^{\infty} \frac{(-x^2/4)^k}{k!(n+k)!}$$

$$= \frac{j^{-n}}{\pi} \int_0^{\pi} e^{jx \cos \theta} \cos (n\theta)\, d\theta$$

$$J_n(xe^{jm\pi}) = e^{jnm\pi} J_n(x)$$

$$I_n(x) = j^n J_n\left(\frac{x}{j} \right)$$

$$= \frac{1}{\pi} \int_0^{\pi} e^{x \cos \theta} \cos (n\theta)\, d\theta$$

Useful Trigonometric Identities

$$\sin (\alpha \pm \beta) = \sin \alpha \cos \beta \pm \cos \alpha \sin \beta$$

$$\cos (\alpha \pm \beta) = \cos \alpha \cos \beta \mp \sin \alpha \sin \beta$$

$$\tan (\alpha \pm \beta) = \frac{\tan \alpha \pm \tan \beta}{1 \mp \tan \alpha \tan \beta}$$

$$\sin \alpha \sin \beta = \tfrac{1}{2} \cos (\alpha - \beta) - \tfrac{1}{2} \cos (\alpha + \beta)$$

$$\cos \alpha \cos \beta = \tfrac{1}{2} \cos (\alpha - \beta) + \tfrac{1}{2} \cos (\alpha + \beta)$$

$$\sin \alpha \cos \beta = \tfrac{1}{2} \sin (\alpha - \beta) + \tfrac{1}{2} \sin (\alpha + \beta)$$

$$e^{\pm j\theta} = \cos \theta \pm j \sin \theta$$

$$\cos \theta = \tfrac{1}{2}(e^{j\theta} + e^{-j\theta})$$

$$\sin \theta = \frac{e^{j\theta} - e^{-j\theta}}{2j}$$

$$\sin^2 \theta + \cos^2 \theta = 1$$

$$\cos^2 \theta - \sin^2 \theta = \cos 2\theta$$

$$\cos^2 \theta = \tfrac{1}{2}(1 + \cos 2\theta)$$

$$\cos^3 \theta = \tfrac{1}{4}(3 \cos \theta + \cos 3\theta)$$

$$\sin^2 \theta = \tfrac{1}{2}(1 - \cos 2\theta)$$

$$\sin^3 \theta = \tfrac{1}{4}(3 \sin \theta - \sin 3\theta)$$

Expansions

$$(1 + x)^n = 1 + nx + \frac{n(n-1)}{2!} x^2 + \cdots \quad |nx| < 1$$

$$e^x = 1 + x + \frac{1}{2!} x^2 + \cdots$$

$$a^x = 1 + x \ln a + \frac{1}{2!} (x \ln a)^2 + \cdots$$

$$\ln (1 + x) = x - \tfrac{1}{2}x^2 + \tfrac{1}{3}x^3 + \cdots$$

$$\sin x = x - \frac{1}{3!} x^3 + \frac{1}{5!} x^5 - \cdots$$

$$\cos x = 1 - \frac{1}{2!} x^2 + \frac{1}{4!} x^4 - \cdots$$

$$\tan x = x + \tfrac{1}{3}x^3 + \tfrac{2}{15}x^5 + \cdots$$

$$\sum_{m=0}^{M} x^m = \frac{(x^M - 1)}{(x - 1)}$$

$$e^{a \cos b} = \sum_{i=0}^{\infty} \epsilon_i I_i(a) \cos (ib), \qquad \epsilon_0 = 1, \quad \epsilon_i = 2, \quad i \geq 1$$

$$\cos (x \sin \theta) = J_0(x) + 2 \sum_{k=1}^{\infty} J_{2k}(x) \cos (2k\theta)$$

$$\sin (x \sin \theta) = 2 \sum_{k=0}^{\infty} J_{2k+1}(x) \sin [(2k + 1)\theta]$$

$$\cos (x \cos \theta) = J_0(x) + 2 \sum_{k=0}^{\infty} (-1)^k J_{2k}(x) \cos (2k\theta)$$

$$\sin (x \cos \theta) = 2 \sum_{k=0}^{\infty} (-1)^k J_{2k+1}(x) \cos [(2k + 1)\theta]$$

Sampling Identities

(a) $$\sum_{k=-\infty}^{\infty} \delta(\omega - kp) = \frac{1}{p} \sum_{n=-\infty}^{\infty} e^{jnp\omega}$$

(b) $$\mathscr{F} \left[\sum_{n=-\infty}^{\infty} \delta(t - nT) \right] = \frac{2\pi}{T} \sum_{k=-\infty}^{\infty} \delta \left(\omega - \frac{2\pi k}{T} \right)$$

(c) $$\mathscr{F} \left[\sum_{n=-\infty}^{\infty} x(t - nT) \right] = \frac{2\pi}{T} \sum_{i=-\infty}^{\infty} X \left(\frac{2\pi i}{T} \right) \delta \left(\omega - \frac{2\pi i}{T} \right)$$

Integrals

$$\int_0^{\infty} \frac{dx}{1 + x^n} = \frac{(\pi/n)}{\sin (\pi/n)}, \quad n > 1 \qquad \int_0^{\infty} \frac{dx}{(a^2 + x^2)^2} = \frac{\pi}{4a^3}$$

$$\int_0^{\infty} \frac{x^u \, dx}{1 + x^n} = \left(\frac{\pi}{n} \right) \csc \left[\frac{(u + 1)\pi}{n} \right] \qquad \int_0^b \frac{dx}{1 + x^2} = \tan^{-1} (b)$$

A.3   CALCULUS OF VARIATIONS

We wish to minimize or maximize the integral

$$\Gamma[H(\omega)] = \int_{-\infty}^{\infty} f[H(\omega)] \, d\omega \qquad (A.3.1)$$

with respect to the function $H(\omega)$, where $f(x)$ is a differentiable functional. To accomplish this, we use the *calculus of variations* [1, 2]. We let $H_0(\omega)$ be the desired solution to be determined. Assuming that this solution exists, then any perturbation from this solution must produce a larger value of Γ (smaller, if $H_0(\omega)$ is a maximizing solution). Thus for any arbitrary function $\eta(\omega)$ and any $\epsilon > 0$, we must have

$$\Gamma[H_0(\omega) + \epsilon\eta(\omega)] \geq \Gamma[H_0(\omega)] \qquad \text{if a minimum}$$
$$(A.3.2)$$
$$\leq \Gamma[H_0(\omega)] \qquad \text{if a maximum}$$

This means $\Gamma[H_0(\omega) + \epsilon\eta(\omega)]$, considered as a function of ϵ, must have an extremal point at $\epsilon = 0$, no matter how $\eta(\omega)$ is selected. This means

$$\frac{\partial \Gamma[H_0(\omega) + \epsilon\eta(\omega)]}{\partial \epsilon}\bigg|_{\epsilon=0} = 0 \qquad (A.3.3)$$

or equivalently,

$$\int_{-\infty}^{\infty} \frac{\partial f[H_0(\omega) + \epsilon\eta(\omega)]}{\partial \epsilon} \, d\omega \bigg|_{\epsilon=0} = 0 \qquad (A.3.4)$$

The only way the integral can be zero for every choice of $\eta(\omega)$ is to have the integrand zero. Hence we require

$$\frac{\partial f[H_0(\omega) + \epsilon\eta(\omega)}{\partial \epsilon}\bigg|_{\epsilon=0} = 0 \qquad (A.3.5)$$

for every $\eta(\omega)$. The preceding is a necessary condition that any solution $H_0(\omega)$ must satisfy. That is, if a solution exists, it must satisfy (A.3.5). The latter equation is a form of the *Euler–Lagrange* equation. Any such extremal solution found must then be tested in (A.3.1) to determine whether it is a valid minimum (or maximum).

To minimize or maximize (A.3.1) subject to the constraint

$$\int_{-\infty}^{\infty} g[H_0(\omega)] \, d\omega = C \qquad (A.3.6)$$

we instead define

$$\Gamma[H(\omega)] = \int_{-\infty}^{\infty} f[H(\omega)] \, d\omega + \lambda \left\{ \int_{-\infty}^{\infty} g[H(\omega)] \, d\omega - C \right\} \qquad (A.3.7)$$

$$= \int_{-\infty}^{\infty} \{ f[H(\omega)] + \lambda g[H(\omega)] \} \, d\omega - \lambda C \qquad (A.3.8)$$

where λ is an arbitrary constant (called the *Lagrange multiplier*). Since extremal points of $\Gamma[H(\omega)]$ must be extremal points of the first integral in (A.3.7) [since the last terms must sum to zero for $H(\omega) = H_0(\omega)$], we can instead proceed to determine the extremal points of (A.3.8). Since the last term is a constant and will not affect the minimization, we seek the function $H(\omega)$ that minimizes (or maximizes) instead

$$\int_{-\infty}^{\infty} \{ f[H(\omega)] + \lambda g[H(\omega)] \} \, d\omega \qquad (A.3.9)$$

Proceeding as in (A.3.1)–(A.3.5), we therefore attempt to find extremal solutions $H_0(\omega)$ to

$$\frac{\partial}{\partial \epsilon} \{ f[H_0(\omega) + \epsilon\eta(\omega)] + \lambda g[H_0(\omega) + \epsilon\eta(\omega)] \} \bigg|_{\epsilon=0} = 0 \qquad (A.3.10)$$

for arbitrary $\eta(\omega)$. Any solution $H_0(\omega)$ will necessarily contain the arbitrary parameter λ. To determine the proper value of λ, we substitute $H_0(\omega)$ back into (A.3.6) and find the necessary λ to satisfy the constraint.

REFERENCES

1. Formin, A. and Gelfang, A. *Calculus of Variations*, Prentice-Hall, Englewood Cliffs, N.J., 1965.
2. Thomas, J. *Statistical Communication Theory*, Wiley, New York, 1967, Appendix H.

RANDOM VARIABLES
AND RANDOM PROCESSES

In this appendix we briefly summarize the basic definitions and conclusions of probability theory, random variables, and random processes needed throughout the text. The prime objective is for this discussion to serve as a fundamental review while achieving the prerequisite level and introducing the symbolic notation used. A reader previously versed in these areas will find this appendix unnecessary. The reader who is somewhat familiar with the material or the reader who may not have kept up with these topics will find this appendix beneficial as a summary or refresher discussion.

B.1 RANDOM VARIABLES

Let x be a random variable, having the probability density $p_x(\xi)$, where $\int_{-\infty}^{\infty} p_x(\xi)\, d\xi = 1$. The probability that x lies in an interval (a, b) is given by

$$\text{Prob}\,[a \le x \le b] = \int_a^b p_x(\xi)\, d\xi \qquad (B.1.1)$$

The *average* (*expected*, *mean*) *value* of any function of x, $f(x)$, is denoted by

$$\mathscr{E}[f(x)] = \int_{-\infty}^{\infty} f(\xi)p_x(\xi)\, d\xi \qquad (B.1.2)$$

where $\mathscr{E}$ is called the expectation operator. The moments of x are defined by

$$m_n = \mathscr{E}[x^n] \qquad (B.1.3)$$

The first several moments are the most important and are denoted

$$m_2 = mean \text{ of } x = \mathscr{E}[x] \qquad (B.1.4a)$$

$$m_2 = \text{\textit{mean squared value}} \text{ of } x = \mathcal{E}[x^2] \qquad \text{(B.1.4\textit{b})}$$

Also important is the *variance* of x defined by

$$\sigma^2 = \mathcal{E}[(x - m_1)^2] = m_2 - m_1^2 \qquad \text{(B.1.5)}$$

The *characteristic function* of x is defined as

$$\Psi_x(\omega) = \mathcal{E}[e^{j\omega x}] = \int_{-\infty}^{\infty} e^{j\omega \xi} p_x(\xi)\, d\xi \qquad \text{(B.1.6)}$$

Note that $\Psi_x(\omega)$ is the Fourier transform of $p_x(\xi)$. Also, the moments of x can be determined directly from its characteristic function by

$$m_n = \frac{1}{j^n} \left[\frac{\partial^n \Psi_x(\omega)}{\partial \omega^n} \right]_{\omega = 0} \qquad \text{(B.1.7)}$$

If x is a random variable, then $y = g(x)$ is also a random variable obtained from x by the transformation g. The probability density of y is related to that of x via

$$p_y(\xi) = \left[\frac{p_x(x)}{|dg/dx|} \right]_{x = g^{-1}(\xi)} \qquad \text{(B.1.8)}$$

where $g^{-1}(\xi)$ is the solution of $\xi = g(x)$ for x.

The most important random variable in engineering is the Gaussian random variable. A Gaussian random variable is one whose probability density is

$$p_x(\xi) = \frac{1}{\sqrt{2\pi}\sigma} e^{-(\xi - m)^2/2\sigma^2} \qquad \text{(B.1.9)}$$

where m is its mean and σ^2 its variance. The density is sketched in Figure B.1. The probability that a Gaussian random variable will lie in the interval $(m - b, m + b)$ about its mean m is then

$$\text{Prob}\,[(m - b) \le x \le (m + b)] = \int_{m-b}^{m+b} \frac{1}{\sqrt{2\pi}\sigma} \exp\left(-\frac{(\xi - m)^2}{2\sigma^2} \right) d\xi$$

$$= \frac{2}{\sqrt{\pi}} \int_0^{b/\sqrt{2}\sigma} e^{-u^2}\, du \qquad \text{(B.1.10)}$$

The function

$$Q(k) = \frac{1}{\sqrt{2\pi}} \int_k^{\infty} e^{-u^2/2}\, du \qquad \text{(B.1.11)}$$

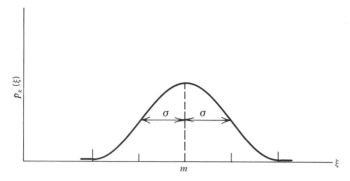

FIGURE B.1. The Gaussian probability density.

called the *Gaussian tail integral*, is shown in Figure B.2 along with some frequently used approximations. Another important function is the *error function*

$$\text{Erf}(a) = \frac{2}{\sqrt{\pi}} \int_0^a e^{-u^2}\, du \qquad (B.1.12)$$

and is tabulated in Table B.1. The *complementary error function* Erfc (*a*) is defined as

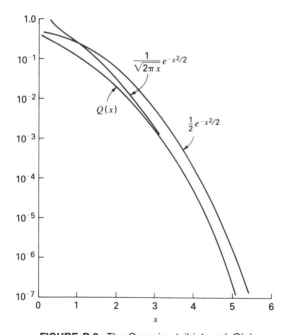

FIGURE B.2. The Gaussian tail integral $Q(x)$.

TABLE B.1.   The Error Function Erf (a)

a	Erf (a)	a	Erf (a)
0.00	0.00000	1.05	0.86244
0.05	0.05637	1.10	0.88021
0.10	0.11246	1.15	0.89612
0.15	0.16800	1.20	0.91031
0.20	0.22270	1.25	0.92290
0.25	0.27633	1.30	0.93401
0.30	0.32863	1.35	0.94376
0.35	0.37938	1.40	0.95229
0.40	0.42839	1.45	0.95970
0.45	0.47548	1.50	0.96611
0.50	0.52050	1.55	0.97162
0.55	0.56332	1.60	0.97635
0.60	0.60386	1.65	0.98038
0.65	0.64203	1.70	0.98379
0.70	0.67780	1.75	0.98667
0.75	0.71116	1.80	0.98909
0.80	0.74210	1.85	0.99111
0.85	0.77067	1.90	0.99279
0.90	0.79691	1.95	0.99418
0.95	0.82089	2.00	0.99532
1.00	0.84270	2.50	0.99959
		3.00	0.99998

$$\text{Erfc } (a) = 1 - \text{Erf } (a) \tag{B.1.13}$$

The Erfc function is accurately approximated by

$$\text{Erfc } (a) \approx \frac{e^{-a^2}}{\sqrt{\pi} a} \tag{B.1.14}$$

and is related to the Q-function by

$$\text{Erfc } (a) = 2Q(\sqrt{2}a) \tag{B.1.15}$$

More extensive tabulations of these functions are available [1].

B.2 JOINT RANDOM VARIABLES

Let x, y be a pair of real random variables having joint probability density $p_{xy}(x, y)$. Then

$$\text{Prob } [x \in I_x \quad \text{and} \quad y \in I_y] = \int_{I_y} \int_{I_x} p_{xy}(x, y) \, dx \, dy \tag{B.2.1}$$

The individual densities $p_x(x)$ can be obtained by integrating the joint density,

$$p_x(x) = \int_{-\infty}^{\infty} p_{xy}(x, y) \, dy \, ; \qquad p_y(y) = \int_{-\infty}^{\infty} p_{xy}(x, y) \, dx \qquad \text{(B.2.2)}$$

The joint expectation operator becomes

$$E_{xy}[f(x, y)] = \int\!\!\int_{-\infty}^{\infty} f(x, y) p_{xy}(x, y) \, dx \, dy \qquad \text{(B.2.3)}$$

for any function f of x and y. The (ij) joint moments are then defined by

$$m_{ij} = \mathcal{E}_{xy}[x^i y^j] \qquad \text{(B.2.4)}$$

and the joint characteristic function is

$$\Psi_{xy}(\omega_1, \omega_2) = \mathcal{E}_{xy}[\exp(j\omega_1 x + j\omega_2 y)] \qquad \text{(B.2.5)}$$

Note that the joint characteristic function is the two-dimensional Fourier transform of $p(x, y)$. An important moment in (B.2.4) is $m_{11} = \mathcal{E}(xy)$. The variables x and y are said to be *uncorrelated* if $m_{11} = \mathcal{E}(x)\mathcal{E}(y)$. The variables are said to be *independent* if the joint density factors as

$$p_{xy}(x, y) = p_x(x) p_y(y) \qquad \text{(B.2.6)}$$

The conditional density of x given $y = \hat{y}$ is denoted

$$p_{x|y}(x|\hat{y}) = \frac{p_{xy}(x, \hat{y})}{p_y(\hat{y})} \qquad \text{(B.2.7)}$$

and can be interpreted as the probability density of x when y has the value $\hat{y}$. The conditional expectation operator is then

$$\mathcal{E}_{x|y}[f(x)] = \int_{-\infty}^{\infty} f(x) p_{x|y}(x|y) \, dx \qquad \text{(B.2.8)}$$

and is a function of y. The joint expectation in (B.2.3) is seen to be equivalent to the sequence of expectations,

$$\mathcal{E}_{xy}[f(x, y)] = \mathcal{E}_y\{\mathcal{E}_{x|y}[f(x, y)]\} \qquad \text{(B.2.9)}$$

and can therefore be obtained by first conditional averaging, and then averaging over the conditioning variable. Conditional moments and characteristic functions can be obtained as in (B.1.3) and (B.1.6) using conditional probabilities.

When dealing with sums of independent random variables $z = x + y$, we note that the characteristic function of z becomes

$$\Psi_z(\omega) = \mathcal{E}_z[e^{j\omega z}] = \mathcal{E}_z[e^{j\omega(x+y)}] = \Psi_x(\omega)\Psi_y(\omega) \qquad (\text{B}.2.10)$$

and is therefore equal to the product of the characteristic function of each term. Being a Fourier transform, the probability density of z is then the convolution of the individual densities of x and y:

$$p_z(z) = \int_{-\infty}^{\infty} p_x(z - u)p_y(u)\, du \qquad (\text{B}.2.11)$$

If (u, v) are two new random variables formed from (x, y) by the transformations

$$u = g_1(x, y)$$
$$v = g_2(x, y) \qquad (\text{B}.2.12)$$

then the joint density of (u, v) is obtained from that of (x, y) by the density transformation

$$p_{uv}(u, v) = \frac{1}{|J|} \left. p_{xy}(x, y) \right|_{(x,y)=g^{-1}(u,v)} \qquad (\text{B}.2.13)$$

where J is the determinant of the Jacobian matrix

$$J = \det \begin{bmatrix} \dfrac{\partial g_1}{\partial x} & \dfrac{\partial g_1}{\partial y} \\ \dfrac{\partial g_2}{\partial x} & \dfrac{\partial g_2}{\partial y} \end{bmatrix} \qquad (\text{B}.2.14)$$

and $(x, y) = g^{-1}(u, v)$ is the inverse solution of the pair of equations in (B.2.12) for (x, y). That is, (B.2.12) is solved for x and y and the results are substituted into the right side of (B.2.13) to yield $p_{uv}(u, v)$.

A pair of random variables are *jointly Gaussian* variables if

$$p_{xy}(x, y) = \frac{1}{2\pi|\mu|^{1/2}}$$

$$\times \exp - \left[\frac{\mu_{02}(x - m_{10})^2 + \mu_{20}(y - m_{01})^2 - 2\mu_{11}(x - m_{10})(y - m_{01})}{2|\mu|} \right] \qquad (\text{B}.2.15)$$

where $|\mu| = \mu_{20}\mu_{02} - \mu_{11}^2$, $\mu_{20} = \sigma_x^2$, $\mu_{02} = \sigma_y^2$, and $\mu_{11} = m_{11} - m_{10}m_{01}$. The corresponding joint characteristic function of (B.2.5) follows as

$$\Psi_{xy}(\omega_1, \omega_2) = \exp \tfrac{1}{2}[\mu_{20}\omega_1^2 + \mu_{02}\omega_2^2 + 2\mu_{11}\omega_1\omega_2] \exp j[m_{10}\omega_1 + m_{01}\omega_2]$$

(B.2.16)

B.3 SEQUENCES OF RANDOM VARIABLES

Let $\mathbf{x} = (x_1, x_2, \ldots, x_N)$ be a sequence (ordered set) of N real random variables. Many of the properties of $\mathbf{x}$ are then simply the N-dimensional extension of the properties of two random variables. In particular, the N-dimensional joint density $p_{\mathbf{x}}(x_1, x_2, \ldots, x_N)$ can be used to derive joint probabilities of being in specific intervals. The joint average over the sequences of any function $f(\mathbf{x})$ is given by

$$\mathscr{E}_{\mathbf{x}}[f(\mathbf{x})] = \int_X f(\mathbf{x})p_{\mathbf{x}}(\mathbf{x})\, d\mathbf{x}$$

$$= \int_{-\infty}^{\infty} \cdots \int_{-\infty}^{\infty} f(x_1, x_2, \ldots, x_N)p_{\mathbf{x}}(x_1, x_2, \ldots, x_N)\, dx_2 \cdots dx_N$$

(B.3.1)

which we see is an N-fold integral over N-dimensional Euclidean space X.

A sequence $\mathbf{x}$ is said to be an *uncorrelated* sequence if every pair of components are uncorrelated. That is, (x_i, x_j) are uncorrelated for every $i \neq j$. The sequence is *independent* if components are pairwise independent.

The N-dimensional joint characteristic function is the N-dimensional extension of (B.2.5) and therefore becomes

$$\Psi_{\mathbf{x}}(\omega_1, \omega_2, \ldots, \omega_N) = \mathscr{E}_{\mathbf{x}}[\exp(j\omega_1 x_1 + j\omega_2 x_2 + \cdots + j\omega_N x_N)] \quad \text{(B.3.2)}$$

A Gaussian sequence has the joint density given by

$$p_{\mathbf{x}}(x_1, x_2, \ldots, x_N) = \frac{1}{(2\pi)^{N/2}\sqrt{|M|}} \exp\left[-\tfrac{1}{2}\chi^{\mathrm{Tr}}M^{-1}\chi\right] \quad \text{(B.3.3)}$$

where

$$\chi = \begin{bmatrix} x_1 - m_{x_1} \\ x_2 - m_{x_2} \\ \vdots \\ x_N - m_{x_N} \end{bmatrix}_{N \times 1}, \qquad M = \begin{bmatrix} \mu_{11} & \mu_{12} & \cdots \\ \mu_{21} & \mu_{22} & \\ \vdots & & \ddots \\ & & & \mu_{NN} \end{bmatrix} \quad \text{(B.3.4)}$$

and m_{x_i} is the mean of x_i, $\mu_{ij} = \mathscr{E}_{\mathbf{x}}[(x_i - m_{x_i})(x_j - m_{x_j})]$, and $|M|$ is the determinant of M. The matrix M is called the *covariance matrix* of the sequence. We see that if the sequence is uncorrelated, $\mu_{ij} = 0$, $i \neq j$, then the quadratic form in the exponent of (B.3.3) expands such that $p_{\mathbf{x}}(x_1 \cdots x_N)$ is a product of terms in each x_i. Hence uncorrelated Gaussian sequences are also independent sequences.

We often must sum the components of a random sequence. Define $z = x_1 + x_2 + \cdots + x_N$ as the sum variable. If the sequence is independent, the characteristic function of z is then

$$\Psi_z(\omega) = \prod_{i=1}^{N} \Psi_{x_i}(\omega) \tag{B.3.5}$$

which is the product of the individual characteristic functions. The corresponding probability density becomes

$$p_z(z) = p_{x_1}(x_1) \otimes p_{x_2}(x_2) \otimes \cdots \otimes p_{x_N}(x_N) \tag{B.3.6}$$

where $\otimes$ denotes convolution. The sum density is therefore obtained by an $(N-1)$-fold convolution of the individual densities. For a Gaussian sequence, z will always be a Gaussian random variable, whose mean is the sum of the means of each component. If the Gaussian sequences is independent, then z will have in addition a variance equal to the sum of the component variances. If the sequence is not Gaussian, then it has been shown that under relatively weak conditions, the density of z will converge to a Gaussian density as $N \to \infty$ (central limit theorem).

B.4 RANDOM PROCESSES

A random process $x(t)$ defines a random variable over a continuum of scalar points t; that is, at each scalar point t, $x(t)$ is a random variable. Random processes are also called *stochastic processes*, and if t denotes time, they are also referred to as *random*, or *noise*, *waveforms*.

If $x(t)$ is a random process, then a sequence of scalar points $(t_1, t_2, \ldots, t_N)$ defines a sequence of random variables. This sequence is then described by the Nth order joint probability density of the random variables involved, which in general depends on the location of the points $\{t_i\}$. A random process is said to be *completely described statistically* if all of its Nth order densities for all $N \to \infty$ are known.

A single point t defines a single random variable, which we denote x_1. The probability density of that variable x_1, $p_{x_1}(\xi)$ is called a *first order density* of the process. From such a density, probabilities, averages, moments, and so on, can be computed for the process at that t_1. If the $p_{x_1}(\xi)$ does not depend on the location of t_1, the process is said to be *first order stationary* and all the resulting averages and probabilities do not depend on t_1.

Two points, t_1 and t_2, define a pair of random variables, $x_1 \overset{\Delta}{=} x(t_1)$ and

$x_2 \overset{\Delta}{=} x(t_2)$. Such a pair is statistically described by its joint second order density $p_{x_1 x_2}(\xi_1, \xi_2)$, which generally depends on the location of t_1 and t_2. From this joint density, joint probabilities, averages, and moments can be determined. An important joint moment of the process $x(t)$ is its *autocorrelation* function

$$R_x(t_1, t_2) = \mathscr{E}[x_1 x_2] \tag{B.4.1}$$

The average on the right is the joint correlation moment m_{11} in (B.2.4) for the random variable x_1 at t_1 and x_2 at t_2. Specifically,

$$\mathscr{E}[x_1 x_2] = \int_{-\infty}^{\infty} \int_{-\infty}^{\infty} \xi_1 \xi_2 p_{x_1 x_2}(\xi_1, \xi_2) \, d\xi_1 \, d\xi_2 \tag{B.4.2}$$

Since this second order density will, in general, depend on both the points t_1 and t_2, the expectation in (B.4.2) will produce a function of both t_1 and t_2; that is, it will depend on the specific two points involved. If $p_{x_1 x_2}(\xi_1, \xi_2)$ does not depend on t_1 and t_2 but only on the separation $\tau = t_2 - t_1$ between the points, the process is said to be *second order stationary*. If the autocorrelation function in (B.4.1) depends only on τ, the process is said to be *wide sense stationary* (WSS). Note that a second order stationary process is always WSS, but the converse is not necessarily true. A WSS process has the basic property that correlation values associated with points of the process depend only on their separation and not on where the points are located— that is, all pairs of points separated by τ have the same correlation value.

For WSS processes, the function in (B.4.1) can be written simply as $R_x(\tau)$. It is obvious that $R_x(\tau)$ is always a real function in τ, since it involves averages of real random variables. In addition, relabeling the points t_1 and t_2 as t_2 and t_1, respectively, does not alter their correlation value in (B.4.2). Hence it follows that

$$R_x(-\tau) = R_x(\tau) \tag{B.4.3}$$

and the autocorrelation function must be even in τ. Furthermore, since it is necessary that

$$\mathscr{E}[(x_1 \pm x_2)^2] \geq 0 \tag{B.4.4}$$

for any t_1 and t_2, we can easily expand out and show

$$\mathscr{E}[x_1^2] + \mathscr{E}[x_2^2] \pm 2\mathscr{E}[x_1 x_2] \geq 0 \tag{B.4.5}$$

Noting that $x_1 = x(t_1)$ and $x_2 = x(t_2)$, and letting $t_2 = t_1 + \tau$, allows us to write (B.4.5) as

$$\mathscr{E}[x^2(t_1)] + \mathscr{E}[x^2(t+\tau)] + 2\mathscr{E}[x(t_1)x(t_1+\tau)] \geq 0 \qquad (B.4.6)$$

For WSS processes, this becomes

$$R_x(0) + R_x(0) \pm 2R_x(\tau) \geq 0 \qquad (B.4.7)$$

for all τ, or

$$|R_x(\tau)| \leq R_x(0) \qquad (B.4.8)$$

Thus the autocorrelation function of a WSS stationary process must take on its maximum magnitude value at $\tau = 0$. This means its slope must be zero at $\tau = 0$.

The *power spectral density* $S_x(\omega)$ of a random process $x(t)$ is formally defined as

$$S_x(\omega) = \underset{T \to \infty}{\text{Lim}} \frac{1}{2T} \mathscr{E}[|X_T(\omega)|^2] \qquad (B.4.9)$$

where $X_T(\omega)$ is the Fourier transform of the random process $x(t)$ restricted to the range $(-T, T)$ and zero elsewhere. Since $X_T(\omega)$ defines an integral of a random process, it is itself a random variable at each ω. The expectation $\mathscr{E}$ in (B.4.9) averages over the probability density of this random variable. The function obtained at each ω in the limit at T is increased and the averaging is repeated in the process spectral density. When the process is wide sense stationary, we can expand as

$$S_x(\omega) = \underset{T \to \infty}{\text{Lim}} \frac{1}{2T} \mathscr{E}\left[\int_{-T}^{T} x(t)e^{-j\omega t}\, dt \int_{-\infty}^{\infty} x(\rho)e^{j\omega\rho}\, d\rho\right] \qquad (B.4.10)$$

Writing the product of integrals as a multiple integral, and inverting the integration and averaging operations, yields

$$S_x(\omega) = \underset{T \to \infty}{\text{Lim}} \frac{1}{2T} \int_{-T}^{T}\int_{-T}^{T} \mathscr{E}[x(t)x(\rho)]e^{-j\omega(t-\rho)}\, dt\, d\rho \qquad (B.4.11)$$

The average in the integrand is recognized as the process autocorrelation function and, because of the stationarity assumption, can be written as

$$\mathscr{E}[x(t)x(\rho)] = R_x(t-\rho) \qquad (B.4.12)$$

With (B.4.12) inserted, (B.4.11) can be rewritten as (see Papoulis [2, p. 325])

$$S_x(\omega) = \lim_{T \to \infty} \int_{-T}^{T} \left(1 - \frac{|u|}{T}\right) R_x(u) e^{-j\omega u} \, du \qquad (\text{B.4.13})$$

In the limit this becomes

$$S_x(\omega) = \int_{-\infty}^{\infty} R_x(u) e^{-j\omega u} \, du \qquad (\text{B.4.14})$$

Thus the spectral density of a wide sense stationary process is simply the Fourier transform of its autocorrelation function. This means the process autocorrelation function and spectral density form a transform pair.

Higher order levels of stationarity can be similarly defined. A process is *Nth order stationary* if the joint probability density between the N random variables $x_1, \ldots, x_N$ at times $t_1, \ldots, t_N$ depends only on the time separations $t_2 - t_1$, $t_3 - t_2$, $t_4 - t_3 \ldots$ of the points, no matter where they are chosen. A process is *strictly stationary* if it is stationary in all orders.

If $x(t)$ is a random process obtained by linearly filtering $x(t)$,

$$y(t) = \int_{-\infty}^{\infty} h(t - \rho) x(\rho) \, d\rho \qquad (\text{B.4.15})$$

is also a random process, where $h(t)$ is the impulse response of the filter. The autocorrelation of the filter output can be derived by direct substitution from (B.4.15):

$$R_y(\tau) = \mathcal{E}[y(t)y(t + \tau)]$$

$$= \mathcal{E}\left[\int_{-\infty}^{\infty} x(t - \rho)h(\rho) \, d\rho \int_{-\infty}^{\infty} x(t + \tau - \alpha)h(\alpha) \, d\alpha\right] \qquad (\text{B.4.16})$$

where $\mathcal{E}$ is the expectation operator over the statistics of the integrals. However, we see that the integrals involve the random variables $x(t - \rho)$ and $x(t + \tau - \alpha)$, and we need only average over their joint density. Interchanging averaging and integration, and moving the expectation operation inside the integrals, allows us to rewrite (B.4.16) as

$$R_y(\tau) = \int_{-\infty}^{\infty} \int_{-\infty}^{\infty} \mathcal{E}[x(t - \rho)x(t + \tau - \alpha)]h(\rho)h(\alpha) \, d\alpha \, d\rho \qquad (\text{B.4.17})$$

We now note that the expectation in the integrand is precisely the definition of the autocorrelation function of the input process $x(t)$ at points $(t - \rho)$ and $(t + \tau - \alpha)$. If $x(t)$ is a stationary input process, this expectation will depend only on the time difference of the points, and we can substitute

$$\mathcal{E}[x(t - \rho)x(t + \tau - \alpha)] = R_x(\tau + \rho - \alpha) \qquad (\text{B.4.18})$$

This simplifies (B.4.17) to

$$R_y(\tau) = \int_{-\infty}^{\infty} \int_{-\infty}^{\infty} R_x(\tau + \rho - \alpha) h(\rho) h(\alpha) \, d\alpha \, d\rho \qquad \text{(B.4.19)}$$

Equation (B.4.19) therefore indicates how the output autocorrelation is related to the input autocorrelation when a stationary process is filetered by a linear filter having impulse response $h(t)$. The output power spectrum is obtained by Fourier transforming (B.4.19). Thus

$$S_y(\omega) = \int_{-\infty}^{\infty} R_y(\tau) e^{-j\omega\tau} \, d\tau = \int \int \int_{-\infty}^{\infty} R_x(\tau + \rho - \alpha) h(\rho) h(\alpha) e^{-j\omega\tau} \, d\alpha \, d\rho \, d\tau$$

$$\text{(B.4.20)}$$

Interchanging the order of integration, and integrating out the τ variable first, changes this to

$$S_y(\omega) = \int_{-\infty}^{\infty} h(\rho) \, d\rho \int_{-\infty}^{\infty} h(\alpha) \, d\alpha \left[\int_{-\infty}^{\infty} R_x(\tau + \rho - \alpha) e^{-j\omega\tau} \, d\tau \right] \qquad \text{(B.4.21)}$$

The last bracket can be recognized as $S_x(\omega) e^{-j\omega(\alpha - \rho)}$, and (B.4.21) reduces to

$$S_y(\omega) = S_x(\omega) \int_{-\infty}^{\infty} h(\rho) e^{j\omega\rho} \, d\rho \int_{-\infty}^{\infty} h(\alpha) e^{-j\omega\alpha} \, d\alpha$$

$$= S_x(\omega) H^*(\omega) H(\omega)$$

$$= S_x(\omega) |H(\omega)|^2 \qquad \text{(B.4.22)}$$

where the asterisk denotes a complex conjugate. Hence the square of the filter gain function filters the power spectral density of the input to produce the output spectral density.

B.5 GAUSSIAN RANDOM PROCESSES

A *Gaussian random process* is a process in which the sequence of random variables defined at any N points $\{t_i\}$ is a Gaussian sequence; that is, at any point t, $x(t)$ is a Gaussian random variable. Hence the general Nth order density of a Gaussian process is given by the general Nth order Gaussian density in (B.3.3). Such a density depends only on the mean and mean squared values at each t and on the correlation $\mathscr{E}[x(t_i)x(t_j)]$ between two points. For wide sense stationary Gaussian processes, this correlation is given precisely by $R_x(t_j - t_i)$ for any pair of points t_i and t_j; that is, the matrix M in (B.3.4) will have the general (i, j) entry

$$u_{ij} = R_x(t_j - t_i) - m_i m_j \qquad \text{(B.5.1)}$$

where m_i is the mean of $x(t_i)$. Since the matrix M is all that is needed to specify a general Nth order Gaussian density for any N, the autocorrelation function $R_x(\tau)$, evaluated at the proper point $\tau = t_j - t_i$, allows us to describe uniquely all joint probability densities of the process. Hence a wide sense stationary Gaussian process is completely described statistically by its mean function and autocorrelation. By the uniqueness of the Fourier transform, it is therefore equivalently described by its power spectral density.

B.6   REPRESENTING GAUSSIAN PROCESSES

In analyzing systems that contain Gaussian random processes, it is often necessary to have a convenient mathematical representation of the process. The representation must be statistically equivalent (have the same N order densities for all N) to the original process, so that all subsequent operations and transformations are identical to dealing with the process itself. From our discussion in the previous section, a WSS Gaussian process has the advantage that any representation is statistically equivalent if it has the same autocorrelation or power spectral density of the process itself. Hence we need only guarantee equivalence in this respect.

Consider a zero mean Gaussian process having the arbitrary two-sided power spectral density $S_n(\omega)$ shown in Figure B.3a. We let ω_c be an

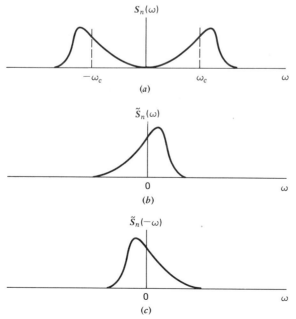

FIGURE B.3. Noise spectral densities: (*a*) two-sided spectrum, (*b*) shifted version, (*c*) image of part *b*.

arbitrary frequency along the frequency axis. We can always write this two-sided density as

$$S_n(\omega) = \tfrac{1}{2}\tilde{S}_n(\omega - \omega_c) + \tfrac{1}{2}\tilde{S}_n(-\omega - \omega_c) \tag{B.6.1}$$

where $\tilde{S}_n(\omega)$ is the shifted version of $S_n(\omega)$, obtained by shifting the one-sided spectrum (the latter obtained by folding the negative frequency part on to the positive frequency part) by an amount ω_c, as shown in Figure B.3b. Equation (B.6.1) is that necessary to regenerate $S_n(\omega)$ from $\tilde{S}_n(\omega)$. When written as in (B.6.1), the corresponding autocorrelation function of the process is then

$$
\begin{aligned}
R_n(\tau) &= \frac{1}{2\pi} \int_{-\infty}^{\infty} S_n(\omega) e^{j\omega\tau}\, d\omega \\
&= \frac{1}{2\pi} \int_{-\infty}^{\infty} \frac{1}{2}\tilde{S}_n(\omega - \omega_c) e^{j\omega\tau}\, d\omega + \frac{1}{2\pi} \int_{-\infty}^{\infty} \frac{1}{2}\tilde{S}_n(-\omega - \omega_c) e^{j\omega\tau}\, d\omega \\
&= \frac{1}{2} [\tilde{R}_n(\tau) e^{j\omega_c\tau} + \tilde{R}_n^*(\tau) e^{-j\omega_c\tau}] \\
&= \mathrm{Real}\, \{\tilde{R}_n(\tau) e^{j\omega_c\tau}\}
\end{aligned}
\tag{B.6.2}
$$

where * denotes complex conjugate and

$$\tilde{R}_n(\tau) = \frac{1}{2\pi} \int_{-\infty}^{\infty} \tilde{S}_n(\omega) e^{j\omega\tau}\, d\omega \tag{B.6.3}$$

Here $\tilde{R}_n(\tau)$ is the transform of the shifted spectrum $\tilde{S}_n(\omega)$. We can further expand

$$\tilde{R}_n(\tau) = \tilde{R}_c(\tau) + j\tilde{R}_s(\tau) \tag{B.6.4}$$

where

$$\tilde{R}_c(\tau) = \frac{1}{2\pi} \int_{-\infty}^{\infty} \tilde{S}_n(\omega) \cos \omega\tau\, d\omega \tag{B.6.5a}$$

$$\tilde{R}_s(\tau) = \frac{1}{2\pi} \int_{-\infty}^{\infty} \tilde{S}_n(\omega) \sin \omega\tau\, d\omega \tag{B.6.5b}$$

When (B.6.5) is substituted into (B.6.2), we have

$$R_n(\tau) = \tilde{R}_c(\tau) \cos \omega_c\tau - \tilde{R}_s(\tau) \sin \omega_c\tau \tag{B.6.6}$$

The preceding represents a general expression for the autocorrelation function of a random process in terms of an arbitrary frequency ω_c. We would now like to find a representation of a random process that has the same autocorrelation as (B.6.6).

Let us consider writing the random process as

$$n(t) = n_c(t) \cos(\omega_c t + \psi) - n_s(t) \sin(\omega_c t + \psi) \qquad (B.6.7)$$

where ψ is an arbitrary phase angle and $n_c(t)$ and $n_s(t)$ are each random processes. We first observe that if $n(t)$ on the left is to be a Gaussian process (a Gaussian variable at any t), the processes $n_c(t)$ and $n_s(t)$ must also be Gaussian processes (sums of Gaussian variables from Gaussian variables). In addition, let $n_c(t)$ and $n_s(t)$ have autocorrelation $R_c(\tau)$ and $R_s(\tau)$, respectively, and cross-correlation function $R_{cs}(\tau)$. The autocorrelation on $n(t)$ in (B.6.7) is then

$$
\begin{aligned}
R_n(\tau) &= \mathscr{E}[n(t)n(t+\tau)] \\
&= \mathscr{E}[n_c(t) \cos(\omega_c t + \psi) - n_s(t) \sin(\omega_c t + \psi)] \\
&\quad \times \{n_c(t+\tau) \cos[\omega_c(t+\tau) + \psi] - n_s(t+\tau) \sin[\omega_c(t+\tau) + \psi]\} \\
&= R_c(\tau) \cos(\omega_c t + \psi) \cos[\omega_c(t+\tau) + \psi] \\
&\quad + R_s(\tau) \sin(\omega_c t + \psi) \sin[\omega_c(t+\tau) + \psi] \\
&\quad - R_{sc}(\tau) \sin(\omega_c t + \psi) \cos[\omega_c(t+\tau) + \psi] \\
&\quad - R_{cs}(\tau) \cos(\omega_c t + \psi) \sin[\omega_c(t+\tau) + \psi] \qquad (B.6.8)
\end{aligned}
$$

We now see that if

$$
\begin{aligned}
R_c(\tau) &= R_s(\tau) = \tilde{R}_c(\tau) \\
R_{cs}(\tau) &= \tilde{R}_s(\tau) = -R_{sc}(\tau)
\end{aligned}
\qquad (B.6.9)
$$

then the autocorrelation in (B.6.9) of $n(t)$ in (B.6.7) is the same as that in (B.6.6) for any ψ. That is, $n(t)$ has the power spectral density $S_n(\omega)$. Note that (B.6.9) states that this occurs only if $n_c(t)$ and $n_s(t)$ have the proper autocorrelation $[\tilde{R}_c(\tau)]$ and cross-correlation $[\tilde{R}_s(\tau)]$. Hence we conclude that if $n_c(t)$ and $n_s(t)$ are Gaussian processes with correlation functions

$$R_c(\tau) = R_s(\tau) = \frac{1}{2\pi} \int_{-\infty}^{\infty} \tilde{S}_n(\omega) \cos \omega\tau \, d\omega \qquad (B.6.10a)$$

$$R_{cs}(\tau) = \frac{1}{2\pi} \int_{-\infty}^{\infty} \tilde{S}_n(\omega) \sin \omega\tau \, d\omega \qquad (B.6.10b)$$

then $n(t)$ in (B.6.7) is also Gaussian with the spectrum $S_n(\omega)$. Thus $n(t)$ can be made to represent any Gaussian process. If the processes involved are not Gaussian, we can only conclude that $n(t)$ will have the same autocorrelation function but will not be statistically equivalent.

The representation in (B.6.7) is called a *quadrature expansion* of the noise, $n_c(t)$ and $n_s(t)$ are called the *quadrature component* processes. Since

(B.6.10a) is the component autocorrelation function, their spectral density follows as

$$S_c(\omega) = \int_{-\infty}^{\infty} \frac{1}{2\pi} \int_{-\infty}^{\infty} \tilde{S}_n(u) \cos{(u\tau)}\, du\; e^{-j\omega\tau}\, d\tau$$

$$= \frac{1}{2} \int_{-\infty}^{\infty} \tilde{S}_n(u)\delta(u - \omega)\, du + \frac{1}{2} \int_{-\infty}^{\infty} \tilde{S}_n(u)\delta(u + \omega)\, du$$

$$= \frac{\tilde{S}_n(\omega) + \tilde{S}_n(-\omega)}{2} \qquad\qquad\text{(B.6.11)}$$

that is, $\tilde{S}_n(\omega)$ must be summed with its mirror image to obtain the component spectrum (Figure B.3c).

We note the quadrature components have the following resulting properties:

1. Since $R_{cs}(0) = 0$, $n_c(t)$ and $n_s(t)$ are always statistically uncorrelated (independent if Gaussian) at the same t.

2. If $\tilde{S}_n(\omega)$ is even in ω [$S_n(\omega)$ is symmetrical about $\pm\omega_c$], then $R_{cs}(\tau) = 0$ for all τ. In this case the components become uncorrelated (independent if Gaussian) processes. This means that if the frequency ω_c (which has been arbitrary to this point) can be selected so that $S_n(\omega)$ is symmetric about it, the quadrature expansion in (B.6.7) will have statistically independent components. This requires $S_n(\omega)$ to be a bandpass spectrum and ω_c its center frequency. Furthermore, under these conditions, $S_c(\omega)$ in (B.6.11) reduces to $\tilde{S}_n(\omega)$, and the shifted (to zero) bandpass spectrum becomes the component spectrum. We often refer to the components here as being the "low frequency equivalent" of the original process. Since bandpass noise is common in communication systems, this quadrature expansion is extremely convenient for manipulation and interpretation, and is used extensively in Chapters 3 and 4 of the text.

3. Since the integral of the spectral density is the power, we see that

$$P_{n_c} = P_{n_s} = \frac{1}{2\pi} \int_{-\infty}^{\infty} \frac{\tilde{S}_n(\omega) + \tilde{S}_n(-\omega)}{2}\, d\omega$$

$$= \frac{1}{2\pi} \int_{-\infty}^{\infty} S_n(\omega)\, d\omega$$

$$= P_n \qquad\qquad\text{(B.6.12)}$$

Hence $n_c(t)$ and $n_s(t)$ have the same power content as $n(t)$ itself. We emphasize that each does not have half the total power, as one may at first conclude.

4. If $S_n(\omega)$ is *narrowband* and symmetrical about ω_c (i.e., bandwidth about ω_c is much smaller than ω_c), then the frequencies of the components will not overlap those of $n(t)$. The components $n_c(t)$ and

$n_s(t)$ are then said to be "slowly varying" functions, and $n(t)$ can be interpreted as a "modulation" of the quadrature components onto the carrier at ω_c. This narrowband condition is typically indicative of communication receiver noise models.

It is often convenient to write the Gaussian noise process in (B.6.7) in terms of its complex expansion

$$n(t) = \text{Real} \{\alpha(t)e^{jv(t)} e^{j(\omega_c t + \psi)}\}$$
$$= \alpha(t) \cos[\omega_c t + \psi + v(t)] \tag{B.6.13}$$

where

$$\alpha(t) = [n_c^2(t) + n_s^2(t)]^{1/2}$$
$$v(t) = \tan^{-1}\left(\frac{n_s(t)}{n_c(t)}\right) \tag{B.6.14}$$

The random process $\alpha(t)$ is the *envelope* of the bandpass noise, whereas $v(t)$ is the *phase* process. The envelope process $\alpha(t)$ is a nonlinear function of the random quadrature components $n_c(t)$ and $n_s(t)$, and therefore is itself not a Gaussian process. Its probability density, at any t, is obtained by converting the Gaussian variables $n_c(t)$ and $n_s(t)$ through the transformation in (B.6.14). This envelope has been shown [2] to have the *Rayleigh* probability density at any t, given by

$$p_{\alpha_t}(\xi) = \frac{\xi}{P_n} e^{-\xi^2/2P_n}, \qquad \xi \geq 0 \tag{B.6.15}$$

We mention that if a sine wave $A \cos(\omega_c t + \psi)$ is added to $n(t)$, then the sum of the two waveforms has an envelope process

$$\alpha(t) = \{[A + n_c(t)]^2 + n_s^2(t)\} \tag{B.6.16}$$

This envelope process, at any t, now has the probability density

$$p_{\alpha_t}(\xi) = \frac{\xi}{P_n} e^{-(\xi^2 + A^2)/2P_n} I_0\left(\frac{A\xi}{P_n}\right), \qquad \xi \geq 0 \tag{B.6.17}$$

The preceding is referred to as a *Rician* density.

REFERENCES

1. Abramowitz, A. and Stegun, I. *Handbook of Mathematical Functions*, National Bureau of Standards, Washington, D.C., 1965, Chap. 26.
2. Papoulis, A. *Probability, Random Variables, and Random Processes*, McGraw-Hill, New York, 1965.

APPENDIX C

ORTHOGONAL EXPANSIONS
OF SIGNALS

C.1 DETERMINISTIC WAVEFORMS

It is convenient in system analysis to represent or express signal waveforms by orthogonal expansions. Given a real, deterministic waveform $s(t)$ of finite energy, defined over an arbitrary time interval (a, b), we wish to represent it as

$$s(t) = \sum_i s_i \Phi_i(t) , \qquad a \le t \le b \tag{C.1.1}$$

where $\{\Phi_i(t)\}$ is a complete set of orthonormal functions over (a, b):

$$\int_a^b \Phi_i(t)\Phi_j(t) \, dt = \begin{cases} 0 & \text{if} \quad i \ne j \\ 1 & \text{if} \quad i = j \end{cases} \tag{C.1.2}$$

and s_i is a scalar. The coefficients $\{s_i\}$ represent the coordinates of the expansion with respect to the orthonormal set $\{\Phi_i(t)\}$. In effect, we are expanding $s(t)$ into a sum of functions with amplitude adjusted coefficients.

We can determine the best choice for the coefficients s_i. Denote the partial sum

$$s_N(t) = \sum_{i=1}^N s_i \Phi_i(t) \tag{C.1.3}$$

and consider the coefficients such that $s_N(t)$ is "closest" to $s(t)$, in an integrated squared sense, for any N. This is obtained by defining

$$\begin{aligned} J &= \int_a^b [s(t) - s_N(t)]^2 \, dt \\ &= \int_a^b \left[s(t) - \sum_{i=1}^N s_i \Phi_i(t) \right]^2 \, dt \end{aligned} \tag{C.1.4}$$

604

and finding the set $\{s_i\}$, $i = 1, \ldots, N$, that minimizes J. Expanding the square, integrating, and using the orthonormality of the $\Phi_i(t)$ functions, yields

$$
J = \int_a^b \left[s^2(t) - 2s(t) \sum_{i=1}^N s_i \Phi_i(t) + \left(\sum_{i=1}^N s_i \Phi_i(t) \right)^2 \right] dt
$$

$$
= \int_a^b s^2(t)\, dt - 2 \sum_{i=1}^N s_i \int_a^b s(t)\Phi_i(t)\, dt + \sum_{i=1}^N s_i^2 \qquad \text{(C.1.5)}
$$

The value of the coordinate s_i minimizing J is that for which $\partial J / \partial s_i = 0$. This yields, for any N,

$$
s_i = \int_a^b s(t)\Phi_i(t)\, dt \qquad \text{(C.1.6)}
$$

Each s_i in the expansion in (C.1.3) should therefore be obtained by integrating $s(t)$ with the corresponding orthonormal function. The scalars in (C.1.6) are called the *Fourier coefficients* of the expansion. No matter how many terms are used in the partial sum, each coefficient used will have the value determined by (C.1.6) for minimizing J in (C.1.4).

If the Fourier coefficients are used in (C.1.4), we obtain the minimal value

$$
J_{\min} = \int_a^b s^2(t)\, dt - 2 \sum_{i=1}^N s_i \int_a^b s(t)\Phi_i(t)\, dt + \sum_{i=1}^N s_i^2
$$

$$
= \int_a^b s^2(t)\, dt - \sum_{i=1}^N s_i^2 \qquad \text{(C.1.7)}
$$

Since $J \geq 0$, and the sum must increase with N, it follows that $J_{\min} \rightarrow 0$ as $N \rightarrow \infty$. If $J_{\min}$ did not converge to zero, then either (1) all s_i, for i exceeding some N are zero, implying that $s(t)$ has components orthogonal to the $\Phi_i(t)$ set (however, the latter was assumed to be a complete set—all orthogonal functions are included in the set as $N \rightarrow \infty$); or (2) $\int_a^b s^2(t)\, dt$ is infinite, violating the assumption of finite energy signals $s(t)$. With $J_{\min} \rightarrow 0$, the partial sum in (C.1.3) with Fourier coefficients therefore converges to $s(t)$ in an integrated squared sense.

We can now show that for any $s(t)$, the Fourier coefficients must be unique; that is, there can be only one distinct sequence $\{s_i\}$. For if there were two such sequences $\{s_i\}$ and $\{s_i'\}$, then it will follow that

$$
\int_a^b [s(t) - s(t)]^2\, dt = 0
$$

$$
= \int_a^b \left[\sum_i s_i \Phi_i(t) - \sum_i s_i' \Phi_i(t) \right]^2 dt
$$

$$
= \sum_i (s_i - s_i')^2 \qquad \text{(C.1.8)}
$$

The last sum can only be zero if each $s_i = s'_i$, proving that the coefficients are identical. Thus there is a unique relation between a given $s(t)$ and its Fourier coefficients, when expanding with a particular orthonormal set. If we select a different orthonormal function set, the Fourier coefficients will change, but the uniqueness is still preserved.

Note that if $J_{min} = 0$, it follows from (C.1.7) that

$$\sum_i s_i^2 = \int_a^b s^2(t)\, dt$$

$$= \text{energy in } s(t) \tag{C.1.9}$$

Hence the sum of the squares of the Fourier coefficients must add to the energy of the waveform. Each squared coefficient gives the energy contribution of that coordinate.

Consider now the class of signals in (C.1.1) represented by only a finite number of terms:

$$s(t) = \sum_{i=1}^{\nu} s_i \Phi_i(t) \tag{C.1.10}$$

where $\{s_i\}$ are again the Fourier coefficients. The coordinate number ν is called the *dimension* of the signal, and $s(t)$ is said to be a signal of ν dimensions. For a particular set of orthonormal functions $\Phi_i(t)$, we can represent all signals of dimension ν by a vector of its Fourier coefficients,

$$\mathbf{s} = (s_1, s_2, \ldots, s_\nu) \tag{C.1.11}$$

Given the vector $\mathbf{s}$ (and the orthonormal set), we can always obtain $s(t)$ by summing (C.1.10). Likewise, given $s(t)$, we can always generate $\mathbf{s}$ by evaluating each Fourier coefficient in (C.1.6). Hence a one-to-one relationship exists between any ν-dimensional signal $s(t)$ and its vector $\mathbf{s}$ in (C.1.11). In this sense, $\mathbf{s}$ is a unique vector representation of the waveform $s(t)$.

A result of this vector representation is that $s(t)$ can be plotted as a single vector point $\mathbf{s}$ in a ν-dimensional Euclidean space (Figure C.1). Likewise each vector point in this space uniquely defines a waveform, since the coordinates can be used to sum in (C.1.10). Hence there is a direct relationship between ν-dimensional waveforms and vector points in a Euclidean space. It must be remembered that a particular orthonormal function set underlies this relationship. If we select a different orthonormal set, the vector point changes.

The following properties of this waveform–vector space relationship can be stated as follows:

1. The waveform energy becomes

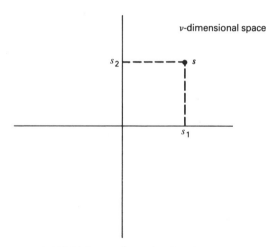

FIGURE C.1. ν-dimensional vector space.

$$\int_a^b s^2(t)\,dt = \int_a^b \left[\sum_{i=1}^{\nu} s_i \Phi_i(t)\right]^2 dt$$

$$= \sum_{i=1}^{\nu} s_i^2$$

$$= (\text{radial distance to vector } \mathbf{s})^2 \qquad (C.1.12)$$

The radial distance to a vector point $\mathbf{s}$ is equal to the square root of the energy in the waveform $s(t)$ that it represents. This means that the greater the energy in a waveform, the further out its vector point. Likewise, all waveforms of equal energy must have vectors that lie equal distance from the origin in vector space (i.e., must lie on a hypersphere).

2. Let $s(t)$ and $y(t)$ be two waveforms with vectors $\mathbf{s} = \{s_i\}$ and $\mathbf{y} = \{y_i\}$, respectively; that is,

$$s(t) = \sum_{i=1}^{\nu} s_i \Phi_i(t)\,,$$
$$\qquad\qquad a \le t \le b \qquad (C.1.13)$$
$$y(t) = \sum_{i=1}^{\nu} y_i \Phi_i(t)\,,$$

Then the correlation of $s(t)$ and $y(t)$ is

$$\int_a^b s(t)y(t)\,dt = \int_a^b \sum_{i=1}^{\nu} s_i \Phi_i(t) \sum_{j=1}^{\nu} y_j \Phi_j(t)\,dt$$

$$= \sum_{i=1}^{\nu} s_i y_i \qquad (C.1.14)$$

The above is called the *inner product* of the vectors **s** and **y**. We can therefore determine correlations between waveforms by computing the inner product of their vectors. A well-known property of inner products in vector space is that it can be related to the angle θ between the vectors by

$$\cos \theta = \frac{\Sigma \, s_i y_i}{[(\Sigma \, s_i^2)(\Sigma \, y_i^2)]^{1/2}} \qquad (C.1.15)$$

Hence the angle between waveform vectors is a measure of their waveform correlations. Waveforms with high positive correlation must have vector angles lying close together ($\theta \approx 0$), while waveforms with zero correlation will have orthogonal vectors ($\theta \approx 90°$). Waveforms with high negative correlation will be opposite each other ($\theta \cong 180°$). Vector angle relationships therefore indicate the waveform correlation values.

3. Suppose that a waveform $s(t)$ is transformed to a new waveform $x(t)$ by a linear operation (filtering, integrating, etc.). Denote this by

$$T[s(t)] = x(t) \qquad (C.1.16)$$

Then, because of the linearity, it must follow that

$$x(t) = T\left[\sum_{i=1}^{\nu} s_i \Phi_i(t) \right] = \sum_{i=1}^{\nu} s_i T[\Phi_i(t)] \qquad (C.1.17)$$

Here $T[\Phi_i(t)]$ is a new waveform obtained by linearly transforming the orthonormal function. Let this new waveform have the orthonormal expansion

$$T[\Phi_i(t)] = \sum_{j=1}^{\nu} a_{ij} \Phi_j(t) \qquad (C.1.18)$$

Here a_{ij} is the jth Fourier coefficient of the waveform $T[\Phi_i(t)]$. Thus the transformation $T[s(t)]$ in (C.1.16) produces a waveform that can be expanded out as

$$x(t) = \sum_{i=1}^{\nu} s_i \sum_{j=1}^{\nu} a_{ij} \Phi_j(t)$$

$$= \sum_{j=1}^{\nu} \left[\sum_{i} s_i a_{ij} \right] \Phi_j(t) \qquad (C.1.19)$$

This is equivalent to a vector expansion with coefficients

$$x_j = \sum_{i=1}^{\nu} a_{ij} s_i \qquad (C.1.20)$$

Alternatively, the column vector **x**, with components in (C.1.20), can be written as

$$\mathbf{x} = A\mathbf{s} \tag{C.1.21}$$

where A is the matrix of a_{ij} values in (C.1.18),

$$A = \begin{bmatrix} a_{11} & a_{21} & a_{31} & \cdots \\ a_{12} & a_{22} & & \cdots \\ \vdots & & & \\ a_{1\nu} & & & \cdots \end{bmatrix} \tag{C.1.22}$$

Thus the linear transformation of the waveform $s(t)$ into the waveform $x(t)$ is equivalent to the matrix transformation of the column vector **s** into the new vector **x** in (C.1.21). Likewise, this means the converse is also true. The matrix transformation of a vector **s** into a new vector **x** is equivalent to some linear transformation of the waveform $s(t)$ to the waveform $x(t)$. It should be pointed out that the relation between the matrix A and the transformation T may not be simple. Although the elements a_{ij} can be determined in a relatively staightforward way from the expansion in (C.1.18), the conversion from an A matrix back to the equivalent T transformation could be much more difficult.

C.2 RANDOM PROCESSES

Orthonormal expansions can be extended to random waveforms. Let $n(t)$, $a \le t \le b$ be a random process, zero mean, with known correlation function

$$R_n(t, \rho) = \mathscr{E}[n(t)n(\rho)] \tag{C.2.1}$$

Consider the expansion

$$n_N(t) = \sum_{i=1}^{N} n_i \Phi_i(t) \tag{C.2.2}$$

with

$$n_i = \int_a^b n(t)\Phi_i(t) \, dt \tag{C.2.3}$$

where $\{\Phi_i(t)\}$ is again an orthonormal set over (a, b). Since $n(t)$ is a random process, the $\{n_i\}$ are random variables, and $n_N(t)$ in (C.2.2) is itself a random waveform. We define the random vector of n_i coefficients as

$$\mathbf{n} = \begin{bmatrix} n_1 \\ n_2 \\ \vdots \\ n_N \end{bmatrix} \qquad \text{(C.2.4)}$$

and its correlation matrix is then

$$R_{\mathbf{n}} = \mathscr{E}[\mathbf{n}\mathbf{n}^{\text{tr}}] = [(\mathscr{E}[n_i n_j])] \qquad \text{(C.2.5)}$$

where tr is transpose.

Consider first the variable $\{n_i\}$. They have the following properties:

1. $\mathscr{E}(n_i) = 0$ for all i, since $n(t)$ in (C.2.3) has a zero mean.
2. If the process $n(t)$ in (C.2.3) is Gaussian, then the n_i are Gaussian (integrals of Gaussian processes are Gaussian), and $\mathbf{n}$ is a random Gaussian vector. The joint density of its components is then

$$p(n_1, n_2, \ldots, n_N) = \frac{1}{(2\pi)^{N/2}[\det R_{\mathbf{n}}]^{1/2}} e^{-\frac{1}{2}\mathbf{n}^{\text{tr}}R_n^{-1}\mathbf{n}} \qquad \text{(C.2.6)}$$

where $R_{\mathbf{n}}^{-1}$ is the inverse of $R_{\mathbf{n}}$ in (C.2.5).

3. The correlation between n_i and n_j is given by

$$\mathscr{E}(n_i n_j) = \mathscr{E}\left[\int_a^b n(t)\Phi_i(t) \, dt \int_a^b n(\rho)\Phi_j(\rho) \, d\rho \right]$$

$$= \int_a^b \int_a^b \mathscr{E}[n(t)n(\rho)]\Phi_i(t)\Phi_j(\rho) \, dt \, d\rho$$

$$= \int_a^b \int_a^b R_n(t, \rho)\Phi_i(t)\Phi_j(\rho) \, dt \, d\rho \qquad \text{(C.2.7)}$$

The correlation of two (n_i) can therefore be computed by integrating the correlation function in (C.2.7). Consider the case where

$$\int_a^b R_n(t, \rho)\Phi_j(\rho) \, d\rho = \lambda_j \Phi_j(t) \qquad \text{(C.2.8)}$$

for every j, where λ_j is any scalar. In other words, the orthonormal functions have the property that they integrate on the left side to reproduce a scaled version of themselves on the right. When this is true, (C.2.7) becomes

$$\mathscr{E}(n_i n_j) = \int_a^b \lambda_j \Phi_j(t)\Phi_i(t)\,dt$$

$$= \begin{cases} 0 & \text{if } i \neq j \\ \lambda_j & \text{if } i = j \end{cases} \qquad (C.2.9)$$

Thus the random variables n_i and n_j will be uncorrelated if the orthonormal set $\Phi_i(t)$ used in (C.2.2) satisfies (C.2.8). [In fact, since $\{\Phi_i(t)\}$ is a complete set, this is the only way (C.2.9) can occur.] Note that this condition specifies a *particular* orthonormal set, that is, the set satisfying (C.2.8) for all j, when the correlation function is inserted. When this particular set is used in (C.2.2), the resulting expansion is called a *Karhunen–Loeve* (KL) expansion. Thus the coefficient variables $\{n_i\}$ will be uncorrelated random variables if, and only if, the expansion in (C.2.2) is a KL expansion. Note that for a KL expansion, the mean squared value of each n_i is given by λ_i, the coefficient in (C.2.8). One may, in fact, recognize (C.2.8) as an eigenfunction equation with respect to the *Fredholm operator* with kernel $R_n(t_1, t_2)$:

$$T[x(t_1)] = \int_a^b R_n(t_1, t_2)[x(t_2)]\,dt_2 \qquad (C.2.10)$$

that is, $\Phi_j(t)$ is an eigenfunction of the equation

$$T[\Phi_j(t)] = \lambda_j \Phi_j(t) \qquad (C.2.11)$$

with λ_j the corresponding *eigenvalue*. The orthonormal set $\{\Phi_i(t)\}$ of a KL expansion must therefore be *eigenfunctions* of the operator in (C.2.10), and the mean squared coordinate values are the associated eigenvalues of this transformation. When a KL expansion is used, the correlation matrix R_n of the coordinate vector $\mathbf{n}$ in (C.2.5) becomes a diagonal matrix

$$R_{\mathbf{n}} = \begin{bmatrix} \lambda_1 & & & & \mathbf{0} \\ & \lambda_2 & & & \\ & & \lambda_3 & & \\ & & & \ddots & \\ \mathbf{0} & & & & \lambda_N \end{bmatrix} \qquad (C.2.12)$$

and if $\mathbf{n}$ is Gaussian, (C.2.6) reduces to

$$p(n_1, n_2, \ldots, n_N) = \frac{1}{(2\pi)^{N/2}[\Pi\lambda_i]^{1/2}}\, e^{-\frac{1}{2}\sum_{i=1}^N n_i^2/\lambda_i} \qquad (C.2.13)$$

4. If we compute the average energy of a random process expanded as in (C.2.2)

$$\text{Average energy of } n_N(t) = \mathscr{E} \int_a^b n_N^2(t) \, dt$$

$$= \mathscr{E} \int_a^b \left[\sum_{i=1}^N n_i \Phi_i(t) \right]^2 dt$$

$$= \sum_{i=1}^N \mathscr{E}(n_i)^2$$

$$= \sum_{i=1}^N \lambda_i \tag{C.2.14}$$

The mean square component value (eigenvalues for a KL expansion) sum to give the average energy. We can therefore consider the λ_i values as representing the average energy of the ith component of $n_N(t)$. For this reason eigenvalues of KL expansions are often referred to as *energy coefficients*.

5. When a KL expansion is used in (C.2.2), the resulting process has the correlation function

$$R_{n_N}(t, \rho) = \mathscr{E}[n_N(t) n_N(\rho)]$$

$$= \mathscr{E} \sum_{i=1}^N n_i \Phi_i(t) \sum_{j=1}^N n_j \Phi_j(\rho)$$

$$= \sum_{i=1}^N \sum_{j=1}^N \mathscr{E}(n_i n_j) \Phi_i(t) \Phi_j(\rho)$$

$$= \sum_{i=1}^N \lambda_i \Phi_i(t) \Phi_j(t) \tag{C.2.15}$$

Let $n(t)$ in (C.2.1) be any random process whose correlation function can be expanded in the specific form

$$R_n(t, \rho) = \sum_j C_j \Phi_j(t) \Phi_j(\rho) \tag{C.2.16}$$

where $\{C_j\}$ are positive scalars and $\{\Phi_j(t)\}$ are orthonormal. Then the resulting process $n_N(t)$ formed from $n(t)$ via (C.2.3) will *always* have $\{\Phi_j(t)\}$ as its KL functions and $\lambda_j = C_j$. This can be seen by using (C.2.3) and noting that

$$\mathscr{E}(n_i n_q) = \int_a^b \int_a^b \mathscr{E}[n(t) n(\rho)] \Phi_i(t) \Phi_q(\rho) \, dt \, d\rho$$

$$= \int_a^b \int_a^b R_n(t, \rho) \Phi_i(t) \Phi_q(\rho) \, dt \, d\rho \tag{C.2.17}$$

Substituting from (C.2.16), and using the orthonormality of the Φ_i functions, we see that (C.2.17) is always zero, except for the case where $i = q = j$. From this latter condition, $\mathscr{E}(n_i^2) = C_i$. This means that $R_{n_N}(t, \rho)$ is given by (C.2.15) with $\lambda_i = C_i$ and $\{\Phi_i(t)\}$ a KL expansion set.

6. For any random process $n(t)$ expandable as in (C.2.16), the partial sum in (C.2.2) converges in a mean squared sense to $n(t)$. This can be proved by evaluating $J = \mathscr{E}[n(t) - n_N(t)^2]$ at any t. Expanding shows

$$J = \mathscr{E}[n^2(t)] - 2\mathscr{E}[n(t)n_N(t)] + \mathscr{E}[n_N^2(t)] \tag{C.2.18}$$

The middle term is

$$\mathscr{E}\left[n(t)\sum_{i=1}^{N} n_i\Phi_i(t)\right] = \sum_{i=1}^{N}\Phi_i(t)\int_a^b \mathscr{E}[n(t)n(\rho)]\Phi_i(\rho)\,d\rho$$

$$= \sum_{i=1}^{N} C_i\Phi_i^2(t) \tag{C.2.19}$$

Hence (C.2.18) reduces to

$$J = R_n(t, t) - 2R_{n_N}(t, t) + R_{n_N}(t, t)$$

$$= \sum_{i=1}^{\infty} C_i\Phi_i^2(t) - \sum_{i=1}^{N} C_i\Phi_i^2(t) \tag{C.2.20}$$

for which $J \to 0$ as $N \to \infty$. The partial expansion (C.2.2) therefore represents any $n(t)$, in a mean squared error sense, whose correlation expands as in (C.2.16). Some well-known theorems in functional analysis (Mercer's theorems) state the conditions for which a correlation function can be so expanded. In general, any correlation function $R_n(t, \rho)$, continuous in t and ρ, can be expanded in the form of (C.2.16), for some $\{\Phi_i(t)\}$ set.

C.3 EXAMPLES OF RANDOM EXPANSIONS

Example C.1 (White Noise). Let $n(t)$ be a white noise process, having the correlation function

$$R_n(t, \rho) = \left(\frac{N_0}{2}\right)\delta(t - \rho) \tag{C.3.1}$$

where N_0 is the one-sided spectral level. Consider the integral

$$\int_{-\infty}^{\infty} R_n(t, \rho)\Phi_j(\rho)\,d\rho = \int_{-\infty}^{\infty}\left(\frac{N_0}{2}\right)\delta(t - \rho)\Phi_j(\rho)\,d\rho$$

$$= \left(\frac{N_0}{2}\right)\Phi_j(t) \tag{C.3.2}$$

This exactly satisfies the KL correlation in (C.2.8) for *any* function $\Phi_j(t)$. Thus any orthonormal function set produces a KL expansion for white noise. Furthermore,

$$\lambda_j = \frac{N_0}{2} \tag{C.3.3}$$

for every j. The mean squared energy value (eigenvalue) is the same for every component. This means that the coefficients of the KL expansion will have correlation matrix

$$R_{\mathbf{n}} = \begin{bmatrix} \dfrac{N_0}{2} & & & 0 \\ & \dfrac{N_0}{2} & & \\ & & \ddots & \\ 0 & & & \dfrac{N_0}{2} \end{bmatrix} \tag{C.3.4}$$

If, in addition, the process $n(t)$ is Gaussian, the joint density of the $\mathbf{n}$ vector follows from (C.2.13) as

$$p(\mathbf{n}) = \frac{1}{(2\pi N_0/2)^{N/2}} e^{-(1/N_0)\sum_{i=1}^{N} n_i^2} \tag{C.3.5}$$

Example C.2 (Bandlimited White Noise). Let $n(t)$ be bandlimited white noise process (Figure 1.9c). Its correlation function is

$$R_n(t, \rho) = N_0 B \left\{ \frac{\sin [2\pi B(t - \rho)]}{[2\pi B(t - \rho)]} \right\} \tag{C.3.6}$$

as shown in Figure 1.9d. The KL condition requires

$$\int_{-\infty}^{\infty} R_n(t, \rho)\Phi_j(\rho)\, d\rho = \lambda_j \Phi_j(t) \tag{C.3.7}$$

We can use the identity

$$\int_{-\infty}^{\infty} \frac{\sin [\pi(x - a)]}{[\pi(x - a)]} \cdot \frac{\sin [\pi(x - b)]}{[\pi(x - b)]}\, dx = \frac{\sin [\pi(a - b)]}{[\pi(a - b)]} \tag{C.3.8}$$

to prove that (C.3.7) is satisfied with

$$\Phi_j(t) = \sqrt{2B}\, \frac{\sin (2\pi Bt - i\pi)}{(2\pi Bt - i\pi)}$$

$$\lambda_j = \frac{N_0}{2} \tag{C.3.9}$$

The orthonormal functions in (C.3.9) are called *sampling functions* (see Section 6.4), and the resultant coordinates in (C.2.3) become

$$n_i = \frac{1}{\sqrt{2B}} \, n\left(\frac{i}{2B}\right) \qquad \text{(C.3.10)}$$

The coordinates $\{n_i\}$ are obtained by samples of the process $n(t)$ at $t_i = i/2B$ [samples of $n(t)$ taken every $1/2B$ sec apart]. The correlation matrix is then

$$
\begin{aligned}
R_{\mathbf{n}} &= \frac{1}{2B} \, [\mathscr{E}[n(t_i)n(t_j)]]_{N \times N} \\
&= \frac{1}{2B} \, [R_n(t_i - t_j)]_{N \times N}
\end{aligned}
\qquad \text{(C.3.11)}
$$

This matrix will be diagonal since $R_n(t, \rho)$ in (C.3.6) passes through zero at all $t - \rho = i/2B$, $i \neq 0$.

C.4   SIGNAL PLUS NOISE WAVEFORMS

The previous expansions can be directly applied to waveforms composed of the sum of a deterministic signal waveform $s(t)$ and a random noise process $n(t)$. Let

$$x(t) = s(t) + n(t) \qquad \text{(C.4.1)}$$

and assume $s(t)$ can be expanded into a ν-dimensional orthonormal set $\{\Phi_i(t)\}$, $i = 1, \ldots, \nu$, as before. We can then expand $x(t)$ into the same orthonormal set

$$x(t) = \sum_{i=1}^{\nu} x_i \Phi_i(t) \qquad \text{(C.4.2)}$$

where

$$
\begin{aligned}
x_i &= \int_a^b x(t)\Phi_i(t) \, dt \\
&= \int_a^b s(t)\Phi_i(t) \, dt + \int_a^b n(t)\Phi_i(t) \, dt \\
&= s_i + n_i
\end{aligned}
\qquad \text{(C.4.3)}
$$

and n_i is the random noise variable formed from the $\Phi_i(t)$. The sequence of x_i values then define the vector $\mathbf{x}$ which, from (C.4.3), will have the form

$$\mathbf{x} = \mathbf{s} + \mathbf{n} \qquad \text{(C.4.4)}$$

where **n** contains the random components of the noise process in the ν dimensions of the signal. That is, it is *only* the noise lying in the ν dimensions of the signal that will appear in the sum vector **x**. This fact is used in Chapters 7–9 where vector expansions of noisy waveforms are inserted.

When the joint probability density of the random noise vector **n** is known, it immediately determines the probability density of the vector **x** in (C.4.4). This can be obtained by a direct transformation from the vector **n** to the vector **x** via the Jacobian of the transformation [see (B.2.14)]. For this transformation, the Jacobian is the identity matrix, and it therefore follows from (B.2.14) that

$$p_x(x) = (1)p_n(n)|_{n=x-s}$$
$$= p_n(n_1 - s_1, n_2 - s_2, \ldots, n_\nu - s_\nu) \qquad (C.4.5)$$

Hence the probability density of the vector **x** is always obtained from that of **n** by the above substitution in (C.4.5). It is to be emphasized that (C.4.5) holds for *any* density $p_n(n)$ on the noise vector **n**.

INDEX